Progress in Food Contaminant Analysis

Progress in Food Contaminant Analysis

Edited by

J. GILBERT
Ministry of Agriculture,
Fisheries and Food
CSL Food Science Laboratory
Norwich
UK

BLACKIE ACADEMIC & PROFESSIONAL
An Imprint of Chapman & Hall
London · Weinheim · New York · Tokyo · Melbourne · Madras

Published by
Blackie Academic & Professional, an imprint of Chapman & Hall,
2–6 Boundary Row, London SE1 8HN, UK

Chapman & Hall, 2–6 Boundary Row, London SE1 8HN, UK

Chapman & Hall GmbH, Pappelallee 3, 69469 Weinheim, Germany

Chapman & Hall USA, 115 Fifth Avenue, New York, NY 10003, USA

Chapman & Hall Japan, ITP-Japan, Kyowa Building, 3F, 2-2-1 Hirakawacho, Chiyoda-ku, Tokyo 102, Japan

DA Book (Aust.) Pty Ltd, 648 Whitehorse Road, Mitcham 3132, Victoria, Australia

Chapman & Hall India, R. Seshadri, 32 Second Main Road, CIT East, Madras 600 035, India

First edition 1996

© 1996 Chapman & Hall

Typeset in 10/12pt Times by Academic & Technical Typesetting, Bristol
Printed in Great Britain by St Edmundsbury Press Ltd, Bury St Edmunds, Suffolk

ISBN 0 7514 0337 7

A catalogue record for this book is available from the British Library

Library of Congress Catalog Card Number: 96–83055

∞ Printed on acid-free text paper, manufactured in accordance with ANSI/ NISO Z39.48-1992 (Permanence of Paper).

Preface

'Analysis of Food Contaminants' was published in 1984 by Elsevier Applied Science Publishers and 10 years later I was asked to consider producing an updated second edition. Surprisingly little has really changed in a decade in terms of the public interest in food safety and the continued vigilance of Government in monitoring the food supply for contaminants. This means that food contamination in itself is still a very relevant topic. However, much has changed in terms of the techniques now employed in trace analysis. The 1984 book used a combination of an analytical technique and a specific food contaminant problem area per chapter (each written by a specialist) which resulted in a multi-authored text which was mostly application based but provided a good introduction to the 'how' in terms of applying techniques to real problems. Rather than producing a second edition of this text, it seemed on reflection more sensible to produce a new and complementary book, using the same formula as before of application plus technique, but to concentrate on contaminant areas of current interest and to highlight recent advances in techniques. Thus, the present book 'Progress in Food Contaminant Analysis' has originated as a follow-up to 'Analysis of Food Contaminants'.

Sampling and the application of appropriate sample plans for food analysis has always been a neglected area, and whilst this topic does not really constitute an advance in food analysis it has been included in the present book simply to emphasise its importance. Invariably the cost of adequate sampling tends to deter when planning food surveillance exercises, and in the laboratory the glamour of sophisticated instrumental techniques and ever pushing to lower limits of detection is inevitably far more attractive than development of sampling plans. The first chapter of this book draws attention to the issues that need to be considered in sampling and hopefully will be read alongside later chapters on instrumental methods. The 1984 book contained a chapter on size exclusion chromatography (gel permeation) for sample clean-up which is still relevant and remains under-utilised as a sample preparation method. In this book the same author has extended this theme and reviewed the wider area of automated clean-up techniques (including size exclusion) which is the way forward for many in terms of meeting demands for increased sample numbers and reduced cost of analysis.

Of the various food contaminants mycotoxins still remain important food contaminants with particular interest today in ochratoxin A and the recently discovered fumonisins. In a chapter on chromatographic analysis of mycotoxins two authors from a world-leading mycotoxin laboratory in

South Africa have critically and comprehensively reviewed the relative merits of different chromatographic approaches to the analysis of mycotoxins. The 1984 book contained a chapter which reviewed analytical approaches to trace element determinations in foods. This is still up-to-date in terms of sample preparation which has essentially remained unchanged and in respect of techniques such as AAS. However, ICP-MS has subsequently revolutionised trace element analysis and in this new book a chapter is devoted exclusively to this topic. In 1984 a chapter covered immunoassay techniques applied to veterinary drug residue analysis—in this book the remarkable advances in immunochemical analysis in the last few years are reviewed but applied to pesticides analysis as opposed to drug residues.

The remaining chapters of this book deviate somewhat from the earlier text but nevertheless are still complementary. Instead of reviewing quantitative GC-MS this present book concentrates on LC-MS which, in less than 10 years, has moved from being a relatively esoteric technique to being routinely applied in many laboratories world-wide. Despite tremendous advances in instrumental techniques, the use of biological-based methods is still, in many areas, irreplaceable at the present time. This is illustrated by a chapter looking at approaches used to monitor the presence of phycotoxins in shellfish. The ultimate state-of-the-art in terms of requirements for sensitivity and specificity for contaminant analysis is in the area of dioxins and PCBs, where high resolution GC-MS remains the only method capable of meeting stringent demands, and a chapter is dedicated to this topic. Perhaps the most specialised chapter concerns testing of food packaging materials for migration into foods and this concentrates on high temperature testing which has become important with changes in lifestyle towards greater use of microwave convenience foods. Finally this book concludes with an insight into some of the developments that are being made internationally in terms of harmonisation of analytical methods, analytical quality assurance and proficiency testing. Developments in this area do represent an advance, and this chapter provides an insight into the workings of frequently impenetrable international committees involved in food analysis.

It is not only a pleasure but also an honour to draw together contributions from friends and colleagues who are renowned specialists in their own fields into the form of an edited book. It is hoped that the end-product will provide for those unfamiliar with the field of food contaminant analysis much stimulating reading, and an overview of trends and developments in this fascinating area.

J.G.

Contributors

T.H. Begley Indirect Additives Section, Division of Product Manufacture and Use, Center for Food and Applied Nutrition, Food and Drug Administration, 200 C Street SW, Washington DC 20204, USA

S.J. Buckland ESR Environmental, PO Box 30 547, Lower Hutt, New Zealand

P.A. Burdaspal Centro Nacional de Alimentación Instituto de Salud Carlos III, Ministerio de Sanidad y Consumo 28220 Majadahonda, Madrid, Spain

H.M. Crews Ministry of Agriculture, Fisheries and Food, CSL Food Science Laboratory, Norwich Research Park, Colney, Norwich NR4 7UQ, UK

N.T. Crosby Laboratory of the Government Chemist, Queen's Road, Teddington, Middlesex TW11 0LY, UK

J. Gilbert Ministry of Agriculture, Fisheries and Food, CSL Food Science Laboratory, Norwich Research Park, Colney, Norwich NR4 7UQ, UK

D.J. Hannah ESR Environmental, PO Box 30 547, Lower Hutt, New Zealand

H.C. Hollifield Indirect Additives Section, Division of Product Manufacture and Use, Center for Food and Applied Nutrition, Food and Drug Administration, 200 C Street SW, Washington DC 20204, USA

B.M. Kaufman Novavax Inc, 12111 Parklawn Drive, Rockville, MD 20852, USA

L.J. Porter ESR Environmental, PO Box 30 547, Lower Hutt, New Zealand

G.S. Shephard Programme on Mycotoxins and Experimental Carcinogenesis, Medical Research Council, PO Box 19070, Tygerberg 7505, South Africa

M.J. Shepherd   Ministry of Agriculture, Fisheries and Food, CSL Food Science Laboratory, Norwich Research Park, Colney, Norwich NR4 7UQ, UK

E.W. Sydenham   Programme on Mycotoxins and Experimental Carcinogenesis, Medical Research Council, PO Box 19070, Tygerberg 7505, South Africa

R. Wood   Ministry of Agriculture, Fisheries and Food, Food Labelling and Standards Division, Norwich Research Park, Colney, Norwich NR4 7UQ, UK

Contents

9 Approaches to evaluating high-temperature food packaging materials as sources of food contamination   332
HENRY C. HOLLIFIELD and TIMOTHY H. BEGLEY

10 Progress in developing European statutory methods of analysis   368
ROGER WOOD

CONTENTS

1 Sampling and sample plans for food surveillance exercises

NEIL T. CROSBY

Summary

Sampling is the most important first stage in the analytical process. Unless the test portion taken for analysis is truly representative of the bulk material, any resulting analytical data are likely to be in error. This may lead to incorrect decisions being taken, resulting, for example, in loss of money, or foods being sold out of specification, containing unacceptable quantities of contaminants, or deficient in vital nutrients. Prosecutions instigated by enforcement authorities could be unsound if sampling is not effected in a satisfactory manner.

Foods are heterogeneous products, being mixtures of proteins, carbo-hydrates and fats with moisture and minerals, and a wide range of additives and contaminants may also be present. They are therefore inherently difficult to sample correctly. This chapter describes the general principles of good sampling practice and identifies some practical problems encountered in the representative sampling of different types of foods. It describes how to devise a sampling plan that fulfils the aims and objectives of the survey, and how that plan is put into practice. It does not, however, specify detailed instructions for any one particular survey or food, since such instructions will vary with the aims and scope required.

1.1 Introduction

Sampling is the act of extracting a small fraction of material from a large bulk in such a way that the fraction removed is representative of the bulk in both properties and characteristics. This process includes the withdrawal of a portion from a bulk material or, equally, the selection of one or more items or articles from a large number of such products ostensibly from the same production batch or consignment. Both approaches may be required in food surveillance exercises.

Most modern methods of analysis for food contaminants are applied to very small test samples, often 1 g or even less, since pretreatment processes such as digestion, solvent extraction or chromatographic clean-up systems require increased time or volume of reagents as the weight of sample is increased. This can increase the cost of the analytical programme as well as the magnitude of 'blank' values, thus increasing the uncertainty of the result when concentrations near to the limit of detection of the method are being measured. Analytical techniques are now so sensitive that only very small quantities of analyte are required to produce a response at the final detection stage. It is vital, therefore, that the final test portion taken for analysis is truly representative of the commodity under examination, otherwise the results of the analysis could be erroneous, however sophisticated the method used and however carefully the analytical manipulations are performed. Youden (1967) stated that once the analytical uncertainty has been reduced to one-third or less of the sampling uncertainty, further improvement to the analytical method is of little consequence to the overall accuracy of the study. This means that, where sampling errors are likely to be high, it is better to use a rapid method of analysis with a high throughput of samples rather than a more accurate but laborious technique that can only cope with a few samples per day or per week. In this way the homogeneity (or degree of heterogeneity) of the product can be established. This is a vital step in working out a sampling protocol that will ensure the success of the surveillance exercise.

Most foods are heterogeneous products and hence difficult to sample in a representative manner. For example, meat pies consist of meat and jelly in a pastry case. The meat itself may well be a heterogeneous mixture of water, vitamins, inorganic salts and several types of connective tissue mixed with fat and carbohydrates. Table 1.1 shows a list of ingredients found in a typical infant feed which would not be obvious from the proximate composition given at the foot of the table, or from a visual examination of the product. Furthermore, several of the ingredients listed are themselves complex mixtures of different constituents.

Other problems are of concern when analysing for contaminants, or even nutrients at low concentrations. The particular analyte under test may not be evenly distributed throughout the food. For example, a pesticide is more likely to be found at higher concentrations on the outside of fresh products and yet outer leaves are often discarded by the consumer. The product may be washed or cooked before consumption. Some veterinary drugs are bound to proteins in the food and so may not be readily extracted by normal processes. Contaminants arising from a food container will be found at a higher concentration in that portion of food in direct contact with the container material than in the portion at the centre of the container.

All the above factors are examples of how the uncertainty in the accuracy of the analytical data can be increased, in addition to the uncertainty arising from the analytical method itself. Further problems may arise during storage of the sample or transport to the laboratory, when degradation or segregation could occur. Other errors may be introduced if the sample received in the laboratory is subjected to pretreatment processes such as grinding, or sub-sampled prior to analysis. The use of unsuitable sampling, storage and

Table 1.1 Ingredients of a typical infant feed

Skimmed milk powder
Electrodialysed whey
Lactose
Oleo oil
Coconut oil
Soya bean oil
Oleic (Safflower) oil
Soya bean lecithin
$CaCl_2$
$NaHCO_3$
Vitamin C
Ca citrate
$FeSO_4$
KOH
$KHCO_3$
$ZnSO_4$
Vitamin E
Nicotinamide
Vitamin A
$CuSO_4$
Ca pantothenate
Vitamins B_1, B_2, B_6
β-Carotene
KI
Folic acid
Vitamins D_3, B_{12}

Proximate analysis: fat 28.0%, carbohydrate 56.0%, ash 2.0%, protein 12.0%, water 2.0% w/w.

pretreatment equipment may lead to contamination of the sample, depending upon the analyte under investigation.

1.1.1 Sampling procedures

The act of sampling consists of two parts:

(i)   the design of a sampling plan;
(ii) the implementation of the agreed plan in a practical situation.

These two activities are quite separate and will normally involve different personnel, although it is advisable that good communications are maintained at all times between those responsible for commissioning the survey, the sampling officers and the analytical scientists who will perform the laboratory tests and provide the final report. It may also be necessary to make use of the specialist services of a statistician to design the study and experts in nutrition or toxicology to interpret the final data.

1.1.1.1 Types of sampling. There are a number of different approaches to sampling depending on the size of the consignment, the characteristics of the product and the objectives of the study. Sampling plans may specify that samples are to be obtained by: (a) random; (b) systematic; (c) stratified; (d) sequential; or (e) *ad hoc* means. Some comments on these different methods with particular reference to the sampling of foodstuffs are now presented. A glossary of some terms used in sampling is given in Appendix 1.1.

(a) Random.   This refers to the drawing of a small number of samples from a large consignment (or bulk material) in a non-biased way. Random does not, therefore, mean haphazard! The process must be carried out in such a way that all constituent parts of the consignment have an equal chance of being included in the sample. This can be achieved using a table of random numbers (or those generated by a calculator) to select packages from a consignment. The packages would be assigned a number in a systematic way and then the random numbers used to isolate those that will be subsequently sampled. This approach can be adapted to a wide range of manufactured products, particularly where large numbers of individual containers (drums, packets or cans) are to be sampled. Large heaps of foods, e.g. cereals, cannot be satisfactorily sampled in this way, especially if the product is heterogeneous, since it is not possible to ensure that each particle has an equal chance of being included in the sample. In heaps, the larger or heavier particles tend to congregate towards the bottom of the heap. Such commodities are best sampled 'on the move' as described below.

(b) Systematic.   This is the most commonly used technique for the sampling of bulk materials. It involves the collection of increments of material at predetermined intervals as defined in the sampling plan (time-weighted averages). Flow-proportional averages in moving systems can also be employed if the speed is constant. Where samples are taken from a conveyor belt, it is important to ensure that incremental samples are removed from across the whole width of the belt at right angles to the direction of movement. Automatic equipment can also be used, but its reliability and performance should be checked where possible.

(c) Stratified.   This approach is best for the sampling of liquids in lakes, reservoirs, large vats or tankers. In the area of food sampling, it is most likely to be employed for taking samples of milk from tankers where there is a layer of cream on the surface. Samples are removed in proportion to the weight (or volume) of the different layers (strata). Alternatively, it may be better to sample the liquid during discharge from the container.

(d) Sequential.   Sampling in this way is used to confirm conformity to a given specification, particularly for manufactured products. Samples are

removed from a production line at predetermined but random intervals to check that the product is within the given specification. Where the results are seen to be approaching the specification limits, it may be necessary to undertake more exhaustive sampling procedures.

(e) Ad hoc. This covers the case where sampling is non-statistically based and deliberately biased towards specific foods, e.g. vegetarian diets, baby foods, imported eggs, etc. Other categories of surveys may be restricted to individual foods, health foods, 'own brand' products, etc. These categories will need to be clearly identified in the sampling plan. This approach can also be referred to as 'biased' or 'worst-case' sampling where a known (or suspected) problem area is targeted specifically.

1.1.1.2 Sample compositing. A limited amount of money is normally available for each food contaminant surveillance exercise. Hence it is seldom possible to adopt an ideal sampling strategy for each exercise. A compromise has to be struck between what is desirable and what is practicable and can be achieved under the limitations of available resources. In particular, it is unlikely that the ideal number of samples can be included in the survey because of the costs of analysis, especially when very low levels of contaminants, such as veterinary residues, are to be determined.

One way around this problem is to use a screening method which permits a higher throughput of samples. This approach can then be used to target only those samples with presumptive positive results by using more accurate, though laborious, analytical methods.

Another approach is to combine portions from a number of different samples of the same product to produce a single analytical sample. This is often known as compositing. In effect this reduces the number of primary analyses required by a factor equal to the number of samples included in the composite sample, thus reducing the cost of the survey by the same factor. In essence this means that larger numbers of samples can be tested for the same cost. Thus advocates of this policy would claim that compositing of samples permits a more efficient use of analytical resources and enables more data to be obtained, thus increasing the significance of the results. However, there are disadvantages to this approach. Combining samples containing contaminants with those that are free from the same contaminants effectively increases the risk of finding false negatives. Hence, the presence of contaminants in certain foods could be missed by this approach. It would also be dangerous to use the data obtained on composite samples to calculate human dietary exposure to a particular contaminant. Where compositing is practised, it is important to obtain agreement about the maximum number of samples that can be combined to produce a composite sample.

1.2 Design of a sampling plan

The principal stages in sampling and analysis operations required for a food surveillance exercise are illustrated in Figure 1.1. These steps are discussed in more detail below.

The design of the sampling programme must ensure that the information obtained from the analytical data produced is sufficient to satisfy the objectives set by the originators of the work. Hence the aims and objectives of the surveillance programme must be clearly stated and agreed by all parties at the outset. These will include the analytes (contaminants) to be determined and the methods to be used for their determination. The sampling plan must fix the number of samples to be obtained. It will include the foods to be taken and from which locations. Any necessary requirements for preservation, pretreatment and transport to the laboratory will also need to be agreed. These conclusions then need to be written down to form a protocol for use by the sampling officers and the analytical team. The protocol may need to be reviewed subsequently in the light of experience, particularly if a pilot study is planned in advance of the main surveillance exercise. The detailed instructions in the protocol must be explicit, comprehensive and

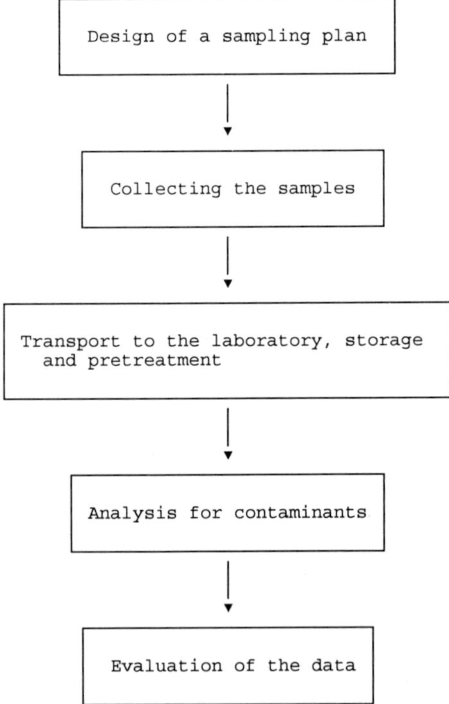

Figure 1.1 The principal stages of sampling and analysis in a food surveillance exercise.

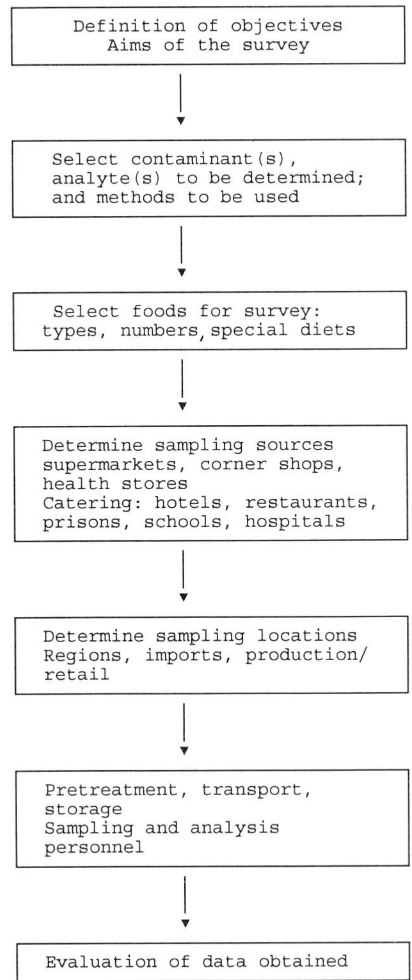

Figure 1.2 Factors involved in the design of a sampling plan for contaminants in foods.

unambiguous, so that sampling officers and analytical staff can follow the agreed plan without introducing further uncertainties or bias into the data produced. Some of the factors involved in the design of a sampling plan for nutrients or contaminants in foods are shown in Figure 1.2. Further comments on these factors are given below.

1.2.1 *Objectives and scope*

The first stage in designing a surveillance exercise is to obtain agreement on why the programme is needed and what results are to be expected. This will entail discussions and consultations as appropriate with experts from the food industry with a specialist knowledge of the commodity to be surveyed. It will be necessary to decide whether the survey will extend to retail products, the catering trade, and establishments such as hotels, hospitals and prisons, or be restricted to selected foods such as imports or special diets. At the end of these preliminary discussions it may be useful to ask some or all of the following questions:

- What do we want to know, i.e. which analytes, which foods?
- Why do we need this information?
- What will happen to the data obtained in the survey?
- What actions/decisions will be taken after interpretation of the data?

Inevitably the resources allocated for food surveys are limited. The money and staff available will only permit a certain number of samples to be purchased. Equally, the funds required for analysis will probably be the key factor in defining the scope of the programme. These limitations will force attention on the objectives of the programme to ensure that they are clearly defined.

1.2.2 Analytes

In any surveillance exercise for food contaminants, the analytes of interest are usually clearly defined. However, in the case of inorganic contaminants it is important to consider whether speciation is important, e.g. Cr^{III} or Cr^{VI}, or simply that total elemental concentration is required to be determined, since this will affect the analytical method to be used. Some elements can occur in both soluble and insoluble forms. The sampling plan must define whether soluble or total elemental concentration is required.

Again, some elements, e.g. mercury, can exist either in the inorganic form or combined organically, e.g. as methyl mercury. The sampling plan must make clear which form(s) are to be determined.

Similar considerations apply also to organic contaminants that can occur either in the free state or combined with a constituent of the foodstuff. In some cases only a single compound may need to be determined, e.g. aflatoxin B_1, since this is the compound of greatest interest. In other surveys, the total content of aflatoxins $(B_1 + B_2 + G_1 + G_2)$ may be required. Similarly, for organochlorine pesticides the determination of a single compound or alternatively of a whole range of different compounds or isomers may have to be specified in the sampling plan. Some food contaminant chemicals are metabolised when introduced into mammalian systems. Any survey of edible tissues would need to make clear whether the target compound and/ or its metabolites are to be determined. For some food contaminants, e.g. the *N*-nitrosamines, it may be necessary to determine also the concentration of nitrates and nitrites present in the food to assist in the interpretation of the data. For contamination arising from packaging materials, specific compounds, e.g. monomers, may need to be determined. Alternatively, a knowledge of the total weight of contaminants that could migrate from the container or package into the food under simulated conditions, e.g. plasticisers or other additives, may be of greater interest.

1.2.3 Methods of analysis

Having decided which analytes are to be determined, it is then necessary to specify the particular analytical methods to be used in the exercise. This is especially important if more than one laboratory is to be used to carry out the analytical studies, in order to reduce the inevitable variation of results from one laboratory to another to as low a level as possible. It is important to ensure that the analytical procedure is written down in sufficient detail that all analysts participating in the study do in fact carry out the analysis in the same way. This can best be achieved by the use of official, statutory or recommended methods wherever possible. A compendium of such methods for use in food and agriculture has recently been published by the Royal Society of Chemistry (1994). Equally, standard and validated methods have been published by AOAC International and by bodies such as CEN (Comité Européen de Normalisation/European Committee for Standardization) and ISO. Wherever possible the analytical methods used should have been subjected to a collaborative study carried out in accordance with an internationally agreed protocol (AOAC, 1989).

Furthermore, laboratories participating in the surveillance exercise should be encouraged to participate in a proficiency testing scheme such as FAPAS (Food Analysis Performance Assessment Scheme) run by the CSL Food Science Laboratory, Norwich (UK), so that their performance can be evaluated against that of other laboratories. Only laboratories that consistently obtain a satisfactory z-score, i.e. within the range $+2.0$ to -2.0, should be used in the survey. In many cases, however, proficiency testing schemes for the particular analyte(s) in question may not be available. As a further check on the comparability of data obtained by different laboratories, it may be useful to circulate trial samples for analysis prior to the main surveillance exercise. Where it is not possible to organise interlaboratory trials before the start of the survey, it may be useful to include in the sampling plan the number of replicates to be determined. Other internal laboratory quality-control criteria may need to be specified if accuracy of the data is of paramount importance.

1.2.4 Types of analytical methods used for surveillance

The analytical methods employed in food surveillance studies for contaminants can be classified into four groups as follows:

- screening
- surveillance
- confirmatory
- reference

(a) Screening.   These methods are extremely rapid procedures permitting a high throughput of samples at low cost. For organic analytes, immunochemical techniques are often used (see Chapter 5). The methods must be sufficiently sensitive to eliminate false negatives (i.e. where an analyte is actually present but is not detected). On the other hand, a small number of false positives (i.e. where the analyte is detected although it is not actually present) can be tolerated, since such samples can then be subjected to further analysis by using other methods. The main purpose of screening methods is to eliminate samples that contain the target analyte at below the limit of detection (or below the concentration of interest, since by dilution some semi-quantitative value for the level present can often be obtained). Screening methods can sometimes be used outside the laboratory.

(b) Surveillance.   Such methods are laboratory-based and yield quantitative results. The throughput of samples is generally less than for screening methods. Automatic equipment is often used but the specificity of the methods is not unequivocal. Some performance data should be available for the method, but full validation in a collaborative study will not generally have been achieved.

(c) Confirmatory.   These methods are used where presumptive positive results have been obtained by using screening or surveillance methods, but the degree of certainty that the observed response is due to the target analyte is insufficient for the purpose of the survey. Normally, an analytical technique based on completely different physicochemical principles is required for this purpose. Thus, for inorganic contaminants, methods that could be used include techniques such as colorimetry (visible spectrophotometry), atomic absorption spectrophotometry and inductively coupled plasma optical emission spectroscopy, either alone or in combination with mass spectrometry (see Chapter 4). For organic analytes a wide range of chromatographic methods coupled with varying detection systems have been described. These include TLC, HPLC and GLC, with mass spectrometry as the preferred detection system for unequivocal identification (see for example Chapters 7 and 8).

Further specificity and certainty in the identification of organic analytes can be achieved by the use of alternative systems of clean-up for the removal of potentially interfering co-extracted substances (see Chapter 2).

(d) Reference.   These are methods that have been validated by collaborative studies and have been used in a large number of laboratories over

a long period of time. Many such methods have been incorporated into legislation. They are generally costly to use and are seldom suitable for food surveillance, as the sample throughput is low and the time taken to complete an analysis often exceeds a working day.

Where several methods of analysis exist for the same analyte(s), various criteria can be used to evaluate the different techniques and procedures available. These include:

- limit of detection
- recovery
- precision (repeatability and reproducibility)
- accuracy
- specificity
- time for a single analysis (or batch of analyses)
- cost
- requirements for expensive equipment or staff training
- need for confirmation of identity of analyte

Some of these factors which are of importance for methods used in food surveillance exercises are discussed in more detail below.

(a)   Limit of detection.   The limit of detection of a method is perhaps the most important criterion in contaminant analysis, since one would expect such analytes to be present only at very low concentrations. In many cases the object of the survey will be to determine whether or not particular contaminants are present in foods above or below the statutory (or recommended) limit. The analytical method used should have a limit of detection in such cases of one-tenth of the appropriate limit. The limit of detection of a method is usually defined in terms of the variation in the blank value. Some workers also define a limit of quantification, i.e. a concentration which can not only be detected with reasonable statistical certainty but can also be measured. The limit of detection of the method has been defined as three times the standard deviation of the blank, whilst the limit of quantification is often defined as five times the standard deviation of the blank. However, these are somewhat empirical definitions, but the issue is nevertheless crucial to surveys for contaminants.

(b)   Recovery.   Analysts should check their ability to use the method by additions of known amounts of the specified nutrients or contaminants to samples assumed to be free of the analyte, thus calculating a recovery factor. In some cases a value greater than 95% can be obtained; in other cases, e.g. for some veterinary residues, a recovery factor of only 50% may be the best that can be achieved. The sampling plan should state whether this recovery factor should be used to correct the values obtained on analysis. Where the recovery factor is very high, it is of little importance whether the

data are corrected or not, particularly where interpretations are made against data obtained in toxicological experiments that are themselves subject to a high degree of variability. The only justification for the use of recovery factors is that, where they are known to be consistently low, reporting the raw data alone could give a false impression of the level of contaminant actually present in the food. Normally, however, when low recovery values are obtained they are variable, and hence should not be used to correct the raw data and so give a false indication of accuracy. Furthermore, addition of known amounts of contaminants in a solution to a heterogeneous matrix such as food seldom mirrors the situation in which the contaminant is normally present in the food. Such experiments tend to give a false and enhanced value of the true recovery factor. In all cases where recovery factors have been used to adjust the raw data, this must be clearly stated along with the numerical factors employed. It is vital to check the recovery factor at concentration levels of contaminants likely to be found in foods included in the survey.

(c) Precision. Where several laboratories are to be used in the surveillance programme it is essential that methods of analysis are used that have been validated by collaborative study and that satisfactory values have been obtained, otherwise the results obtained by the separate laboratories will not be directly comparable. In any case, it is probably wise to include some duplicate samples to act as a cross-check between the performances of the different laboratories. The individual laboratories should establish and declare their own repeatability criteria; reproducibility values can only be established by collaborative studies or proficiency testing schemes. Coefficients of variation vary inversely with the concentration level. The Horwitz curve (Horwitz, 1982) should be checked to ensure that the methods used have reproducibility values that fall within an acceptable range. Some typical values derived from the curve are shown in Table 1.2. As these data have been compiled from AOAC studies over the years, they probably represent the best that can be achieved. Most participants in such studies take the work

Table 1.2 Variation of coefficient of variation (CV) with analyte concentration

Analyte concentration		Recommended maximum value for CV
1 ppb	(10^{-9})	45%
0.01 ppm	(10^{-8})	32%
0.1 ppm	(10^{-7})	23%
1 ppm	(10^{-6})	16%
0.001%	(10^{-5})	11%
0.01%	(10^{-4})	8%
0.1%	(10^{-3})	5.6%
1%	(10^{-2})	4%
10%	(10^{-1})	2.8%
100%		2%

seriously and do their best to obtain optimum results. Equally, participants in food surveillance exercises would also be expected to perform at their best. Nevertheless, the Horwitz curve (and the values in Table 1.2) provide a useful baseline standard to aim at.

(d)   Accuracy and specificity.   The accuracy of a method can only be assessed by the analysis of certified reference materials (CRMs) where the concentration of the analyte is known with a high degree of certainty. Unfortunately, there is a limited supply of CRMs and they are costly to purchase for routine use. As a typical food surveillance exercise is likely to encompass a wide range of different food types, it will not be possible to test the method against a CRM for each individual foodstuff. However, one or two CRMs of appropriate composition should be included in the sampling plan when the highest degree of accuracy is required.

The specificity of a method relates to the ability of the detection system to differentiate between the target analyte and other constituents present in the matrix which have not been removed by earlier clean-up procedures. For inorganic analytes the main interferences are spectral (see Chapter 4), whereas for organic analytes less specific detection systems (e.g. flame ionisation detection, FID, or ultraviolet spectroscopy, UV) are generally employed. Hence, it may be necessary to investigate the specificity of the method by using matrix materials known to be free from the target analyte wherever possible. Where specificity is important, it may also be necessary to check those samples with a presumptive positive for the target analyte by another method based on a fundamentally different scientific principle. Thus mass spectrometry, if available, is often used for this purpose. Alternatively, different detection systems can be employed in parallel, e.g. FID and electron capture in gas–liquid chromatography. Some workers are content to use columns of different polarity (hence producing different retention times) or to use measurements at different wavelengths (UV, diode array detection) to obtain further evidence of specificity.

(e)   Other factors.   In today's commercially orientated world, cost is nearly always a limiting factor. This will determine to a large extent the scope of the survey and the number of samples submitted for analysis. However, there is little point in carrying out a survey only to find that the data obtained are not statistically significant. Hence, this factor must always be an important consideration when drawing up the sampling plan.

The availability of laboratories with the necessary equipment and analytical skills to undertake the survey may also be a limiting factor, particularly if the work has to be completed within a short timescale. The scope of the survey (number of samples, analytes and methods to be used) should be agreed with the participating laboratories before the survey is started. The timing of submission of samples, including batch size, should also be agreed, together with the necessary documentation for the reporting of results. Some savings in analysis time and costs may be achievable by sample compositing, although at the expense of less detailed information (see section 1.1.1.2).

1.2.5 Selection of foods

After agreement has been reached on the specific nutrients and/or contaminants to be included in the surveillance exercise, it is then necessary to decide which foods are to be sampled and analysed. In most cases background knowledge from the literature and from contacts with industry will suggest the principal foods likely to contain the analyte under investigation. Priority should be accorded in the sampling plan to such foods, along with staple foods, those consumed by infants and those which form a significant proportion of the diet of special interest groups, e.g. diabetics, old people, ethnic groups or vegetarians. It may also be wise to carry out a limited survey of other foods to check that the contaminant really is absent from such products.

Decisions then have to be taken about the number of samples of each foodstuff to be included in the survey. This will vary depending upon the objectives of the exercise, the accuracy required and whether factors such as seasonal or geographical variations, special diets and the need to analyse individual foods as opposed to diets (mixed foods) need to be considered, or whether samples can be composited. The budget agreed for the exercise will impose further limitations on sample numbers and the design of the sampling plan. Other factors relating to the source of the samples include:

- home-produced or imported foods
- retail outlets, e.g. supermarkets, corner shops, health food stores
- catering, e.g. hotels, restaurants, hospitals, prisons, schools
- presentation of product, e.g. peas can be fresh, frozen, canned, processed or cooked
- how many different brands are to be examined
- are several points of the food chain to be checked?

Consideration of the above factors should then result in a clear scheme for sampling, covering types, sources and numbers of samples to be included in the survey. Further points to be taken into account include: (a) the time-scale for sampling, since many foods are fresh and perishable and all the analytical operations cannot be completed in a short time if the sample numbers are large; (b) who will collect the samples and will training be required? It may be advisable to use local authority Trading Standards Department personnel, since these officers are experienced in the purchasing of appropriate foodstuffs and can cover wide regions of the country. Otherwise, staff from participating laboratories may be a satisfactory alternative, particularly if regional variations are not to be studied systematically. Where imported foods are to be included in the survey, the co-operation of the port health authorities will be required.

1.2.6 Pretreatment, transport and storage of the samples

The sampling plan must include instructions for the pretreatment of the foods to be included in the survey. This is particularly important for fresh foods. Considerations may include: the removal of outside leaves of lettuces, cabbages, etc.; the washing of root vegetables to remove soil; the removal of excess fat from meats; the removal of skin and bones from fish; any cooking operations to be carried out, how and by whom. After collection, samples must be packed into suitable containers for transport and storage before analysis. If the samples are to be analysed for metallic contaminants, plastic containers are to be preferred. Where organic contaminants are of interest, plastic containers must be selected with care to ensure that transfer of additives, e.g. plasticisers, does not occur and thereby cause problems at the analytical stage. Polythene or polytetrafluorethylene products are preferred in many cases to poly(vinyl chloride). Some foods will need to be kept cool or even frozen during transport and storage before analysis. Freeze-drying is also used for some products. This must be clearly specified in the sampling plan. Finally, full documentation and records must be kept of all samples collected. An example of the information required is shown in Table 1.3. Not all the details may be appropriate for every food sample. This should be discussed during the preparation of the sampling plan and the proforma adjusted accordingly. A checklist of factors to be considered when planning a food surveillance exercise is shown in Table 1.4.

Table 1.3 Food sampling survey

Sample number	a unique code
Name of food	including brand name, variety, etc.
State	dried, frozen, etc.
Batch number	
Producer/manufacturer	
Copy of label	nutritional information, use by date
Place sample obtained	location and description of premises (supermarket, corner shop, restaurant, etc.)
Time of sampling	
Date	
Name of sampling officer	
Office address	
Telephone number	

Other comments: condition of product, any treatment, dispatched to.

Table 1.4 Checklist: food contaminants surveillance

Which contaminants are to be determined?
Which foods are likely to contain such contaminants?
Should foods in which the contaminants are not expected to be present be examined in the
  survey?
How many samples of each food are required?
Where will the samples be collected from?
Do seasonal variations need to be taken into account?
Do special diets or sectors of the population need to be considered?
Are regional variations of importance?
Are catering foods to be included in the survey?
Who will take the samples and arrange transport?
What pretreatment of the foods is required?
Which laboratories will take part in the analytical programme?
Have the methods of analysis been agreed?
Who will collect and collate the data produced?
Who will interpret the data, and how?

1.3 Food surveillance exercises

1.3.1 Total diet studies

The organisation of total diet studies in the UK was originally described by
Harries *et al.* (1969). Major changes to the scheme (Buss and Lindsay, 1978;
Peattie *et al.*, 1983) have been made since that time. For ease of analysis it
was considered necessary to divide foodstuffs into groups of commodities
similar in chemical composition. Originally seven such groups were described
but this has now been increased to 20 groups. Changes to such a sampling
scheme should be made as rarely as possible, since the main objective of
total diet studies is to identify trends in exposure to contaminants over a
number of years, rather than to determine a single concentration in an individual food. For this reason, it is also advisable not to change the methods of
analysis too frequently, or the laboratories undertaking the analytical work,
so that any perceived changes in the data produced are real and not artefacts
resulting from variations in analytical operations or equipment used.

The types and quantities of foods purchased for inclusion in a total diet
survey are based on information obtained from the National Food Surveys
undertaken and published annually by the Ministry of Agriculture, Fisheries
and Food (MAFF), UK. The foods, following purchase, are prepared and
cooked as appropriate and then sent for analysis.

This approach inevitably means that any food containing a relatively high
residual level of a contaminant will be masked by the dilution effect of
admixture with other foods which do not contain the same contaminant.
Contaminant levels may thus appear reassuringly low and may even be
missed altogether. However, the total diet study does provide the maximum
information on mean population exposure to given contaminants from a
relatively small number of analyses and, hence, at minimum cost. Therefore,

it is essential in most cases to supplement total diet studies by analyses of individual foods thought to be more likely to be subject to contamination by the target analyte.

In the USA, total diet studies are known as Market Basket surveys. The organisation of the FDA pesticides monitoring programme has been described by Reed *et al.* (1987). The objectives of the programme are (a) to monitor domestic and imported food to check compliance with the tolerances imposed under the Food, Drug and Cosmetic Act, and (b) to gather information on the incidence and levels of pesticides in the food supply. The survey is carried out four times each year at 12 locations, sampling table-ready food. Each 'basket' contains 234 individual foods chosen to represent the typical diet in the USA. Separate samples of infant and toddler food are prepared.

However, sampling is rarely a problem in this type of work since relatively large samples are taken and homogenised before analysis. Furthermore, the primary aims of such surveys are to bring to light changes in the concentration of contaminants rather than the accurate determination of absolute levels of contamination.

1.4   Sampling of commodities for aflatoxins

Perhaps the most extreme example of a sampling problem can be found in the determination of aflatoxins in various food commodities. Aflatoxins (and mycotoxins in general) are produced by mould growth and are therefore found in high concentrations but only at isolated locations in the bulk product, where micro-organisms have invaded the product and where conditions of temperature and humidity have been favourable to growth. This results in an heterogeneous (skewed) distribution which makes it difficult, if not impossible, to obtain a representative (or accurate) sample for analysis. The problem is usually minimised by the taking of very large samples and, where possible, comminuting the product to a small particle size, thus increasing the number of particles in the test sample taken for analysis (i.e. number of particles per unit mass). Errors can occur at the sampling and sub-sampling as well as at the analytical stage. In a study on peanuts, Whitaker *et al.* (1994) showed that sampling, sample preparation and analysis accounted for 92.7, 7.2 and 0.1% of the total variability respectively, when a 2.27 kg sample and a 100 g sub-sample were taken of a product containing 100 μg/kg of aflatoxins.

Some sampling procedures recommended by the Food and Drug Administration, USA, are shown in Table 1.5. This table shows that the smaller the particle size of the product to be analysed, the smaller the total sample size required to obtain a representative sample since, for example, a 100 g sample of flour will contain more particles than a 10 kg sample of grain. It

Table 1.5 Recommended sampling procedures for mycotoxin analyses (USA)

Product	Number of sample units (minimum)	Unit size (lb) (minimum)	Total sample size (minimum) (lb)
Peanuts (shelled)	48	1	48
Peanut butter (smooth)	24	0.5	12
Peanut butter (crunchy)	48	1	48
Corn: meal, flour	10	1	10
Cottonseed	15	4	60
Oilseed meals	20	1	20
Milk powder	10	1	10
Figs	50	1	50

is important to ensure also that the sample units are collected from as many sites as possible in the lot (consignment), selected at random.

In the UK, slightly lower minimum sample sizes are recommended for sampling nuts and figs (see Table 1.6 and Statutory Instrument, 1992). These regulations limit the concentration of aflatoxin to $4\,\mu g/kg$. With such a stringent limit it is vital that sampling and analysis of suspect products are carried out carefully and accurately. The skewed distribution of aflatoxins in figs is illustrated in Table 1.7 following work by MAFF (1993). A 10-tonne consignment of whole dried figs was obtained via the Suffolk Port Health Authority in 1990. Two hundred 12-kg boxes were selected at random from the 850 in the total consignment and analysed individually for total aflatoxins. The results obtained are shown in Table 1.7. Whilst in this case

Table 1.6 Sampling requirements for aflatoxins in nuts, nut products, dried figs and dried fig products

Product	Minimum number of samples	Minimum sample size (kg)
Peanut butter, smooth	24	5
Peanut butter, crunchy	24	5
Peanuts, shelled raw	30	10
roasted	30	10
Peanuts, in shell	30	20*
Cashews	20	3
Brazil nuts, shelled	20	3
in shell	20	6*
Pistachios, shelled	20	1.5
in shell	20	3*
Hazelnuts, shelled	20	3
in shell	20	6*
Almonds, shelled	20	3
in shell	20	6*
paste	20	3
Figs, dried whole	20	20
paste	20	5

Samples shall be taken as randomly as possible from throughout the consignment. *Weight of nuts in shell.

Table 1.7 Distribution of aflatoxins within a consignment of whole dried figs

Total number of boxes analysed	Total aflatoxin level (μg/kg) found						
	Number of boxes in the range						
	<4.0	4.0–9.9	10.0–19.9	20.0–49.9	50.0–99.9	100–199	>200
200	75	59	29	23	8	5	1

the contamination level found (mean 33 μg/kg) was well above the statutory limit, it does illustrate the difficulties encountered in sampling products of this type for low levels of contaminants distributed non-homogeneously throughout the product.

1.5 Some other practical problems in food sampling

1.5.1 Trace metals in packeted teas

Since foods are such heterogeneous mixtures of inorganic and organic constituents, some aqueous, some non-aqueous, it is not surprising that problems can arise when small quantities are withdrawn for analysis. However, nobody would expect to experience difficulty in sampling tea for the determination of trace elements. Yet problems have been reported by Michie and Dixon (1977). They found that tea is in fact a mixture of leaf and dust and that higher concentrations of trace elements occur in the dust than in the tea leaf. This explains the variations observed in Table 1.8. Checks on the digestion and determination procedures showed that recovery of added lead fell in the range 95–108%. Hence, the variation in the results observed in Table 1.8 could only be ascribed to a non-uniform

Table 1.8 Variation between repeat lead determinations of non-retail leaf teas

Sample	Lead content (mg/kg)	
A	6.2	1.1
B	1.2	6.6
C	3.9	3.9
D	25.0	6.5
E	3.9	9.5
F	4.2	2.4
G	0.2	0.6
H	0.6	0.4
I	0.2	1.4
J	0.6	0.4

Samples A–F, tea bags; samples G–J, loose tea; data from Michie and Dixon (1977).

Table 1.9 Distribution of lead between leaf and dust fractions of teas

Sample	Total wt of leaves fraction (g)	Total wt of dust fraction (g)	% of leaves in sample	% of dust in sample	Pb (mg/kg)	
					Leaves	Dust
A	6.0	2.5	44	18	0.2	17.2
B	7.6	2.4	55	18	5.9	14.6
C	4.7	2.7	33	19	0.4	8.5
D	5.0	3.2	33	21	0.2	11.7
E	4.8	2.4	34	17	8.6	10.0
F	4.4	2.6	33	19	0.2	6.0
H	84.0	6.0	78	6	0.2	3.7

Dust fraction: passed by 50 mesh sieve; broken leaves and dust fraction passed by 30 mesh, retained by 50 mesh sieve—not analysed.

distribution of lead in the sample of tea. This hypothesis was confirmed by separating the dust from the leaf by using a sieve. Determinations of lead were then made on the separate leaf and dust fractions (Table 1.9). This table shows clearly that the dust fraction contains much higher concentrations of lead than are found in the leaves. Similarly, other metals were also found at higher concentrations in the dust fraction than in the leaf (Table 1.10).

1.5.2 Trace metals in canned foods

Until recently most cans for foods were made from steel and contained a soldered seam. The inner surface was protected from corrosion by a thin layer of tin and in some cases a film of lacquer as well. Hence, products such as canned fruits and vegetables were potential sources of contamination by metals such as lead, tin and iron.

Despite the fact that foods are usually stored in cans for relatively long periods of time (6 months to several years) when one might have expected equilibrium to be attained, analysis of the contents showed that there is a distribution of trace elements between the liquid and the solid contents of the can; generally higher concentrations are found in the solid portion. In the case of canned fruits (Table 1.11) this may not be important since the consumer would normally eat both the fruit and the syrup. The sampling plan would therefore describe how the whole contents of the can were to be homogenised and small portions removed for analysis.

A similar distribution is observed in the case of canned vegetables (Table 1.12). However, in this case most consumers would drain the brine away from the solid before heating in clean water. In this instance the sampling plan must indicate whether or not the whole contents of the can are to be sampled, or, if only the solid portion is sampled, how this operation is to be carried out and whether any subsequent cooking procedures are to be performed before samples are taken for analysis.

Table 1.10 Distribution of metals between leaf and dust fractions of teas

Sample	Total wt of leaves fraction (g)	Total wt of dust fraction (g)	% of leaves in sample	% of dust in sample	Metal content (mg/kg)											
					Cd		Ni		Fe		Zn		Mn		Cu	
					Leaves	Dust	Leaves	Dust	Leaves	Dust	Leaves	Dust	Leaves	Dust	Leaves	Dust
A	6.0	2.5	44	18	<0.1	<0.1	5.2	3.4	150	1300	35	105	750	500	17	32
B	7.6	2.4	55	18	<0.1	<0.1	5.9	3.5	350	1500	40	140	750	500	20	33
C	4.7	2.7	33	19	<0.1	<0.1	5.2	3.4	150	1350	35	100	800	500	19	33
D	5.0	3.2	33	21	<0.1	<0.1	5.2	3.4	300	1400	35	125	750	500	20	33
E	4.8	2.4	34	17	<0.1	<0.1	5.6	2.0	200	1200	140	120	750	500	20	31

Dust fraction: passed by 50 mesh sieve; broken leaves and dust fraction passed 30 mesh, retained by 50 mesh sieve—not analysed.

Table 1.11 Trace metals in canned fruit

	Pb (mg/kg)	Sn (mg/kg)	Fe (mg/kg)	
Grapefruit (P)	0.18	75	5.3	S
	1.0	130	6.9	F
Blackcurrants (La)	1.0	45	1300	S
	10	160	2600	F
Pineapple (P)	0.12	55	31	S
	0.35	105	3.0	F

S = syrup; F = fruit; P = plain; La = lacquered.

Table 1.12 Distribution of trace metals in canned vegetables

Vegetable	Can		Metal content (mg/kg)		
			Lead	Tin	Iron
Green beans	La	L	0.10	5	2.8
		S	0.70	10	4.8
Haricot beans	La	L	0.02	5	9.8
		S	0.15	10	26
Petit pois	La	L	0.04	10	10
		S	0.55	20	12
Celery hearts	La	L	0.13	10	4.0
		S	1.50	20	3.4
Sweetcorn	La	L	0.04	10	1.0
		S	0.30	20	6.4
Mushrooms	P	L	0.01	15	5.1
		S	0.04	55	16

P = plain; La = lacquered; L = liquid; S = solid.

1.5.3   Vitamin D in cereals

Whilst a new method for vitamin D was being developed, it was found that the concentrations detected in cereals were far lower than the manufacturer's declared value. Known additions of the pure chemical to the sample were recovered satisfactorily. Subsequent work established that the dust at the bottom of the packet contained far higher concentrations of vitamin D than was present in the product as a whole. This is shown in Table 1.13. It is possible that this situation arose as a result of addition of vitamin D by

Table 1.13 Vitamin D in cereals

Sample	Wt taken (g)	Vitamin D (μg/100 g)	Vitamin D in sample (μg)
1	40	0.5	0.2
2	50	0.5	0.25
3	41	0.5	0.2
4	44	0.9	0.4
5 (fines)	44	7.5	3.3

spraying a solution over the product. On drying, the vitamin would be present in powdered form and so could fall off into the dust at the bottom of the packet during transport to the shop and to the home

1.5.4 Monomers in plastics

When it was first realised that vinyl chloride monomer was carcinogenic, there was a need to measure low levels of this compound in food contact materials as well as the residues of it that are transferred to foods. A report has been published (MAFF, 1978). In developing the method for vinyl chloride monomer in bottles used to contain foods, e.g. soft drinks, alcoholic drinks and cooking oils, it was found that the monomer content varied from place to place. This is illustrated in Figure 1.3. Subsequent improvements in packaging technology have resulted in substantially reduced levels of the monomer occurring in the plastic.

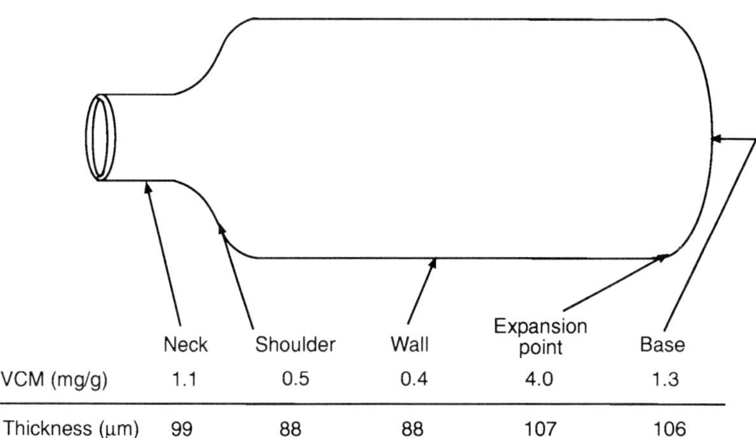

	Neck	Shoulder	Wall	Expansion point	Base
VCM (mg/g)	1.1	0.5	0.4	4.0	1.3
Thickness (μm)	99	88	88	107	106

Figure 1.3 Variation of vinyl chloride monomer in a PVC bottle.

1.6 Sample pretreatment

Once the primary samples of foods have been obtained, it is necessary to consider whether any pretreatment is required before analysis. Some examples of possible processes that may have to be employed are shown in Table 1.14. The application of any or all of these processes must be described fully in the sampling plan.

Removal of inedible parts is normally desirable in food surveys. This may include the de-stoning of fruit, the filleting of fish and the jointing of meat carcasses. Closely associated is the removal of parts of the food not normally consumed by the average consumer. This can include the removal of outer

Table 1.14 Possible pretreatment processes prior to analysis

1.	Removal of inedible parts
2.	Removal of outer leaves, peel
3.	Washing
4.	Cooking
5.	Grinding and sieving
6.	Homogenisation
7.	Sample division
8.	Compositing

leaves from vegetables and the peel from fruit. Whilst most people would peel bananas and oranges, in the case of grapes and apples the decision is not quite so straightforward. It is also necessary to specify whether the core of apples should be included in the analytical sample. Many people consume both the skins and the starchy part of potatoes. Depending on the contaminant and the objectives of the survey, it might be necessary to test both fractions separately. Removal of the outer layers of a food is often not desirable when a 'worst-case' scenario is being investigated, since chemicals such as pesticides or atmospheric pollutants are likely to be found in a higher concentration on the outside of the product. Similar considerations apply to any washing operations used before analysis. It is normally a good idea to wash off any soil from root vegetables, and again the practice of washing fruit prior to consumption varies from household to household. Cooking processes (e.g. time and temperature) should be described in detail where foods normally consumed after cooking are to be analysed. Both contaminants and nutrients may be destroyed by the cooking process; alternatively, concentrations may apparently be increased by reason of the removal of moisture and other volatiles from the product.

The inhomogeneity of most foodstuffs has been referred to several times already in this chapter. Where only small portions are taken for analysis, it is necessary to homogenise the product as much as possible. For dry foods this is best achieved by grinding and sieving. Some grinding equipment where a high degree of mechanical action is used produces frictional heat which will change the moisture content of the sample and may also degrade certain sensitive constituents. It may be necessary to check the moisture content before and after grinding and then to recalculate any analytical data obtained back to the original state. Where excessive heat must be avoided, dry ice can be added to counteract the production of frictional heat. For foods containing larger amounts of moisture which have to be stored before analysis, an alternative approach is to freeze-dry the product. Volatiles can also be collected in some cases. Homogenisation using a high-speed blender is often appropriate for samples consisting of a mixture of solid and liquid fractions, e.g. canned fruit or meat products. The latter commodities may benefit from passage through a mincer (sometimes more than once).

Following homogenisation of the foodstuff, it will be necessary to specify how the test portion for analysis is to be removed from the bulk material. Alternatively, sample division may be more appropriate prior to homogenisation. Again, this forms an essential part of the sample plan. Sample compositing has already been discussed (see section 1.1.1.2), and may need to be considered at this point.

During the analysis of samples for contaminants it is usual to employ a number of quality-control procedures. Check recoveries are a vital part of this process. After homogenisation and removal of one or more test portions of the sample for analysis, it is normal practice to 'spike' or fortify the portion with a known amount of analyte to check on the recovery through the analytical method. This has already been described (see section 1.2.4, *(b) Recovery*). The exact point at which the 'spike' is added must be stated in the sampling plan, as well as the mode of addition.

1.7 Sampling equipment for foodstuffs

Foodstuffs to be sampled can vary from a large heap of grain, through canned or packaged products, to entire meals, as well as individual products available in the shops. The last-named products usually present no problems as far as physical abstraction of the sample is concerned. The main problems centre on which particular packets should be selected, and how many. These questions have been addressed in earlier parts of this chapter.

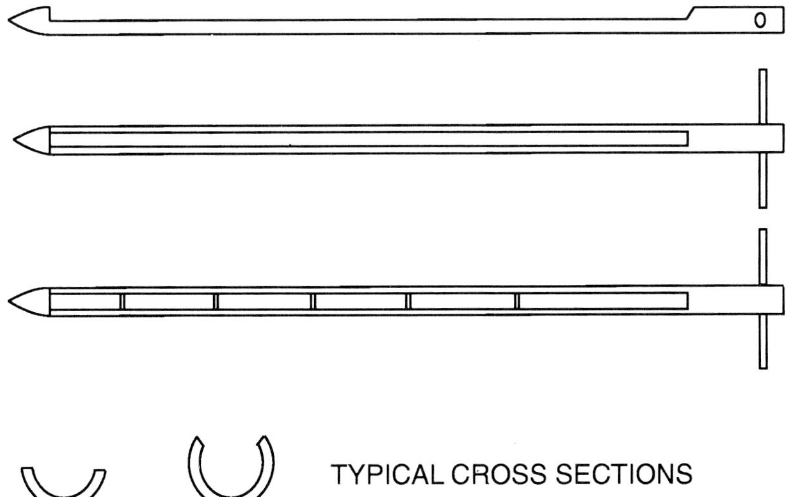

Figure 1.4 Open-sided sampling spear and divided spear for dry free-running powders.

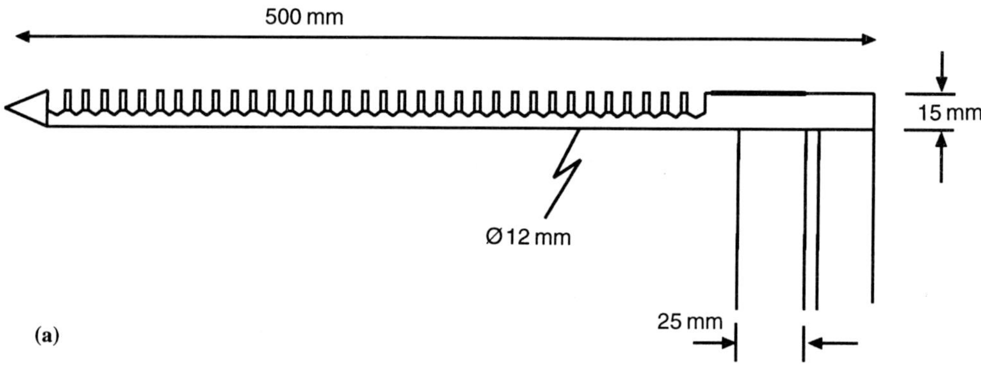

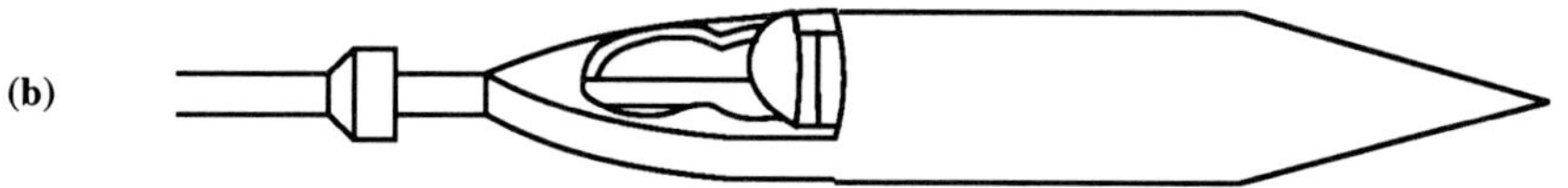

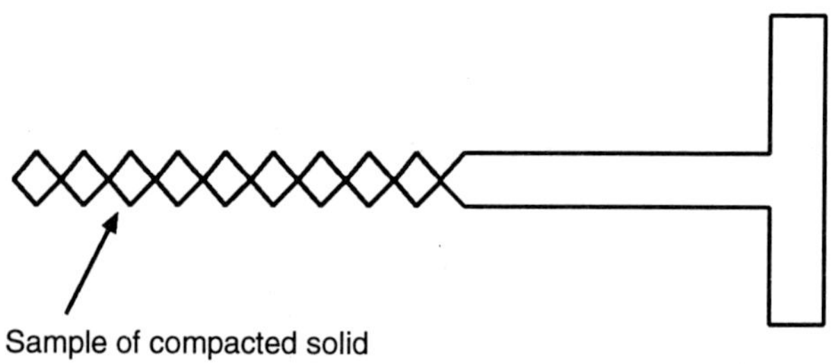

Figure 1.5 (a) Shuttered sampling spear for dry free-running solids; (b) grain probe; (c) auger.

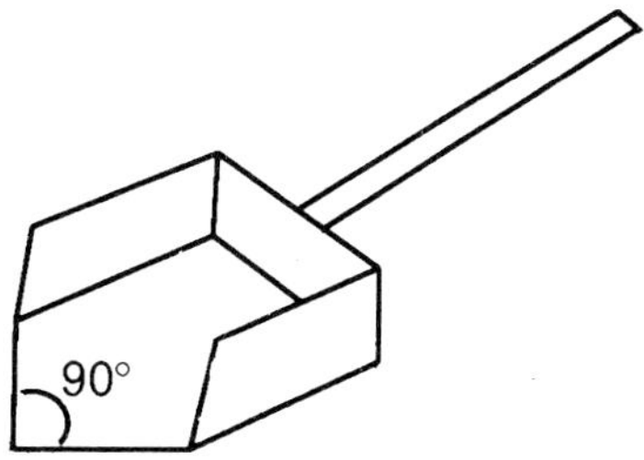

Figure 1.6 Sampling scoop for solids.

From time to time it is necessary to sample bulk quantities of foods. These may be heaps of grains or large bulks of butter or cheese. In such cases sampling spears or augers are available. Some examples of the types of equipment available are shown in Figures 1.4 and 1.5. The use of sampling spears can require a considerable amount of force to penetrate into the bulk material. It is usual then to twist the spear through 180° before withdrawing a core sample of the product. Augers may be more suitable for compacted products that are difficult to penetrate. Where the product is thought to be heterogeneous, several incremental samples should be drawn and then mixed to form an aggregate sample of adequate size. Spears generally only remove small amounts of the product. Static heaps are almost impossible to sample satisfactorily if the product is heterogeneous. The best that can be achieved is by using a scoop or shovel (Figure 1.6) of suitable dimensions to remove a number of incremental samples from different parts of the heap, mixing in between. Heterogeneous heaps are best sampled 'on the move'. Automatic sampling equipment is available to remove samples from conveyor belts, but such equipment needs to be checked to ensure that representative portions are being abstracted.

1.7.1 Sub-division of solids

To ensure that the sample taken is truly representative of the bulk material, it is usually necessary to take a fairly large number of incremental samples which are subsequently mixed together to form the aggregate sample. In many cases the aggregate sample is then too large to send to the laboratory for analysis, or alternatively, it is often necessary to split the aggregate sample into several portions so that it can be analysed separately by more than one laboratory. Furthermore, on arrival at the laboratory the sample has normally to be sub-sampled prior to removal of a test portion for analysis.

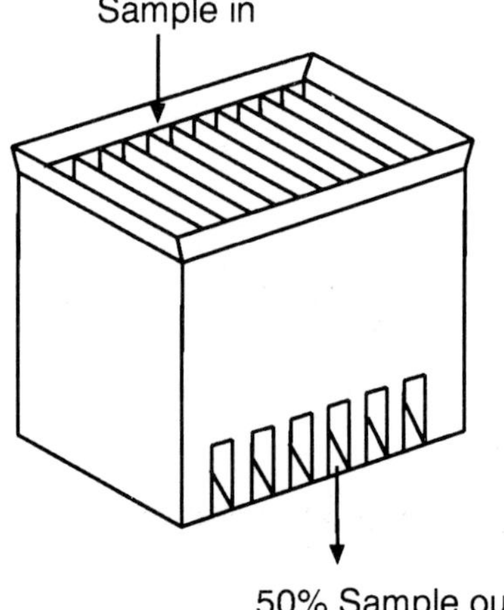

Figure 1.7 Schematic diagram of a static riffler.

Where the aggregate sample is very large owing to the amalgamation of a large number of increment samples, the heap should first be well mixed using a shovel on a clean surface. The heap is then flattened and divided into four roughly equal parts. Two diametrically opposite quarters are removed and the two remaining parts re-mixed. This process can be continued until a reduced sample of suitable size is obtained. Alternatively, a riffler can be used. A static version is shown in Figure 1.7 and a dynamic type in Figure 1.8. The latter equipment is better when representative portions of heterogeneous solids are required, since segregation is less likely to occur.

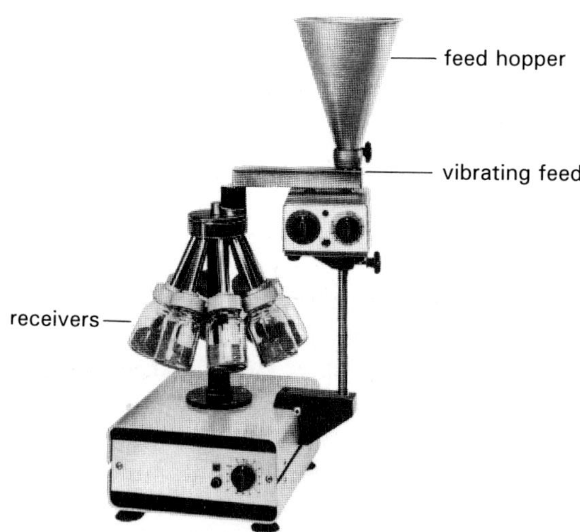

Figure 1.8 Schematic diagram of a dynamic riffler.

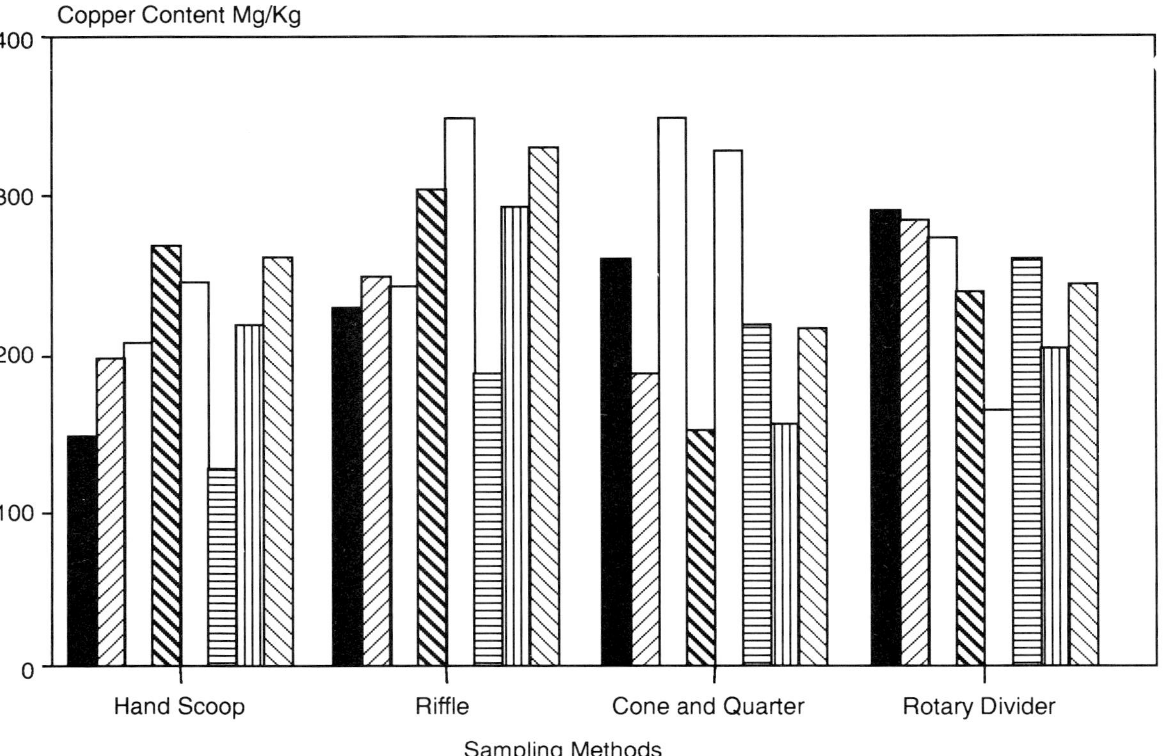

Figure 1.9 A comparison of sub-sampling methods for copper content in turkey feeds (histogram bars show replicate sampling and analysis by various methods).

Figure 1.9 shows the results obtained when a sample of turkey feed containing copper was sampled by a number of different procedures. The spinning riffler clearly gave the best results, although the other procedures, if carried out carefully, can also give acceptable and representative sample portions in this particular case.

1.8 Conclusions

Sampling is an important first step in any food surveillance programme. It is vital that all aspects of sampling and analysis are discussed and agreed at the start of any surveillance exercise. In turn this requires that the aims and objectives of the programme are clear and unambiguous. A sampling plan can then be drawn up to ensure that these aims and objectives will be satisfied following sampling and analysis.

References

AOAC (1989) Guidelines for collaborative study procedure to validate characteristics of a method of analysis. *J. Assoc. Off. Anal. Chem.*, **72**, 694–704.

Buss, D.H. and Lindsay, D.G. (1978) Reorganisation of the UK total diet study for monitoring minor constituents of food. *Food Cosmet. Toxicol.*, **16**, 597–600.

Harries, J.M., Jones, C.M. and Tatton, J.O'G. (1969) Pesticide residues in the total diet in England and Wales, 1966–7. 1—Organisation of a total diet study. *J. Sci. Food. Agric.*, **20**, 242–245.

Horwitz, W. (1982) Evaluation of analytical methods used for regulations of food and drugs. *Anal. Chem.*, **54**, 67A–76A.

MAFF (1978) Food Surveillance Paper no. 2, Survey of vinyl chloride content of polyvinyl chloride for food contact and of foods. Ministry of Agriculture, Fisheries and Food. HMSO, London.

MAFF (1993) Food Surveillance Paper no. 36, Mycotoxins: 3rd Report, Ministry of Agriculture, Fisheries and Food. HMSO, London, 12–13.

Michie, N.D. and Dixon, E.J. (1977) Distribution of lead and other metals in tea leaves, dust and liquors. *J. Sci. Food Agric.*, **28**, 215–224.

Peattie, M.E., Buss, D.H., Lindsay, D.G. and Smart, G.A. (1983) Reorganisation of the British Total Diet Study for monitoring food constituents from 1981. *Food Chem. Toxicol.*, **21**, 503–507.

Reed, D.V., Lombardo, P., Wessel, J.R., Burke, J.A. and McMahon, B. (1987) The FDA pesticides monitoring programme. *J. Assoc. Off. Anal. Chem.*, **70**, 591–595.

Royal Society of Chemistry (1994) *Official and Standardized Methods of Analysis*, 3rd edn, C.A. Watson (ed.). Royal Society of Chemistry, Cambridge.

Statutory Instrument (1992) Aflatoxins in Nuts, Nut Products, Dried Figs and Dried Fig Products Regulations 1992 (S.I. [1992] No 3236). HMSO, London.

Whitaker, T.B., Dowell, F.E., Hagler, W.M., Giesbrecht, F.G. and Wu, J. (1994) Variability associated with sampling, sample preparation, and chemical testing for aflatoxin in farmers' stock peanuts. *J. Assoc. Off. Anal. Chem. Intern.*, **77**, 107–116.

Youden, W.J. (1967) The role of statistics in regulatory work. *J. Assoc. Off. Anal. Chem.*, **50**, 1011–1013.

Appendix 1.1   Glossary of terms used in sampling

Bias	an error which on replication does not tend to zero.
Comminution	the reduction of particle size by grinding, crushing, milling, etc.
Consignment	the quantity of material sent by the vendor. Other similar terms are 'batch' or 'sampled portion' or 'lot', i.e. a quantity of a material constituting a unit and having characteristics presumed to be uniform.
Characteristic	a chemical or physical property of a material.
Division	splitting of a sample into one or more identical portions. Also known as reduction.
Error	the difference between the experimental value and the true value. Does not include bias. Can include random variations from manufacturing processes, sampling and analysis.
Heterogeneous material	a material with a wide variation between increments. Similar to blends.
Sample incremental	means a portion removed from one point of the consignment, e.g. a shovelful.
aggregate	means a mixture of incremental samples.
reduced	means a representative part of the aggregate sample. Not to be confused with grinding, etc.
final	means a representative part of the reduced sample, or aggregate sample if no reduction is required. The final sample is the portion sent for analysis.
grab or spot	a single increment taken for a rough check of analyte levels.
random	a sample taken by selecting increments from all parts of the consignment with equal probability.
systematic	a sampling scheme where increments are selected on the basis of a fixed-term pattern.
stratified	a sampling scheme where the consignment is divided into strata or layers, each being sampled separately and then combined.

2 Automated clean-up techniques for trace component analysis in complex biological matrices including foods

MARTIN J. SHEPHERD

Summary

Chromatography-based techniques for automating trace component determination in biological matrices are examined. Once samples are brought into solution or extracted, a range of fluid handling systems may be applied to automate most analyses. However, extraction remains a limiting step and the development costs and time required to automate any method successfully must be carefully considered. These and other factors which will influence the decision to automate any analytical method are explored and the problems of trace analysis discussed. Methods based on solid-phase extraction, dialysis and column switching approaches are evaluated in detail and interesting approaches described.

Abbreviations

AASP	Advanced automated sample preparation
ASPEC	Automated sample preparation by extraction and concentration
ASTED	Automated sample treatment by extraction and dialysis
CAS	Chemical Abstracts Service
CE	Capillary electrophoresis
CV	Coefficient of variation
ECD	Electron capture detector
GPC	Gel permeation chromatography
IAC	Immunoaffinity chromatography
IR	Infra-red
LC–GC	Coupled liquid chromatography/gas chromatography
LC–LC	Coupled liquid chromatography/liquid chromatography
LIMS	Laboratory information management systems
MSTFA	Monosilyltrifluoroacetamide (silylation derivatising reagent)
MTBE	Methyl t-butyl ether

OC	Organochlorine
PIONA	Paraffin-iso/normal olefin–naphthalene–aromatic
SEC	Size exclusion chromatography
SFE and SFC	Supercritical fluid extraction and chromatography
SPE	Solid-phase extraction
THF	Tetrahydrofuran

2.1 Introduction

This chapter examines the automation of trace analysis methods which employ chromatography for the preparation or final measurement of analytes. The cited literature covers the trace analysis of complex biological matrices, including clinical and environmental samples and foods. The focus is on solid-phase extraction, dialysis and column switching methods. Automation of other techniques such as immunoassay, flow injection analysis or robotic systems is not covered. No attempt has been made to present a comprehensive review of automation applications, but examples have been selected to highlight current practice and interesting approaches. General reviews and articles covering aspects of the subject include those by Cortes (1990), Grob (1991), Gere *et al.* (1993), van de Merbel *et al.* (1993), Brinkman (1994) and Vreuls *et al.* (1994).

For solid samples, analysis may be considered to fall into three stages: initial extraction of the sample from the matrix into a fluid medium; purification of the extract; and quantification of the analyte by chromatographic measurement. The work covered here explores the combination of the last two processes. Sample extraction is a major consideration in this sequence of events but is difficult to automate, particularly given the large sample size often necessary to achieve required detection limits, and will not be considered further except briefly in relation to supercritical fluid extraction.

Initial extracts usually require processing to bring the residue mass and analyte concentrations within the dynamic ranges of the final separation and measurement systems. Biological samples present problems associated with the protein likely to be present. Proteins bind to most chromatographic supports, often irreversibly, and must be taken into account when developing systems where columns are to be re-used. Re-use will always be the case for analytical columns and is usually desirable for precolumns because it simplifies instrument design. However, systems for automated solid-phase extraction (SPE) are commercially available and present one option for dealing with the problem. Dialysis may be employed for the removal of proteins and other macromolecules. This is not often used off-line because it is relatively slow and because small molecules become diluted in a large volume of dialysate, which is likely to be difficult to handle. Both of these problems are partially overcome by on-line systems, as discussed later in this chapter.

The other major difficulty presented together with elimination of protein is the probable complexity of the sample. The typical analyte is a low-molecular-weight middle-polarity molecule. This class of compound represents, after removal of lipids, 0.2–0.3% of the mass of a matrix such as cattle liver, and fractions isolated on this basis may contain many thousands of components at trace analysis concentrations. Current chromatographic techniques, despite being termed 'high performance', are unable to resolve the complex mixture of components typical of crude extracts from biological matrices such as food or environmental samples.

This is shown clearly in the work of Davis and Giddings (1983) and Martin *et al.* (1986). In order to achieve a 95% probability of unit resolution of a randomly selected mixture of 20 components, chromatographic efficiencies in excess of 10^7 theoretical plates are required. This may be contrasted with current typical column efficiencies of about 2.5×10^4 plates for HPLC and 5×10^5 for capillary GC. Although more plates may be obtained by coupling several columns, this is rarely done in practice as it is at the expense of long run times—proportional to column length, whereas the increase in efficiency is proportional to the square root of length—and particularly for HPLC because longer columns create higher back pressures which often require lower flow rates, further increasing analysis time.

It seems likely that any extract from a complex biological sample will contain a chemically heterogeneous mixture of components, and Giddings (1995), via the concept of sample dimensionality, has shown that such a mixture will always be distributed randomly in terms of retention. The efficiency problem is likely to be even more severe in practice, as preliminary clean-up will ensure that the components of a partially purified extract are not selected at random but have similar gross physicochemical properties, while still likely to be of high sample dimensionality. Further chromatographic resolution would demand a separation mechanism unlike any of the preceding stages, yet the number of available mechanisms is quite limited. Thus the 'brute-force' approach of employing high separation efficiency by itself in order to achieve unambiguous determination of a specific analyte is usually not tenable, and it is necessary to choose methods which combine efficiency with selectivity, either of separation, or of detection, or often both. Although high selectivity of detection may in rare instances by itself accomplish the required determination, susceptibility to interferences even when these are not visualised is likely to reduce confidence in the results. For example, co-elution of non-responding sample components may seriously affect responses of many detectors to target molecules, including both fluorimeters and mass spectrometers.

The limitations of chromatographic resolution of sample components are indicated in Figure 2.1. The original peak (a) was taken from an organophosphorus pesticide standard separated on a capillary GC system with an efficiency of about 250 000 theoretical plates. The peak shape is close to

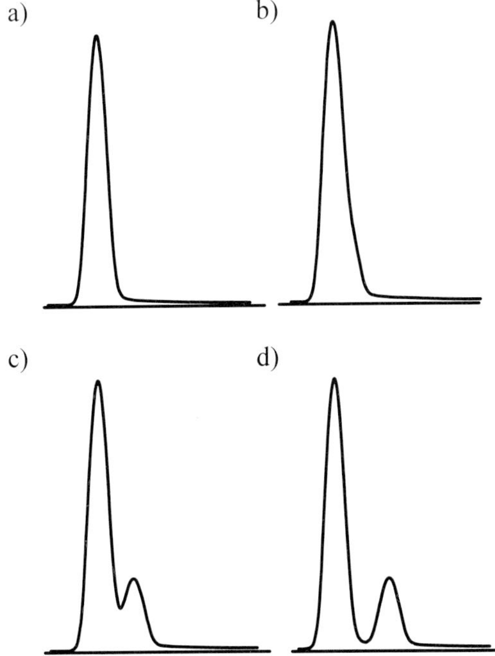

Figure 2.1 Superposition of digitised GC peaks to demonstrate the separation achieved at different resolutions. a: original peak; b–d: original plus one-quarter copy at 2, 4 and 6σ resolution respectively.

ideal, having an asymmetry at 10% height of only 1.001. The digitised peak was manipulated in a computer spreadsheet to simulate potential interferences. A copy of the peak was reduced to 25% of the area of the original and superimposed on its tail at resolutions of 0.5 (b), 1.0 (c) and 1.5 (d). A value of 1.0 corresponds to unit (4σ) resolution, which has widely been regarded as a satisfactory separation. However, it is clear that unit resolution is almost the minimum acceptable resolution of the 'interference' peak. Separation at half-unit resolution could conceivably be missed by inexperienced staff, particularly if (as is likely) the peak shape were worse. It is interesting to compare peak height and area 'integrations' of the two peaks. Height measurements are much less susceptible to poorly-resolved interferences such as the half-unit resolution example, where integrators would show a 25% increase in peak area, but only a 4% increase in peak height. For the unit resolution example, both peak area and peak height measurements for the major component are accurate to within 0.1%, assuming 4σ limits for area integration.

In order to attain required detection limits at the measurement stage, it may be necessary to inject large amounts of sample or sample equivalent of residue, which emphasises the requirement for preliminary clean-up, because the large bulk of the crude extract usually obtained with biogenic samples would cause problems with most chromatographic techniques. Capillary GC columns can accept only microgram amounts of material, while even HPLC columns are capable at the best of coping with sample at perhaps 0.01% of the stationary phase mass before becoming unacceptably overloaded. For typical analytical reverse-phase columns, this implies a sample mass of less than 1 mg, but even then resolution is likely to be compromised. Separations based on partition or size exclusion chromatography (SEC) provide an exception to this rule because phase surface area is not a limiting factor, and analytical SEC columns for example can accept 10–20 mg of low-molecular-weight compounds such as lipids in true SEC mode. Substantially higher capacities are possible when using these columns in mixed SEC–partition mode. However, except for pesticide analysis, SEC has not received the attention it perhaps deserves as the initial stage in clean-up of crude extracts.

Thus effective clean-up will usually require several stages of chromatographic separation. Manual clean-up essentially consists of several relatively low-resolution chromatographic operations to reduce the bulk of the extract in a manner that eliminates potential interferences with the final analytical separation. Fluid flows are simple to manipulate, and one major goal of analytical automation is to combine clean-up with analysis by coupling different modes of column chromatography on-line (multi-dimensional chromatography). This imposes a number of major system requirements:

(a)   High loading capacity of initial stages, together with the ability to accept large sample volumes and remove protein
(b)   Different retention mechanisms for the columns
(c)   Reconcentration of analyte before loading on to the second column.

Once a sample is in solution it is very easily controlled by pumps and valves, and the time and tedium penalties of manual clean-up actively promote method automation. However, a decision to automate a method should be taken only after all the factors have been considered. There are a number of potentially serious drawbacks involved.

2.2   Rationale for automating trace analysis

Manual sample throughput is inherently limited. In addition, manual methods are usually carried out in batches, which means that if something goes wrong, the whole batch may have to be discarded.

It seems reasonable to suppose that automated methods should be capable of greater precision than manual approaches. Opportunities for contamination are generally reduced, while each sample is treated identically in terms of both the manipulations employed and the time taken for each stage in the process. The results obtained, however, do not consistently reflect this expectation. One possible explanation is that human operators are able to evaluate the various stages of the clean-up process critically, and to compensate for variations in method performance.

2.3   Criteria for selecting methods for automation

Sample numbers

Automation is applied very successfully in clinical chemistry, where the typical workload often consists of monitoring physiologically active drugs in long runs of biological fluid samples. It is also often the case that analyte concentrations in these samples are considerably higher than residues or other target compounds in more complex matrices such as foods or environmental samples.

Matrix variability

Many methods are very specific for a single type of sample, and automation will generally be easier where very similar samples are to be analysed. One exception to this is immunoaffinity chromatography, where the high selectivity of antibodies is employed to isolate analytes from any of a wide range of sample matrices.

Analyte concentration

The lower the detection limit required the more difficult trace analysis becomes, as demonstrated empirically by the Horwitz equation (Horwitz, 1982) relating CV to analyte concentration. There is likely to be more incentive to automate the complex methods typically required to achieve high sensitivity, both because of the increased cost of manual operation and because the repetitive but demanding nature of those operations increases the demands on staff vigilance.

2.4   Advantages of automation

Identical handling conditions

It is clear that automated operation should increase the reproducibility of each stage in a method. However, the loss of the 'critical eye' of the human operator can sometimes offset this.

Continuous operation—sample throughput

Manual sample throughput is inherently limited.

Sample protection

For samples susceptible to degradation by light or oxygen, coupled column methods provide some protection from exposure, although holding time in sample carousels (see below) may introduce complications.

Operator protection

There is less exposure of personnel to toxic solvent vapours.

2.5 Disadvantages of automation

Analyte stability

Samples may be held for long periods of time in carousels before analysis. Exposure to room temperature, oxygen, solvent impurities or buffer pH may be important and must be checked.

Sample heterogeneity

Not all operations can be successfully automated using normal laboratory instrumentation. Extraction remains a labour-intensive stage—automation of the handling of large samples requires substantial investment in industrial robotics. Processes to be avoided include those where there is a possibility of considerable variation between samples, such as emulsion formation in liquid–liquid extraction.

Critical samples

Can these be trusted to unattended operation?

Urgent samples

The difference between batch and automated sequential working has implications for processing urgent samples.

Staff approach

(a) A 'black box' mentality must be avoided in all aspects of automation, but particularly with regard to integration systems and laboratory information management systems (LIMS). All results need critical evaluation; this may be no more than simple inspection to ensure that the expected pattern of peaks has been found and that peak areas reported for individual peaks are reasonable. It is important to plot the baselines being generated by integrators.

(b)   Lack of the manual operator's critical eye.

Cost

(a)   Automated systems are inherently more complex than manual ones; thus capital costs will be higher.
(b)   Automated systems contain more components and there will be greater dependence on trouble-free operation, so additional preventive maintenance may be required.
(c)   Development costs—considerable time may be required to produce an effective and reliable method, particularly if it is necessary to handle a range of different sample matrices or to perform multi-analyte analysis.

2.6   Methods for automation

2.6.1   *Headspace GC and purge and trap*

Although these older approaches to automated sample preparation have not seen much development recently, they provide excellent solutions for the analysis of very volatile analytes and should always be considered for such compounds. They will not be covered in any detail here. Headspace methods may be exemplified by the determination of halogenated fumigants in cereals, nuts and seeds at concentrations down to 1 μg/kg (Norman, 1991). Purge and trap procedures provide potentially higher sensitivities, but suffer from breakthrough problems and system interferences. They may be combined with headspace methodology as demonstrated by Gan *et al.* (1995), who carried out 300 determinations per day of methyl bromide in air, with a detection limit of 17 ng/m^3.

2.6.2   *Auto-samplers*

These have become ubiquitous on chromatographs. Older machines tend to be pneumatically operated and, although cheaper, have a major disadvantage for trace analysis because substantial volumes of sample are required simply to flush out the system. Often, only one injection may be made from several hundred microlitres of sample. More modern positive displacement syringe instruments permit 1 μl injections to be made from samples of less than 10 μl. One factor to consider with some auto-injectors is that sample vials may be restricted in volume. This may prove limiting as some approaches to trace analysis require the injection of large samples. Some commercial instruments combine the function of auto-injection with sophisticated liquid handling abilities, permitting programmed precolumn derivatisation or other manipulations of samples. This approach has been developed further to afford automation of solid-phase extraction. Advantage has been taken of the programmability of current auto-injectors to automate procedures such as liquid–liquid extraction for determination by capillary GC of organophosphorus pesticides in water (van der Hoff *et al.*, 1993). Aspiration and re-injection of sample and methyl *t*-butyl ether (MTBE) (1.5 ml each) within closed vials followed by withdrawal of the upper phase after settling provided extraction of 1.5 ml water with 70–100% recovery for most pesticides. Injection of 500 μl MTBE was accomplished by use of a retention gap and early vapour exit. Detection limits of 0.01–0.1 μg/l were attained.

2.6.3   Liquid handling systems

Automation of solid-phase extraction.   Solid-phase extraction (SPE) is essentially low-resolution column chromatography employed in binary mode— analytes and interferences are either completely retained or completely eluted, and packings and wash and elution solvents are chosen to give as much selectivity as possible in this process. SPE offers advantages in terms of speed and the range of separation modes available (including standard HPLC phases, metal chelation and immunoaffinity), and it also offers a unit operation which is widely employed in a standard format. This makes it ideal for automation, and a number of manufacturers produce dedicated instruments. These fall into two categories: those (such as the Gilson ASPEC) which use standard commercial SPE cartridges, and those (such as the Varian AASP—now discontinued—and the Spark-Holland PROS-PEKT) which use proprietary cartridges containing very small amounts of packing (about 20 mg). Sensitivity is maximised by eluting the cartridge directly on to an HPLC column, and the small size permits this without any compromise in analytical efficiency. Their limitations include the relatively restricted range of packings available and the greater significance of sample breakthrough due to the small amount of stationary phase. The ASPEC is a very flexible system built round an *xyz* Cartesian co-ordinate liquid handling device, and permits a wider choice of packing material and higher capacities. It also has the potential for more sophisticated sample manipulations, including evaporation steps. However, the use of standard SPE cartridges also results in greater dilution of the final eluate. This may necessitate a more complex LC method for the end analysis if the highest sensitivity is required. All automated SPE systems permit better control of flow rates than manual methods using vacuum manifolds, and this is an important advantage as under many circumstances it leads to improved method reproducibility.

Since the systems are based upon standard SPE principles, all the usual advantages and disadvantages of SPE apply. In particular, it is imperative that any new lot of cartridges employed should be checked for performance. It cannot be assumed that the chemistry of apparently identical cartridges will be the same. Packings are made in batch processes, and although manufacturers are now more aware of the need to ensure reproducible properties between batches, this cannot be guaranteed for every single analyte. It may also be prudent, before committing to it, to check any newly developed method with cartridges from a number of different batches. Use of appropriate internal standards is also highly recommended, both to improve variance and to provide simple checks for quality control.

The performances of similar manual and automated methods have been compared by several workers and the automated procedures are generally found to be at least as reliable. Pratt *et al.* (1992) found an excellent correlation between manual and ASPEC methods for pyridinoline and deoxypyridinoline in urine and a significant improvement for analysis of hydrolysed urine due to the more reproducible timing of the method. The method also relied upon a special SPE packing; this is undoubtedly more readily achievable with systems capable of employing standard SPE cartridges rather than the miniaturised and sealed versions required for the PROSPEKT, where customised cartridges would have to be specially made.

Hotter *et al.* (1993) showed that for prostanoid analysis the PROSPEKT provided improved recoveries (65–90%) and better CVs (3%) than did manual processing (recoveries of 60–80% with CVs of 5–8%). The method included a cartridge wash with water, followed by a wash with petroleum ether and drying with air. The air was then replaced with water before elution. Retention time reproducibility was good, with CVs in the range 0.8–1.4%, similar to those expected from standard HPLC injections.

Strutton (1992) found that for a nootropic drug the ASPEC method initially gave a worse performance than a manual analogue. Volumetric accuracy and repeatability of the ASPEC were found to be 2% and 1% respectively, so this was not the source of the problem. An increase in interferences was observed due to a change in the lot of Bond-Elut SPE cartridge used; a second problem was the instability of the analyte in pH 3 buffer. An overall cycle time of about 11 h was required to process a 60-sample batch, and there was progressive degradation of analyte over this period.

A previously developed manual method for *N*-methyl carbamate pesticides was automated, using an ASPEC, by de Kok and Hiemstra (1992) and employed for routine analysis over four years. The cleaned-up pesticides were detected by postcolumn hydrolysis using a solid-phase catalyst and fluorescent derivatisation using *o*-phthalaldehyde. Detection limits of 5–50 μg/kg were obtained, with sample throughputs of 20 per day. One significant advantage of the automated method was control of the relatively rapid degradation of the pesticides in aqueous media. Use of the ASPEC ensured that the time of contact with water was both minimised and constant between samples. Recoveries for automated SPE were in the range 75–100% and CVs were between 1 and 3%. These workers consider that the major advantage of the ASPEC is that clean-up is no longer the rate-determining step in the analytical procedure.

van der Hoff *et al.* (1991) used an ASPEC as an injection device on to a capillary GC. After SPE clean-up, sample was loaded into a 200 μl loop and directly injected using concurrent solvent evaporation. Organochlorine (OC) and pyrethroid pesticides were detected using an ECD. Care was taken to remove electron-capturing impurities from the silica SPE cartridge during preconditioning. The SPE stage could be used to separate OC and pyrethroid pesticides, a procedure which required careful control of the SPE flow rates—readily achieved using an automated apparatus but difficult to attain manually. Detection limits of about 10 μg/l were obtained.

Sharman and Gilbert (1991) and Sharman *et al.* (1992) have reported the application of the ASPEC to mycotoxin analysis, employing immunoaffinity SPE cartridges. Sample application and elution flow rates are major factors in obtaining reproducible results with immunoaffinity chromatography (IAC), and the ASPEC afforded a considerable improvement over manual procedures. IAC gels were packed into standard SPE cartridges by the manufacturer. Manual preparation of nut and fig samples for aflatoxin analysis consisted simply of blending with acetonitrile–water and diluting the filtrate with buffer (Sharman and Gilbert, 1991). The remainder of the analysis was completely automated using the ASPEC and HPLC with postcolumn derivatisation. One interesting feature of the ASPEC method was automated sub-sampling of the cleaned-up extract for retention in case confirmation was required. For naturally contaminated dried fig slurry, a CV of 5.1% for aflatoxin B_1 ($n = 20$) was found. Excellent results were also obtained for aflatoxin M_1 in a certified milk powder reference material. The method was employed for analysis of 2000 samples over six months and the only instrumental problems encountered were with needle seals. A batch of 20 extracts could be analysed in 11 h. Ochratoxin A in cereals and pig tissues was determined by using a similar system (Sharman *et al.*, 1992). It was stated that significant time savings were achieved over manual analysis. For both mycotoxins, use of IAC permitted a single method to be applied to a wide range of sample types.

2.6.4 Dialysis

Membrane methods (including ultrafiltration) provide another approach to separation of sample macromolecules from low-molecular-weight analytes.

They have been reviewed by van de Merbel *et al.* (1993). Dialysis is widely used to separate proteins from small molecules, usually to purify the protein. Dialysis is an equilibrium diffusion process, which has two significant consequences. It tends to be slow, and also it may be difficult to achieve good recoveries of analytes migrating through the membrane. In the normal configuration, protein is held in a limited volume in a sealed dialysis bag, and repeated changes of the water or buffer in which the bag is immersed, often involving litres of fluid, are required to complete the separation. Trace analysis requires the reverse operation, eliminating protein from extracts, which demands a different approach. In order to achieve good recoveries in a minimum time it is usual to employ two long channels of minimum depth separated by a membrane. Sample (donor) is placed in one channel, and cleaned-up analyte diffuses into the medium in the other (recipient). Dialyser blocks to this design are available from Technicon (Aerts *et al.*, 1988), while the ASTED instrument from Gilson has been optimised for HPLC methods and can produce efficient dialysis of small volumes. Either or both of the donor and recipient streams may be pumped, usually in countercurrent mode. Pumping again may be continuous or pulsed, and donor and recipient are then left static in contact with the membrane for a period of equilibration. Obviously, if both liquid streams are held motionless in contact with the membrane indefinitely, the final distribution of analyte between the two channels will simply be in proportion to their volumes, assuming no partition effects. Because dialysis is a diffusion process, the depth of the dialyser channels has a significant bearing on the time required to achieve equilibrium. One further influence on the process is the pore size of the membrane. Typical analytes have molecular weights in the range 200–500 daltons. Thus the ideal membrane would have a cut-off at about 1000 daltons. However, such membranes are not commonly available and would also significantly increase equilibration times because of the entropic constraints imposed by narrow pores. The ASTED is supplied with membranes having a nominal exclusion point of about 15 000 daltons, but the effective cut-off point under normal operating conditions is about 2000 daltons (Gilson, personal communication).

Two dialysis blocks are available for the ASTED, with 100 and 370 μl volumes for the channels. The channel length for the larger block is about 65 cm. With a width of 1.1 mm, the average depth is thus about 0.5 mm. It will be seen that the size of the dialysis channels severely limits sample volumes, and this may be a significant constraint if sensitivity is an issue. Although more than one block volume of donor may be dialysed per analysis, clean-up times increase proportionally. Since the volume of dialysate is usually large, it is essential to re-concentrate the analyte before the final analytical separation. Most methods carry this out directly on the analytical column by solvent strength control, but dialysate may also be re-focused on a precolumn. Recoveries tend to be in the range 20–60%, depending upon dialysis conditions, with a trade-off between recovery and dialysis time (typically 5–20 min). Recovery of aflatoxin M_1 at 1 μg/kg in 90% milk was 65% (Tuinstra *et al.*, 1989); the detection limit was 50 ng/kg. It was necessary to purge the dialyser membrane with 1:1 water:acetonitrile periodically to prevent fouling giving rise to reduced recoveries. For dialyser blocks not optimised for HPLC, recoveries tend to be lower, at 10–20%. Mattern *et al.* (1990) found 17% recovery of meticlorpindol in egg extract and achieved a detection limit of 20 ng/g.

Cooper *et al.* (1994) employed this approach to determine 1-(beta-D-arabinofuranosyl)-5-(1-propynyl)uracil and its metabolite 5-propynyluracil in urine. The precolumn was configured to permit heart-cutting on to the analytical column, which reduced analysis time. Washing with solvents both before and after elution minimised interferences and eliminated waiting for late-eluting peaks. The mean within- and between-run CVs at three different urine analyte concentrations were 3.6 and 3.6% and 1.7 and 3.3% for 5-propynyluracil and 1-(beta-D-arabinofuranosyl)-5-(1-propynyl)uracil, respectively. An interesting point is that the ASTED transfer tubing was chemically treated to reduce both surface tension and protein binding. This permitted a sixfold increase in maximum liquid flow rate, which in the ASTED is normally limited by the fragility of the liquid–air interface used to separate liquid streams and reduce dispersion. This modification necessitated adding a stop-flow valve at the donor channel outlet to prevent sample movement within the dialyser block, which otherwise gave rise to higher CVs.

Temperature fluctuations affect both HPLC k' values and diffusion coefficients; temperature control is therefore useful (as it is for all automated chromatographic systems). Within-run CVs for a nucleoside were 1.3–3.0%; between-run CVs were 2.0–7.4% ($n = 15$) depending upon analyte concentration (1–30 μM). Absolute recovery of analyte was 20%, restricted by its high polarity which affected retention on the loop cartridge. Elimination of late-eluting interferences reduced analysis time from 30 to 10 min; ASTED preparation was concurrent with HPLC analysis. Monochloroacetic acid was added to samples to minimise protein binding.

Dialysis gave excellent results for the determination of flumequine and oxolinic acid in fish extracts obtained by blending and defatting, and compared very favourably with a method involving automated precolumn clean-up without dialysis. Detection limits of 3 and 2 μg/kg respectively were obtained (Thanh *et al.*, 1990); CVs of 1.9–4.3% ($n = 6$) were found at concentrations of 50–100 ng/g. Absolute recoveries of drug standards of about 50% were achieved.

Other membrane-based approaches.   Jackson and Jones (1991) described a hollow perfluorosulphonate fibre used to remove high concentrations of halide ions from samples prior to ion chromatography of other inorganic anions. With use of the pretreatment device, the limit of detection of nitrite was about 0.1 mg/kg in brine containing 5000 mg/kg of chloride. This system was operated in manual form but could readily be automated.

Electrodialysis overcomes the principal drawback of conventional dialysis, the dependence upon passive diffusion. It has its own limitations, as it is applicable only to charged analytes; this, however, also has benefits in terms of increased selectivity. Brewster and Piotrowski (1991) employed the method manually for sample preparation in the analysis of sulphamethazine in milk. Automation would be straightforward.

The ASTED was modified to carry out liquid–liquid extraction by exchanging the dialysis unit for a porous hydrophobic membrane containing immobilised dihexyl ether (Jönsson *et al.*, 1994). Basic compounds in a high-pH donor stream were partitioned into the membrane and then transferred to an acidic recipient medium, where they were protonated and thus inhibited from back-diffusion. However, maximum recovery from water was only about 60%. See also the review by van de Merbel *et al.* (1993).

2.6.5 Column switching

General. Reliable and effective column switching systems may be constructed having regard to a few general considerations. The component columns of the system should have separation mechanisms as different as possible, having due regard to the physicochemical properties of the target analyte. It is not usually effective, for example, to couple two reversed-phase columns of similar chemistry, such as C_8 and C_{18}. Although in some individual cases this may give the desired separation, combinations such as normal-phase HPLC/capillary GC, or ion-exchange/reversed-phase HPLC are far more likely to be satisfactory. To exclude potential interferences when fractions are transferred, high column efficiencies are desirable in all dimensions in addition to the final analytical separation. Alternatively, preliminary stages with high selectivity such as IAC may be employed. It is important to take care about the quality of the solvents employed. When fractions of 1–20 ml are transferred between columns, solvent impurities can give rise to significant interferences.

Retention times must be consistent. As the ambient temperature in many laboratories drops significantly overnight, column thermostatting is desirable. Also, columns should not be overloaded. All columns introduce dispersion, which must be eliminated before peaks are transferred to the next dimension. HPLC peaks produced from a small injection ($\sim$10 μl) onto a standard 250×4.7 mm analytical column are typically 0.5–1 ml. If resolution is not to be compromised the analyte must be re-focused in some way at the head of the next column. The need for re-focusing imposes severe limitations on column combinations that may be coupled directly. Some are straightforward,

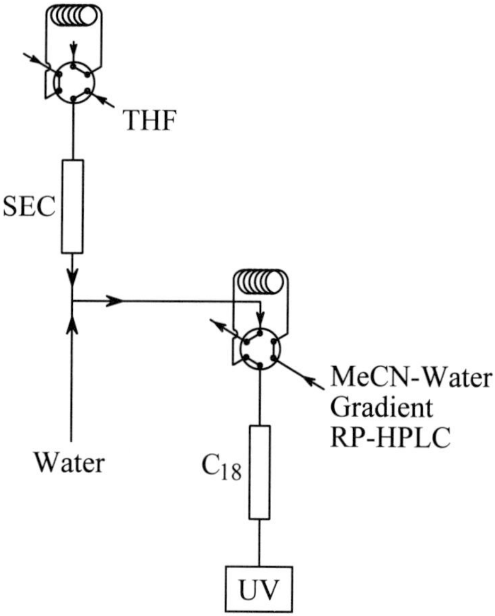

Figure 2.2 Non-aqueous SEC–RP-HPLC. Peak from SEC column is diluted with water in reverse-plumbed holding loop to ensure compression on RP column. Re-drawn from Williams *et al.* (1989).

such as ion-exchange and reversed-phase HPLC. Other combinations may require specialised interfaces. Obvious problem areas include immiscible mobile phases. Interfaces involving evaporation steps must take into account the volatility of analytes. Even relatively high-molecular-weight compounds may be lost when present in nanogram amounts spread over a large surface area. The only water-miscible non-aqueous SEC mobile phase is tetrahydrofuran (THF), but this is a strong eluent in reversed-phase terms. Thus to couple these two chromatographic modes, it was necessary to incorporate an interface (Figure 2.2) which diluted the THF peak with water before injection on to a reversed-phase column (Williams *et al.*, 1989). The 5 ml holding loop was plumbed so that instead of sample entering and leaving by the same port, as is usual, it was swept through the whole loop during injection. This was done to encourage mixing of the THF peak with diluent water, ensuring that the final solvent strength was sufficiently low that analytes were retained at the head of the C_{18} column.

Disadvantages of column switching include the difficulty of simultaneously determining multiple analytes. It is unlikely that a group of analytes will present identical retention times in all dimensions except the last. With the exception of systems incorporating an SEC step, coupled column systems primarily target single compounds. Substantial additional complexity of system design (Figure 2.3) was required, for example, to determine the two hormones trenbolone and 19-nortestosterone in cattle liver extracts

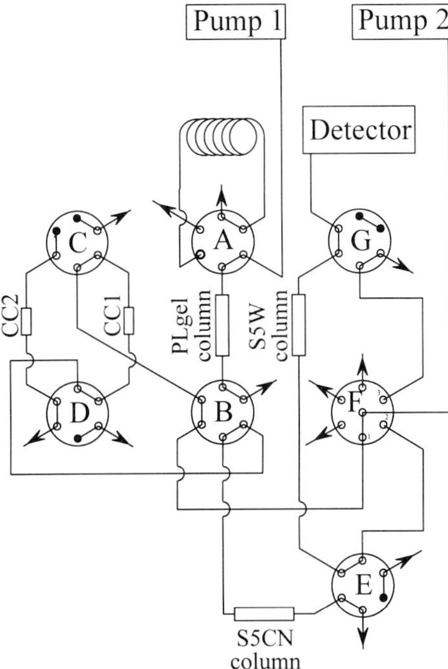

Figure 2.3 Multiple-column LC. Two steroids from SEC column are held individually on trapping columns and then separately chromatographed in second and third dimensions. Re-drawn from Stubbings and Shepherd (1993).

(Stubbings and Shepherd, 1993). An initial SEC column was employed, but because of $\pi-\pi$ interactions involving trenbolone, the two compounds were resolved. Each was therefore transferred to a holding column before the determination was completed on coupled cyano-silica HPLC.

2.6.6   LC–LC column switching

Injection loop substitution.   The simplest form of LC–LC may be carried out by replacing the injection loop with a small precolumn, which retains sample components of interest while washing interferences to waste. The precolumn must have sufficient porosity to enable samples to be injected on to it readily. Typically this is achieved by using $40\,\mu m$ particle size stationary phase in a $10–50\,mm \times 2\,mm$ column. The precolumn is washed as necessary and eluted on to the analytical column for quantification. After the precolumn has been reconditioned, preferably by backflush, the cycle is repeated. There are two major constraints: analyte loss by breakthrough from the precolumn, and strongly retained sample components interfering with subsequent analyses. Precolumn lifetime is wholly dependent upon the cleanliness

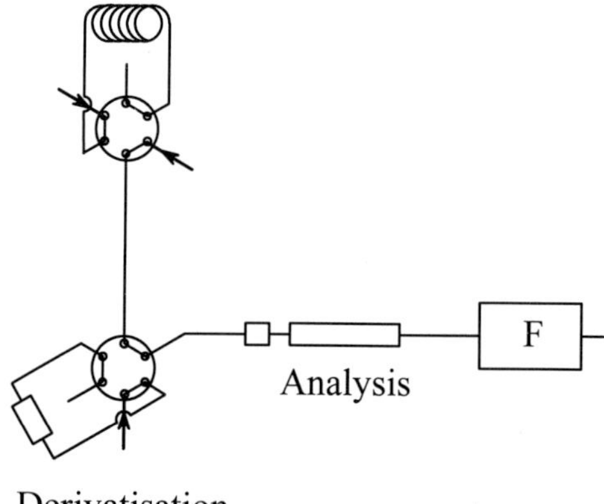

Figure 2.4 On-line solid-phase derivatisation. F = fluorescence detector. Redrawn from Zhou *et al.* (1992).

and protein content of the sample. The use of coupled LC–LC techniques for determination of drugs in biological samples was reviewed by Campíns-Falcó *et al.* (1993), including the use of internal-surface reversed-phase and protein-coated columns as the first dimension, for elimination of biological macromolecules. It is possible to carry out reaction chemistry on the pre-column, such as derivatisation for improved detection. Zhou *et al.* (1992) described the preparation and use of a solid-phase reagent for 9-fluorene-acetyl derivatisation of amphetamine, based on a Merrifield-type resin (Figure 2.4). The size-exclusion properties of the resin, together with incorporation of sodium dodecylsulphonate in the mobile phase, permitted direct injection of plasma. Sample, trapped on the derivatisation column by valve switching, was heated at 75°C for 8 min, giving a conversion efficiency of about 70%. A CV of 5.9% ($n = 40$) was achieved at a concentration of 10 μg/l with a detection limit of 0.2 μg/l. The capacity of the derivatisation column depended upon the rate of competing hydrolysis by mobile phase, which itself was dependent upon temperature, mobile-phase composition and time of reaction. This procedure seems promising, but applications require careful selection to minimise reagent losses by hydrolysis, which reduces the capacity of the system, and reaction with other sample components, which increases the risk of interferences.

Low-pressure LC.   Few areas require the clean-up of such large samples as dioxin analysis. Because sensitivities of at least one part in 10^{12} are required, sample sizes of 100 g are routinely employed. Low-pressure column switching systems have been devised to automate part of the clean-up process. Often the whole sample is ground with sodium sulphate (to bind water) and packed into the first of a series of columns. This procedure uses large amounts of solvent, and for some stages may be replaced by HPLC column switching (Thompson *et al.*, 1991). However, low-pressure systems provide the greatest degree of automation and are widely employed (Turner *et al.*, 1992; Dettmer and Stieglitz, 1994).

Size-exclusion chromatography.   SEC has many advantages as the initial stage in multi-dimensional LC–LC or LC–GC systems (Shepherd, 1984, 1988). The mechanism of separation is entropy-based, requiring no transfer between physically different phases and therefore completely independent of other modes of chromatography. This fulfils one of the key requirements of column-switching systems. The most common application is elimination of lipid, using cross-linked polystyrene packings. SEC methods for aqueous samples are not so well developed. The choice of small-pore packings is restricted, and the packings themselves often superimpose on to the SEC mechanism secondary interactions which may cause undesirably long retention of some sample components. However, secondary interactions may be employed to advantage, as Huber and Lamprecht (1995) showed with an ODS precolumn used in combined SEC and adsorption modes for preliminary clean-up of serum for neopterin analysis. Gel permeation chromatography (GPC), which is widely used for lipid removal in pesticide analysis, depends for its effectiveness on a superposition of SEC with adsorption or partition (Shepherd, 1984).

Columns have an extremely high loading capacity: 5–10 mg of lipid for conventional 7–8 mm internal diameter high-performance analytical columns. With the appropriate choice of mobile phase, highly reproducible results may be obtained and long-retained peaks avoided. Automation of these systems is very straightforward, and columns have a considerable life-time if pressure pulsing is eliminated. Their major disadvantage is the restricted resolution achievable, because all the retention takes place within what would be the column void volume for any other form of chromato-graphy. Thus selectivity is relatively poor, and SEC is commonly employed primarily to separate low-molecular-weight analytes from larger molecules such as lipid or protein (Shepherd, 1984). GPC is frequently automated as a stand-alone application because of the sample size involved, while HPLC column switching applications frequently involve elimination of protein. Szuna and Blain (1993) demonstrated the determination of an antibacterial agent in plasma in this way (Figure 2.5), and found that determination of its major metabolite was simplified because the automated procedure minimised the artefactual formation by hydrolysis experienced during manual sample preparation. The system was stated to be reliable, with no change in performance during the course of 500 injections, and this is typical of SEC-based clean-up.

Immunoaffinity chromatography.   Farjam *et al.* (1991) determined 19-nor-testosterone in human urine by on-line IAC–capillary GC. The major

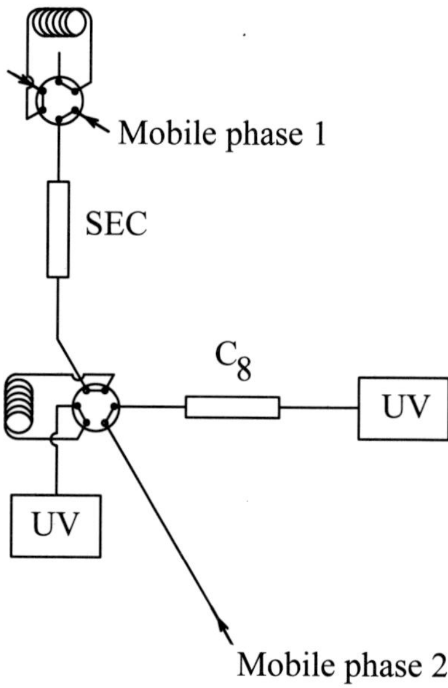

Figure 2.5 Use of coupled aqueous SEC to eliminate protein from plasma samples prior to RP-HPLC. Redrawn from Szuna and Blain (1993).

problem with this approach is that most current IAC packings are based upon soft gels which cannot withstand the pressure changes to which they would be subjected in normal on-line systems. Thus the immunoaffinity column was isolated from the high-pressure system (Figure 2.6). The methanol–water (95–5, 2 ml) IAC eluent was diluted with water to an organic content of 8%, and the hormones were reconcentrated on a small (10 mm × 2 mm) C_{18} column. This in turn was eluted with ethyl acetate, 75 μl of which was introduced to the GC via a retention gap at 25 μl/min, after the first 26.25 μl was diverted to waste to purge water. For a urine volume of 5 ml, the detection limit was 0.1 μg/ml, with a CV of 6% ($n = 6$). Moretti *et al.* (1992) described the preparation of an IAC column for the drug chloramphenicol in milk and pig muscle, based on a rigid tresyl-activated matrix. This permitted direct coupling of IAC with analytical HPLC. Fouling of the IAC column by irreversible binding of sample components was not a problem; 150 samples were analysed over a period of three months.

Micellar clean-up. Posluszny *et al.* (1990) employed micellar clean-up for the determination of tricyclic antidepressant drugs in plasma by direct injection with fluorescence detection. Filtered plasma was injected in a micellar solvent onto an extraction precolumn (Figure 2.7) which retained the analyte. After the precolumn had been washed free from the micellar

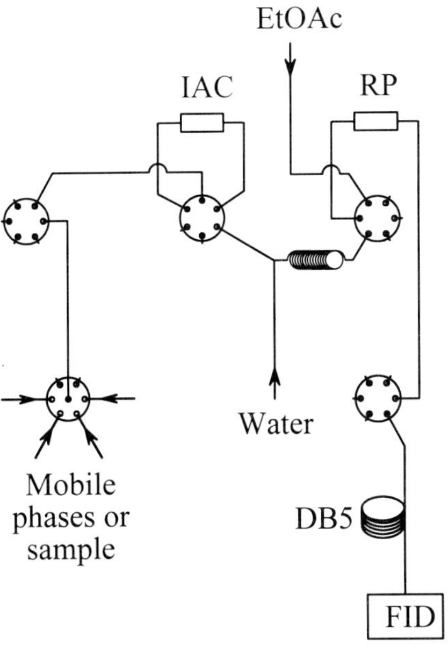

Figure 2.6 On-line immunoaffinity chromatography–GC employing a column of soft immuno-affinity gel isolated by valves from pressure damage, with a trapping column to re-focus eluted analytes. Re-drawn from Farjam *et al.* (1991).

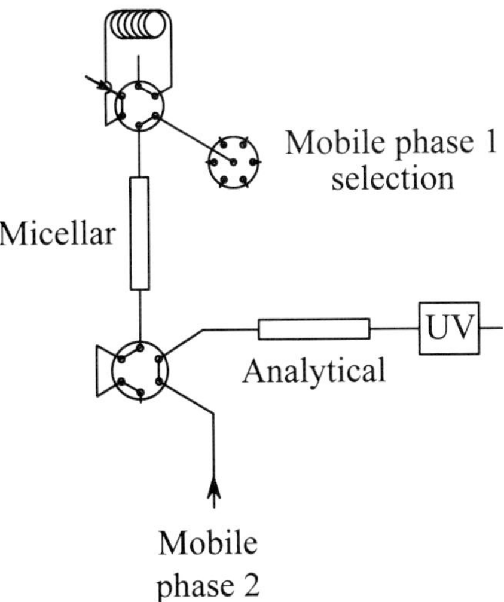

Figure 2.7 Improved efficiency via separation of micellar and RP-HPLC systems by on-line trapping. Re-drawn from Posluszny *et al.* (1990).

mobile phase, drugs were eluted by using an intermediate mobile-phase strength which permitted reconcentration on the analytical column, where the final analysis was carried out by reversed-phase chromatography. The use of the precolumn was found to provide three advantages over direct injection: no serum protein comes into contact with the analytical column; mobile-phase restrictions and baseline shifts are circumvented; and many potential interferences may be excluded by appropriate washing of the precolumn. For propranolol, a tenfold improvement in sensitivity was achieved by using a precolumn compared with direct injection on to the analytical column. Throughput was stated to be typically 50 samples per day, with 450 samples analysed in a period of three weeks. Micellar separations tend to have rather low efficiency, and this coupled system employed the strengths of each separation mode to advantage.

Ionic interactions.   Many drugs are basic in character and thus ion exchange is an attractive mechanism for initial clean-up. Tarbin and Shearer (1993) described the determination of nitroxynil in cattle muscle tissue with on-line anion-exchange clean-up. The instrumentation is shown in Figure 2.8. Concentrated extract was injected on to a weak anion-exchange precolumn, and analyte was eluted with aqueous 1% trifluoroacetic acid:acetonitrile (1:1). The eluate was switched on to a polymer reversed-phase column for analysis with detection by UV at 273 nm. Average recoveries of nitroxynil at four levels from 0.005 to 1 mg/kg added to cattle muscle were >88% with CVs <8% ($n = 5$–7). The limit of determination was 5 μg/kg. Anionic compounds may be treated similarly. White *et al.* (1991) used ion pairing and direct injection of filtered citrus juice on to a precolumn to determine folate at concentrations of 0.1–0.4 mg/l.

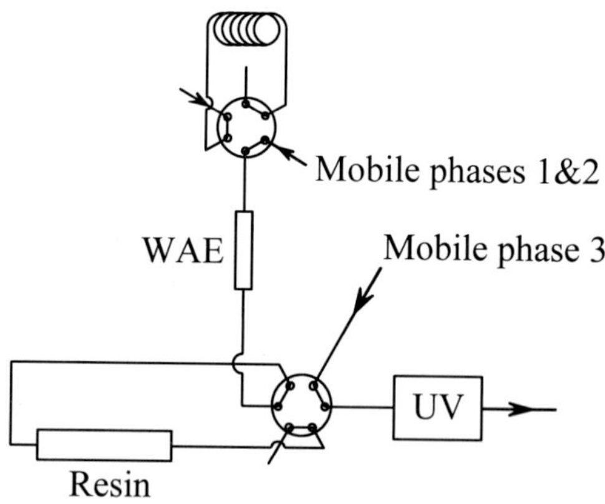

Figure 2.8 On-line ion exchange–RP-HPLC. Re-drawn from Tarbin and Shearer (1993).

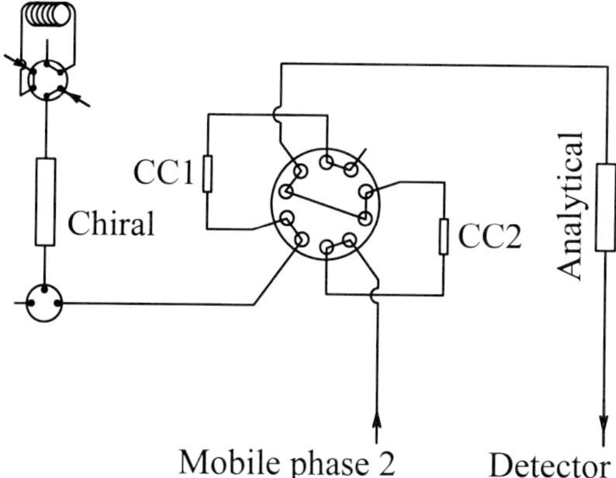

Figure 2.9 Separation of protein column and RP-HPLC systems by on-line trapping. CC1 = concentration column 1; CC2 = concentration column 2. Re-drawn from Walhagen and Edholm (1989).

Chiral separations. Chiral stages may be incorporated into LC–LC systems. Chiral columns tend to be less efficient than other common types, and thus chiral separation followed by peak recompression and a non-chiral analytical separation affords overall improvements in detection limits. Kelly *et al.* (1994) described the resolution of four geometrical isomers of a platelet-activating factor antagonist on coupled silica–aminopropyl phases, with subsequent on-line separation of the enantiomers of one of the isomers by switching to a chiral column. Walhagen and Edholm (1989) separated drug enantiomers in plasma initially on an α_1-acid glycoprotein column (Figure 2.9), trapping the two compounds on C_{18} concentration columns and carrying out the final analysis by C_8 or C_{18} phases. Peak trapping on the concentration columns was facilitated because the glyco-protein separation was carried out in highly aqueous medium.

Other LC–LC separation mechanisms. Sample clean-up should always take into account the physicochemical properties of the analyte, and this may entail employing less-common separation mechanisms, such as metal chelation. Irth *et al.* (1989) described the analysis of $3'$-azido-$3'$-deoxythymidine (AZT) in 1 ml serum by initial isolation on to cross-linked polystyrene, followed by elu-tion with aqueous methanol at pH 11.6, and chelation on Ag(I)–thiol resin (Figure 2.10). Chelation depended upon the presence of the pyrimidino NH–CO group. Protonating this group by injection of perchloric acid enabled the facile elution of AZT on to the analytical column. The Ag(I) column could be re-used at least 25 times under these conditions. Direct injection of plasma resulted in substantial interference from sulphur-containing components. The PRP-1 column could be used only once, as it was irreversibly fouled with plasma components. It would seem likely that the method could be improved

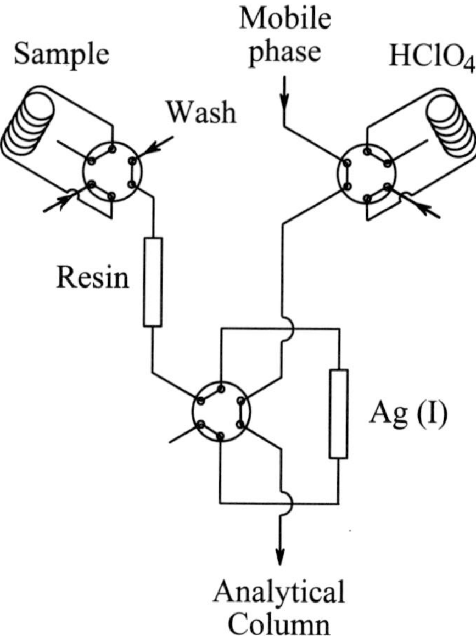

Figure 2.10 On-line RP-HPLC and metal chelate chromatography. Re-drawn from Irth *et al.* (1989).

by adopting a micellar approach as described above, or by employing automated SPE as the initial stage in the clean-up.

2.6.7 GC–GC column switching

Multi-dimensional GC is a mature technique. Bertsch (1990) gives a good summary of methodology. Again, the major consideration is re-focusing of analytes on the second column. This is perhaps most readily carried out by using a dual oven system, but where this is not practicable cryogenic traps are very effective. Multiple LC and GC stages may readily be concatenated as shown in Figure 2.11 (Chappell *et al.*, 1993).

PIONA (Paraffin-iso/normal olefin–naphthalene–aromatic) analysis of petroleum naphthas cannot be described as trace analysis, but provides an excellent example of column switching employed to permit the automated analysis of many components in a highly complex sample (Buchanan and Nicholas, 1994). The instrument is shown in Figure 2.12. Aromatics and large non-aromatic hydrocarbons are retained on the OV-275 column; other components pass through to the olefin trap. Normal paraffins continue to the 5A molecular sieve and branched paraffins and naphthenes are then resolved by carbon number on the 13X column. Thermally-desorbed *n*-paraffins are also resolved on the 13X column. After desorption, olefins are hydrogenated and then determined as *n-/iso*-paraffins. Aromatics are

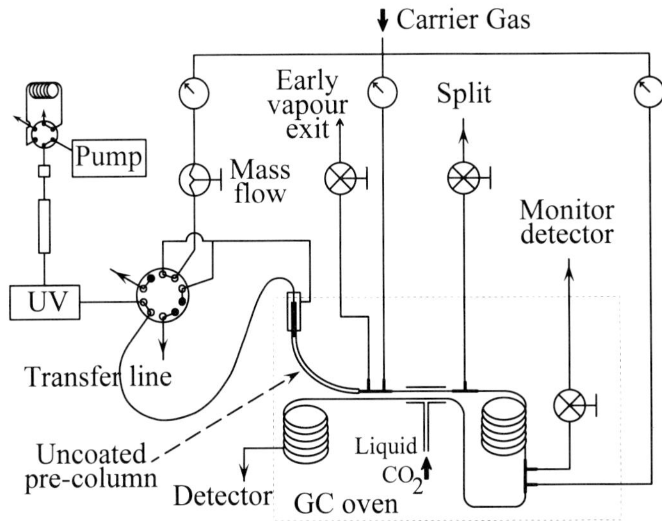

Figure 2.11 On-line LC–GC–GC employing single GC oven with cryogenic re-focusing. Re-drawn from Chappell *et al.* (1993).

desorbed in three fractions from the OV-275 column, and each is re-focused using Tenax and separated on the OV-101 column. Drawbacks to use of the PIONA instrument were considered to be the increased cost of maintenance, problems with isomerisation of olefins, and limitations on the carbon-number range of components analysable.

Along with enhanced efficiency, a major advantage of GC is the wealth of gas-phase detectors commercially available. Furthermore, cryogenic trapping offers a facile method for re-focusing analytes which may be used with particular advantage with non-destructive detectors. Krock *et al.* (1994) used both these aspects of GC–GC to advantage with the separation of essential-oil components (Figure 2.13). GC fractions up to and including the whole sample emerging from an IR detector were re-focused and re-chromatographed on alternative phases.

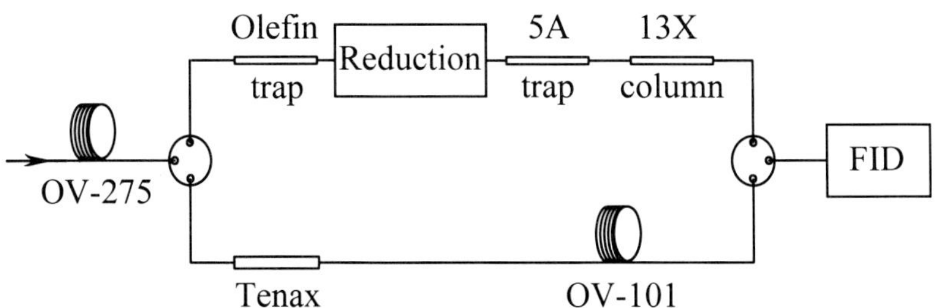

Figure 2.12 On-line fractionation and GC–GC separation of a complex petroleum sample. Re-drawn from Buchanan and Nicholas (1994).

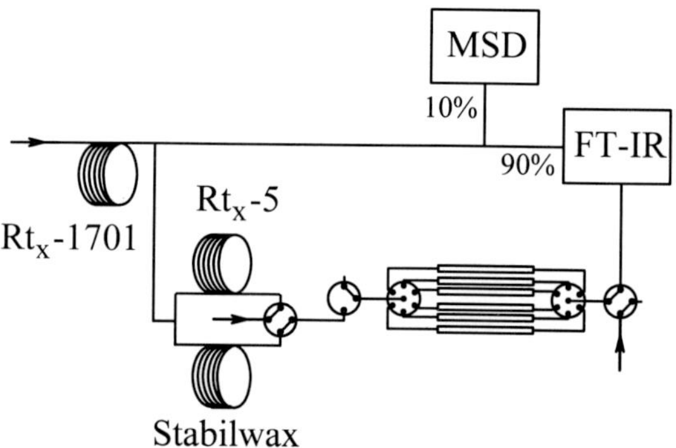

Figure 2.13 On-line GC–GC by sample collection and recycling. Re-drawn from Krock *et al.* (1994).

2.6.8 *LC–GC column switching*

In order for coupled LC–GC to be viable, the analyte must be sufficiently volatile to pass through a GC column. Although only about 5% of CAS registry compounds are amenable to GC, combined LC–GC is of interest as it affords all the advantages of liquid handling, together with the efficiency of capillary GC. It also provides a very straightforward and well-developed route for the introduction of separated compounds into mass spectrometers and other gas-phase detectors (Chappell *et al.*, 1992).

Davies *et al.* (1989) reviewed early applications of LC–GC that focus on fossil fuels. Vreuls *et al.* (1994), in a more recent review, described the three types of interface (auto-injector, direct injection, loop) employed to transfer liquid sample on to a gas-phase column. Another method of phase transfer is by SPE (Vreuls *et al.*, 1991). The loop interface is widely employed because of its simplicity and ease of use. In all cases it is necessary to take into account the large volume increase, about 300-fold, which occurs on evaporation of liquids. In addition, large amounts of solvent are likely to have deleterious effects on analyte separation and detector performance and must be voided beforehand. Grob (1991) has pioneered the use of retention gap and other techniques for elimination of solvent. Highly polar solvents and particularly water do not wet the hydrophobic phases used in capillary GC to coat retention gaps and other precolumns. This severely limits the application of techniques such as concurrent solvent evaporation, which are very effective for injection of large volumes of relatively non-polar solvents. Furthermore, most GC phases (including those employed to de-activate precolumns) are susceptible to damage by hydrolysis. The problems of introducing aqueous HPLC mobile phases directly into GC columns have not yet been fully overcome, and an interface between LC and GC incorporating phase transfer of separated analytes into a non-polar solvent provides perhaps the most satisfactory approach.

Capillary GC performance is easily degraded by non-volatile substances present in the sample and by oxidation or hydrolysis, and consequently great care should be taken over the purity of the LC mobile phase. Chromatographic interferences from solvent impurities may be significant and these can arise from apparently trivial sources of contamination. Biedermann and Grob (1991) described artefact GC peaks caused by finger-prints due simply to touching the glass stopper of an eluent reservoir.

It is possible to extend the range of analytes by on-line derivatisation methods. Goosens *et al.* (1992) described the analysis of propionic and 2,6-difluorobenzoic acids in aqueous solution by extraction into dichloromethane with a phase transfer agent followed by alkylation with pentafluorobenzyl bromide. Wessels *et al.* (1993) carried out derivatisation on the GC column by sequential injection of sample and derivatising reagents, including diazo-methane, bromine and MSTFA. Both groups employed sandwich-type phase separators to interface between aqueous reversed-phase HPLC mobile phases and the GC column, an approach also followed by van Zoonen *et al.* (1990) for pesticide analysis.

SEC is an effective first stage in the clean-up process (see discussion above) and offers particular advantages for preliminary clean-up of plastics before determination of low-molecular-weight additives. In this instance it is very straightforward to isolate the required analytes for transfer to capillary GC. Blomberg *et al.* (1994) employed a size-exclusion column operated with THF containing up to 5% *n*-decane as mobile phase. Decane was required as a co-solvent to retain the more volatile sample components. THF is an excellent solvent for polymers, but experience in this laboratory suggests that it should be avoided wherever possible, because it is likely to damage plastic and leach HPLC components and because it is difficult to obtain in a state of high purity. Oxidation products, including THF oligomers, may give rise to interference problems. A ten-port valve with dual loops was employed to permit direct injection of standards onto the GC, circumventing the HPLC system (Figure 2.14). The interface was modified by inclusion of a nitrogen purge to remove pre-existing solvent before loading with sample, to overcome the losses normally encountered due to mixing arising from laminar flow dispersion in the loop. Visualisation and optimisation of the LC–GC transfer was achieved by using a glass GC observation door. Transfer of hydrocarbons was 100% efficient for C_{13}–C_{38} alkanes. The method was applied to the determination of UV stabiliser additives (200–400 μg/g) in a polystyrene reference material; the CVs were 0.9–5.7%. This separation was facile because of the large difference between the molecular weights of the sample components.

Biological samples will have a more uniform molecular-weight distribu-tion. Although SEC remains an effective initial clean-up, removing the bulk of unwanted co-extractives, additional steps may be required to ensure satisfactory elimination of interferences. Thus de Paoli *et al.* (1992)

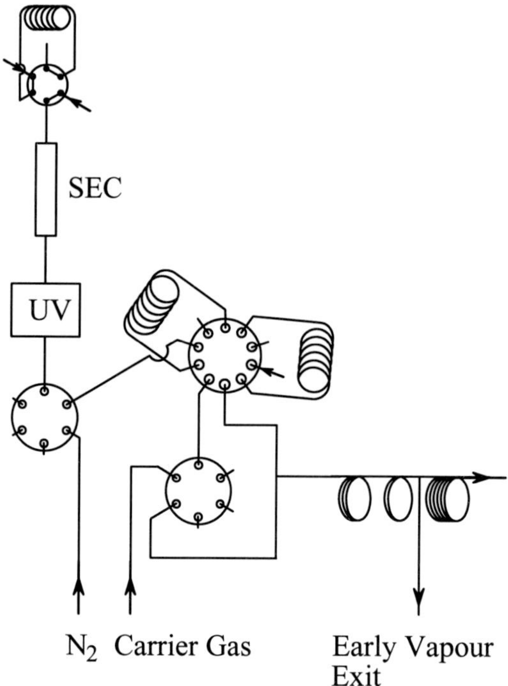

Figure 2.14 On-line SEC–GC for plastics additives; interface includes nitrogen purge to minimise within-loop mixing. Re-drawn from Blomberg *et al.* (1994).

found it necessary to include an intermediate silica column stage for the determination of 22 organophosphorus pesticides in fruit by capillary GC (Figure 2.15). After each analysis the silica column was backflushed with methyl *t*-butyl ether. Recoveries of 80–95% were obtained, with limits of detection of 1 μg/kg.

2.6.9   *Supercritical fluid switching*

There is considerable interest in the use of supercritical fluids both for sample extraction (SFE) and chromatography (SFC), driven to some extent by the decreasing acceptability of chlorinated and other solvents and the cost of disposal. Carbon dioxide is by a considerable margin the most widely used supercritical fluid because of its excellent physicochemical properties, ready availability and non-toxic nature (Bøwadt and Hawthorne, 1995). Problems remain in extending the applicability of SFE to polar analytes because potential alternative supercritical fluids are either too toxic or require unacceptably high operating pressures. McNally (1995) and Taylor (1995) have described the theoretical background to SFE and its implications for analysis. McNally (1995) considered that SFE results may permit a better

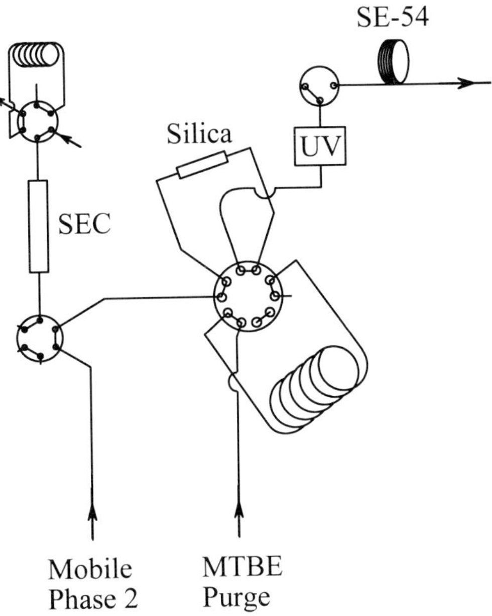

Figure 2.15 SEC–GC for pesticide analysis, with on-line silica column removal of interferences. Re-drawn from de Paoli *et al.* (1992).

understanding of matrix–analyte interactions, which would represent a major advance for trace analysis. Although organic modifiers, pressure and temperature may be varied to gain some degree of selectivity, super-critical fluids extract a fairly wide range of components from samples. Some SFE applications require relatively long extractions, compared to blending, but may provide improvements over Soxhlet methods in terms of both time and recoveries. Analyte trapping and interfacing with other instruments have been improved by the use of SPE cartridges (Sandra *et al.*, 1995) but trapping efficiency remains a critical aspect of performance (Bøwadt and Hawthorne, 1995). Initially, the methodology tended to be applied to fossil fuels and other relatively non-polar analytes, but it is being extended to more polar compounds by approaches including deriva-tisation *in situ* before extraction (Hawthorne *et al.*, 1992; Hillmann and Bachmann, 1994). Greibrokk (1995) has reviewed the use of SFE and SFC in automated systems.

2.7 Electrophoresis

Because of the plug flow characteristics of capillary electrophoresis (CE), very high efficiencies are potentially available. Although electrophoresis may appear to be restricted to charged compounds, the induced electrolyte flow causes uncharged molecules to migrate too. The major disadvantage for trace analysis is the poor loadability of CE. It is carried out in capillaries of around 50 μm diameter, which restricts sample volumes to a few nano-litres. Sensitivity is normally very poor for biological matrices because the large amount of co-extractives precludes loading an adequate mass of the extract without extensive preliminary clean-up. Generally the disparity between the dimensions of CE and other LC columns has inhibited the development of on-line coupled systems. The on-line combination of isotachophoresis with CE (Kaniansky and Marák, 1990; Monnig and Kennedy, 1994) may provide one answer. Kaniansky and Marák (1990) found that an initial isotachophoretic step permitted high-sensitivity detection of an analyte even in the presence of a 10^5-fold excess of another sample component. Many instruments incorporate programmable auto-injectors and these may be employed, for example, for therapeutic drug monitoring in serum (Evenson and Wiktorowicz, 1992).

2.8 Data reduction

In addition to sample preparation, operations such as data reduction, reporting and archiving may be extremely time-consuming activities if carried out manually, and the development of Laboratory Information Management Systems (LIMS) has been the focus of much attention. Typically raw data are quantitated by integration, often using PC-based software, and the results transferred to a database where reports are generated. There are many texts dealing with LIMS and the subject will not be dealt with further here. However, it is important to ensure that appropriate quality control measures are in place to check the performance of the system. In particular, integration must not be relegated to 'black box' status. Baselines should be plotted out and seen to be reasonable, and the resolution of fused peaks verified. The text by Dyson (1990) is recommended reading for anybody using integrators. Most data from automated systems will be collected and stored electronically and this has an impact on accreditation and data security. If one is operating under formal accreditation, it will be important to ensure that original data files are stored for an appropriate period of time in a form that cannot be edited. Hard copies might also be required.

2.9 Conclusions

It is quite possible to devise an automated system for essentially any analysis involving chromatography. However, before doing so it is important to carry out a suitable cost-benefit analysis. Commercial liquid-handling instruments are flexible and reliable, and offer the easiest route to automation. Many column-switching instruments described in the literature have been assembled and employed essentially as feasibility studies. Relatively few have been tested on extended sample runs. Probably the guiding principle is to keep matters as simple as possible. On-line LC–LC systems are versatile and easy to assemble and are likely to find a place in most laboratories, particularly those analysing extended runs of biological fluid samples. The ever-increasing driving force is the need to carry out more analyses for less cost. This will ensure the continued development of automated methods.

References

Aerts, M.M.L., Beek, W.M.J. and Brinkman, U.A.Th. (1988) Monitoring of veterinary drug residues by a combination of continuous flow techniques and column-switching high performance liquid chromatography. I. Sulphonamides in egg, meat and milk using post-column derivatization with dimethylaminobenzaldehyde. *J. Chromatogr.*, **435**, 97–112.

Bertsch, W. (1990) Multidimensional GC. *Chromatogr. Sci.*, **50**, 74–144.

Biedermann, M. and Grob, K. (1991) GC ghost peaks caused by fingerprints. *J. High Resolut. Chromatogr.*, **14**, 558–559.

Blomberg, J., Schoenmakers, P.J. and Van den Hoed, N. (1994) Automated sample clean-up using on-line coupling of size-exclusion chromatography to high-resolution gas chromatography. *J. High Resolut. Chromatogr.*, **17**, 411–414.

Bøwadt, S. and Hawthorne, S.B. (1995) Supercritical fluid extraction in environmental analysis. *J. Chromatogr. A*, **703**, 549–571.

Brewster, J.D. and Piotrowski, E.G. (1991) Rapid electrodialytic clean-up of biological samples for HPLC. *J. Chromatogr.*, **585**, 213–218.

Brinkman, U.A. Th. (1994) On-line sample treatment for or via column liquid chromatography: a review. *J. Chromatogr. A*, **665**, 217–231.

Buchanan, J.S. and Nicholas, M.E. (1994) Analysis of olefinic gasolines with multidimensional GC. *J. Chromatogr. Sci.*, **32**, 199–203.

Campíns-Falcó, P., Herráez-Hernández, R. and Sevillano-Cabeza, A. (1993) Column switching techniques for HPLC of drugs in biological samples (Review). *J. Chromatogr.*, **619**, 177–190.

Chappell, C.G., Creaser, C.S., Stygall, J.W. and Shepherd, M.J. (1992) On-line high performance liquid chromatographic/gas chromatographic/tandem ion trap mass spectrometric determination of levamisole in milk. *Biol. Mass Spectrom.*, **21**, 688–692.

Chappell, C.G., Creaser, C.S. and Shepherd, M.J. (1993) On-line high performance liquid chromatography–multidimensional gas chromatography and its application to the determination of the stilbene hormones in corned beef. *J. High Resolut. Chromatogr.*, **16**, 479–482.

Cooper, J.D.H., Sheung, C.T.C.F. and Buick, A.R. (1994) Automated sequential trace enrichment of dialysates combined with HPLC and automated heart-cutting for the determination of the nucleoside 1-(beta-D-arabinofuranosyl)-5-(1-propynyl)uracil and its metabolite 5-propynyluracil in urine. *J. Chromatogr. B*, **652**, 15–21.

Cortes, H.J. (1990) *Multi-dimensional chromatography. Techniques and applications.* Dekker, New York.

Davies, I.L., Markides, K.E., Lee, M.L., Raynor, M.W. and Bartle, K.D. (1989) Applications of coupled LC–GC: a review. *J. High Resolut. Chromatogr.*, **12**, 193–207.

Davis, J.M. and Giddings, J.C. (1983) Statistical theory of component overlap in multi-component chromatograms. *Anal. Chem.*, **55**, 418–424.

de Kok, A. and Hiemstra, M. (1992) Optimization, automation, and validation of the solid-phase extraction clean-up and on-line liquid chromatographic determination of *N*-methyl-carbamate pesticides in fruits and vegetables. *J. AOAC Int.*, **75**, 1063–1072.

de Paoli, M., Taccheo Barbina, M., Mondini, R., Pezzoni, A., Valentino, A. and Grob, K. (1992) Determination of organophosphate pesticides in fruit by on-line SEC–LC–GC–FPD. *J. Chromatogr.*, **626**, 145–150.

Dettmer, K. and Stieglitz, L. (1994) PARC—an automated apparatus for clean-up procedures used in routine-analysis for PCDD/PCDF and related compounds. *Chemosphere*, **29**, 1789–1796.

Dyson, N. (1990) *Chromatographic integration methods.* Royal Society of Chemistry, Cambridge, UK.

Evenson, M.A. and Wiktorowicz, J.E. (1992) Automated capillary electrophoresis applied to therapeutic drug monitoring. *Clin. Chem.*, **38**, 1847–1852.

Farjam, A., Vreuls, J.J., Cuppen, W.J.G.M., Brinkman, U.A.Th. and de Jong, G.J. (1991) Direct introduction of large volume urine samples into an on-line immunoaffinity sample pretreatment capillary gas chromatographic system. *Anal. Chem.*, **63**, 2481–2487.

Gan, J., Yates, S.R., Spencer, W.F. and Yates, M.V. (1995) Optimisation of analysis of methyl bromide on charcoal sampling tubes. *J. Agric. Food Chem.*, **43**, 960–966.

Gere, D.R., Knipe, C.R., Castelli, P., Hedrick, J., Randall Frank, L.G., Schulenberg-Schell, H., Schuster, R., Doherty, L., Orolin, J. and Lee, H.B. (1993) Bridging the automation gap between sample preparation and analysis: an overview of SFE, GC, GC–MS, and HPLC applied to environmental samples. *J. Chromatogr. Sci.*, **31**, 246–258.

Giddings, J.C. (1995) Sample dimensionality: a predictor of order–disorder in component peak distribution in multidimensional separation. *J. Chromatogr. A*, **703**, 3–15.

Goosens, E.C., Broekman, M.H., Wolters, M.H., Strijker, R.E., de Jong, D., de Jong, G.J. and Brinkman, U.A.Th. (1992) A continuous two-phase reaction system coupled on-line with capillary chromatography for the determination of polar solutes in water. *J. High Resolut. Chromatogr.*, **15**, 242–248.

Greibrokk, T. (1995) Applications of supercritical fluid extraction in multi-dimensional systems. *J. Chromatogr. A*, **703**, 523–536.

Grob, K. (1991) *On-line coupled LC–GC.* Hüthig, New York.

Hawthorne, S.B., Miller, D.J., Nivens, D.E. and White, D.C. (1992) SFE of polar analytes using *in situ* chemical derivatisation. *Anal. Chem.*, **64**, 405–412.

Hillmann, R. and Bachmann, K. (1994) On-line supercritical fluid derivatisation and extraction-capillary GC of polar compounds. *J. High Resolut. Chromatogr.*, **17**, 350–352.

Horwitz, W. (1982) Evaluation of analytical methods used for regulation of foods and drugs. *Anal. Chem.*, **54**, 67A–76A.

Hotter, G., Ramis, I., Bioque, G., Sarmiento, C., Fernandez, J.M., Rosello-Catafau, J. and Gelpi, E. (1993) Application of totally automated on-line sample clean-up for prostanoid extraction and HPLC separation. *Chromatographia*, **36**, 33–38.

Huber, J.F.K. and Lamprecht, G. (1995) Assay of neopterin in serum by means of two-dimensional HPLC with automated column switching using three retention mechanisms. *J. Chromatogr. B*, **666**, 223–232.

Irth, H., Tocklu, R., Welten, K., de Jong, G.J., Brinkman, U.A.Th. and Frei, R.W. (1989) Trace-level determination of 3′-azido-3′-deoxythymidine in human plasma by preconcentration on a silver (I)-thiol stationary phase with on-line reversed-phase HPLC. *J. Chromatogr.*, **491**, 321–330.

Jackson, P.E. and Jones, W.R. (1991) Hollow-fibre membrane-based sample preparation device for the clean-up of brine samples prior to ion-chromatographic analysis. *J. Chromatogr.*, **538**, 497–503.

Jönsson, J.Å., Mathiasson, L., Lindegård, B., Trocewicz, J. and Olsson, A.-M. (1994) Automated system for the trace analysis of organic compounds with supported liquid membranes for sample enrichment. *J. Chromatogr. A*, **665**, 259–268.

Kaniansky, D. and Marák, J. (1990) On-line coupling of capillary isotachophoresis with capillary zone electrophoresis. *J. Chromatogr.*, **498**, 191–204.

Kelly, J.W., Aggarwal, N.D., Murari, R. and Stewart, J.T. (1994) HPLC separation of the ZZ, ZE, EZ and EE geometric isomers and EE isomer enantiomers of a substituted pentadienyl carboxamide using achiral/chiral column switching. *J. Liq. Chromatogr.*, **17**, 1433–1442.

Krock, K.A., Ragunathan, N. and Wilkins, C.L. (1994) Multidimensional gas chromatography coupled with infra-red and mass spectrometry for the analysis of eucalyptus essential oils. *Anal. Chem.*, **66**, 425–430.

Martin, M., Herman, D.P. and Guiochon, G. (1986) Probability distributions of the number of chromatographically resolved peaks and resolvable components in mixtures. *Anal. Chem.*, **58**, 2200–2207. [Erratum: Idem (1987) ibid., **59**, 384.]

Mattern, E.M., Kan, C.A. and van Gend, H.W. (1990) An automated HPLC determination of meticlorpindol in eggs with UV absorbance detection using on-line dialysis and pre-concentration as sample clean-up occurrence in and carry over to eggs. *Z. Lebensm.-Unters. Forsch.*, **190**, 25–30.

McNally, M.E.P. (1995) Advances in environmental SFE. *Anal. Chem.*, **67**, 308A–315A.

Monnig, C.A. and Kennedy, R.T. (1994) Capillary electrophoresis. *Anal. Chem*, **66**, 280R–314R.

Moretti, V.M., van de Water, C. and Haagsma, N. (1992) Automated HPLC determination of chloramphenicol in milk and swine muscle tissue using on-line immunoaffinity sample clean-up. *J. Chromatogr.*, **583**, 77–82.

Norman, K.N.T. (1991) A rapid method for the determination of liquid fumigant residues in food commodities using automated headspace analysis. *Pesticide Sci.*, **33**, 23–34.

Posluszny, J.V., Weinberger, R. and Woolf, E. (1990) Optimization of multidimensional high-performance liquid chromatography for the determination of drugs in plasma by direct injection, micellar clean-up and photodiode array detection. *J. Chromatogr.*, **507**, 267–276.

Pratt, D.A., Daniloff, Y., Duncan, A. and Robins, S.P. (1992) Automated analysis of the pyridinium crosslinks of collagen in tissue and urine using solid-phase extraction and reversed-phase HPLC. *Anal. Biochem.*, **207**, 168–175.

Sandra, P., Kot, A., Medvedovici, A. and David, F. (1995) Selected applications of the use of supercritical fluids in coupled systems. *J. Chromatogr. A*, **703**, 467–478.

Sharman, M. and Gilbert, J. (1991) Automated aflatoxin analysis of foods and animal feeds using immunoaffinity column clean-up and HPLC determination. *J. Chromatogr.*, **543**, 220–225.

Sharman, M., MacDonald, S. and Gilbert, J. (1992) Automated liquid chromatographic determination of ochratoxin A in cereals and animal products using immunoaffinity column clean-up. *J. Chromatogr.*, **603**, 285–289.

Shepherd, M.J. (1984) Size exclusion and gel chromatography: theory, methodology and applications to the clean-up of food samples for contaminant analysis. In: *Analysis of Food Contaminants*, Ed. Gilbert, J. Chapman & Hall, London, pp. 1–72.

Shepherd, M.J. (1988) Gel permeation chromatography for sample clean-up. *Internat. Analyst*, **2**, 14–15, 17, 19.

Strutton, A. (1992) The isolated ASPEC applied to the assay of a nootropic agent, compared with manual operation. *Methodol. Surv. Biochem. Anal.*, **22**, 321–324.

Stubbings, G.W.F. and Shepherd, M.J. (1993) A multi-dimensional liquid chromatography method for determination of androgen hormone residues in cattle liver. *J. Liq. Chromatogr.*, **16**, 241–255.

Szuna, A.J. and Blain, R.W. (1993) Determination of a new antibacterial agent (Ro 23-9424) by multi-dimensional HPLC with UV detection and direct plasma injection. *J. Chromatogr.*, **620**, 211–216.

Tarbin, J.A. and Shearer, G. (1993) High-performance liquid-chromatographic method for the determination of the anthelmintic nitroxynil in cattle muscle tissue with on-line anion-exchange clean-up. *J. Chromatogr.*, **613**, 347–353.

Taylor, L.T. (1995) Strategies for analytical SFE. *Anal. Chem.*, **67**, 364A–370A.

Thanh, H.H., Andresen, A.T., Agasoster, T. and Rasmussen, K.E. (1990) Automated column-switching high-performance liquid chromatographic determination of flumequine and oxolinic acid in extracts from fish. *J. Chromatogr.*, **532**, 363–373.

Thompson, T.S., Kolic, T.M. and MacPherson, K.A. (1991) Dual-column high-performance liquid-chromatographic clean-up procedure for the determination of polychlorinated dibenzo-*p*-dioxins and dibenzofurans in fish tissue. *J. Chromatogr.*, **543**, 49–58.

Tuinstra, L.G.M.Th., Kienhuis, P.G.M., Traag, W.A., Aerts, M.M.L. and Beek, W.M.J. (1989) Fully automated column-switching HPLC determination of aflatoxin M1 in milk using dialysis as sample preparation. *J. High Resolut. Chromatogr.*, **12**, 709–713.

Turner, W.E., Isaacs, S.G. and Patterson, D.G. (1992) Automated sample clean-up apparatus used in the procedure for measuring polychlorinated dibenzo-*p*-dioxins, dibenzofurans and ortho-unsubstituted (planar) biphenyls in human serum and adipose tissue. *Chemosphere*, **25**, 805–810.

van de Merbel, N.C., Hageman, J.J. and Brinkman, U.A.Th. (1993) Membrane-based sample preparation for chromatography. *J. Chromatogr.*, **634**, 1–30.

van der Hoff, G.R., Gort, S.M., Baumann, R.A., van Zoonen, P. and Brinkman, U.A.Th. (1991) Clean-up of some organochlorine and pyrethroid insecticides by automated solid-phase extraction cartridges coupled to capillary GC–ECD. *J. High Resolut. Chromatogr.*, **14**, 465–470.

van der Hoff, G.R., Baumann, R.A., Brinkman, U.A.Th. and van Zoonen, P. (1993) On-line combination of automated micro liquid–liquid extraction and capillary gas chromatography for the determination of pesticides in water. *J. Chromatogr.*, **644**, 367–373.

van Zoonen, P., van der Hoff, G.R. and Hogendoorn, E.A. (1990) RP LC–GC with a NPD using a loop-type interface combined with a sandwich-type phase separator. *J. High Resolut. Chromatogr.*, **13**, 483–488.

Vreuls, J.J., Brinkman, U.A.Th., de Jong, G.J., Grob, K. and Artho, A. (1991) On-line SPE–thermal desorption for introduction of large volumes of aqueous samples into a GC. *J. High Resolut. Chromatogr.*, **14**, 455–459.

Vreuls, J.J., de Jong, G.J., Ghijsen, R.T. and Brinkman, U.A.Th. (1994) LC coupled on-line with GC: state of the art. *J. AOAC Int.*, **77**, 306–327.

Walhagen, A. and Edholm, L.E. (1989) Coupled column chromatography on immobilised protein phases for direct separation and determination of drug enantiomers in plasma. *J. Chromatogr.*, **473**, 371–379.

Wessels, P., Ogorka, J., Schwinger, G. and Ulmer, M. (1993) Elucidation of the structure of drug degradation products by on-line coupled reversed phase HPLC–GC–MS and on-line derivatisation. *J. High Resolut. Chromatogr.*, **16**, 708–712.

White, D.R., Lee, H.S. and Kruger, R.E. (1991) Reversed-phase HPLC–EC determination of folate in citrus juice by direct injection with column switching. *J. Agric. Food Chem.*, **39**, 714–717.

Williams, R.A., Macrae, R. and Shepherd, M.J. (1989) Non-aqueous size exclusion chromatography coupled on-line to reversed phase high performance liquid chromatography. Interface development and applications to the analysis of low molecular weight contaminants and additives in food. *J. Chromatogr.*, **477**, 315–325.

Zhou, F.X., Krull, I.S. and Feibush, B. (1992) 9-Fluoreneacetyl-tagged, solid phase reagent for derivatisation in direct plasma injection. *J. Chromatogr.*, **609**, 103–112.

3 Chromatographic and allied methods of analysis for selected mycotoxins

ERIC W. SYDENHAM and GORDON S. SHEPHARD

Summary

This chapter discusses and reviews analytical methods that have been developed (over the past decade) for the determination of several selected, naturally occurring mycotoxins in human foods. The mycotoxins selected include the following: aflatoxins B_1, B_2, G_1, G_2 and M_1; fumonisins B_1, B_2 and B_3; ochratoxin A; patulin; several trichothecenes (deoxynivalenol, diacetoxyscirpenol, nivalenol and T-2 toxin) and zearalenone. Emphasis is placed on new techniques that are being applied, while analytical trends (related to various aspects associated with the analysis of mycotoxins in different foodstuffs) are also discussed. Areas of particular concern include selective extraction of target toxins from substrates, sample purification, (where necessary) derivatization, detection and confirmation. Method and analyst performance criteria are also presented.

3.1 Introduction

Fungi are members of the lower plant species that do not contain chlorophyll. Accordingly, fungi are unable to synthesize carbohydrates via photosynthesis as do members of the higher plant species, having instead to obtain their nutrients by growing either parasitically on the living tissues of plants, animals or man, or saprophytically on dead organic matter. It is due to the process by which they obtain their nutrients that fungi in general pose a potential threat to both animal and human health. During their growth stage (either in the field or under storage conditions), numerous fungi have the ability to produce a quite diverse range of *secondary metabolites* (inclusive of mycotoxins) which can be toxic and/or carcinogenic if ingested by animals or humans. The term *secondary metabolite* was introduced into microbial biochemistry in order to differentiate those compounds such as alkaloids, terpenes, flavonoids and other plant products that may be considered as non-essential for the growth of the plants themselves. Conversely, amino acids, fatty acids, saccharides, nucleic acids and proteins—compounds essential for all living organisms—were termed *primary metabolites*.

Although the mechanisms involved in the formation of several mycotoxins by various fungal species have been documented, the reason for their production is less well understood, although it has been postulated that their production may form part of intrinsic defence mechanisms. However, it is clear that many mycotoxins are able to exhibit an equally diverse range of biological activities in various animal species, including humans, at relatively low concentrations (i.e. in the mg/kg [ppm] to μg/kg [ppb] range). The isolation and characterization of the aflatoxins, and their subsequent implication as potential human carcinogens, may be regarded as the catalyst for modern mycotoxin research programmes. Although major research efforts initially centred on the aflatoxins, attention rapidly turned to alternative mycotoxins and their possible cause/effect relationship with other idiopathic diseases. The worldwide concern for aflatoxins and aflatoxicoses therefore became a broader one of mycotoxins and mycotoxicoses.

The natural occurrence of mycotoxins in cereal grains can, in addition to health problems, also have substantial economic implications for numerous commercial sectors including the crop, livestock and poultry producers, as well as for food and feed processors. Losses may originate from poor grain quality, low crop yields, reduced animal performance, the downgrading of crops, and/or the necessity to decontaminate affected grains. The perceived human health risks associated with the natural occurrence of selected mycotoxins has resulted in some degree of regulatory control in a number of countries, although the rationales for the selection of permitted tolerance levels are often vague (Van Egmond, 1989; Stoloff *et al.*, 1991). Alternatively, limited control is achieved through the establishment of non-legally binding action levels.

Irrespective of the extent of control, human exposure to mycotoxins has primarily been restricted by the application of chemical screening/monitoring programmes of suspect commodities. These programmes have required the development of analytical methods for the accurate determination of relatively low concentrations of specific mycotoxins, which often occur in chemically complex matrices. The Association of Official Analytical Chemists International (AOAC Int.) has, under its General Refereeship on Mycotoxins, recommended a number of analytical methods for several mycotoxins in various matrices (Stoloff and Scott, 1984). This chapter, however, summarizes some of the more recent developments in the application of chromatographic and allied analytical techniques developed over the past ten years for the determination and confirmation of several important mycotoxins which occur in various food matrices intended for human consumption. Given the wide scope of topics covered in this book, it has been necessary to restrict the number of mycotoxins discussed so as to include only those which are considered to be of major toxicological concern. Accordingly, the contents of this chapter should not be considered as an extensive literature review of the field of mycotoxins and mycotoxicoses.

Individual toxins (or groups of structurally related toxins) will be preceded by a brief introduction, which will include pertinent information related to their source and unique biological activities. Thereafter, the chromatographic and allied analytical techniques developed for their individual estimation in host substrates will be covered, while multi-mycotoxin methods (those methods which incorporate the determination of several chemically dissimilar toxins) are considered separately.

3.2   Aflatoxins in food commodities

3.2.1   Introduction

The aflatoxins (AF) were first discovered in the early 1960s as a result of investigations into the outbreak of so-called 'turkey-X' disease in the United Kingdom. They are highly fluorescent heterocyclic compounds consisting of a dihydrodifurano or tetrahydrodifurano moiety fused to a substituted coumarin (Figure 3.1) and are produced by *Aspergillus flavus* and *A. parasiticus*. The main members of this group of mycotoxins are the four naturally occurring toxins, AFB_1, AFB_2, AFG_1, and AFG_2, together with their metabolic products (AFM_1, AFM_2, AFP_1, AFQ_1 and aflatoxicol) produced by microbial and animal metabolic systems (Jones, 1977).

Aflatoxins have been shown to be acutely toxic to a host of animal species, with the liver being the primary target. AFB_1 is a powerful teratogen, mutagen and hepatocarcinogen, via its metabolic activation to the epoxide form which can then bind covalently to DNA. Its association with primary liver cancer in humans has led to its being classified by the International Agency for Research on Cancer (IARC) as carcinogenic to humans (Group 1 carcinogen) (Smith *et al.*, 1994).

The aflatoxins have been found in a wide variety of foodstuffs susceptible to contamination by *A. flavus* or *A. parasiticus*, either in the field or during

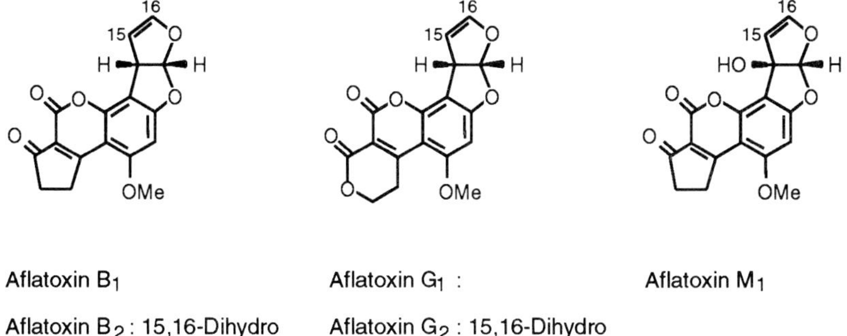

Figure 3.1 Chemical structures of aflatoxins B_1, B_2, G_1, G_2 and M_1.

storage, including cereals such as maize, oats and barley, ground and other nuts, oilseeds, and spices such as pepper and chilli.

Of the various mycotoxins for which legal limits of contamination have been set, aflatoxins are the most widely regulated. Van Egmond (1989) reported 50 countries worldwide with some form of aflatoxin regulation. Where regulations specify only the AFB_1 level, a range of levels from zero detectable to $50\,\mu g/kg$ have been set by various countries, although the action level most frequently specified is $5\,\mu g/kg$. Apart from AFB_1 levels, some countries regulate total aflatoxin levels within this same concentration range.

3.2.2 Application of TLC techniques

The chemical identification of the aflatoxins led to a vast amount of analytical research and the development of a number of TLC methods, some of which, such as the BF and CB methods for peanuts and peanut products, received official recognition (Stoloff and Scott, 1984). Despite the appearance of methodologies based on new concepts (e.g. HPLC and immunochemical methods), TLC remains a widely used tool for routine aflatoxin analysis. Developments over the previous decade have involved the collaborative study of existing methods and the application of instrumental TLC and high-performance thin layer chromatography (HPTLC) to achieve greater accuracy and precision. Improved TLC methods have been applied to various foodstuffs, including palm kernels (Nawaz et al., 1992), vegetable oils (Miller et al., 1985) and raisins (Boyacioglu and Gönül, 1988).

Tosch et al. (1984) compared instrumental HPTLC and HPLC for the determination of aflatoxins in peanut products. Although the resolution achieved by HPTLC was not as good as that from HPLC, the two methods appeared to be equivalent with respect to precision, accuracy and sensitivity. HPTLC could achieve considerable savings in analysis time and solvent consumption for a series of samples. A more comprehensive study, involving 300 peanut butter samples, compared HPTLC both with HPLC using trifluoroacetic acid (TFA) precolumn derivatization, and a commercial enzyme-linked immunosorbent assay (ELISA) method (Dell et al., 1990). Results indicated that the HPTLC method gave lower coefficients of variation than the other methods, while the ELISA method was found to be subject to matrix interferences at low toxin levels. Statistical comparison of analytical results indicated good agreement between HPTLC and ELISA but identified a measurable bias between HPTLC and HPLC, which the authors ascribed to the TFA derivatization procedure.

A bidirectional HPTLC method for determination of aflatoxins in maize has been developed which, when compared with two-dimensional methods, offers a 15–30-fold improvement in sample capacity per plate (Tomlins et al., 1989b). This method employs a prechromatographic development in diethyl ether to eliminate interfering compounds. After removal of the top 2 cm of the plate, it is developed twice in the opposite direction using the same mobile phase (chloroform:xylene:acetone, 6:3:1). The initial diethyl ether stage has been improved by the design of a continuous linear development chamber, which affords greater accuracy and reduced analysis times (Bradburn *et al.*, 1990).

A solvent-efficient method for determination of aflatoxins in maize and peanut butter was introduced which combined aspects of the BF and CB methods and used 40-μm diameter silica packing to improve the column-purification step, and hence reduce solvent consumption (Trucksess *et al.*, 1984). Average recoveries for all aflatoxins from both maize and peanut butter were above 82%. This method was subsequently studied collaboratively for maize, raw peanuts and peanut butter (Park *et al.*, 1994). Average recoveries for the aflatoxins at all levels determined by densitometry were 95.3, 95.6 and 139.0% from the three matrices, respectively. The high recovery from peanut butter appeared to be due to contamination of the blank material.

3.2.3 Application of HPLC techniques

HPLC has been widely used for the determination of aflatoxins in foodstuffs (Holcomb *et al.*, 1992; Frisvad and Thrane, 1993). AFB_1, AFB_2, AFG_1 and AFG_2 are readily separated by both normal-phase (NP) and reversed-phase (RP) columns, although in a European interlaboratory comparison of compound-feed reference materials, 19 of the 20 laboratories using HPLC reported isocratic RP separations with mobile phases of methanol, acetonitrile and water mixtures (Van Egmond and Wagstaffe, 1990). Although RP separations of aflatoxins are generally performed on C_{18} columns, Dorner and Cole (1988) described the use of a phenyl column with a mobile phase of water:tetrahydrofuran (4:1). The natural fluorescence of these toxins provides a useful mode of detection, although for the most sensitive analysis, chemical enhancement is required. Most of the advances in methodology over the past decade have involved extraction and purification methods in order to provide cleaner extracts and to reduce organic solvent consumption, while a number of methods have been investigated to improve detection limits.

Aflatoxins are routinely extracted with chloroform:water and methanol:water mixtures (Frisvad and Thrane, 1993). For the extraction of peanuts, Whitaker *et al.* (1986) and Cole and Dorner (1994) have both investigated optimum methanol concentration at various solvent:peanut ratios. The latter workers concluded that at low ratios (used to avoid excessive waste) the best choice was 80:20 methanol:water (3:1 ml/g). However, if extracts were purified by a Mycosep® multifunctional clean-up column, an acetonitrile:water mixture (optimum of 90:10, 2:1 ml/g) was needed for compatibility with this technique. These columns were developed to remove, in a single step, analytical interferences from the extract (Wilson and Romer, 1991), and have been the subject of an international collaborative study for the determination of aflatoxins in maize, almonds, Brazil nuts, peanuts and pistachio nuts (Trucksess *et al.*, 1994). Average recoveries were above 90%, within-laboratory repeatability was 6.0–23.2% and between-laboratory reproducibility was 12.0–69.4%. Other extraction solvents investigated were mixtures of acetone and water, which, together with a purification step using phenyl solid-phase extraction (SPE) cartridges, was applied to the analysis of groundnut meal and maize (Bradburn *et al.*, 1990; Roch *et al.*, 1992). Although phenyl SPE cartridges have found some use in aflatoxin analysis (Tomlins *et al.*, 1989a; Dell *et al.*, 1990), the majority of laboratories using SPE clean-up that participated in a recent intercomparison of methods reported the use of Florisil® followed by C_{18} SPE cartridges (Van Egmond and Wagstaffe, 1990), while others have used a combination of amino and C_{18} columns in series (Hurst *et al.*, 1987a). Hetmanski and Scudamore (1989) have reported a purification method for cereal extracts based on gel permeation chromatography.

A collaborative study of a proposed official method for use within the European Community revealed a number of potential sources of error in aflatoxin analysis, including inadequate protection from daylight, use of inappropriate glassware and temperature fluctuations in the fluorimetric detector (Van Egmond *et al.*, 1991). The method, which involved chloroform extraction and subsequent purification on Florisil® and C_{18} SPE cartridges, gave within-laboratory repeatability of 11% and between-laboratory reproducibility of 18%. Beaver (1990) has shown that aflatoxins dissolved in common reversed-phase solvent mixtures are subject to degradation, especially when exposed to light, but that addition of 0.5% acetic acid yields adequate stability. Lázaro *et al.* (1988) have investigated flow injection analysis for screening peanut and maize samples for aflatoxin contamination. This system can be integrated with HPLC for accurate determination of the individual toxins.

Although the aflatoxins are naturally strongly fluorescent, various members of the group exhibit solvent-dependent quenching in HPLC systems. Detection systems for aflatoxins have recently been reviewed (Kok, 1994). In chlorinated solvents used for NP-HPLC, the fluorescence intensities of AFB_1 and AFB_2 can be one to two orders of magnitude lower than those of AFG_1 and AFG_2. This effect was originally widely overcome by adoption of a fluorimetric cell packed with silica gel, which improved the fluorescence intensities of AFB_1 and AFB_2 to similar levels to AFG_1 and AFG_2 (Panalaks and Scott, 1977). However, with the development of RP-HPLC, the use of NP columns for aflatoxin analysis has greatly declined. In the aqueous systems of methanol and acetonitrile used for RP-HPLC, AFB_1 and AFG_1 fluorescence intensities are significantly quenched. This problem was initially overcome by precolumn derivatization with TFA, which results in the hydration of the double bond of the dihydrofuran moiety and the formation of the hemiacetals, AFB_{2a} and AFG_{2a}. These derivatives have similar intensities to AFB_2 and AFG_2. However, the relative instability of these derivatives and the potential advantages for automation led to the development of postcolumn derivatization techniques. Although TFA was unsuitable because of corrosion problems, iodination using saturated solutions of iodine and subsequent reaction at elevated temperatures (60–75°C) in a reaction coil provided enhanced sensitivity for AFB_1 and AFG_1. The postcolumn reaction conditions were optimized (Shepherd and Gilbert, 1984; Thiel et $al.$, 1986) and the detection limit for AFB_1 was found to be about 20 pg/injection. Confirmation of peak identity could be provided by switching off the iodine flow and observing the reduction in AFB_1 and AFG_1 peak heights.

Figure 3.2 shows a HPLC analysis, with and without postcolumn addition of iodine, of a sorghum malt extract in which this confirmatory technique is clearly demonstrated (Thiel, 1986). Similar postcolumn derivatization can be achieved using bromine, generated electrochemically (Kok et $al.$, 1986; Traag et $al.$, 1987). Potassium bromide is incorporated in an acidic mobile phase

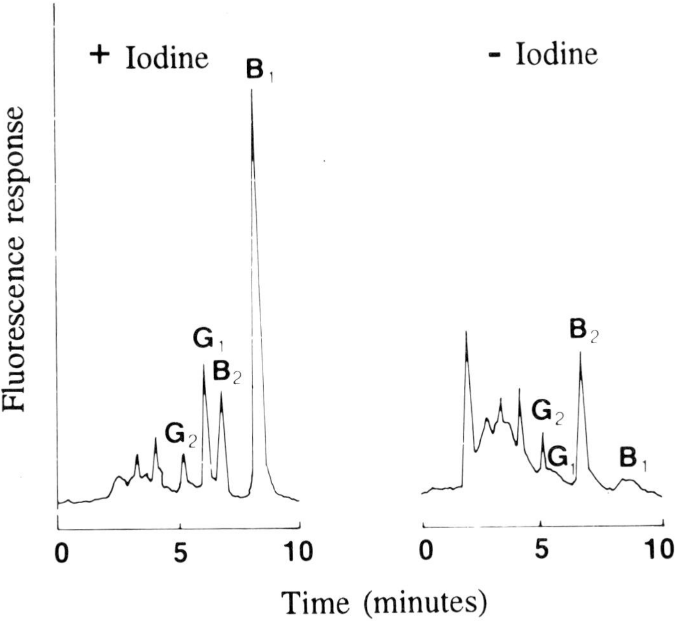

Figure 3.2 Determination of aflatoxins B_1, B_2, G_1 and G_2 at levels of 36, 12, 7 and 2 µg/kg, respectively, in naturally contaminated sorghum malt extracted and cleaned-up by the CB method and analyzed by RP-HPLC both with and without postcolumn iodination. Comparison of the two chromatograms demonstrates the ability of the technique to confirm the presence of aflatoxins B_1 and G_1 (Thiel, 1986).

and bromine is generated in an electrochemical cell by using a constant oxidative current. The resulting derivatives have about the same fluorescence intensity as the iodine derivatives, but the reaction with bromine is much faster (about 4 s at room temperature) and thus requires a shorter reaction coil, with a resultant reduction in band broadening. This derivatization also produces fewer matrix interferences and requires less expensive equipment. Francis *et al.* (1988) have shown that postcolumn derivatization of aflatoxins with β-cyclodextrin also enhances the fluorescence intensity of AFB_1 and AFG_1 in aqueous systems. The application of electrochemical detection using a differential-pulsed amperometric detector at a dropping mercury electrode has been investigated, but was found to give reduced sensitivity with the detection limit of about 5 ng underivatized standards (Duhart *et al.*, 1988). A postcolumn photochemical derivatization, followed by fluorescence detection, permits AFB_1 detection in maize down to 1 μg/kg (Joshua, 1993). Chromatographic evidence suggests that photolysis causes hydration of AFB_1 and AFG_1 to form their hemiacetals, AFB_{2a} and AFG_{2a}.

3.2.4   Application of immunochemical techniques

Immunochemical methods for aflatoxin determination have found application in ELISA tests and in immunoaffinity columns for sample extract clean-up prior to HPLC. The ELISA techniques can provide a screening method in the form of a single yes/no answer by means of the appearance or non-appearance of colour in the test, or they can give a semi-quantitative result through measurement of the developed colour. Both polyclonal and monoclonal antibodies to aflatoxins have been raised for incorporation in direct and indirect ELISA. Due to the regulatory limits that have been imposed and the commercial importance of aflatoxin contamination of foodstuffs, a number of commercial kits have been extensively marketed. Koeltzow and Tanner (1990) found that Afla-20-cup[®], Aflatest[®], EZ-Screen[®] and Oxoid[®] methods in maize were not statistically different from the conventional Holaday–Velasco mini-column method. Dorner and Cole (1989) compared the EZ-Screen Quick Card Test[®] and Afla-10 Cup Test[®] with HPLC, using groundnut extracts, and concluded that both screening methods identified 95% of samples containing no detectable aflatoxin as negative and >97% of samples containing >10 μg/kg aflatoxin as positive. The Cite Probe[®] screening test for aflatoxins in maize was compared with HPLC analysis and determined to be a reliable screening method for $\geq$5 μg/kg aflatoxins in maize (Beaver *et al.*, 1991).

New antibodies to aflatoxin have continued to be reported and their use in food analysis demonstrated (Morgan *et al.*, 1986; Chu *et al.*, 1987; Dixon-Holland *et al.*, 1988; Kawamura *et al.*, 1988). Using an ELISA based on that of Morgan *et al.* (1986), a comparison between ELISA and HPLC conducted during a survey of peanut butter in the UK showed good agreement between methods for smooth peanut butter in the range 3–24.4 μg/kg (Mortimer *et al.*, 1987b). Ram *et al.* (1986) compared a direct ELISA based on polyclonal antibodies with the CB TLC method and showed a correlation coefficient of 0.95 for contaminated maize in the range 7–420 μg/kg. Hongyo *et al.* (1992) compared results on maize extracts obtained by a direct ELISA using monoclonal antibodies with those obtained by TLC and HPLC and found correlation coefficients of 0.93 and 0.95, respectively, for contamination levels up to about 200 μg/kg.

Since 1988 a number of collaborative studies on commercial ELISA tests have been reported. Mortimer *et al.* (1988) reported that in a 14-laboratory UK study on peanut butter samples contaminated at 8, 25 and 75 μg/kg, the Quantitox B_1[®] ELISA kit could be used effectively to indicate AFB_1 levels. An international collaborative study concluded that the Agri-Screen[®] ELISA test (or other ELISA meeting the same criteria) was suitable for use as a screening method to determine the presence or absence of AFB_1 at a concentration of ≥ 20 μg/kg in maize and roasted peanuts (Park *et al.*, 1989). Another international collaborative study gave approval to the ImmunoDot Screen[®] method for aflatoxins at 20 μg/kg in peanut butter and ≥ 30 μg/kg in maize and raw peanuts (Trucksess *et al.*, 1989). A modification of this screening test has been studied and found acceptable as a positive/negative test at a level of 20 μg/kg aflatoxins in maize (Trucksess and Stack, 1994). Patey *et al.* (1992) conducted an international collaborative study of the Biokits[®] total aflatoxin ELISA kit for analysis of peanut butter naturally contaminated at levels of 9, 30 and 89 μg/kg total aflatoxin, and spiked at levels of 13 and 22 μg/kg. Recoveries from the spiked samples were 84 and 89% respectively, while the naturally contaminated samples gave between-laboratory reproducibility of 28–37%. The method was adopted first action by the AOAC Int. for aflatoxins in peanut butters in the range 9–90 μg/kg.

Immunoaffinity columns provide rapid and efficient clean-up of sample extracts, but, like other immunochemical methods, their use places limitations on the choice of extraction solvent, which must be compatible with the antibody. Trucksess *et al.* (1990) used the Aflatest-P[®] affinity column to clean up extracts of maize and peanuts prior to HPLC or solution fluorimetry for estimation of total aflatoxins. This work was tested by international collaborative study and both the HPLC and the solution fluorimetry showed acceptable recoveries of above 80% for the HPLC determination and 105–123% for the solution fluorimetry. Within- and between-laboratory reproducibilities at spiking levels of 10, 20 and 30 μg/kg were acceptable (Trucksess *et al.*, 1991). Other collaborative studies using immunoaffinity columns for clean-up of extracts of peanuts and other foodstuffs have shown acceptable reproducibilities, but low recoveries, the causes of which were not known (Patey *et al.*, 1991; Barmark and Larsson, 1994).

3.2.5   Application of alternative/confirmation techniques

The potential use of supercritical fluid extraction (SFE) for extraction of aflatoxins has been investigated. Engelhardt and Haas (1993) were unable to achieve high recoveries of AFB_1 from naturally contaminated peanut meal. Aflatoxins are only weakly soluble in carbon dioxide and require organic modifiers for efficient extraction, with resultant decrease in selectivity. Extraction kinetics from naturally contaminated peanut meal were also unfavourable when compared with conventional organic solvent extraction. However, Taylor et al. (1993) achieved over 90% extraction efficiencies from maize when compared with conventional solvent extraction. They utilized 15% acetonitrile:methanol (2:1) as modifier at 5000 psi and 80°C. However, for determination of AFB_1 in the low $\mu g/kg$ range, the resulting extract required further clean-up.

Initial attempts to determine aflatoxins by gas chromatography (GC) using glass columns were unsuccessful due to decomposition or binding of these compounds. However, the development of fused silica capillary columns has overcome these problems and has thus enabled GC–MS to be developed as a confirmatory procedure for aflatoxin contamination. Rosen et al. (1984) developed a method based on GC coupled to high resolution MS for confirmation of AFB_1 and AFB_2 in peanuts. Peanut extracts were prepared by the BF or CB methods, subjected to a rapid silica SPE clean-up with benzene and analysed by GC–MS with selected ion monitoring (SIM). The high-resolution MS was required to separate AFB_1 and AFB_2 from co-eluting impurities, while AFG_1 and AFG_2 did not elute from the GC column. AFB_1 in maize and peanut butter has been confirmed by GC–MS with negative-ion chemical ionization after the AFB_1 had been eluted from a TLC plate (Trucksess et al., 1984). A similar MS method, using solid probe introduction and full mass scans, was subject to an international collaborative study for use as a confirmation procedure for AFB_1 in roasted peanuts, cottonseed and ginger root (Park et al., 1985). The identity of AFB_1 was confirmed in 19.5, 90.9 and 100% of samples containing <5, 5–10 and >10 $\mu g/kg$, respectively. Separation of AFB_1, AFB_2, AFG_1 and AFG_2 by capillary GC with on-column injection and flame ionization detection (FID) has been reported and applied to culture filtrates (Goto et al., 1988). Aflatoxins in peanuts have been determined by HPLC coupled to a thermospray MS detector (Hurst et al., 1991). Detection limits with selected ion monitoring were found to be 100 pg or less.

The need for simple, reliable, portable and cost-effective screening methods for use under field conditions has been addressed by the development of a pressure minicolumn technique which was shown to give results comparable with those obtained by TLC (Sudershan et al., 1992). Sashidhar (1993) has reported the development of an adsorbent-coated dipstick for determination of AFB_1 down to 10 $\mu g/kg$ levels.

3.3 Aflatoxin M₁ in milk

3.3.1 Introduction

AFM_1 is a hydroxylated derivative of AFB_1 (Figure 3.1) and has been shown to be produced in the milk of lactating animals consuming AFB_1-contaminated feed. It has been reported to be excreted at a level of 0.03–1% of the level of AFB_1 in the feed (Heathcote and Hibbert, 1978). The toxicological concern over AFM_1 arises from its close chemical similarity to AFB_1 and the fact that infants consume relatively large quantities of milk. Direct evidence for its carcinogenicity is limited to studies in rats (Cullen *et al.*, 1985) and rainbow trout (Canton *et al.*, 1975). It is listed by IARC as a Group 2B carcinogen (possibly carcinogenic to humans) (Smith *et al.*, 1994).

Van Egmond (1989) reported 14 countries with regulations for AFM_1 in milk and dairy products. These limits have mostly been set either at $0.05\,\mu g/kg$ or $0.5\,\mu g/kg$, with little apparent rationale for these disparate values.

3.3.2 Application of TLC and HPLC techniques

The original methods developed for the determination of AFM_1 were based, like other aflatoxin methods, on TLC techniques. More recently, however, other methodologies based on HPLC and immunochemical advances have come to the fore. Scott (1989) has recently reviewed a wide range of TLC and HPLC methods that have reported detection limits for AFM_1 in milk down to 5 ng/l, and which have been collaboratively studied or used in check sample programmes. It was concluded that determination of AFM_1 in milk down to low ng/l concentrations was practical using TLC, HPLC or ELISA for quantitation. The reviewed methods used various clean-up procedures, including immunoaffinity column and C_{18} and silica SPE cartridges.

Shepherd *et al.* (1986) evaluated six previously published methods for the determination of AFM_1 in milk. The methods chosen involved chloroform, Extrelut® or C_{18} SPE extraction, silica gel or liquid–liquid partition for clean-up and determination by TLC (one or two developments) or RP-HPLC (AFM_1 itself or, for greater sensitivity, its TFA derivative). The authors concluded that the method of Takeda (1984) was the most suitable, as it provided the cleanest chromatograms and a rapid analysis time due to the use of C_{18} SPE cartridges for extraction and clean-up. This method has subsequently been developed into an automated robotic method capable of performing multiple unattended determinations (Gifford *et al.*, 1990). A collaborative study for the determination of AFM_1 and AFM_2 has been conducted on a variation of the Ferguson-Foos and Warren (1984) method, which involved C_{18} SPE extraction of a diluted milk sample and subsequent clean-up on a silica column (Stubblefield and Kwolek, 1986). The original method involved NP-HPLC, but was modified for the study by adoption of RP-HPLC of the TFA derivatives of AFM_1 and AFM_2. The method was found satisfactory with average recoveries for AFM_1 of 93.7%, within-laboratory repeatability of 27.9% and between-laboratory reproducibility of 44.5% for AFM_1.

An alternative approach to AFM_1 extraction and clean-up is the development of an automated method utilizing on-line dialysis, which was reported to achieve over 50% recovery and a detection limit of 0.02 μg/kg (Tuinstra *et al.*, 1990). The problem of analytical recoveries has been addressed by the use of AFB_2 as an internal standard (Carisano and Torre, 1986), while Qian *et al.* (1984) described a method for retention of AFM_1 on a C_{18} cartridge, which was then shipped unrefrigerated to a suitable analytical laboratory for elution and determination.

Van Egmond and Wagstaffe (1987) have reported on the development of milk powder reference materials certified for AFM_1 content. In their report, they commented on potential sources of error in AFM_1 analytical methods and noted that TLC–densitometry (but not visual TLC) can be as precise as HPLC determinations. The following sources of analytical difficulty were listed: correct concentration of standard solutions, formation of emulsions in methods using organic solvents, variations in the activity of silica and impurities (e.g. water) in solvents used for clean-up, adequate chromatographic resolution of AFM_1 and relative instability of AFM_1 in RP-HPLC solvents such as water:acetonitrile.

HPLC and TLC methods have also been developed for the determination of AFM_1 in cheese. These methods involve prior extraction of the cheese with organic solvents such as chloroform (Hisada *et al.*, 1984) or acetone:water (3:1) (Bijl *et al.*, 1987) followed by clean-up on silica and reversed-phase C_{18} cartridges.

3.3.3 *Application of immunochemical techniques*

AFM_1 has been determined in dairy products (milk, yoghurt and cheese) by using direct ELISA based on antibodies raised in rabbits immunized with AFM_1-bovine serum albumin (BSA) conjugate (Fremy and Chu, 1984). Antibodies raised to other aflatoxin analogues had little affinity for AFM_1. Although milk can be used directly in this assay, an additional clean-up with C_{18} SPE cartridges enables the assay to detect AFM_1 down to 10 ng/l for raw milk and 5 ng/l for reconstituted dry skimmed milk. For yoghurt and cheese extracts, a silica SPE cartridge was used for purification. For levels of contamination >0.1 μg/kg, samples must be diluted prior to assay. Another ELISA has been developed using polystyrene beads coated with antibody which is suitable for measurement in the range 0–30 ng/l (Nieuwenhof *et al.*, 1990), while hybridoma technology has been used to generate monoclonal antibodies for use in a direct ELISA for the detection of AFM$_1$ in milk (Dixon-Holland *et al.*, 1988).

A number of laboratories have reported the use of immunoaffinity columns as a purification step prior to HPLC determination of AFM$_1$. Mortimer *et al.* (1987a) reported a considerable saving in analysis time by the use of a commercially available immunoaffinity column which had been developed for aflatoxin analysis. Aliquots of milk were defatted by centrifugation prior to direct loading on to the clean-up column. The detection limit of the method was 0.05 ng/l, being governed mostly by the volume of milk sampled, while recoveries averaged 85.7%. This method for milk was later extended to determine AFM$_1$ in cheese (Sharman *et al.*, 1989), also with considerable saving of analysis time. The cheese was extracted with chloroform, which was then subjected to liquid–liquid extraction before loading on to the immunoaffinity column. Recoveries varied between different cheeses (55–80%), while the detection limit was 0.005 μg/kg. Another commercial immunoaffinity column containing monoclonal antibodies against AFM$_1$ has been collaboratively tested over the range 0.080–0.600 μg/kg milk

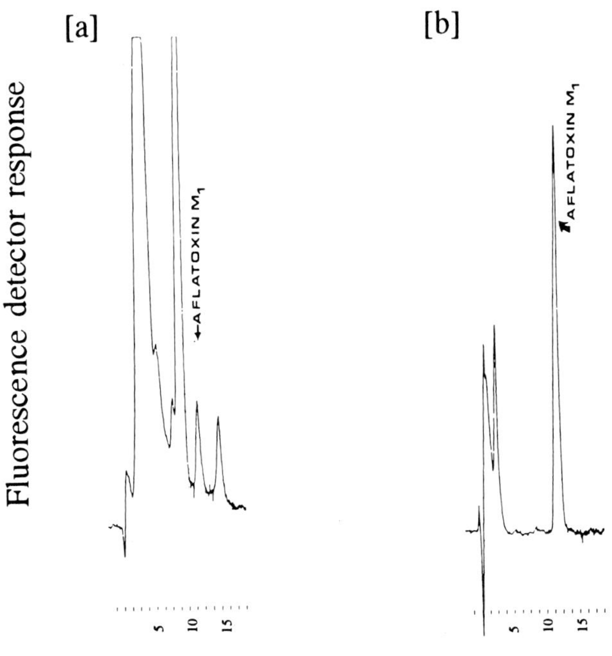

Figure 3.3 Comparison between two methods of AFM$_1$ purification for RP-HPLC analysis of naturally contaminated Camembert cheese. The purified samples were concentrated to the same volume. (a) Using a silica-gel SPE cartridge: 0.34 ng of AFM$_1$ injected in 50 μl; (b) using an immunoaffinity column (Aflaprep M$^®$): 1.4 ng of AFM$_1$ injected in 50 μl (Dragacci *et al.*, 1995).

powder, in 16 laboratories from eleven countries (Tuinstra *et al.*, 1993). Fourteen laboratories produced acceptable results with between-laboratory reproducibility in the range 11–23%. This immunoaffinity column has also been used as the clean-up step in the determination of AFM_1 in cheese, with an overall recovery of about 75% and a limit of quantification of 0.02 μg/kg (Dragacci *et al.*, 1995). Figure 3.3 shows the cleaner chromatograms achieved with this column compared with a clean-up based on silica gel SPE cartridges.

Farjam *et al.* (1992) investigated the use of a dialysis-based clean-up of milk prior to loading on the immunoaffinity column. Of the systems tested, a hollow-fibre unit was found to give better repeatability (3%) and lower detection limit (10 ng/l) than a flat membrane system. Figure 3.4 shows a chromatogram of a blank and spiked (50 ng/l) milk extract using this

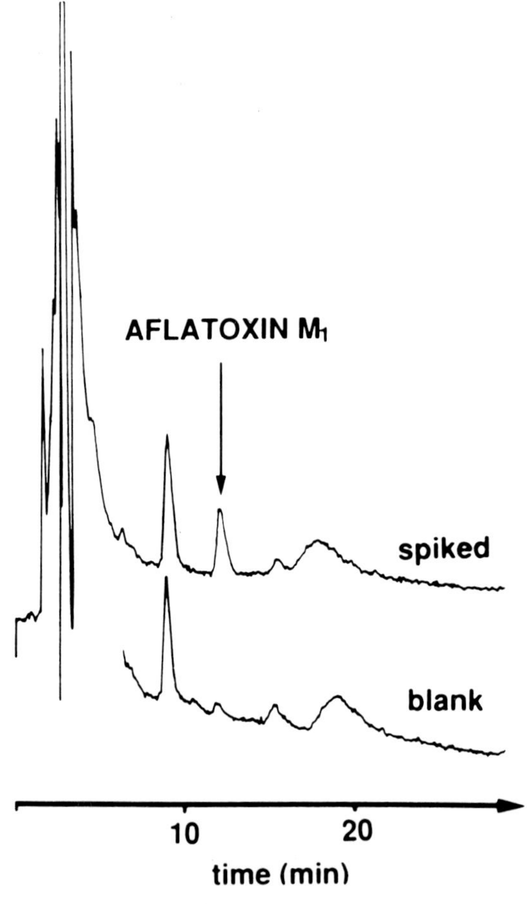

Figure 3.4 RP-HPLC determination of AFM_1 in a spiked (50 ng/l) crude milk sample and its corresponding blank after sequential clean-up by on-line hollow-fibre dialysis followed by an immunoaffinity column (Farjam *et al.*, 1992).

purification system. By removing impurities which destroy the activity of the antibodies, this prepurification allowed a single column to be re-used for over seventy milk analyses. Other authors, using immunoaffinity columns for the clean-up of defatted milk, have reported that regenerated columns can be satisfactorily re-used five times (Toyoda *et al.*, 1994).

3.3.4 *Application of alternative confirmation techniques*

The confirmation of the presence of AFM_1 in milk and milk products has long relied on the reaction of AFM_1 with TFA to form a derivative which has been assumed to be AFM_{2a}. This derivative can be formed *in situ* on the TLC plate which is then developed (Stoloff and Scott, 1984; Kamimura *et al.*, 1985), or prepared *in vitro* prior to HPLC analysis (Bijl *et al.*, 1987). The optimum conditions for this preparation have been studied (Stubble-field, 1987). The reaction proceeds best with non-polar solvents such as hexane or iso-octane, and requires silylated glass for complete reaction of standards if no milk residue is present. Another confirmatory method involves preparation of an acetyl derivative of AFM_1 and the subsequent use of two-dimensional TLC (Bijl *et al.*, 1987).

3.4 Fumonisins

3.4.1 *Introduction*

The fumonisins are secondary fungal metabolites produced by *Fusarium moniliforme, F. proliferatum* and other morphologically-related species, which are pathogens of maize worldwide (Gelderblom *et al.*, 1988; Scott, 1993). Although various analogues have been isolated from fungal culture (Cawood *et al.*, 1991; Branham and Plattner, 1993), the forms present at significant levels in naturally contaminated maize are fumonisin B_1 (FB_1), fumonisin B_2 (FB_2) and fumonisin B_3 (FB_3) (Sydenham *et al.*, 1991; Murphy *et al.*, 1993). They are all diesters of propane-1,2,3-tricarboxylic acid (tricarballylic acid) and various 2-amino-12,16-dimethylpolyhydroxy-eicosanes in which the hydroxyl groups on C_{14} and C_{15} are esterified with the terminal carboxyl group of the tricarballylic acid (Figure 3.5).

The major naturally occurring analogue, FB_1, has been shown to cause the syndromes of leukoencephalomalacia in horses (Kellerman *et al.*, 1990) and pulmonary oedema in swine (Harrison *et al.*, 1990). Studies in laboratory rats have shown FB_1 to be both hepatotoxic and hepatocarcinogenic (Gelder-blom *et al.*, 1991). The fumonisins have been shown to have toxic effects in broiler chicks (Ledoux *et al.*, 1992) and turkey poults (Weibking *et al.*, 1993), while exhibiting both cytotoxicity to various mammalian cell lines (Shier *et al.*, 1991; Gelderblom *et al.*, 1993) and phytotoxicity to tomato

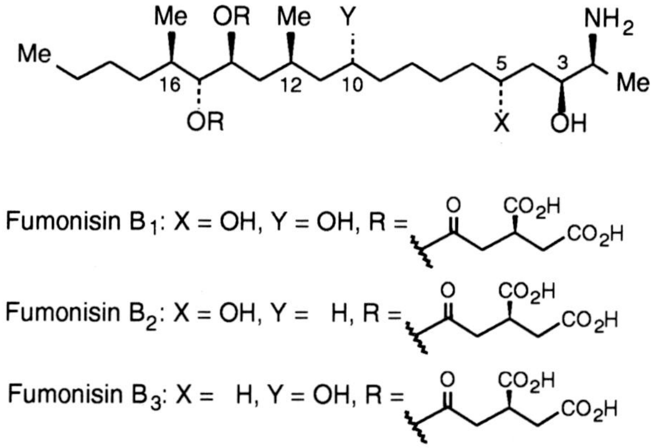

Fumonisin B$_1$: X = OH, Y = OH, R =

Fumonisin B$_2$: X = OH, Y = H, R =

Fumonisin B$_3$: X = H, Y = OH, R =

Figure 3.5 Chemical structures of fumonisins B$_1$, B$_2$ and B$_3$.

and maize seedlings (Lamprecht *et al.*, 1994). The role of fumonisins in the aetiology of human disease is not certain, but its occurrence in home-grown maize has been statistically associated with the prevalence of oesophageal cancer, both in the Transkei region of southern Africa (Rheeder *et al.*, 1992) and in Linxian county, China (Chu and Li, 1994).

3.4.2 Application of TLC techniques

The original isolation of the fumonisin mycotoxins in 1988 from maize cultures of *F. moniliforme* MRC 826 was aided by the development of techniques for both RP-TLC (C$_{18}$ plates) using methanol:water (3:1) as solvent, and NP-TLC on silica plates using chloroform:methanol:water:acetic acid mixtures as solvent (Gelderblom *et al.*, 1988; Cawood *et al.*, 1991). After development, the fumonisins were visualized on the plates by spraying either with ninhydrin or with *p*-anisaldehyde solution. Although useful for the analysis of fungal cultures, the detection limit of the method (500 mg/kg) precluded its use for the analysis of naturally contaminated maize (Sydenham *et al.*, 1990a). Further development of the method for maize culture material utilized a two-stage development of the TLC plate with removal of the lower section of the plate between stages and instrumentalized spectrophotodensitometry for fumonisin detection, which was reported to be suitable for levels above 100 mg/kg (Dupuy *et al.*, 1993).

Improved sensitivity and selectivity that enabled fumonisins at low-ppm levels in naturally contaminated maize to be determined by TLC was achieved by using fluorescamine as spray reagent (Rottinghaus *et al.*, 1992). The fumonisins were extracted with acetonitrile:water (1:1) and an aliquot of the extract was diluted with 1% KCl prior to application to a C$_{18}$ SPE cartridge. After washing the cartridge with 1% KCl and acetonitrile:1% KCl (1:9) to remove some matrix interferences, the fumonisins were eluted with acetonitrile:water (7:3). The purified extract was chromatographed by RP-TLC using methanol:4% aqueous KCl (4:1) as solvent, and the fumonisins were visualized by the successive spraying of 0.1 M sodium borate buffer (pH 8–9), a solution of 0.04% fluorescamine in acetonitrile and, after 1 min, a solution of 0.01 M boric acid:acetonitrile (4:6). Fumonisins were quantified by comparison of spot intensities under long-wavelength UV with those of standards. Although a detection limit of 100 μg/kg was reported for this method in maize-based feeds, considerable matrix interferences are frequently present at levels below 1 mg/kg (in maize) which hamper the visual discernment of the individual chromatographic bands. These interferences can be greatly reduced by replacing the C_{18} cartridge with a clean-up method based on strong anion exchange (SAX) SPE cartridges (Stockenström *et al.*, 1994).

Another reported RP-TLC method for the determination of FB_1 in maize relies on a spray reagent of 0.5% vanillin in 97% sulphuric acid:ethanol (4:1) which, after heating at 120°C for 10 min, reveals FB_1 as a blue–purple spot with a detection limit of 250 μg/kg (Pittet *et al.*, 1992).

3.4.3 Application of HPLC techniques

RP-HPLC is a technique ideally suited to the sensitive determination of fumonisins in maize and maize-based foods and has found extensive use worldwide for investigating the occurrence of these mycotoxins, and in assessing fumonisin exposure of animals and humans. Extraction of these polar compounds from food matrices is achieved by using either acetonitrile:water (1:1) (Wilson *et al.*, 1990; Bennett and Richard, 1994) or methanol:water mixtures containing 70–80% methanol (Shephard *et al.*, 1990; Bennett and Richard, 1994). Aqueous acetonitrile (with 60 min shaking) has been reported to give superior extraction efficiency compared to aqueous methanol (Bennett and Richard, 1994) whereas Sydenham *et al.* (1992) reported slightly better efficiencies for methanol:water (3:1) using homogenization for 1–5 min. Different workers have also reported using various sample:solvent ratios from 1:2 to 1:5, without rationalization of these differences. It has been noted that fumonisins in naturally contaminated samples are more difficult to extract than fumonisin standard added as a spike to maize, possibly due to an association with other sample matrix components (Bennett and Richard, 1994). In this regard, Scott and Lawrence (1994) have reported that in certain substrates, such as maize bran flour and mixed baby cereal, poor recoveries are obtained when using these extraction solvents, and that for the bran flour, but not the mixed cereal, the use of methanol:borate buffer (pH 9.2) (3:1) as extractant gave acceptable recoveries.

Prior to HPLC analysis, the crude extracts need purification to remove matrix impurities and to concentrate the fumonisins. This step has been achieved either by SPE on reversed-phase (C_{18}) or SAX cartridges, or by immunoaffinity columns. These latter columns have the disadvantage of limited capacity and are discussed more fully below in section 3.4.5. Of the two SPE techniques, SAX cartridges achieve superior purification (Bennett and Richard, 1994; Stockenström *et al.*, 1994), but require monitoring of the pH of the sample extract (above 5.8 for adequate retention on the SAX cartridge), and careful control of elution flow rates at not more than 1 ml/min for reproducible recoveries (Sydenham *et al.*, 1992). In addition, they cannot be used for the determination of the hydrolysed aminopolyol moieties of the fumonisins, which lack the anionic carboxylate groups needed for the SAX clean-up procedure. Large variations in recovery from C_{18} cartridges have been noted, which were surmised to be due to interaction of the fumonisins with active sites on the sorbent (Bennett and Richard, 1994). In addition to the reversed-phase and SAX cartridges, SPE sorbents containing both these functionalities have been reported to yield good recoveries of fumonisins over a wide pH range (Miller *et al.*, 1993), while the sequential use of first a C_{18} and then a SAX cartridge has been applied to the clean-up of rodent feed (Holcomb *et al.*, 1993). C_{18} SPE cartridges have been applied to the purification of the aminopentol produced by the hydrolysis of FB_1 in maize (Hopmans and Murphy, 1993; Sydenham *et al.*, 1995a) and milk (Scott *et al.*, 1994), while wort samples in beer brewing were also purified on C_{18} media, after a prior clean-up on SAX SPE cartridges (Scott *et al.*, 1995).

As the fumonisins are not fluorescent and lack a significant UV chromophore for sensitive detection by conventional spectrophotometric methods, analysts have described various precolumn derivatization techniques involving reaction of the primary amine group. The first HPLC method reported was developed for fumonisin quantitation in culture extracts (Alberts *et al.*, 1992). It involved the formation of the maleyl derivative and the separation of the derivatives by RP-HPLC with UV detection, which gave a detection limit of 10 mg/kg and was hence unsuitable for use in naturally contaminated maize (Sydenham *et al.*, 1990a). Much more sensitive results were obtained with fluorescamine and fluorescence detection (Wilson *et al.*, 1990; Ross *et al.*, 1991) but reaction with this reagent results in the formation of two reaction products (an equilibrium mixture of a lactone and its hydrolysed analogue) and this was considered undesirable (Sydenham *et al.*, 1990a). *o*-Phthaldialdehyde (OPA) has proved a sensitive reagent for the fumonisins by using precolumn derivatization and isocratic RP-HPLC with fluorescence detection (Figure 3.6) (Shephard *et al.*, 1990; Sydenham *et al.*, 1992).

The derivatization reaction with OPA and 2-mercaptoethanol is rapid and reproducible at room temperature in a borate buffer (pH 9–10) but the method suffers from the disadvantage of the limited stability of the fluorescent reaction product. These derivatives were found to be stable for a period

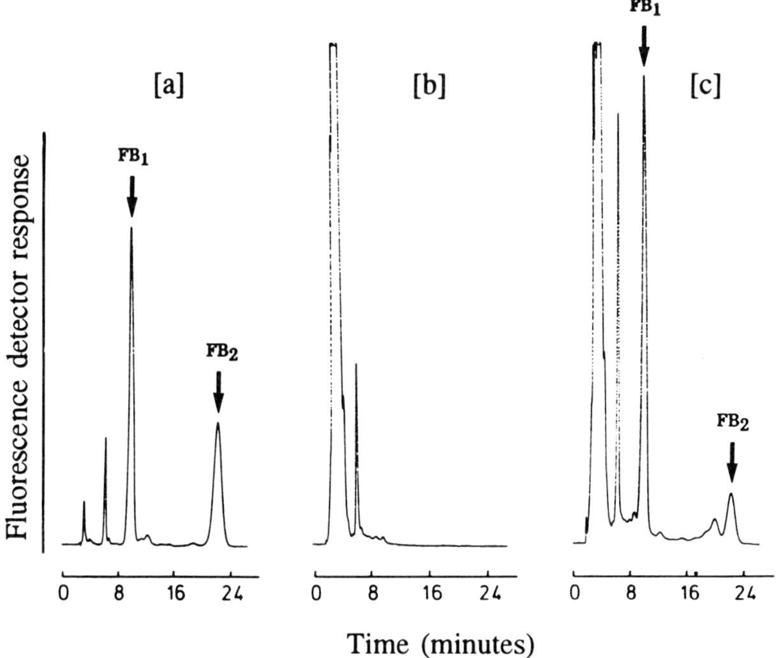

Figure 3.6 RP-HPLC determination of the OPA derivatives of FB_1 and FB_2. (a) FB_1 and FB_2 standards (100 ng of each injected); (b) extract of fumonisin-free maize purified on a SAX SPE cartridge; (c) similarly purified extract of maize naturally contaminated at levels of 4.4 and 1.3 mg/kg, respectively (Shephard *et al.*, 1990).

of 4 min after preparation, whereafter a decrease (5%) in fluorescence response of the FB_1 derivative was noted after 8 min, with further decreases thereafter. This problem is readily overcome by standardizing the time (about 2 min) between reagent addition and HPLC injection. Detection limits for the method are of the order of 50 μg/kg or better.

Naphthalene-2,3-dicarboxaldehyde with KCN formed a highly fluorescent derivative which was relatively stable over 24 h and which allowed detection of 50 pg of FB_1 standard (Ware *et al.*, 1993; Bennett and Richard, 1994). This reagent has been incorporated in a method for the determination of FB_1 and FB_2 in milk which has a sensitivity of 5 μg/l (Figure 3.7) (Maragos and Richard, 1994).

Scott and Lawrence (1992) reported the use of 4-fluoro-7-nitrobenzo-furazan (NBDF) with a detection limit of 100 μg/kg and found it a useful alternative reagent to other published methods, although the derivative also exhibited limited stability. These authors also found that 1-dimethyl-aminonaphthalene-5-sulphonyl chloride (dansyl chloride) formed a good derivative but was not useful for maize due to analytical interferences. 9-Fluorenylmethyl chloroformate (FMOC) has been used as a sensitive reagent

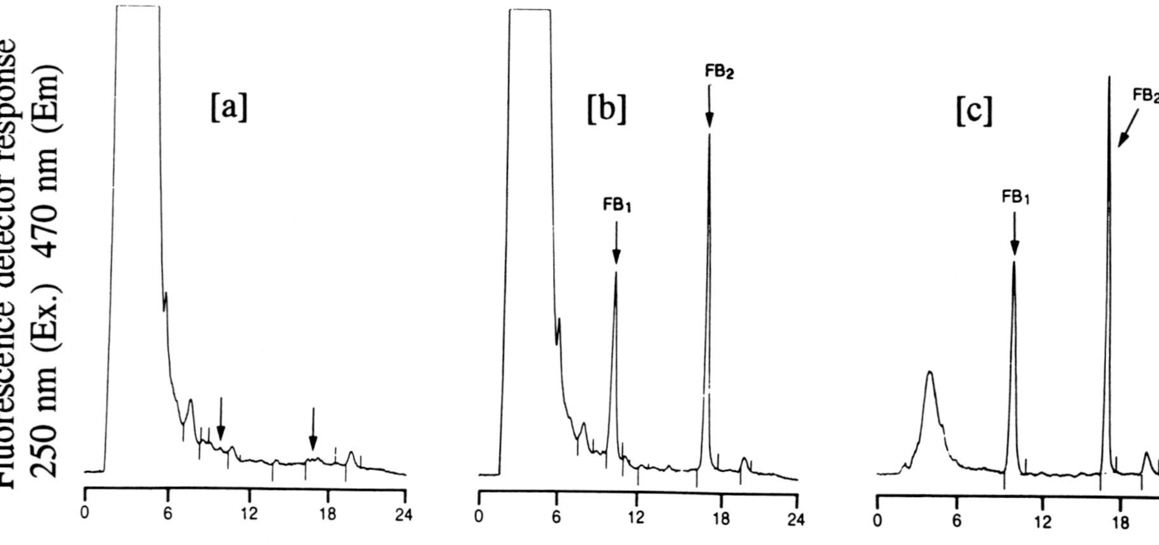

Figure 3.7 Analysis of fumonisins in raw (unpasteurized, unhomogenized) milk by HPLC as their NDA derivatives following clean-up on SAX SPE cartridges. (a) Toxin-free control; (b) the same milk as in (a), to which 50 ng each of FB$_1$ and FB$_2$ was added per ml; (c) standard FB$_1$ and FB$_2$ at a level equivalent to 50 μg/l of milk (Maragos and Richard, 1994).

for fumonisin determination in rodent feed, forming derivatives which were stable for at least 72 h and which had a detection limit of 200 μg/kg (Holcomb *et al.*, 1993).

Soon after the publication of the first analytical techniques for fumonisins in maize, a small collaborative study between four laboratories was conducted to test the fluorescamine procedure. In addition, two laboratories included GC–MS and OPA methods in their results, which showed acceptable agreement in the range 4–1800 mg/kg (Plattner *et al.*, 1991). The OPA method using methanol:water (3:1) for extraction and SAX cartridges for clean-up was studied collaboratively under the auspices of the Commission of Food Chemistry of IUPAC (Thiel *et al.*, 1993). Naturally contaminated maize samples were analysed by eleven laboratories from six countries. For FB_1 concentrations ranging from 2000 to 200 μg/kg, within-laboratory repeatability varied from 7.7 to 25.5% RSD and between-laboratory reproducibility from 18.0 to 26.7% RSD. Corresponding figures for FB_2 were 12.5–36.8% RSD for within-laboratory repeatability and 28.0–45.6% relative standard deviation (RSD) for between-laboratory reproducibility. A follow-up collaborative study using 'blank' maize spiked with FB_1, FB_2 and FB_3 standards at concentrations ranging from 100 to 8000 μg/kg produced mean recoveries of between 75.8 and 86.8% for the individual toxins (Sydenham *et al.*, 1996b).

3.4.4 Application of GC techniques

Prior to the development of accurate and sensitive HPLC techniques, fused-silica capillary GC methods were used to confirm the natural occurrence of these newly discovered mycotoxins. These methods involved hydrolysis of the fumonisins to cleave the ester bonds, followed by esterification with isobutanol for tricarballylic acid determination (Sydenham *et al.*, 1990a) or trimethylsilylation following isolation of the hydrolysed aminopolyol backbone on XAD-2 resin (Plattner *et al.*, 1990). This latter method, coupled to MS with chemical and electron ionization, was used for the quantitative analysis of heavily contaminated maize samples associated with field cases of leukoencephalomalacia in horses. Subsequent work showed that the hydrolysed backbones of FB_2 and FB_3 could not be resolved by capillary GC after silylation but required trifluoroacetylation for separation from each other and from FB_1 (Plattner *et al.*, 1992). The accuracy and precision of this GC–MS analytical method was improved by the use of deuterium-labelled FB_1 as an internal standard added to the sample extract prior to hydrolysis (Plattner and Branham, 1994). The labelled FB_1 was produced in liquid culture using methyl-d_3-labelled methionine as precursor of the branch methyl groups at C_{21} and C_{22}. Since the fragment ion in the main peak of the electron ionization spectrum did not contain any of the labelled atoms, electron capture negative chemical ionization of the trifluoroacetyl derivative was used for quantification as it gave rise to abundant molecular ions. Detection level for this method were better than $10\,\mu g/kg$.

Although initially used for fungal cultures and heavily contaminated maize associated with animal toxicoses, these methods have found only limited use in food analysis due to their reliance on sophisticated MS and the development of more facile HPLC methods. Nevertheless, together with other MS methods, they are important confirmatory techniques.

3.4.5 Application of immunochemical techniques

The success of immunochemical techniques for the determination of other mycotoxins, and the economic and health implications of this recently discovered group of mycotoxins, led to an immediate interest in developing antibodies to the fumonisins for use in immunochemical-based analysis systems. The free amino group of FB_1 enables it to be rendered immunogenic by conjugation, via glutaraldehyde, to protein carriers. Murine polyclonal antibodies to FB_1 were produced by using cholera toxin (CT) as both the hapten carrier protein and the adjuvant, after initial attempts using BSA as the carrier and Freund's adjuvant proved ineffective (Azcona-Olivera et al., 1992a). The antiserum produced cross-reacted with FB_2 (87%) and FB_3 (40%), but not with the hydrolysed backbones of the fumonisins or with tricarballylic acid. This suggested that the antibody recognized a section of the molecule near the union of the tricarballylic acids and the fumonisin backbone, enabling the antibody to be used for detection of all three fumonisin analogues simultaneously. Using an indirect ELISA, the detection limit for FB_1 was 100 ng/ml. This sensitivity limit was improved to 50 ng/ml by using a direct ELISA based on monoclonal antibodies produced from stabilized hybridomas formed by fusing NS-1 myeloma cells with splenic lymphocytes from mice immunized with the same FB_1–CT conjugate (Azcona-Olivera et al., 1992b).

This direct ELISA, based on monoclonal antibodies, was used to compare the immunochemical technique with other chemical methods (HPLC and GC–MS) for the determination of fumonisins in maize and maize-based products (Pestka et al., 1994). Although the two chemical techniques were well correlated ($r = 0.946$) with each other, correlation of each with ELISA was poorer ($r = 0.512$ and 0.478, respectively), with ELISA generally producing higher results than the other methods. Similar higher levels have been found in other studies based on these antibodies (Sydenham et al., 1996a). However, determination of fumonisins in a series of spiked maize samples using these antibodies resulted in an excellent correlation ($r = 0.996$) between the total fumonisin concentration added to the maize and the levels determined by ELISA (Sydenham et al., 1996a). Hence it seems likely that the higher levels found in naturally contaminated maize may result from the presence of fumonisin precursors or metabolites. A recently developed ELISA, based on polyclonal antibodies, has been reported to yield improved correlations between ELISA and HPLC methods (Sydenham *et al.*, 1996c).

Another direct ELISA has been developed based on polyclonal antibodies produced in rabbits after immunization with FB_1–keyhole limpet haemocyanin conjugate. This ELISA was reported to be highly sensitive, with a detection limit for FB_1 of about $10\,\mu g/kg$, but showed significant non-specific binding due to matrix effects at low fumonisin concentrations (Usleber *et al.*, 1994).

The results reported to date on the use of immunochemical techniques for fumonisin determination show that they are rapid and effective in screening maize for fumonisin contamination. Research on these methods is ongoing in order to improve their accuracy in comparison to chemical methods.

Antibodies reactive with fumonisins have also been incorporated in a commercial immunoaffinity column for the purification of fumonisins in contaminated food. In its application to maize, fumonisins were extracted using methanol:water (4:1) after the addition of sodium chloride to the sample (Ware *et al.*, 1994). The filtered extract was diluted with a salt solution and then applied to the immunoaffinity column. This was washed with the saline dilution solution and the fumonisins eluted with 0.05 M $Na_2B_4O_7$ in water:methanol (1:4). The purified extract was then dried and derivatized with NDA for HPLC determination of the fumonisins. The column had equal affinity for FB_1 and FB_2 and gave recoveries of 85.4 and 87.1%, respectively. Of importance in the use of these columns are the sample size and level of contamination. These commercial columns at present are reported to have a total capacity of $1.2\,\mu g$ for FB_1 and FB_2. Should determination of fumonisins in a sample indicate that this limit has been exceeded, then the sample has to be re-analysed after dilution. Besides maize, these columns have been used to develop an analytical method for the determination of FB_1 and FB_2 in milk which gave recoveries of 79–109% (Scott *et al.*, 1994).

3.4.6 *Application of alternative/confirmation techniques*

The main confirmatory technique for the presence of fumonisins, apart from repeat analysis using an alternative chemical or immunochemical method, is mass spectrometry. Mass spectral analyses have been reported by several laboratories using a number of sample introduction and ionization techniques. The introduction of fumonisins into the MS can be broadly divided into those methods using direct sampling and those relying on either GC or HPLC interfaces.

Initial comparison of thermospray (TS–MS) using flow injection analysis for sample introduction, electrospray (ES–MS) with infusion of the sample at a flow of $1\,\mu l/min$ and fast-atom bombardment (FAB–MS) using Xe as the primary atom beam and thioglycerol as the matrix showed that both ES–MS and FAB–MS are useful techniques for characterization of nanogram quantities of FB_1 (Korfmacher *et al.*, 1991). Both these techniques provided strong signals from the molecular ion and little fragmentation. By contrast, TS–MS was reported to give multiple fragment ions and lacked sensitivity at sub-microgram levels. A more recent study has reported that the detection limit of TS–MS can be decreased to 2 ng by using negative-ion mode and an aqueous ammonium acetate and acetonitrile mixture for the mobile phase (Thakur and Smith, 1994). In this format, the method was reported to be suitable for detection of fumonisins in maize.

FAB–MS was originally used for the confirmation of the presence of fumonisins in maize associated with field cases of leukoencephalomalacia (Plattner *et al.*, 1990). It has been developed into an analytical method by the use of deuterium-labelled FB_1 as an internal standard to account for variations in the yield of molecular ions which can be suppressed by other sample matrix components (Plattner and Branham, 1994). Naturally contaminated maize was extracted with acetonitrile:water (1:1) and the resulting extract was spiked with the internal standard and purified on SAX cartridges prior to FAB–MS. The approximate detection limit for this method is 0.1 mg/kg, with precision restricted near the detection limit by the fact that FAB–MS by its nature gives a signal at each m/z value.

Mass spectral analysis using GC–MS requires hydrolysis of the fumonisins and derivatization prior to GC analysis. The mass spectral data can be used both for quantification and confirmation. The methods available have been discussed in section 3.4.4.

In contrast to GC–MS, LC–MS can be used without hydrolysis or derivatization to provide direct confirmation of fumonisins. Initial methods were developed for analysis of FB_1 standard on a reversed-phase (C_{18}) column eluted with solvent mixtures of water:acetonitrile:acetic acid (Mirocha *et al.*, 1992). The column effluent was passed into an ES–MS which gave strong signals for the molecular ion and its alkali salts. Sensitivity was shown to be in the 100 pg range. Other workers have used a particle-beam interface for confirmation of fumonisin production in liquid cultures after separation of their methyl esters by RP-HPLC (Young and Lafontaine, 1993). The advantage of the particle-beam interface is its ability to provide electron and chemical ionization spectra, which provide greater structural information than atmospheric-pressure ionization techniques such as electrospray. This latter technique coupled to LC, using a polymeric hydrophobic column packing and volatile mobile phase (acetonitrile and aqueous ammonium acetate), has been used for the determination of FB_1, FB_2 and FB_3 in commercial maize meal (Doerge *et al.*, 1994). Using selected ion monitoring in the positive mode, the detection limits for this method were 100, 20 and 50 μg/kg for the three fumonisin analogues respectively, making this method suitable for analysis of contaminated foods.

FB_1, FB_2 and the hydrolysed backbone of FB_1 have been separated by capillary zone electrophoresis (CZE) as their fluorescein isothiocyanate (FITC) derivatives and detected by laser-induced fluorescence, with a detection limit of 0.025 pg FB_1 injected (Maragos, 1995).

In contrast to the above sophisticated techniques, which require expensive laboratory layout and equipment, simple and inexpensive field tests for the screening of maize for fumonisin contamination are also required. This need is being addressed by the development of a minicolumn-based technique known as chemiselective immobilization and detection (CSID), which relies on derivatization with fluorescamine and adsorption on a minicolumn packed with various sorbents (Phillips *et al.*, 1992). Similarly, it is expected that further development of immunochemical techniques will provide a dipstick assay format for rapid screening of fumonisins (Usleber *et al.*, 1994) similar to methods reported for other mycotoxins (Schneider *et al.*, 1991).

3.5 Ochratoxin A

3.5.1 Introduction

The ochratoxins are a group of mycotoxins consisting of a dihydroisocoumarin linked to phenylalanine, and are produced by various *Penicillium* and *Aspergillus* species. Of this group, ochratoxin A (OA; Figure 3.8) is the main analogue found in naturally contaminated foods, such as maize, wheat, barley and coffee beans. The presence of OA is strongly region-dependent, with lower levels of contamination generally being found in the USA compared to northern Europe and the Balkans. High levels in animal feeds have led to its presence being detected in fluids and tissues of chickens, hens and pigs (Pohland *et al.*, 1992).

OA is a potent nephrotoxin in all animals studied apart from mature ruminants, which are capable of hydrolysing OA in the forestomach (Steyn, 1984). Acute or chronic administration of OA to pigs leads to the syndrome of porcine nephropathy (Albassam *et al.*, 1987). A number of

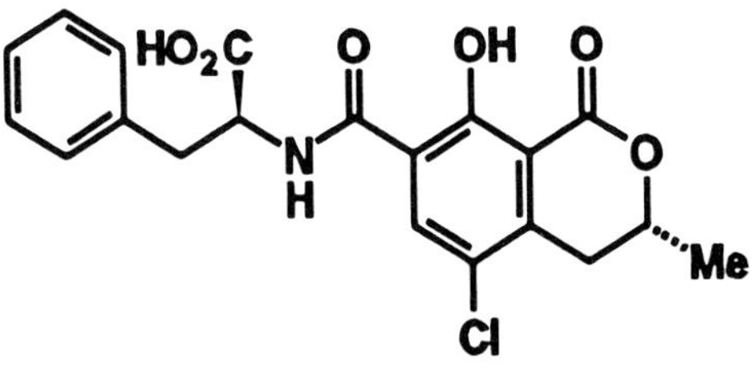

Figure 3.8 Chemical structure of ochratoxin A.

studies in rodents have shown that OA is also a renal carcinogen (Bendele *et al.*, 1985). OA has not been shown to be mutagenic in Ames-type assays, but has been shown to induce single-strand DNA breaks in liver and kidney tissues of rats and mice (Kane *et al.*, 1986). OA has shown teratogenic effects in rats, mice and hamsters and has been reported to be a more potent teratogen than aflatoxin B_1 (Kuiper-Goodman, 1991). OA is immunotoxic to certain elements of the immune system (Pohland *et al.*, 1992) and produces a variety of metabolic effects, most notable being its inhibition of protein synthesis (Creppy *et al.*, 1984).

OA has been linked to the human syndrome, Balkan Endemic Nephropathy (BEN), through similarities in the renal pathology between BEN and porcine nephropathy induced by exposure to OA, and through the occurrence of OA in food and in human blood in endemic areas (Pohland *et al.*, 1992). A limited number of countries have instituted regulatory limits for OA. Van Egmond (1989) reported that these limits ranged from 50 μg/kg for rice, barley, beans and maize in Brazil to 1 μg/kg for infant food in the former Czechoslovakia.

3.5.2 *Application of TLC techniques*

TLC analytical methods for OA have been reviewed by several authors (Steyn, 1984; Betina, 1993). However, with the development of more sophisticated HPLC and immunochemical techniques, few advances have been recorded on established TLC methodology. Nesheim and Trucksess (1986) have reviewed the necessity for using acid modifiers in TLC mobile phases to prevent streaking of OA on silica gel and the precautions needed to obtain reliable R_f values in such systems. Problems with the rapid fading of the fluorescence intensity on the plate can be overcome by exposure to ammonia vapour, which converts the OA to its ammonium salt. If the plate is subsequently covered, then the enhanced fluorescence induced by this reaction can be retained for several days.

A method has been developed for the screening and quantitation of OA in corn, peanuts, beans, rice and cassava based on methanol:aqueous 4% KCl (9:1) extraction and clean-up with ammonium or cupric sulphate (Soares and Rodriguez-Amaya, 1985). After addition of these clarifying agents to the extract and subsequent filtration, the OA is extracted with aliquots of chloroform. These extracts can be screened for the presence of OA (detection limit 80 μg/kg) by the use of a minicolumn of silica gel and neutral alumina. The column is developed with toluene:ethyl acetate:acetic acid (50:49:1) and the OA forms a tight band at the silica gel–alumina interface as it does not migrate into the latter layer under these conditions. The OA can then be estimated by comparison of its blue fluorescence compared to that of standards similarly treated.

A RP-TLC method has been developed for the purpose of sample clean-up in which the OA is extracted from the spot for quantitation by direct fluorimetry or by subsequent HPLC analysis (Frohlich *et al.*, 1988). The grain was extracted by shaking with phosphoric acid and chloroform. An aliquot of the extract was applied to the RP-TLC plate which was initially developed with hexane, then dried and subsequently developed with methanol:water (70:30) for chromatographic separation of OA. The OA spot can be visualized under long-wavelength UV and the adsorbent removed with a suction device on to a glass frit. The OA was then eluted from the adsorbent with methanol. Residues as low as $100\,\mu g/kg$ can be detected by this procedure, being limited by the need to visualize the OA on the plate.

3.5.3 Application of HPLC techniques

HPLC, using the natural fluorescence of OA for detection, has been widely applied to the determination of OA in various food matrices. Extraction of OA is generally achieved with organic solvents after addition of small amounts of acid (generally H_3PO_4) to suppress the ionization of OA and facilitate its dissolution in the organic medium. El-Banna and Scott (1984) described a method for extraction of OA from cooked faba beans and polished wheat with H_3PO_4 and dichloromethane, followed by clean-up on a silica SPE cartridge and separation by RP-HPLC. Mean recoveries were 71.5 and 70.2% for the cooked beans and wheat, respectively, and the detection limit was $0.7\,\mu g/kg$. Another RP-HPLC technique was described by Lepom (1986) for the simultaneous determination of OA and citrinin in wheat and barley using an acid–base solvent partition for sample clean-up, with a detection limit of $40\,\mu g/kg$. Cohen and Lapointe (1986) described a two-stage SPE clean-up for animal-feed and cereal-grain extracts, first with a silica cartridge and then with a cyano cartridge. Overall recoveries were above 90% with the method capable of detecting OA levels down to $5\,\mu g/kg$. Purification of pig-feed extracts has been achieved on C_{18} SPE cartridges (Takeda *et al.*, 1991), while Langseth *et al.* (1989) reported a simultaneous chloroform extraction of OA and zearalenone from wheat, barley, oats and mixed feed with subsequent separation of these two mycotoxins on a silica SPE cartridge and separate quantification by RP-HPLC. Recoveries (77–96%) and detection limits (0.1–$0.3\,\mu g/kg$) varied between the different matrices.

Procedures being used within Europe were compared in a project undertaken by the European Community Bureau of Reference Materials (BCR) as a preliminary stage in improving methodology and preparation of a suitable reference material for OA determination by European laboratories (Hald *et al.*, 1993). A total of 24 laboratories analysed two wheat samples, one being naturally contaminated at a level of $13\,\mu g/kg$ and the second a 'blank' ($<1\,\mu g/kg$). Participants used a supplied standard and an analysis method of their own choice. Extraction solvents used by participants included chloroform, methanol, toluene and ethyl acetate together with a variety of acids for ionization suppression (H_3PO_4, CH_3COOH, HCl). Clean-up methods chosen included liquid–liquid extraction, silica and RP-SPE cartridges and immunoaffinity columns. All participants used RP-HPLC with aqueous acetonitrile or methanol for the quantitation step, except one who used NP-HPLC (aminopropyl column) and another who chose TLC. Analytical recoveries achieved by the participants ranged from 25 to 100%. Coefficient of variation from all results (corrected for recovery) was reported as 23%, with the variability being influenced by clean-up method rather than by extraction method.

A collaborative study, under the auspices of AOAC, IUPAC and the Nordic Committee on Food Analysis, has been conducted on spiked samples of barley, maize and kidney (Nesheim *et al.*, 1992). The principle of the studied method was extraction with chloroform:0.1 M H_3PO_4 (9:1), partitioning with bicarbonate solution, clean-up on C_{18} SPE cartridge and quantitation by HPLC with fluorescence detection. Results were reported by 16 laboratories on the grain samples spiked at 5, 10 and 20 μg/kg. Mean recoveries ranged from 72 to 82% for the grain samples. Blind duplicates at the 20 μg/kg level gave within-laboratory precision of 7.9% for barley and 20.1% for maize, while between-laboratory precision was 20.7–31.7% for the grains. A number of collaborators reported false positives on the blank samples, which may have been due to matrix interferences.

A number of food commodities other than grains have been analysed by HPLC techniques for contamination by OA. Methods have been published for cocoa beans (Hurst and Martin, 1983), coffee beans and coffee products (Terada *et al.*, 1986; Tsubouchi *et al.*, 1988; Micco *et al.*, 1989) and human and cow's milk (Gareis *et al.*, 1988; Breitholtz-Emanuelsson *et al.*, 1993). Analytical methods were similar in principle to those used for grains. Aliquots of milk were acidified and extracted with ethanol:chloroform, partitioned with sodium bicarbonate solution and subsequently purified on silica (Breitholtz-Emanuelsson *et al.*, 1993). This method gave a recovery of 75–85% with a detection limit of 10 ng/l.

3.5.4 *Application of immunochemical techniques*

As for most of the economically important mycotoxins, considerable interest has been shown in recent years in the development of rapid screening and quantitative methods for OA by immunochemical methods. A non-competitive, double-antibody ELISA was developed for determination of OA in barley with a detection limit of 0.06 μg/kg (Morgan *et al.*, 1983). The assay was based on OA–keyhole limpet haemocyanin conjugate coated to the microtitration plates and the use of antisera to OA–BSA conjugate raised in rabbits. These polyclonal antibodies had little cross-reactivity with ochratoxins other than OA. The final enzyme reaction was based on alkaline phosphatase. OA was extracted in chloroform and could be used without clean-up once the chloroform had been evaporated and the OA redissolved in assay buffer. The analysis time for such an assay could be reduced by use of a peroxidase colour reaction and by simultaneous incubation of both antibodies (Sidwell *et al.*, 1989). Lee and Chu (1984) reported the development of a direct, competitive ELISA using antiserum from immunized rabbits and horseradish peroxidase conjugated to OA as the marker enzyme. Sample extracts were purified on RP-SPE cartridges to enable determination at the $1-2\,\mu\text{g/kg}$ level in wheat.

More recently, ELISAs have been developed based on monoclonal antibodies, which have the potential to be produced in larger amounts and with more consistent properties than polyclonal antibodies. Hybridoma technology was used to produce a high-affinity, specific antibody for indirect competitive ELISA determination of OA in barley with a detection limit of $5\,\mu\text{g/kg}$ (Candlish *et al.*, 1988). OA was extracted from barley with chloroform and then partitioned into bicarbonate solution for use in the ELISA. This assay was further developed so as to use a single barley extract for determination of OA, AFB_1 and T-2 toxin (Ramakrishna *et al.*, 1990). Kawamura *et al.* (1989) reported the development of a series of monoclonal antibodies which, with solvent partition clean-up of extracts of wheat flour, gave a detection limit of $0.1-1\,\mu\text{g/kg}$.

Rousseau *et al.* (1985) developed a radioimmunoassay for OA in barley based on polyclonal antibodies raised in rabbits and a $[^{14}\text{C}]\text{OA}$ tracer with a specific activity of $130\,\text{Ci/mol}$. Barley extracts required clean-up on Extrelut 3® columns to achieve a detection limit of $2.5\,\mu\text{g/kg}$. Another radioimmunoassay using rabbit antiserum against OA–BSA conjugate and $[^{125}\text{I}]$ochratoxin prepared by ^{125}I-labelling ochratoxin B (dechloro-OA) as tracer was used to survey cereals and feedstuffs in Czechoslovakia with a detection limit of $1\,\mu\text{g/kg}$ (Fukal, 1990).

Antibody technology has also been applied to the clean-up of cereal and animal sample extracts by utilizing immunoaffinity columns to isolate OA selectively (Sharman *et al.*, 1992). Cereals were extracted with phosphate buffered saline (PBS):methanol (1:1) and directly applied to the affinity column after dilution. OA was recovered from the column by elution with methanol. Total capacity of the column was greater than $2.7\,\mu\text{g}$ OA with quantitative recovery when OA was applied as a standard solution in PBS. A reduced recovery of 74% was obtained in wheat at a level of $10\,\mu\text{g/kg}$, the difference being assumed to result from matrix effects.

Figure 3.9 shows a chromatogram obtained from an extract of naturally contaminated ($13.7\,\mu\text{g/kg}$) wheat. Monoclonal antibody affinity columns have also been shown to be effective for the clean-up of OA from coffee beans and coffee products, with recoveries of more than 98% (Nakajima *et al.*, 1990).

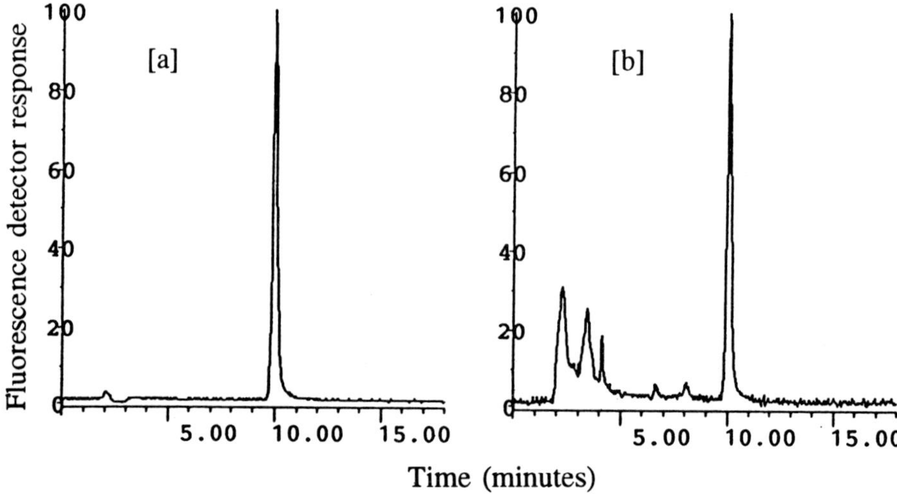

Figure 3.9 RP-HPLC analysis of OA in extracts of naturally contaminated wheat purified by immunoaffinity column. (a) Standard equivalent to 20 μg/kg; (b) wheat sample contaminated with 13.7 μg/kg (Sharman *et al.*, 1992).

3.5.5 Application of alternative/confirmation techniques

The presence of OA in contaminated foods determined by TLC has generally been confirmed by the formation of methyl esters, using BF_3 as catalyst, and further TLC of the derivatized extract (Nesheim and Trucksess, 1986). Other confirmatory tests include exposure of the spot on the TLC plate to acetic anhydride or pyridine or mixtures of the two followed by re-chromatography (Goliński and Grabarkiewicz-Szczęsna, 1984). The methyl ester has also been used for confirmation in HPLC analysis or for shifting the chromatographic peak away from co-eluting impurities (Cohen and Lapointe, 1986; Micco *et al.*, 1989; Nesheim *et al.*, 1992). Terada *et al.* (1986) used methyl, ethyl and *n*-propyl esters for this purpose.

Both LC–MS and GC–MS methods have been published for OA determination. RP-HPLC was used to separate OA with aqueous 2% formic acid:acetonitrile (2:3) as volatile acidic mobile phase. The column effluent was split so as to feed 2.5% into the MS via a direct liquid introduction interface (Abramson, 1987). Using negative-ion chemical ionization, OA could be detected in barley down to 3 μg/kg and, using selected ion monitoring, even lower levels would be possible. OA has also been confirmed in foods by conversion to its *O*-methylochratoxin A methyl ester derivative and subsequent GC–MS analysis using negative-ion chemical ionization (Jiao *et al.*, 1992). The method can be made quantitative by using the hexadeuterated *O*-methyl-d$_3$-ochratoxin A methyl-d$_3$ ester as internal standard, giving a detection limit down to that achieved by HPLC (0.1 μg/kg).

3.6 Patulin

3.6.1 Introduction

Patulin (Figure 3.10), an α, β-unsaturated lactone (4-hydroxy-4H-furo[3,2-c] pyran-2(6H)-one) is one of a group of mycotoxins that contain a five-membered cyclic ring system.

It is a toxic secondary metabolite produced by a wide range of *Penicillium and Aspergillus* species, of which *P. expansum* is the most important (Jimènez *et al.*, 1988, 1991). Under laboratory conditions, patulin may be produced on different food substrates such as fruits, grains, cheese and cured meats. However, under natural conditions the occurrence of patulin is restricted (almost exclusively) to fruits in general and apples and their processed products in particular (Harrison, 1989). Although several chemical and physical procedures may be used to reduce residual patulin levels in apple commodities, the toxin can remain relatively stable under acidic conditions (Damoglou and Campbell, 1986). The presence and extent of patulin contamination in apple products may, therefore, be indicative of the quality of the fruit used in their production (Burda, 1992).

Patulin has been reported to possess mutagenic properties, to exhibit adverse effects on the developing foetus in rats and to cause immunotoxic, neurotoxic and gastrointestinal effects in rodents (Hopkins, 1993). Present recommendations suggest that the provisional maximum tolerable daily intake (PMTDI) level for patulin be 0.4 μg/kg per week (FAO/WHO, 1995), while various national bodies have recommended that products intended for human consumption should not contain residual patulin concentrations greater than 50 μg/kg (Smith *et al.*, 1994). International interest in patulin dramatically increased during 1993, following a report published by the UK Ministry of Agriculture, Fisheries and Food (MAFF), in which patulin levels in excess of the recommended 50 μg/kg were detected in several apple juice samples (MAFF, 1993). However, the carcinogenic and mutagenic potential of patulin, and the validity of the data used to set current recommendations for residual patulin levels in fruit products have been seriously questioned (Hopkins, 1993).

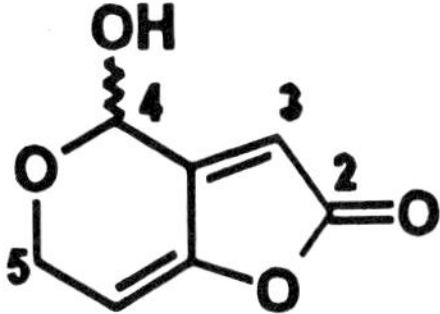

Figure 3.10 Chemical structure of patulin.

3.6.2   Application of TLC techniques

The application of TLC to the determination of patulin in food commodities has been limited, although Lin *et al.* (1993) reported one method that differed from others, in that it was developed for the simultaneous determination of patulin and zearalenone in maize. Purified extracts prepared from maize by standard column chromatography techniques, were separated on NP-TLC plates, developed in iso-octane followed by a chloroform–acetone mixture. The TLC plates were screened in an absorbance–reflectance mode at $\lambda = 275\,nm$, and quantification of patulin was achieved by comparison of the resultant chromatographic spot intensities against a calibration curve, which was shown to be linear over the range 5–75 ng patulin/spot. The detection limit for the method was 3 ng patulin/chromatographic spot, with recoveries of >80% for levels between 70 and 300 ng patulin (Lin *et al.*, 1993).

3.6.3   Application of HPLC techniques

HPLC, coupled with UV detection, is particularly well suited to the determination of patulin, since the toxin is relatively polar in nature and exhibits a strong absorption spectrum. Forbito and Babsky (1985) reported a rapid method for the quantitative determination of patulin in apple juice, in which the purification procedure incorporated two solvent partitioning steps, initially using ethyl acetate for the extraction of patulin from apple juice followed by the removal of some potentially interfering phenolic compounds by partitioning with a solution of sodium carbonate. However, the method was similar to those previously published by Möller and Josefsson (1980) and Tanner and Zanier (1976), with only minor modifications made to the solvent volumes used. In its original form, the method formed part of a collaborative study in which the performance of two RP-HPLC methods for the determination of patulin in apple juice were evaluated in twelve laboratories from ten countries (Kubacki and Goszcz, 1988). The second method (Stray, 1978) differed from the former (Tanner and Zanier, 1976), in that the purification of patulin from a crude ethyl acetate extract was effected by a silica gel column chromatographic step. Recoveries were determined using apple juice spiked at three concentrations (5, 50 and 250 μg/l), although the lowest level (5 μg/l) was considered to represent the detection limits of the methods (Kubacki and Goszcz, 1988). Mean recoveries for patulin spiked at the higher concentrations ranged from 78.4% (Stray, 1978) to 81.4% (Tanner and Zanier, 1976), with mean between-laboratory coefficients of variation of 15.8% and 8.5% for the two methods, respectively (Kubacki and Goszcz, 1988). The RP-HPLC method recommended by the International Standards Organization (ISO, 1993) is also based on extraction of patulin with ethyl acetate and partitioning with sodium carbonate. The method formed the basis of a collaborative study involving 22 participants who analysed twelve test samples artificially spiked with 20, 50, 100 and 200 μg patulin/l. In addition, a naturally contaminated test sample (containing about 31 μg patulin/l) was included. Recoveries of patulin ranged from 91 to 108%, with a mean of 96%. Within-laboratory repeatability values (RSD$_r$) ranged from 10.9 to 53.8%, while between-laboratory reproducibility values (RSD$_R$) ranged from 15.1 to 68.8% (Brause *et al.*, 1996).

The use of dialysis membranes for the separation of high- and low-molecular-weight compounds has been extended to incorporate the determination of fungal metabolites including patulin in apple juice, using ethyl acetate for initial extraction purposes (Dominguez *et al.*, 1992). The mean recovery for patulin, spiked into apple juice at 100 μg/l, was found to be only 62.5%. Prieta *et al.* (1992) applied the dialysis extraction technique to apple juice samples by incorporating the TLC detection of the patulin as its 3-methyl-2-benzothiazolinone hydrazone hydrochloride (MBTH) derivative, but concluded that the method was at best semi-quantitative. Factors such as dialysis extraction temperature, extraction time and sample to solvent ratios, together with additional purification of apple juice extracts on silica gel cartridges and separation by RP-HPLC, were further assessed by Prieta *et al.* (1993). The silica gel cartridge purification step incorporated the use of relatively low solvent volumes, with interfering compounds being removed by washing with chloroform and two sequential chloroform:ethyl acetate mixtures (at ratios of 8:2 and 1:1, respectively), prior to elution of the patulin-containing fraction with a mixture of chloroform:ethyl acetate (2:8). Alteration of the sample to extraction solvent ratio from 1:1 to 1:2, and extended extraction times, improved the analytical recovery for patulin to >85% at a spiking level of 20 μg/l with a detection limit estimated to be of the order of 1 μg/l (Prieta *et al.*, 1993).

Rovira *et al.* (1993) combined an ethyl acetate solvent partitioning step for the initial extraction of patulin from dilute apple juice with the silica gel cartridge purification step described by Prieta *et al.* (1993). Recoveries of patulin from apple juice ranged from 80.0 to 87.7% (spiked at levels of between 3 and 100 μg/l), with a detection limit for the method of 2 μg/l (Rovira *et al.*, 1993). The dissolution of patulin (present in the initial ethyl acetate extracts) into the chloroform phase, prior to silica gel purification, was assisted by the addition of water. However, subsequent removal of the water with sodium sulphate prior to purification on silica gel media and application of a low inert gas flow rate during the final concentration of residues were cited as important aspects of the method that could influence patulin recovery (Rovira *et al.*, 1993).

Of particular importance in the HPLC analysis of processed apple juices is the separation of patulin from intrinsic phenolic compounds in general, and 5-hydroxymethyl-2-furaldehyde (HMF) in particular. HMF is a cyclic aldehyde derived from the reaction between reducing sugars and amino acids, and its presence in fruit juices may be indicative of the use of excessive heat treatment during either the concentration or the pasteurization process used during the production of fruit juice concentrates. Packaging and storage conditions as well as reactions with other compounds may also influence residual HMF concentrations in juice products (Cilliers and Van Niekerk, 1984).

The separation of patulin from HMF is crucial (Figure 3.11a) since concentrations of the latter can often exceed significantly those of the

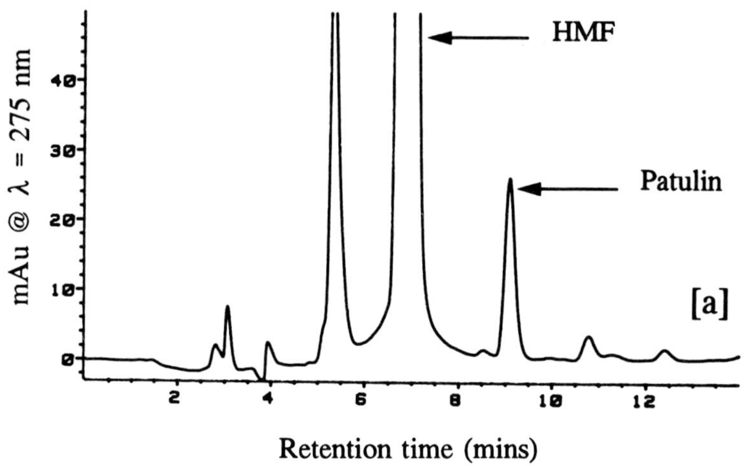

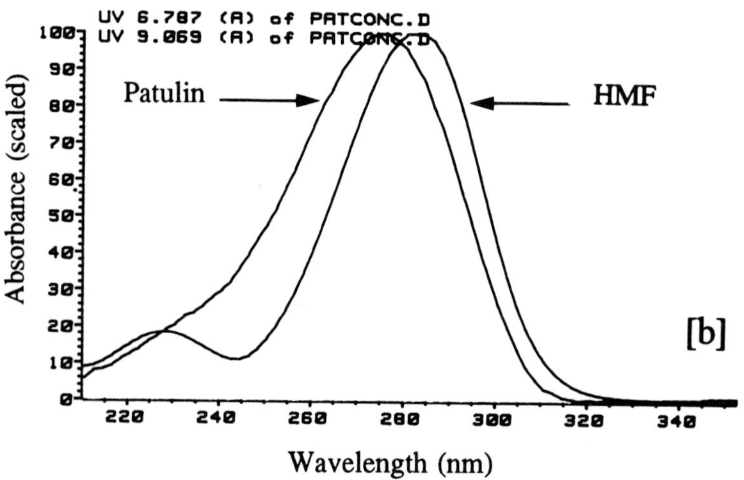

Figure 3.11 RP-HPLC analysis of a purified apple juice sample naturally contaminated with 15 μg/l patulin. Chromatogram (a) illustrates the separation of patulin from HMF, while (b) compares the UV spectra of patulin and HMF over the range 210–350 nm.

former, while in addition HMF exhibits a UV spectrum similar to that of patulin, with a λ_{max} of 282 nm as opposed to a corresponding λ_{max} of 275 nm for patulin (Figure 3.11b). Insufficient chromatographic separation of the two moieties can, therefore, give rise to the incorrect identification of chromatographic peaks.

3.6.4 Application of GC techniques

Tarter and Scott (1991) reported the formation and capillary GC separation (coupled with electron capture detection [ECD]) of the heptafluorobutyrate (HFB) derivative of patulin. The method was applied to the analysis of apple juice samples, which were purified in accordance with ethyl acetate extraction and silica gel column chromatographic techniques (Stoloff and Scott, 1984). The derivatization procedure, which involved the reaction between patulin and heptafluorobutyrylimidazole (HFBI), was similar to that previously used for the analysis of the type B trichothecenes, except that hexachlorobenzene was incorporated as an internal standard. The resultant HFB–patulin derivative was separated on a nonpolar 30 m × 0.32 mm i.d. fused silica capillary column (coated with a 0.25 μm film of 5% phenyl methyl siloxane DB-5) with helium as carrier gas and nitrogen as the make-up gas (Figure 3.12).

The derivatization procedure was reported to be highly reproducible, with a coefficient of variation of 4.0% (at 0.2 ng patulin injected), while the derivative was found to be stable in an *n*-heptane solution for up to 35 h at room temperature (Tarter and Scott, 1991). The detection limit of the method was reported to be $\leq$10 μg/l with an 83% recovery for patulin spiked into apple juice at 100 μg/l (CV = 6.3%, based on six determinations).

3.6.5 Application of alternative/confirmation techniques

The possibility of using RP-TLC for the confirmation of patulin previously detected by NP-TLC screening methods has been suggested, whereby patulin can be detected as its MBTH derivative (Abramson *et al.*, 1989). The authors compared the retention values obtained for patulin (and other mycotoxins) chromatographed on RP-TLC plates produced by two manufacturers, developed in 60:40 ratios of methanol:water, acetonitrile:water and tetrahydrofuran:water. However, the studies were conducted using pure mycotoxin standards and the techniques were not applied to the confirmation of naturally contaminated samples (Abramson *et al.*, 1989).

Preliminary identification of the LC chromatographic peak corresponding to patulin may be obtained by using diode array detection (DAD) as the detection method of choice. DAD permits the simultaneous collection of absorbance data for a range of (user-specified) wavelengths, thereby allowing the chromatographer to record the UV spectra of compounds eluting from

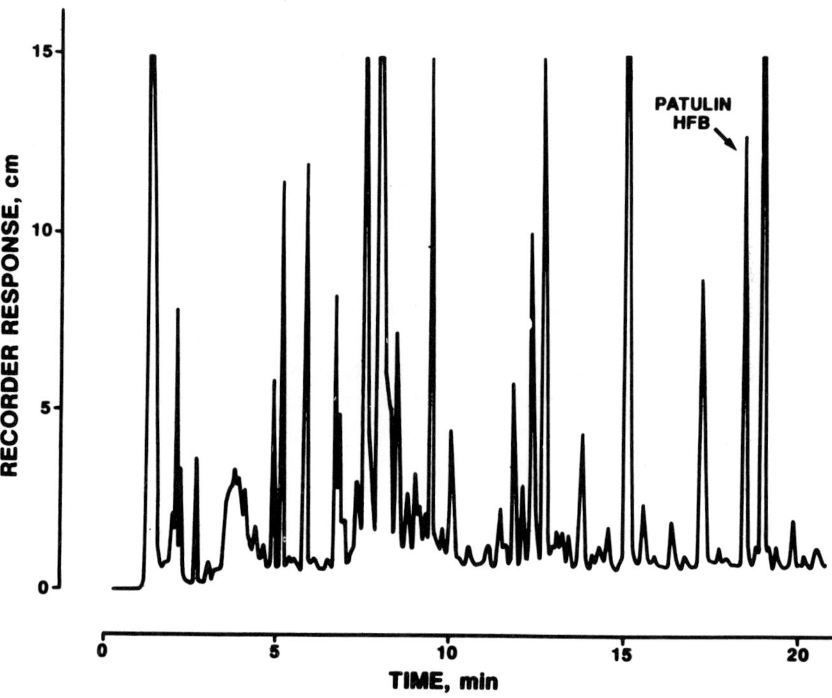

Figure 3.12 Capillary GC separation of the HFB-derivatized extract of apple juice contaminated with 38 μg/l patulin (Tarter and Scott, 1991).

the HPLC column. Subsequent identification of patulin may therefore be obtained by comparing its UV spectrum (Figure 3.11b) against that recorded for an authentic patulin standard (Bartolomé *et al.*, 1994). The spectral characteristics of patulin have been compared against those obtained for a range of phenolic compounds present in apples (Bartolomé *et al.*, 1994) and other common low-molecular-weight compounds in vegetable substrates (Bartolomé *et al.*, 1993). The use of absorbance ratios at alternative wavelengths and the observed maxima in second-order derivative spectra were also suggested as confirmatory methods for the identification of patulin (Bartolomé *et al.*, 1993). Paterson and Kemmelmeier (1990) published the UV spectra of patulin and several other secondary metabolites, obtained in neutral and alkaline solvents. Similarities between the UV spectra of several metabolites (derived in neutral solvents) may inhibit confirmation; however, Paterson and Kemmelmeier (1990) suggested that the metabolites could be differentiated by comparing their corresponding spectra derived in alkaline solvents, or by comparison of their difference spectra. However, it is likely that the technique would be more applicable to chemotaxonomic studies involving fungi, than to the confirmation of observations on the natural occurrence of patulin.

The unequivocal confirmation of patulin was demonstrated by using an LC–electrospray MS technique (Sydenham *et al.*, 1995b). Under alkaline conditions (in a 2% ammonia solution), patulin was observed as a doubly charged species yielding a molecular ion at $m/z\,76.9$. A major fragment ion of a singly charged species was also observed at $m/z\,122.8$, but this fragment ion could not be structurally assigned. Electron ionization mass spectrometry was used by Tarter and Scott (1991) for the GC–MS confirmation of patulin as its HFB derivative, with an observed series of ions between $m/z\,153$ and $m/z\,53$ being consistent with HFB–patulin fragment ions.

Commercial immunoassays for patulin have not yet been developed, although McElroy and Weiss (1993) have reported the production of polyclonal antibodies against patulin as its hemiglutarate derivative. Indirect ELISA studies indicated that the polyclonal antibodies could be used for the qualitative estimations of patulin derivatives, but the authors concluded that affinity purification techniques and/or the production of monoclonal antibodies could improve both the selectivity and sensitivity characteristics of the technique, which ultimately could be developed for the quantitative determination of patulin (McElroy and Weiss, 1993).

3.7 Trichothecenes (deoxynivalenol, nivalenol, diacetoxyscirpenol and T-2 toxin)

3.7.1 Introduction

The trichothecenes consist of a group of approximately 140 structurally related sesquiterpenoids, produced primarily by various species of imperfect fungi including strains of *Cephalosporium, Fusarium, Myrothecium* and *Trichoderma*. Production of the trichothecenes is not, however, restricted to fungi since the baccharinoids (a series of macrocyclic trichothecenes) have been isolated from extracts of *Baccharis megapotamica*, a Brazilian shrub and member of the higher plant species (Jarvis *et al.*, 1983). As a series of structurally related toxins, all trichothecenes contain a 12,13-epoxytrichothec-9-ene ring system, but the toxins may be divided into two major groups, one encompassing the alcoholic derivatives of the trichothecene skeletal nucleus and their related esters, and the second incorporating the chemically complex macrocyclic di- and tri-esters.

In accordance with their individual chemical characteristics, the trichothecenes have been sub-divided into four basic groups or 'types'. The first of these (termed type 'A' trichothecenes—Figure 3.13) is mainly characterized by the presence of either a hydrogen atom or a hydroxyl group at the C_8 position, while type 'B' trichothecenes (Figure 3.14) may be distinguished by the presence of a ketone group at the same position. Conversely, trichothecenes classified as types 'C' and 'D' contain a second

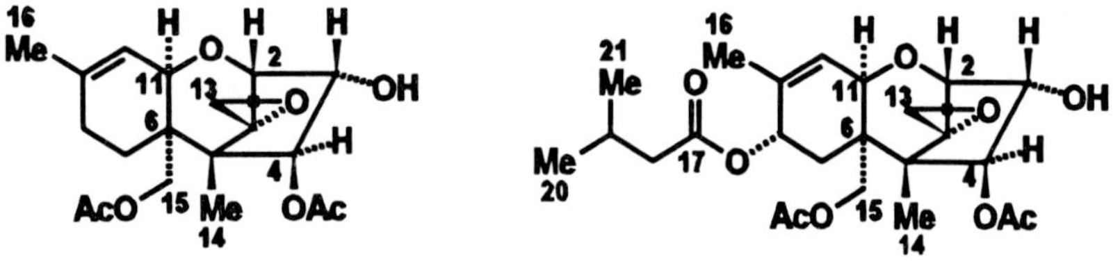

Diacetoxyscirpenol

T-2 Toxin

Figure 3.13 Chemical structures of DAS and T-2 toxin (type A trichothecenes).

epoxy function at the C_7/C_8 or C_9/C_{10} positions, or are macrocyclic in character, respectively.

Additional classifications of toxin variants within each type have also been identified (Betina, 1993). Although numerous trichothecenes and their derivatives have been characterized, relatively few occur as natural contaminants of food commodities intended for human consumption. Accordingly, this section will focus on four of the major naturally occurring trichothecenes: diacetoxyscirpenol (DAS) and T-2 toxin (characterized as type A trichothecenes—Figure 3.13); and deoxynivalenol (DON) and nivalenol (NIV) (identified as type B trichothecenes—Figure 3.14).

DAS (3-hydroxy-4,15-diacetoxy-12,13-epoxytrichothec-9-ene) and T-2 toxin (3-hydroxy-4,15-diacetoxy-8-[3-methyl-butyryloxy]-12,13-epoxytrichothec-9-ene) exist as white crystalline solids. Neither toxin exhibits UV absorption characteristics, nor do they fluoresce. Both DAS and T-2 toxin are regarded as amongst the most toxic trichothecene variants, with experimental exposure to the toxins resulting in skin necrotization and oedema

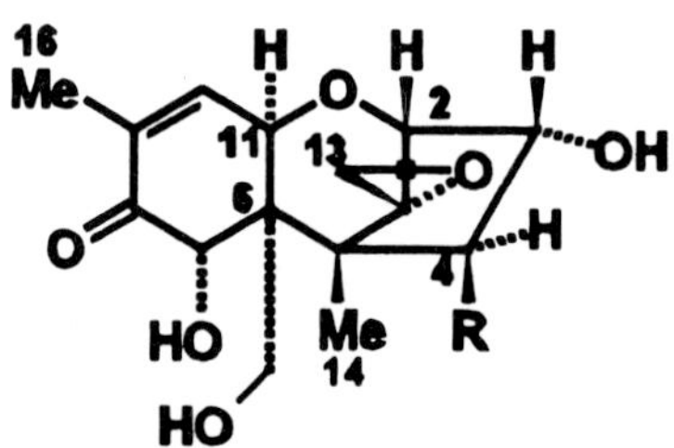

Nivalenol: **R = OH**

Deoxynivalenol: R = H

Figure 3.14 Chemical structures of DON and NIV (type B trichothecenes).

(Ueno, 1983). A recent review on the toxic potential of the trichothecenes to various animal species has been published (Wannemacher *et al.*, 1991).

In comparison to other mycotoxins (notably the aflatoxins, patulin and zearalenone), DON and NIV exhibit relatively weak UV characteristics with absorption maxima at about $\lambda = 220$ nm. DON has been associated with decreased feed consumption and weight gain, vomiting, feed refusal and diarrhoea in livestock, and is characterized by haemorrhaging, skin irritation and immunosuppression in laboratory animals (Ueno, 1983). DON often occurs in cereal grains and their processed foods together with other *Fusarium* mycotoxins, notably NIV and zearalenone (see section 3.8). The natural occurrence of NIV is of particular importance, since its toxicity is higher than that of DON, and is considered to be of the same order as that of T-2 toxin (Ueno, 1984). Therefore, the co-occurrence of NIV and DON may potentially lead to combined or synergistic toxicity in animal species. Regulatory limits for DON in wheat and other grains (and their products) destined for both animal and human consumption have been established in several countries including Canada, the former Soviet Union and the USA (between 500 and 4000 μg/kg; Van Egmond, 1989). However, although the toxicity of the trichothecenes is beyond doubt, little or insufficient evidence exists regarding their carcinogenicity (Smith *et al.*, 1994).

3.7.2 *Application of TLC techniques*

Those methods published from 1984 onwards that describe the TLC determination of the trichothecenes in foods have primarily involved the detection of DON in selected grains and their processed commodities. Most have incorporated solvent extraction of matrices with acetonitrile:water (84:16), followed by purification on short disposable chromatographic columns containing charcoal:alumina:celite media (7:5:3). Extracts have been separated on NP-TLC plates, either pre-impregnated or subsequently sprayed with an aluminium chloride solution to yield DON as a blue fluorescent spot (Eppley *et al.*, 1986; Trucksess *et al.*, 1986, 1987; Fernandez *et al.*, 1994). At least two methods have reported the use of densitometry for the quantitative estimation of DON by TLC, with the plates screened at $\lambda = 366$ nm (Fernandez *et al.*, 1994), or for increased selectivity at $\lambda = 313$ nm (excitation) and $\lambda = 440$ nm (emission) (Trucksess *et al.*, 1987).

In a collaborative study involving 15 participants, recoveries of DON from wheat were reported to range from 78 to 96% at spiking levels of between 50 and 1000 μg/kg (Eppley *et al.*, 1986). Trucksess *et al.* (1986) modified the basic method by adding a secondary purification step, using C_8 media, for the determination of DON in high-sugar-containing breakfast cereals, maize syrup and beer, with a limit of determination of 50 μg/kg. Further modifications to the method were made to enable the co-determination of DON, NIV and fusarenon-X (FUS-X, a structurally similar type B trichothecene) in barley, maize and wheat (Trucksess *et al.*, 1987). These modifications included a lead acetate precipitation step, following initial solvent extraction of the matrix, and additional sample purification on C_{18} media, prior to separation using high-performance NP-TLC. Average recovery for the three toxins was 83% at spiking levels of 100 and 200 $\mu g/kg$, and the detection limit was estimated to be 50 $\mu g/kg$ (Trucksess *et al.*, 1987).

Conversely, Shannon *et al.* (1985) developed a rapid method for the determination of DON, in which samples of maize, wheat and barley were extracted with a mixture of acetonitrile:4% potassium chloride (9:1), followed by purification on C_{18} media. Excess water was removed on a hydrophilic matrix and extracts were separated by NP-TLC, followed by spraying with an aluminium chloride solution. A two-stage TLC development system was reported to decrease the extent of matrix interference, while the use of aluminium chloride-impregnated TLC plates was considered to contribute to 'streaking' on the TLC plate, an effect that was not apparent when the plate was sprayed with the solution following development. The method was, however, less sensitive than others with a detection limit for DON of approximately 1000 $\mu g/kg$ (in maize, barley and wheat) and 1500 $\mu g/kg$ (in oats) (Shannon *et al.*, 1985).

The potential use of 2-(diphenylacetyl)-1,3-indanedione-1-hydrazone (DIPAIN) as a spray reagent for the enhanced fluorescence of several type A trichothecenes has been demonstrated (Novak and Quinn-Doggett, 1991). A reported advantage in the use of DIPAIN is that the response time is based on the rate of the formation of a molecular association complex between the toxin and the detector reagent, rather than on the chemical reactivity of the toxin. Between 50 and 400 ng of T-2 toxin, HT-2 toxin (a structurally similar type A trichothecene) and DAS could be observed under long-wave UV light as orange or yellow-orange spots, while the solid support material also influenced the intensity of the observed fluorescence. The authors also illustrated the effect that alternative DIPAIN analogues had on fluorescence sensitivity and selectivity, and concluded that the use of DIPAIN analogues in which the NH_2 group is replaced with a substituted NH or imine group was necessary for the effective enhancement of fluorescence of the trichothecenes (Novak and Quinn-Doggett, 1991).

3.7.3 Application of HPLC techniques

Trenholm *et al.* (1985) evaluated alternative solvent blends and techniques for the initial extraction of DON from naturally and artificially contaminated grain products. Although HPLC was used as the separation method, their studies should be applicable to other separatory techniques. The results confirmed that the recovery of DON from various substrates was highly dependent on both the matrix to be analysed and the extraction times used. For some naturally contaminated samples, up to 120 min were required for the complete extraction of DON, while in a comparison of three alternative extraction techniques, high-speed homogenization yielded faster solvent extractions (Trenholm *et al.*, 1985). Acetonitrile:water (84:16) was considered to be a more selective extraction solvent blend than methanol:water (1:1), resulting in fewer matrix-related co-extractives.

Lauren and Greenhalgh (1987) reported a RP-HPLC method for the co-determination of NIV and DON in cereals. Acetonitrile:water extracts were initially purified by passage through a combined cation-exchange resin:alumina–carbon (20:1) column. Additional purification was achieved on a carbon minicolumn with the toxins being separated by RP-HPLC and the eluate monitored by UV at $\lambda = 222\,$nm. Recoveries of NIV and DON ranged from 83 to 94% with relative standard deviations of 5% or less. Detection limits of 15 and 50 μg/kg were achieved for NIV and DON, respectively (Lauren and Greenhalgh, 1987).

Sano *et al.* (1987) proposed a two-step successive high-temperature (115°C) postcolumn derivatization method, for the simultaneous determination of NIV, DON and FUS-X by RP-HPLC. The separated toxins were initially subjected to alkaline decomposition with sodium hydroxide to liberate formaldehyde. This was followed by a modified Hantzsch reaction, in which the formaldehyde was reacted with methyl acetoacetate and ammonium acetate to yield fluorophores which could be detected using excitation and emission wavelengths of 370 nm and 460 nm, respectively. The method was applied to cereal samples, which were extracted with acetonitrile:water, partitioned with *n*-hexane and purified on Florisil® and cyano-SPE columns. Recoveries from maize, wheat and barley were found to range from 61.4 to 96.9% for NIV and DON, spiked at levels of 50 and 1000 μg/kg. The detection limits of the method were estimated to be between 20 and 50 μg/kg for each toxin (Sano *et al.*, 1987).

The formation of several alternative derivatives for the UV and fluorescence detection of both type A and type B trichothecenes has been investigated. Derivatives investigated included diphenylindenone sulphonyl esters (Yagen *et al.*, 1986), *p*-nitrobenzoyl (Maycock and Utley, 1985) and coumarin-3-carbonyl chloride (Cohen and Boutin-Muma, 1992). Most of these procedures have, however, been applied to the analysis of toxin standards, and their applicability to the determination of naturally contaminated food commodities has not been fully demonstrated.

3.7.4 *Application of GC techniques*

Although the group B trichothecenes can be detected by UV, the type A trichothecenes do not exhibit UV characteristics, nor do they naturally fluoresce. Consequently, capillary GC techniques have been used extensively for the detection of trace levels of both type A and type B trichothecenes.

However, the application of GC to the analysis of trichothecenes necessitates alteration of their physicochemical characteristics, prior to analysis, by means of derivatization to their corresponding trimethylsilyl (TMS) ethers or trifluoroacetyl (TFA), pentafluoropropionyl (PFP) or heptafluorobutyryl (HFB) esters. Gilbert *et al.* (1985) suggested that an empirical approach had been adopted for the preparation of TMS trichothecene derivatives, leading to the use of a diverse range of reagents and reaction conditions. The authors established optimum conditions for the trimethylsilylation of DON, and identified potential sources of error (Gilbert *et al.*, 1985). The authors used a range of commercially available reagents, including bis-TMS-trifluoroace-tamide (BSTFA), trimethylchlorosilane (TMCS), TMS-imidazole (TMSI) and a commercial combination of all three (Regisil 323®). The results of the study illustrated that the use of a number of combinations of the TMS reagents (together with cited thermal treatments) could result in the incomplete derivatization of DON, yielding a combination of mono-, di- and tri-TMS derivatives. TMSI (on its own or in combination with others) was shown to be the active component leading to the complete conversion of DON to its corresponding tri-TMS derivative (even after reaction for 30 min at room temperature). Following derivatization, removal of excess TMSI reagent is essential (though difficult), since direct injection of the reagent was found to degrade GC column performance. It was concluded that the use of the Regisil 323® reagent (at room temperature) resulted in the quantitative conversion of DON to its tri-TMS derivative, while the reagent was also shown to be suitable for the similar conversion of both NIV and FUS-X to their corresponding TMS derivatives (Gilbert *et al.*, 1985). Kientz and Verweij (1986, 1987) also discussed the problems of derivatization of type A and type B trichothecenes, but extended their studies to include the trifluoroacetylation of the trichothecenes. They concluded that trifluoroacetic anhydride (TFAA) in combination with an acid acceptor (sodium bicarbonate) was an acceptable choice for derivatization purposes. However, the removal of residual water (in addition to excess derivatization reagent) was recommended, while the stability of some TFA-trichothecene derivatives (e.g. neosolaniol, NEO) could be cause for concern (Kientz and Verweij, 1986, 1987).

Tanaka *et al.* (1985a) reported a method for the co-determination of DON and NIV in cereals that yielded average recoveries of 87 and 86% respectively, from maize, wheat and polished rice spiked at levels of 300 μg/kg, with a limit of detection of approximately 2 μg/kg. The method involved extraction of the matrices with acetonitrile:water (3:1) followed by removal of intrinsic fats by partitioning with *n*-hexane. Extracts were further purified in a two-step chromatographic procedure using Florisil and silica gel columns. Levels of DON and NIV were determined by packed-column GC–ECD analysis of their TMS derivatives (Tanaka *et al.*, 1985a). Conversely, Scott *et al.* (1986) used methanol:water (7:3) for the extraction of

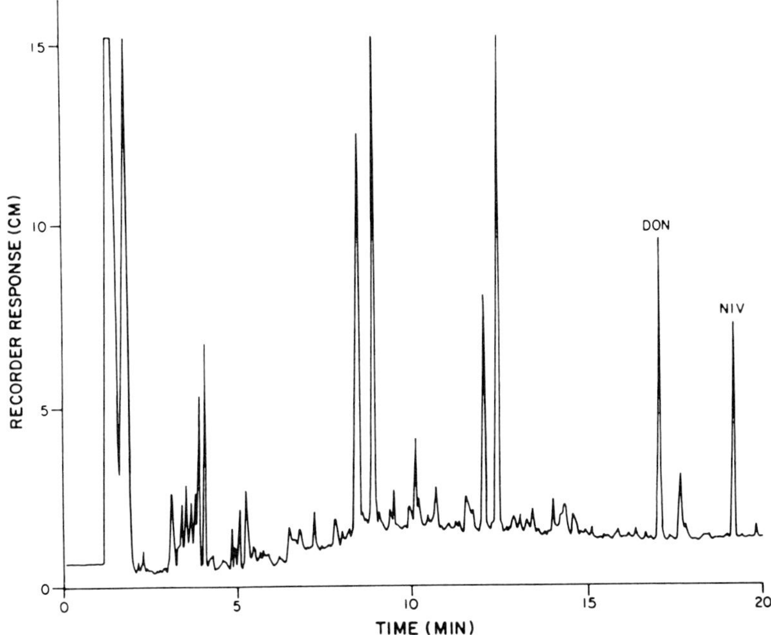

Figure 3.15 Capillary GC–ECD chromatogram of the TMS derivative of a purified wheat sample spiked with DON and NIV at 500 µg/kg (Scott *et al.*, 1986).

DON and NIV from various cereals. Extracts were treated with an ammonium sulphate solution and an aliquot was partitioned into ethyl acetate on a hydrophilic matrix. Further purification was achieved using silica gel chromatography, with NIV and DON derivatized to their TMS derivatives and subsequently separated and detected by GC–ECD, using both packed and capillary GC (Figure 3.15).

The method was evaluated for wheat, maize flour, rye flour and barley, with average recoveries of 72–90% for NIV and 82–106% for DON, spiked at levels of 200–1000 µg/kg. Levels as low as 20 µg/kg could be readily detected (Scott *et al.*, 1986).

Cohen and Lapointe (1984) proposed a method for the co-determination of DAS, T-2 toxin and HT-2 toxin in cereal grains, at levels ranging from 50 to 100 µg/kg. Methanol:water extracts were purified on silica gel and cyano columns, prior to the detection of the toxins by capillary GC–ECD as their HFB derivatives. The method was applied to wheat, oats and barley, with average recoveries of between 65% (for T-2 toxin from barley) and 99% (for DAS from oats) (Cohen and Lapointe, 1984). An alternative method based on the capillary GC analysis of DAS and T-2 toxin in maize and oats samples has also been reported (Sydenham and Thiel, 1987). The method involved extraction with aqueous methanol, followed by a two-stage sample clean-up procedure using a hydrophilic matrix and a silica gel

SPE cartridge. Purified extracts were converted to their corresponding HFB derivatives using neosolaniol monoacetate (a structurally similar type A trichothecene) as an internal standard. The derivatives were separated on an apolar capillary column and monitored by ECD. Average recoveries from maize were found to be 88% for both toxins, with a detection limit of approximately 200 μg/kg (Sydenham and Thiel, 1987).

Langseth and Clasen (1992) used HPLC for the automation of a sample purification step, prior to the determination of NIV, DON, FUS-X and deoxynivalenol triacetate (3-Ac-DON) in cereal samples by capillary GC–ECD. The clean-up column consisted of a cartridge (50 mm × 3 mm i.d.) packed with charcoal:alumina:celite (2.33:1.67:1). Crude acetonitrile:water extracts were filtered and 0.5 ml (corresponding to a 100 mg sample equivalent) was applied to the HPLC system. The trichothecenes were eluted with acetonitrile:water (84:16; 5 ml), and the cartridge was then conditioned for re-use by washing with acetonitrile. The authors indicated that up to ten cereal samples could be purified using a single clean-up cartridge. Extracts were evaporated to dryness and residual water was removed azeotropically with benzene. Trichothecene levels were then determined by capillary GC–ECD as their TMS derivatives. Recoveries were found to be 94% for NIV, DON and FUS-X and 81% for 3-Ac-DON, with detection limits of about 20 μg/kg for each toxin variant (Langseth and Clasen, 1992).

3.7.5 Application of immunochemical techniques

The problems associated with the production of antibodies to small molecules have been highlighted (Zhang *et al.*, 1986). Casale *et al.* (1988) reported the conjugation of DON to carrier proteins by converting DON to 3-*O*-hemisuccinyl-DON, following the initial protection of two of the three available hydroxyls by formation of a cyclic boronate ester. DON could then be detected at 10–250 ng/assay in a direct ELISA, and 10–150 ng/assay in an indirect ELISA. The monoclonal antibody cross-reacted with several structurally related trichothecene analogues (Casale *et al.*, 1988). Xu *et al.* (1986) prepared an antibody to 3-Ac-DON and incorporated it into an RIA for application to wheat, using tritiated 3-Ac-DON as tracer. The samples were initially extracted with acetonitrile:water, defatted with *n*-hexane and then reacted with acetic anhydride to form 3-Ac-DON. Excess reagents and impurities were removed on a reversed-phase C_{18} cartridge, the 3-Ac-DON was eluted with methanol:water and then directly analysed by RIA, utilizing antibodies against 3-Ac-DON. Recoveries of DON, added to wheat at levels ranging from 50 to 5000 μg/kg averaged 86%, with a limit of detection of around 20 μg/kg (Xu *et al.*, 1986). Analyses of naturally contaminated wheat, maize and mixed feed samples for DON indicated that the results obtained by the RIA were in good agreement with those determined by TLC. However, since the method was based on the detection of 3-Ac-DON, the authors confirmed that the RIA technique would also ultimately detect mono-acetyl DON analogues, since they would also be acetylated to 3-Ac-DON (Xu *et al.*, 1986). The same antibody was also incorporated into direct and indirect ELISAs, where recoveries from maize and wheat were found to range from 100 to 121.5%, at spiking levels of between 10 and 1000 μg/kg (Xu *et al.*, 1988). In a comparative study using wheat and maize samples naturally contaminated with DON at levels ranging from <50 to 2000 μg/kg, the ELISA results were shown to agree well with those determined by RIA and TLC. Mills *et al.* (1990) reported the preparation of polyclonal anti-DON antisera. The immunogen was synthesized by the enzyme treatment of 3-Ac-DON hemiglutarate and coupling of the product to BSA. The specificity of the polyclonal antibody was high, with negligible cross-reactivity with NIV, 3-Ac-DON, DAS and T-2 toxin. The ELISA was applied to the analysis of wheat samples, extracted with methanol:water (60:40), and the sensitivity of the assay was considered to be approximately 100 μg/kg (Mills *et al.*, 1990).

Antibodies to DON, prepared by immunizing rabbits with 3-*O*-hemiglutaryl-DON–human serum albumin conjugate, were incorporated into an indirect competitive ELISA for application to liquid matrices occurring in the brewing process, including beer (Niessen *et al.*, 1993). Recoveries ranged from 45 to 83% for DON spiked at levels ranging from 50 to 1000 μg/l. The performance of the indirect competitive ELISA for the detection of DON in beer was compared against HPLC, using the method of Lauren and Greenhalgh (1987), with modifications. When applied to naturally contaminated beer samples, the results obtained by the two techniques compared favourably (over the range <50–500 μg/l) (Niessen *et al.*, 1993).

The production of antibodies against nivalenol tetraacetate (4-Ac-NIV) has been reported (Teshima *et al.*, 1990a; Wang and Chu, 1991). The antibodies were incorporated into a competitive RIA for the determination of NIV in barley samples, in accordance with a method similar to that proposed by Xu *et al.* (1986, 1988) for the determination of DON as its 3-Ac-DON derivative. Extracts were prepared using acetonitrile:water, which were then defatted with *n*-hexane and subsequently reacted with acetic anhydride in order to convert NIV to its 4-Ac-NIV derivative. Additional clean-up was achieved using C_{18} media, and the RIA was conducted directly on methanol:water eluates from the C_{18} column. Recoveries for NIV added to whole barley extracts were between 123.5 and 81.5%, over the spiking range of 50–5000 μg/kg. Good agreement between the RIA and GC results was obtained for naturally contaminated barley samples containing between 140 and 3000 μg/kg NIV (as determined by GC) (Teshima *et al.*, 1990a).

Antisera against DAS have been produced, incorporated into a competitive indirect ELISA and applied to the analysis of wheat (Mills *et al.*, 1988). In comparison to DAS, the cross-reactivities of the antisera to NEO, T-2 toxin, NIV and DON were 1, 0.2, <0.001 and <0.001%, respectively, indicating a high degree of specificity. Wheat samples were extracted with methanol:water, filtered and diluted with a solution of PBS containing 0.05% Tween-20. Recovery of DAS from wheat samples spiked with levels ranging from 300–1000 μg/kg were between 94 and 112%, while the sensitivity of the assay was considered to be 300 μg/kg (Mills *et al.*, 1988).

T-2 toxin was converted to neosolaniol triacetate (3-Ac-NEO) and then to its hemisuccinate prior to conjugation to BSA (Wei and Chu, 1987), and used to raise antibodies following immunization of rabbits. The resultant antibody showed extensive cross-reactivity to most of the group A trichothecenes. The practical application of the antibody for RIA of trichothecenes was demonstrated by testing maize samples spiked with T-2 toxin. The samples were extracted with acetone, purified on a reversed-phase C_{18} cartridge and eluted with acetone:water. The eluates were then diluted with PBS and analysed by RIA. Recoveries of T-2 toxin from maize averaged 94.2% at spiking levels ranging from 10 to 50 μg/kg (Wei and Chu, 1987). Chiba *et al.* (1988) produced several monoclonal antibodies to T-2 toxin. Cross-reactivities of the antibodies with HT-2 toxin and 3′-hydroxy-T-2 toxin were 0.5% of that observed for T-2 toxin, with lower cross-reactivities recorded for other T-2 analogues. One of the antibodies was incorporated into an indirect ELISA and applied to the analysis of wheat flour samples, which were extracted with acetonitrile:water, defatted with *n*-hexane and further purified by partitioning with chloroform and 0.5% aqueous sodium chloride. Recoveries of T-2 toxin from spiked wheat flour ranged from 100 to 119%, at levels of between 0.55 and 50 μg/kg (Chiba *et al.*, 1988). A competitive indirect ELISA, able to detect 0.2 1 ng/ml of T-2 toxin was developed for application to the analysis of various biological fluids, including milk (Fan *et al.*, 1984). Samples of milk spiked with T-2 toxin were subjected to a simple clean-up procedure by passing them through C_{18} media. Recoveries of T-2 toxin from milk ranged from 80 to 83% at spiking levels of 0.2–10 μg/l (Fan *et al.*, 1984).

Chu and Lee (1989) proposed an alternative use for ELISA as a post-column monitoring system, following RP-HPLC analysis of various group A trichothecenes. Fractions eluted from the HPLC column were analysed by competitive direct ELISA using generic antibodies against group A trichothecenes. The technique (referred to as immunochromatography) could be applied to both the identification and the quantitation of the toxin variants (Chu and Lee, 1989). The method required that all fractions eluting from the HPLC column be subjected to alkaline hydrolysis (with methanolic potassium hydroxide) followed by acetylation, prior to analysis by ELISA. Although immunochromatography has not been applied to the analysis of food samples, its potential has been demonstrated by application to the analysis of fungal cultures of *F. sporotrichioides* (Chu and Lee, 1989).

Laamanen and Veijalainen (1992) identified factors that can affect the results of ELISA assays for T-2 toxin, although their observations might well have implications for the analysis of other mycotoxins using ELISA-based techniques. It was suggested that the low efficiency of selected carrier solvents and the presence of natural peroxidases in food and feeds could account for inaccurate reactions, leading to the observed underestimations by ELISA. During the analyses of some processed foodstuffs, false positives were obtained by ELISA (as verified by GC–MS analysis), resulting from the presence of substances in the crude extracts which destroyed or decreased enzyme activity. It was concluded that false positive reactions could be detected by retesting extracts at alternative dilutions (Laamanen and Veijalainen, 1992). Similar observations were also reported by Barna-Vetró *et al.* (1994).

3.7.6 *Application of alternative/confirmation techniques*

The application of selective adsorption minicolumns as rapid screening assays for the determination of DON in cereals has been reported. Ramakrishna *et al.* (1989) detailed the production of the individual minicolumns (packed with separate chromatographic beds of celite, sodium sulphate, Florisil® and silica gel), which were used as the determinative step following purification of crude extracts on charcoal:alumina:celite cartridges. Purified acetonitrile:water extracts of maize, wheat and wheat flour were applied to prepared minicolumns, which had been activated prior to use by heating at 100°C for 1 h. The columns were then eluted with a mixture of hexane:petroleum ether (9:1) to remove residual contaminants, followed by a 20% aluminium chloride solution for the derivatization of DON. Colour development was achieved by drying and heating the minicolumns at 140°C for 10 min; they were then viewed under longwave UV light, where DON was observed as a narrow blue fluorescent band at the Florisil® layer interface. Quantitation was achieved by comparing the colour intensity against similar columns prepared with DON standards over the range 100–1000 ng. The minimum detection limit was assessed to be 200 μg/kg DON. In a comparative study, estimates of DON by the minicolumn tended to be 26–43% higher than those determined by HPLC (Ramakrishna *et al.*, 1989). Gordon and Gordon (1990) used commercially available preparative and analytical (detector) minicolumns for the determination of DON in maize and wheat. In addition to DON, the preparative column was also shown to give excellent recoveries (between 80 and 100%) for NIV, FUS-X, T-2 toxin and DAS. The procedure was estimated to detect DON at levels of 500 μg/kg or higher, and was applied to the analysis of over forty naturally contaminated wheat samples containing DON at levels ranging from 60 to 6300 μg/kg. The results were compared with GC–MS data and showed that 91% of the observations by both techniques compared favourably. The adsorption assay also compared favourably when compared with a TLC method (Gordon and Gordon, 1990).

King *et al.* (1984) developed an effective method for the confirmation of DON in cereals, in which extracts were oxidized by treatment with aqueous periodate or lead dioxide, to form a seven-membered ring lactone DON analogue. The resultant analogue could be quantitatively determined by GC–ECD as its TFA derivative, at levels as low as 50 μg/kg (King *et al.*, 1984).

Capillary GC–MS has been used extensively for the confirmation and/or quantitative estimation of various trichothecenes in cereals, as their TMS, TFA or HFB derivatives. Alternative techniques used have included electron ionization MS for DON, DAS and T-2 toxin (Wreford and Shaw, 1987) and chemical ionization for DON, 3-Ac-DON, NIV, T-2 toxin, HT-2 toxin and DAS (Brumley *et al.*, 1985; Schwadorf and Müller, 1991), while Kostiainen *et al.* (1989) applied chemical ionization tandem MS to the analysis of a range of type A and type B trichothecenes. The application of HPLC–MS has also been demonstrated for the confirmation of the trichothecenes. TS–MS (Voyksner *et al.*, 1987; Kostiainen, 1991; Kostiainen and Kuronen, 1991), dynamic FAB–MS (Kostiainen, 1991; Kostiainen and Kuronen, 1991) and plasmaspray (Kostiainen, 1991) have all been applied to the analysis of both type A and type B trichothecenes. Kalinoski *et al.* (1986) demonstrated the application of supercritical fluid extraction (SFE) with MS for the rapid identification of DAS, DON and T-2 toxin in wheat. The capillary SFC–MS detection of the SFE extract was also applied to wheat samples, with detection limits in the μg/kg range (Kalinoski *et al.*, 1986).

Lanin and Nitikin (1991) studied the effect of the HPLC eluent composition on the retention characteristics of several type B trichothecenes. Ethanol, acetonitrile and tetrahydrofuran were selected as the organic components (modifiers) of the binary water–organic mobile phases that were studied. The retention mechanisms and separation selectivity were evaluated, using a reversed-phase C_{18} column, and optimal conditions were determined for the isocratic reversed-phase separation of five type B trichothecene mycotoxins. The chromatographic system was shown to exhibit high selectivity when eluents incorporating tetrahydrofuran were used (Lanin and Nitikin, 1991). A homologous series of 1-[*p*-(2,3-dihydroxypropoxy)phenyl]-1-alkanones were used as retention index standards in the detection of type A and type B trichothecenes by reversed-phase gradient elution HPLC–MS. Use of the retention indices was reported to provide independent identification of the trichothecene mycotoxins (Kostiainen and Kuronen, 1991).

3.8 Zearalenone

3.8.1 Introduction

Zearalenone (ZEA) ([6-(10-hydroxy-6-oxo-*trans*-1-undecenyl)-β-resorcyclic acid lactone]), also commonly known as F-2 toxin, is a mycotoxin

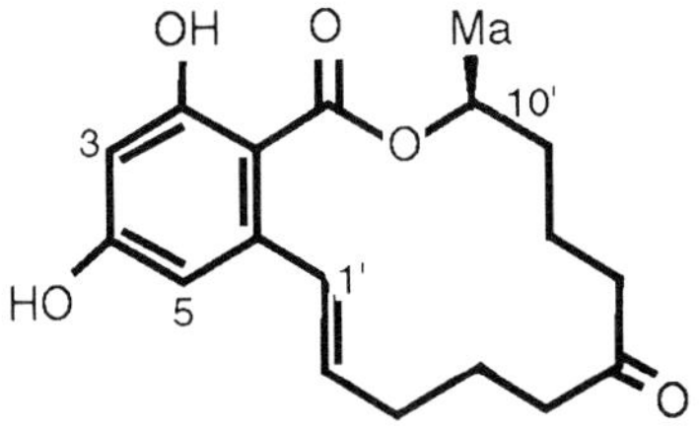

Figure 3.16 Chemical structure of *trans*-zearalenone.

synthesized by several *Fusarium* species. *F. graminearum* is possibly the most common ZEA-producing fungal contaminant of cereal-based foods intended for human consumption. ZEA exhibits a characteristic UV spectrum and also fluoresces. Since *F. graminearum* also produces a number of trichothecene metabolites, the natural co-occurrence of several structurally unrelated mycotoxins in *F. graminearum*-infected foods should be considered. ZEA can exist in both the *cis* and *trans* forms, but it predominantly occurs and is recorded as the latter (Figure 3.16).

Since ZEA possesses strong oestrogenic properties, its natural occurrence is more commonly associated with animal health problems, being the causative agent of hyper-oestrogenism, infertility and other problems in various livestock, notably swine (Mirocha and Christensen, 1974). Feeds contaminated with about 150 μg/kg ZEA might be considered to represent the approximate threshold for hyper-oestrogenism in young swine (Warner *et al.*, 1986). Based on the evidence of the carcinogenic potential of ZEA in rodents, a risk assessment study estimated a safe ZEA intake for humans of 0.05 μg/kg body weight/day (Kuiper-Goodman *et al.*, 1987).

Other oestrogenic substances have been identified, together with ZEA, in various *Fusarium* isolates (Richardson *et al.*, 1985). *Trans*-α-zearalenol (α-ZOL) and *trans*-β-zearalenol (β-ZOL) are two ZEA metabolites which can occur naturally in cereals as diastereomers (Richardson *et al.*, 1985). The oestrogenic potential of α-ZOL has been estimated to be 3–4 times that of ZEA, while the corresponding potential of the β-ZOL metabolite is similar to (or slightly lower than) that of ZEA (Hagler *et al.*, 1979). Correct identification of the specific zearalenol (ZOL) diastereomer is therefore important, given the differences in their relative oestrogenic potentials (Sydenham *et al.*, 1988), although on many occasions references are made only to ZOL without further clarification. All three metabolites may therefore contribute to the oestrogenic activity of implicated feeds, while there is also evidence for the transmission of ZEA, α-ZOL and β-ZOL into milk, following consumption by animals of ZEA-contaminated feeds (Scott and Lawrence, 1988). The possibility of residual ZEA metabolites in animal fats intended for human consumption also exists (Roybal *et al.*, 1988). Human exposure to ZEA and its metabolites may therefore be increased and the contribution of several sources, including cereals, milk and meats should be taken into consideration.

3.8.2 Application of TLC techniques

A rapid TLC method for the co-determination of ZEA and ZOL in various grains and animal feeds was published by Swanson *et al.* (1984). The method, which incorporates the use of low solvent volumes, involves extraction of the matrix with methanol:water (75:25) and precipitation of pigments by the use of a lead acetate solution. Residual fats are removed with petroleum ether, and the toxins are then partitioned into toluene:ethyl acetate (90:10). The presence of additional interferences, especially during the analyses of complex matrices, can be significantly reduced by incorporating a final Florisil® minicolumn clean-up step (Swanson *et al.*, 1984). Purified extracts are then applied to high-performance NP-TLC plates (incorporating a preconcentration zone), which are developed in either chloroform:ethanol (95:5) or benzene:acetone (90:10). ZEA and ZOL may then be observed under shortwave UV light as fluorescent bands, which turn pink after spraying with a solution of Fast Violet B. Additional (partial) confirmation may be achieved by spraying the TLC plate with a sulphuric acid solution, which turns the toxin bands a violet colour. The method was applied to maize, wheat, barley, millet and pelleted feeds, which were spiked with ZEA at levels ranging from 100 to 1000 µg/kg. Recoveries for ZEA were between 87 and 108% (mean 97%), although corresponding recoveries for ZOL were much lower at 40–60%. Swanson *et al.* (1984) considered the method to be quantitative for ZEA but only qualitative for ZOL, with detection limits of 80 and 200 µg/kg, respectively.

Quiroga *et al.* (1994) proposed the use of a 90:10 mixture of aceto-nitrile:4% potassium chloride solution for the initial extraction of ZEA from contaminated maize. Extracts were then purified further by partitioning with iso-octane and a 20% solution of lead acetate. The subsequent transfer (partitioning) of ZEA into toluene was reported to reduce the degree of matrix interference, allowing the toxin to be detected by NP-TLC using a mixture of chloroform:acetone (90:10) as mobile phase. ZEA may then be quantified by observing its fluorescent band intensity under shortwave UV light. Recoveries in excess of 95% were achieved for ZEA spiked into maize at levels ranging from 100 to 300 µg/kg, although these decreased to 40–60% at spiking levels of 20 and 60 µg/kg (Quiroga *et al.*, 1994). The authors reported a detection limit for the method of 50 µg/kg, but indicated that this could be reduced by observing ZEA as its Fast Violet B derivative.

The limited chromatographic resolution achieved by TLC was demonstrated following the analysis of sorghum-based mixed feeds (Sydenham *et al.*, 1988). Using benzene:acetone (95:5) as mobile phase, NP-TLC analysis of purified extracts failed to separate the chromatographic bands corresponding to ZEA and a co-occurring *Alternaria* mycotoxin, alternariol monomethyl ether (AME), adequately. Under the prevailing conditions AME and ZEA exhibited similar R_f and colour characteristics, both prior to and after spraying with a solution of aluminium chloride. Three methods specifically developed for the determination of ZEA in cereal commodities, which differed considerably from each other in their purification procedures, co-extracted AME, with one method (Bagneris *et al.*, 1986) exhibiting a recovery for AME of $91.8 \pm 1.68\%$ (spiked into maize at $90\,\mu g/kg$; Sydenham *et al.*, 1988).

3.8.3 Application of HPLC techniques

The major analytical technique applied to the analysis of ZEA in foods has been HPLC. Bagneris *et al.* (1986) proposed the use of chloroform for the extraction of ZEA and α-ZOL from cereal grains and mixed feeds, followed by a base–acid liquid–liquid partitioning purification step. The toxin levels were then determined by RP-HPLC coupled with fluorescence detection, with the eluate being monitored at $\lambda = 236\,nm$ (excitation) and $\lambda = 418\,nm$ (emission). Average recoveries for ZEA and α-ZOL, spiked into maize, oats, barley, sorghum and mixed feeds were 84 and 69%, respectively, although recoveries from maize tended to be superior to those from other substrates (Bagneris *et al.*, 1986). The detection limit of the method was assessed to be $10\,\mu g/kg$ for each toxin. The method was a modification of that tested collaboratively by 13 participants in the USA, which was conducted using both spiked and naturally contaminated maize (Bennett *et al.*, 1985). Spiking levels ranged from 50 to $200\,\mu g/kg$ for α-ZOL and from 50 to $4000\,\mu g/kg$ for ZEA, with recoveries of 79.6–91.8% and 78.7–103.0% for the two toxins, respectively. For the spiked maize samples, the between-laboratory coefficients of variation were 24.2% for α-ZOL and 33.4% for ZEA, although as expected these figures increased to 47.0 and 37.7%, respectively, for the naturally contaminated samples (Bennett *et al.*, 1985).

A rapid method, described by Chang and DeVries (1984), incorporated cupric carbonate and dichloromethane in the initial extraction of ZEA and α-ZOL from maize and mixed feeds. Filtered extracts were purified by being evaporated to dryness, redissolved in acetonitrile and partitioned with petroleum ether. The purified extracts were separated by RP-HPLC and monitored with fluorescence detection at $\lambda = 280\,nm$ and $\lambda = 465\,nm$ for excitation and emission, respectively. Recoveries ranged from 65.6 to 91.1% (spiked into maize at $10–2000\,\mu g/kg$) with detection limits of about $2\,\mu g/kg$ for each toxin (Chang and DeVries, 1984). In a comparative study, the method was reported to exhibit a superior recovery for α-ZOL than that observed with the method prescribed by AOAC Int. (Chang and DeVries, 1984).

The method of Trenholm *et al.* (1984) for the co-determination of ZEA, α-ZOL and β-ZOL in wheat incorporated the use of zearalenone oxime (ZOX) as internal standard. Wheat samples, buffered with a phosphate solution of pH 7.8, were extracted with water:ethanol:acetonitrile. Crude extracts were subsequently purified by several partitioning steps before being analysed by RP-HPLC and detected either by UV (at $\lambda = 254$ nm) or fluorescence (at $\lambda = 236$ nm excitation and $\lambda = 418$ nm emission). Recoveries for all toxins ranged from 87.5 to 101.0% (spiked at 100–3000 μg/kg) with detection limits of 20 μg/kg each for ZEA and α-ZOL, and 160 μg/kg for β-ZOL (using fluorescence detection) (Trenholm *et al.*, 1984).

The use of ZOX as an internal standard was also recommended for the determination of ZEA and α-ZOL in barley and Job's-tears (Tanaka *et al.*, 1993). Dichloromethane was used for toxin extractions; however, purification was achieved on a short chromatographic column packed with piperidinohydroxypropyl-modified Sephadex® LH-20 media. Separated by RP-HPLC with fluorescence detection, average recoveries for ZEA and α-ZOL (added to barley at levels of 5–250 μg/kg) ranged from 95 to 103% (CV = 3.3%) and from 96 to 102% (CV = 3.6%) for the two toxins, respectively, while the limit of detection was considered to be about 2 μg/kg (Tanaka *et al.*, 1993).

The application of an analytical method developed for the determination of type B trichothecenes to the analysis of ZEA has also been demonstrated (Tanaka *et al.*, 1985b). Applied to the analysis of various cereals, the method involved extraction with acetonitrile:water followed by partitioning with *n*-hexane. Further purification was achieved by Florisil® column chromatography. The major difference in the method was the separation of purified extracts using NP-HPLC, with chloroform:cyclohexane:acetonitrile:ethanol as mobile phase. Tanaka *et al.* (1985b) reported recoveries for ZEA of 85, 93 and 92% from wheat, barley and maize, respectively, spiked at levels of 50 and 500 μg/kg. The detection limit of the method was reported to be 1 μg/kg. Tanaka *et al.* (1985b) and Sydenham *et al.* (1988) demonstrated that detector selectivity and sensitivity to ZEA could be influenced significantly by the incorrect selection of monitoring wavelengths.

An enhancement of the fluorescence response of ZEA was achieved by the incorporation of postcolumn derivatization with aluminium chloride (Hetmanski and Scudamore, 1991). The magnitude of the enhancement was dependent on both the temperature of the reaction coil and the concentration of the aluminium chloride solution. Various reaction coil temperatures were evaluated, with the optimum for ZEA being observed at 50–60°C, while for the derivatization of ZOL the optimum temperature was slightly lower than that for ZEA. The fluorescence response of ZEA increased markedly when using a 0.1 M aluminium chloride solution as the postcolumn reagent. Although the response increased linearly with an increase in aluminium chloride concentration (>0.1 M), it did so at a slower rate. Accordingly, a 0.25 M aluminium chloride solution was selected for the method, since a significantly larger increase in concentration would have been required for a limited further increase in fluorescence response.

Use of the postcolumn derivatization procedure yielded a fivefold increase in the fluorescence response of ZEA, without significantly affecting the level of background interference from co-extractives. However, in the case of maize and feed samples, it was suggested that the presence of a chromatographic peak appearing approximately 1 min prior to that of ZEA (Figure 3.17b) could possibly interfere with the co-determination of ZOL (Hetmanski and Scudamore, 1991).

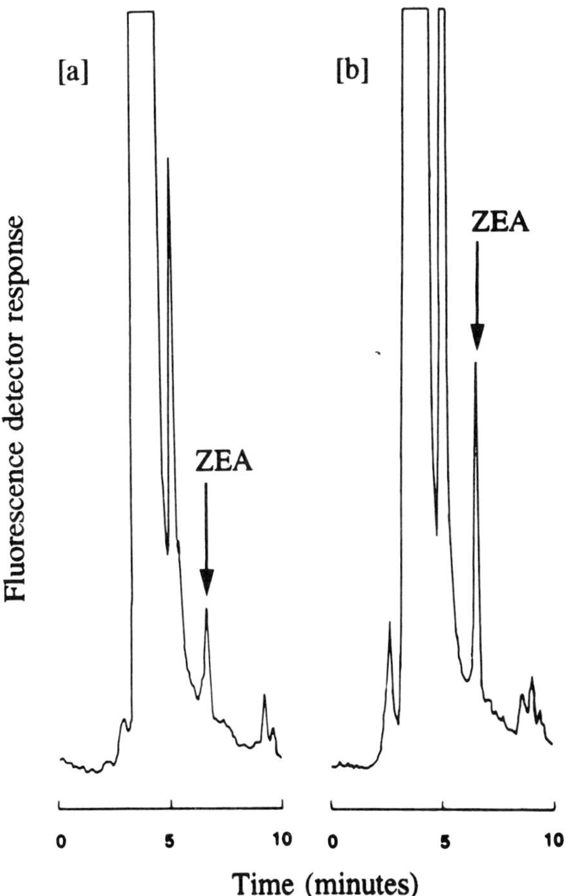

Figure 3.17 RP-HPLC analysis of a maize sample naturally contaminated with 50 μg/kg ZEA, (a) without derivatization, and (b) with postcolumn derivatization with a 0.25 M solution of aluminium chloride (Hetmanski and Scudamore, 1991).

Electrochemical detection has also been applied to the determination of ZEA and ZOL in maize (Ware *et al.*, 1989), extracts of which were prepared in accordance with previously published methods (Bennett *et al.*, 1985; Bagneris *et al.*, 1986). ZEA and ZOL were resolved by RP-HPLC using acetonitrile:methanol:water:0.02 M sodium acetate (buffered to pH 6.5) as mobile phase. The electrochemical detector was operated at an applied potential of +0.95 V versus a silver/silver chloride reference electrode. Phenolic compounds (including ZEA and ZOL) have the potential to form polymeric membranes over an electrode surface (especially at high concentrations), resulting in a reduction in the electrochemical detector performance. Although this 'electrode poisoning' was apparent during studies involving ZEA standards, the authors virtually eradicated the effect by a reduction of the applied maize sample equivalent weight to less than 0.25 g. Average recoveries of ZOL and ZEA, spiked into maize at 25–1000 μg/kg, were 104 and 97.5%, respectively. The sensitivity of the electrochemical detection technique was illustrated by recording a signal to noise ratio of 2:1 for approximately 20 pg each of ZEA and ZOL. However, the authors did not specify a realistic limit of detection for the method as applied to naturally contaminated samples (Ware *et al.*, 1989). Roybal *et al.* (1988) also demonstrated the application of electrochemical detection for the determination of ZEA, α-ZOL and β-ZOL in animal tissues. The method involved maceration of the tissue followed by enzymatic hydrolysis. The toxins were then isolated and purified by sequential partitioning steps. Recoveries of ZEA spiked into chicken and beef muscle and beef liver averaged 75.1, 69.6 and 71.5%, respectively (spiked at 1–10 μg/kg; Roybal *et al.*, 1988).

The possible transmission of ZEA, α-ZOL and β-ZOL into the milk of cows and other animals has prompted the development of suitable rapid and sensitive methods for the determination of these metabolites in milk. The method of Scott and Lawrence (1988) incorporated the extraction of milk with an alkaline acetonitrile solution followed by partitioning of the toxins into dichloromethane, on a kieselgel hydrophilic matrix. Extracts were further purified on an aminopropyl SPE cartridge and subsequently analysed by RP-HPLC with fluorescence detection. Recoveries from milk averaged 84% (ZEA), 93% (α-ZOL) and 90% (β-ZOL), spiked at levels ranging from 0.5 to 20 μg/l. Detection limits of 0.2 μg/l (for ZEA and α-ZOL) and 2 μg/l (for β-ZOL) in milk were reported (Scott and Lawrence, 1988). The method also incorporated the use of an enzyme step for the hydrolysis of potential conjugates prior to extraction. Prelusky *et al.* (1989, 1990) also reported a method for the determination of ZEA and its metabolites in various biological fluids, including milk. The toxins were isolated from milk and purified using a series of pH-controlled solvent extractions, with the toxin levels ultimately determined by RP-HPLC with fluorescence detection. The authors reported detection limits of 1 μg/l for ZEA and α-ZOL and about 5 μg/l for β-ZOL. Recoveries from milk were found to be between 75 and 84%, while the use of ZOX as an internal standard improved the quality of the data. The careful control of solution pH during the extraction phase of the method was cited as a critical point leading to relatively high toxin recoveries, while a mixture of 2-propanol and diethyl ether was considered to be a highly selective extraction solvent (Prelusky *et al.*, 1989).

3.8.4 *Application of GC techniques*

The application of GC techniques to the determination of ZEA in food matrices has received only limited attention. However, the method of Schwadorf and Müller (1992) merits attention since it combines capillary GC with ion trap detection, for the co-determination of ZEA, α-ZOL and β-ZOL in various cereals, including wheat, barley, oats and maize. Cereal samples were extracted and purified in accordance with a method previously described by Mirocha *et al.* (1974), with modifications. Crude ethyl acetate extracts were prepared by shaking for 2 h. The solvent was evaporated and the residue dissolved in chloroform. Following the addition of saturated sodium chloride, the chloroform extract was partitioned with a solution of sodium hydroxide. The toxins were then partitioned back into chloroform, following adjustment of the solution pH by the addition of phosphoric acid. TMS derivatives were prepared and separated on an apolar fused-silica capillary column, and chromatographic peaks were monitored over the scan range $300-540 \, m/z$ (Figure 3.18).

Quantitation was based on selected ion monitoring at $m/z \, 462$ for ZEA and $m/z \, 536$ for α-ZOL (Schwadorf and Müller, 1992). No significant improvement in molecular peak intensities was observed despite the preparation of alternative derivatives or the operation of the detector in the chemical ionization mode. Recoveries for the three toxins ranged from 82 to 86% and the limit of detection was reported to be approximately 1 μg/kg (Schwadorf and Müller, 1992).

3.8.5 *Application of immunochemical techniques*

Warner *et al.* (1986) developed a competitive direct ELISA which was applied to maize samples extracted with a mixture of methanol:PBS:dimethylformamide. The antibodies incorporated into the ELISA were produced against ZOX–BSA conjugates in rabbits. Average recoveries of ZEA from maize, spiked at 57, 151 and 307 μg/kg, were 50, 119 and 123%, respectively. The antibody exhibited greater specificity for α-ZOL than for ZEA (2.8 times), but only limited cross-reactivities were observed for β-ZOL and the zearalanol diastereomers. Analysis of naturally ZEA-contaminated maize samples by both ELISA and HPLC techniques (according to the method of Bennett *et al.*, 1985) gave comparable recoveries, although it was conceded that the natural occurrence of the α-ZOL metabolite could bias total ZEA estimations (Warner *et al.*, 1986).

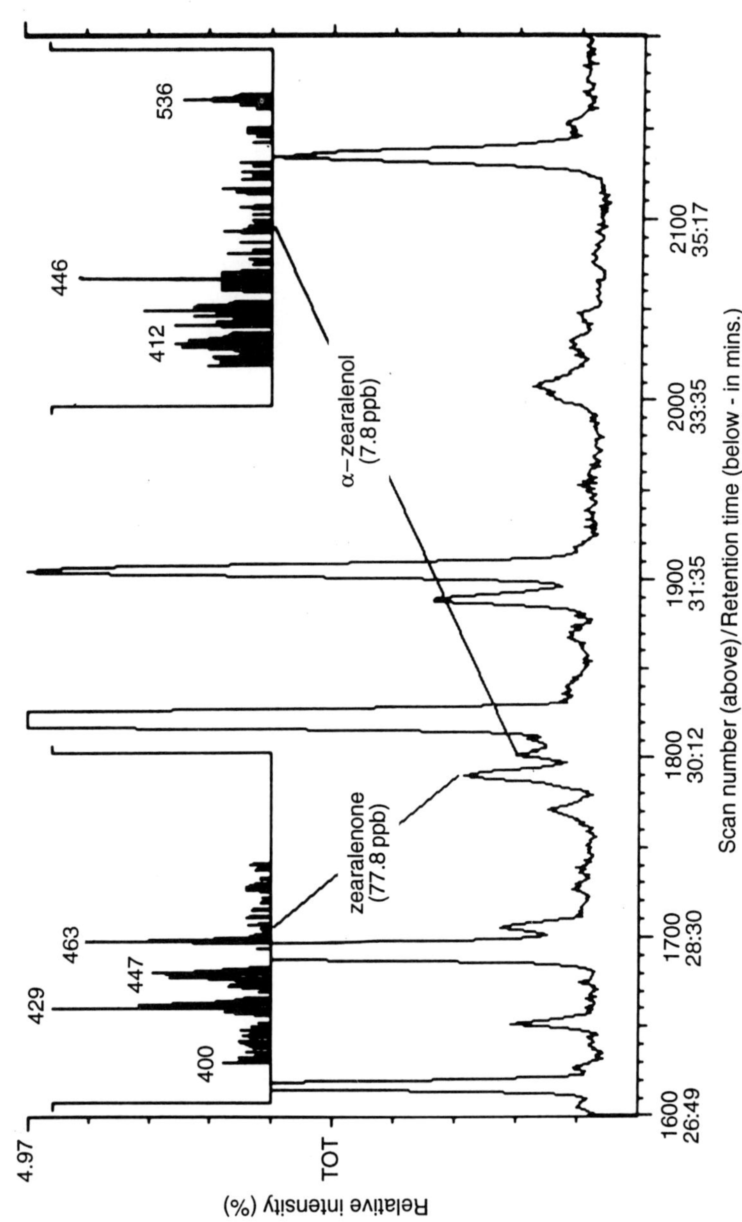

Figure 3.18 Capillary GC—MS (zoomed) total ion chromatogram of a purified wheat sample (as its TMS derivative) naturally contaminated with 78 μg/kg ZEA and 8 μg/kg α-ZOL (β-ZOL was not detected) (Schwadorf and Müller, 1992).

Polyclonal antisera against ZEA were produced in rabbits after immunization with ZOX coupled to human serum albumin (Usleber *et al.*, 1992). The antibodies and a ZOX horseradish peroxidase conjugate were incorporated into a direct ELISA with a detection limit for ZEA of 70 pg/ml. The cross-reactivity of the antibodies to α-ZOL was 37.3% (significantly lower than that previously reported by Warner *et al.*, 1986), with lower cross-reactivities recorded for other ZEA metabolites (with the exception of zearalanone which exhibited a cross-reactivity of 59.2%; Usleber *et al.*, 1992). The direct ELISA was applied to the analysis of wheat samples with a detection limit of 20 μg/kg, and to biological fluids (including cow's milk).

Bennett *et al.* (1994) reported the results of an international collaborative study of a competitive direct ELISA method for the determination of ZEA in maize, wheat and animal feed matrices. The study was conducted by 23 collaborators from five countries, using 18 samples of spiked and naturally contaminated commodities, which were analysed in duplicate. ZEA concentrations were assessed both visually and spectrophotometrically. The ELISA method gave an overall incidence of false negatives of 6.0%, with an incidence of false positives of 22.7%; however, only one false negative was reported for samples containing 800 μg/kg ZEA. Application of the spectrophotometric assessment method yielded average RSD_r and RSD_R values of 11.6 and 25.1%, respectively, for the spiked samples, and 11.7 and 33.1% for the naturally contaminated samples. The method was subsequently adopted first action by AOAC Int. for ZEA at $\geq$800 μg/kg in corn, wheat and pig feed (Bennett *et al.*, 1994).

An indirect ELISA method was conducted by simultaneously incubating ZEA with ZEA antiserum over ZOX poly-L-lysine solid phase, and then determining the bound rabbit immunoglobulin with goat anti-rabbit peroxidase conjugate (Liu *et al.*, 1985). Cross-reactivities of the antisera to α-ZOL, β-ZOL and the zearalanol diastereomers ranged from 3 to 50% of that recorded for ZEA. The detection limit of the method was assessed to be 1 μg/l, equivalent to 0.05 ng per assay. The authors concluded that the ELISA could be used semi-quantitatively to screen for ZEA in methanol extracts of maize, wheat and pig feeds, and that the results correlated with those obtained by TLC (Lui *et al.*, 1985). A stabilized hybridoma cell line secreting anti-ZEA monoclonal antibodies of subclass IgG_1 was prepared by Dixon *et al.* (1987). The antibodies were initially incorporated into an indirect ELISA which detected ZEA at 0.5 mg/ml. Cross-reactivities of the antibody to α-ZOL, β-ZOL and the corresponding diastereomers of zearalanol ranged from 25 to 107%. The antibody was subsequently also incorporated into a direct ELISA, which was used to analyse maize spiked with ZEA at between 50 and 500 μg/kg, with recoveries ranging from 99 to 117% (mean = 110%). Mean inter-well and inter-assay coefficients of variation for the spiked samples were 3.3 and 4.0%, respectively. The authors concluded that the use of hybridoma technology could yield antibodies of enhanced specificity and sensitivity (Dixon *et al.*, 1987). A highly selective monoclonal anti-ZEA antibody was developed by Teshima *et al.* (1990b). Whereas previous anti-ZEA antibodies were obtained by immunization with ZOX–BSA conjugate, Teshima *et al.* (1990b) prepared a novel type of immunogen based on preparation of 5-amino-ZEA, thereby changing the position of conjugation of the ZEA molecule to the carrier. The cross-reactivities of the antibody to other ZEA metabolites ranged from <0.1 to 2.5%, with α-ZOL giving only a 0.9% cross-reactivity. The monoclonal antibody was applied to the indirect ELISA analysis of ZEA in spiked barley, with an average recovery of 102% (spiking range 20–2000 μg/kg). ZEA was also detected in ten naturally contaminated samples, while the quantitative results were comparable to those determined in the same samples by HPLC (Teshima *et al.*, 1990b).

Tanaka *et al.* (1995) combined purification of extracts on an affinity column with detection using indirect ELISA, for the determination of ZEA in barley and Job's-tears. Using different extraction solvent blends, average recoveries of ZEA from barley ranged from 98 to 106%, and were 96–102% from Job's-tears, with a detection limit of 10 μg/kg. The method was applied to the analysis of 200 grain samples, 54 of which were found to be contaminated with ZEA at levels of 10–1395 μg/kg. The 54 ZEA-positive samples were also analysed by HPLC, and a correlation coefficient of $r = 0.971$ was obtained for the two methods.

Monoclonal antibodies to ZEA have, in addition, been incorporated into immunoaffinity cartridges, and these have been used in a tandem affinity chromatography–ELISA method for the determination of ZEA in milk (Azcona *et al.*, 1990). A ZEA-specific monoclonal antibody was attached to Sepharose® for the initial purification of milk samples. Thereafter, ZEA was eluted with methanol and quantified by competitive direct ELISA. The mean recovery from reconstituted skimmed milk, spiked with ZEA at levels ranging from 1–50 μg/l, was 126%. The apparent high recovery may have been indicative of the presence of trace levels of ZEA or ZEA metabolites in the milk, or from partial interference from the sample matrix. The mean inter-well and inter-assay coefficients of variation were found to be 14.5 and 9.1%, respectively (Azcona *et al.*, 1990).

3.8.6 *Application of alternative/confirmation techniques*

The application of a direct chemical confirmation procedure for ZEA developed on silica gel TLC plates has been reported (Goliński and Grabarkiewicz-Szczęsna, 1984). TLC plates previously spotted and chromatographed with toluene:ethyl acetate:90% formic acid (6:3:1) as mobile phase, were subsequently treated by either (a) exposing the chromatographic plate to the vapours of a 1:1 mixture of pyridine:acetic anhydride, or (b) over-spotting the chromatographic band corresponding to ZEA with acetic anhydride.

Following either procedure, the TLC plates were re-chromatographed in the same mobile phase and the fluorescent properties of products of the confirmation procedure were visually assessed. The vapour treatment partially transformed ZEA on the TLC plate to a blue fluorescent compound with a higher R_f value than that of the non-treated species. Direct application of a drop of acetic anhydride esterified ZEA to yield two blue fluorescent chromatographic bands, corresponding to two ester products. Fluorescent intensity of ZEA and its chemical confirmatory products was enhanced by spraying the chromatographic plate with a 20% solution of aluminium chloride (Goliński and Grabarkiewicz-Szczęsna, 1984).

ZEA has a distinctive UV spectrum with maxima at $\lambda = 236\,nm$, $274\,nm$ and $316\,nm$ (Cole and Cox, 1981). Provided the chromatographic peak corresponding to ZEA (emanating from an HPLC analysis) is pure, the ratio of the intensities observed for the three maxima should remain constant. By recording the HPLC profiles obtained at the alternative wavelength maxima, the observed ratio for the peak corresponding to ZEA may be compared with a similar ratio obtained, under similar conditions, for the pure toxin (Sydenham, 1989). With the advent of DAD, chromatographic peaks corresponding to ZEA may be scanned between defined wavelengths, thereby generating a full or partial UV spectrum of the compound. This technique has been applied to the presence of ZEA in both naturally contaminated cereals (Sydenham, 1989) and fungal cultures (Shrivastava and Ansari, 1992).

However, the use of DAD for the confirmation of ZEA in foods may be restricted to those samples containing relatively higher ZEA concentrations, since fluorescence is both a more selective and a more sensitive technique than UV for the detection of ZEA. In addition, at the higher UV wavelength ($\lambda = 316\,nm$) the detectability of ZEA is approximately 3.25 times less than at $\lambda = 236\,nm$ (Shrivastava and Ansari, 1992).

The unequivocal confirmation of the presence of ZEA and its metabolites in animal tissues by GC–MS, as their corresponding TMS derivatives, has been reported (Roybal *et al.*, 1988). The derivatives were separated on a $20\,m \times 0.25\,mm$ i.d. DB-1 fused silica capillary column, with a two stage temperature programme: 80°C to 150°C at 12°C/min followed by 150°C to 235°C at 8°C/min. Under these conditions the molecular ions corresponding to di-TMS ZEA ($m/z = 462$) and tri-TMS α- and β-ZOL ($m/z = 536$) were clearly observed (Roybal *et al.*, 1988).

3.9 Multi-mycotoxin methods

3.9.1 Introduction

Since many fungal species are able to produce (under natural conditions) a diverse range of secondary metabolites, the co-contamination of foods with several diverse mycotoxins may be expected (Sydenham *et al.*, 1990b; Chu and Li, 1994). In addition, various toxins may be derived from the concurrent contamination of foods with several different fungal species. The possibility therefore exists for synergistic effects to be evident between the naturally occurring mycotoxins and/or with other environmental contaminants. The possible biological effects emanating from the co-contamination of foods with multiple toxins should therefore not be underestimated. Since many of the naturally occurring mycotoxins are both structurally and chemically diverse, the successful development of true multi-mycotoxin methods (that are able to yield high recoveries for all toxins regardless of matrix) is problematical. However, the need for such methods is apparent, and hence a number of analytical procedures have been developed for the simultaneous determination of two or more dissimilar mycotoxins in selected matrices. Some of the multi-mycotoxin methods developed over the past ten years are highlighted in sections 3.9.2 to 3.9.6.

An important (though sometimes neglected) potential source of error in mycotoxin analysis is the quality and accuracy of analytical standards. Fortunately, a number of mycotoxins exhibit strong UV characteristics. It is therefore possible to determine the concentration of standards accurately by recording their absorbance at specified wavelengths and calculating their concentration from published coefficients of extinction (ϵ) (although ϵ values published in the literature can differ). Bennett and Shotwell (1990) conducted a short collaborative study to establish the purity of commercially available standards, and to determine their UV maxima and λ values (in methanol). The mycotoxins investigated included α- and β-ZOL, DON, T-2 toxin, HT-2 toxin, DAS, NIV, FUS-X and NEO (Bennett and Shotwell, 1990).

3.9.2 *Application of TLC techniques*

The use of short disposable charcoal–alumina columns for the purification of mycotoxins from food, feed and grain samples, prior to determination by TLC was demonstrated by Romer (1986), with recoveries ranging from 90 to 99%. Levels in the μg/kg range for DON, FUS-X, NIV, T-2 toxin, HT-2 toxin, NEO, and DAS could be detected, using different spray reagents for visual assessment of the NP-TLC plates. The application of clean-up columns to the analysis of the toxins by other chromatographic techniques was also discussed (Romer, 1986).

Soares and Rodriguez-Amaya (1989) reported a NP-TLC method for the co-determination of the aflatoxins, OA, sterigmatocystin (STE) and ZEA, which was applied to the analysis of cereals and processed foods, with detection limits of 2, 5, 15 and 55 μg/kg, respectively. The method involved the extraction of matrices with a mixture of methanol:4% KCl, liquid–liquid partitioning with chloroform followed by several successive solution clarification steps. Toxin quantification was achieved by developing the NP-TLC plates in various solvent systems prior to selective spraying and visual assessment under UV light. Recoveries for AFB_1, STE and ZEA spiked into maize, cassava flour, rice, dried beans and peanuts (spiked at levels from <5 to $2680\,\mu g/kg$) had ranges of 90–100%, 98–117% and 96–107% for the three toxins, respectively.

3.9.3 Application of HPLC techniques

An on-line sample clean-up and postcolumn derivatization method has been proposed for the automated purification and determination of the aflatoxins, OA and ZEA in grains, oilseeds and animal feeds (Chamkasem et al., 1989). On-line sample purification was achieved using a short precolumn ($4.6\,mm$ i.d. $\times$ $50\,mm$) packed with 10-μm C_{18} material, while the analytical column was packed with 5-μm C_{18} material. The use of two in-line fluorescence detectors (prior to and following a heated reaction chamber) allowed for the RP-HPLC determination of OA and ZEA prior to iodine derivatization, and the detection of aflatoxins both prior to and following postcolumn derivatization with iodine. Limits of detection were found to be $5\,\mu g/kg$ for the aflatoxins and OA and $30\,\mu g/kg$ for ZEA. Recoveries for the aflatoxins, OA and ZEA, spiked into maize and sorghum at levels ranging from 20 to $1000\,\mu g/kg$, ranged from 84 to 120% (Chamkasem et al., 1989).

The simultaneous determination of patulin, penicillic acid, STE and ZEA in cocoa beans has been described by Hurst et al. (1987b). Samples were extracted with an acidic acetonitrile solution and sequentially partitioned, first with hexane and then with chloroform. Additional interferences were removed by purification on a silica-gel SPE cartridge. Purified extracts were separated on a cyano column using a mixture of hexane:1-propanol:acetic acid as mobile phase, with dual-channel UV detection at $\lambda = 245\,nm$ and $\lambda = 280\,nm$. Recoveries for the four toxins, spiked into cocoa beans at 30–$1800\,\mu g/kg$, ranged from 75.0 to 112% (Hurst et al., 1987b).

A method for the co-determination of DON, patulin, DAS, HT-2 toxin, T-2 toxin, ZEA and OA was developed for application to wheat samples, using RP-HPLC coupled with DAD (Rajakylä et al., 1987). Initial extracts were partitioned with hexane for the removal of fats and an aliquot was retained for the determination of patulin, OA and ZEA. The remaining extract was purified further by using an aminopropyl SPE cartridge, followed by gradient HPLC analysis. Monitoring wavelengths were altered throughout the HPLC run, in order to optimize the UV detection for specific toxin variants. Toxin recoveries from wheat varied from 75.4 to 90.0%, with the limit of detection being between 250 and $1000\,\mu g/kg$. In addition, the authors demonstrated the use of TS–LC–MS for the determination and confirmation of the toxins. Gradient HPLC was again used for the LC–MS analysis, with the mobile phase modified by the addition of ammonium acetate. By varying the electrolyte concentration, it was possible to utilize the technique either qualitatively or quantitatively (Rajakylä *et al.*, 1987). Stratton *et al.* (1993) reported a HPLC method for the determination of five mycotoxins (DAS, DON, HT-2 toxin, T-2 toxin and ZEA) in wheat and barley samples. The method employed a single extraction step (using acetonitrile:water) and two purification procedures: an aminopropyl-SPE cartridge for the purification of T-2 toxin and ZEA, and a charcoal–alumina column for the purification of the remaining toxins. Toxin levels were determined by using selective isocratic or gradient HPLC programmes with UV detection. Recoveries for all toxins spiked into wheat and barley at levels of between 100 and 2000 μg/kg were greater than 85%, with detection limits of between 20 and 150 μg/kg (Stratton *et al.*, 1993).

Gel-permeation chromatography was utilized for the purification of maize, palm and wheat extracts, prior to the simultaneous RP-HPLC determination of aflatoxins, OA and ZEA (Dunne *et al.*, 1993). Dichloromethane:1 M hydrochloric acid was used for the initial extraction of the matrices prior to clean-up. The toxins were detected by using fluorescence detection at excitation/emission wavelengths which were changed throughout the gradient RP-HPLC cycle, in order to optimize detection. Acceptable recoveries for the aflatoxins were recorded from maize, palm and wheat (between 73.1 and 102.6%), with decreased recoveries for both OA and ZEA, especially from palm (i.e. only 12.5 and 12.9%, respectively) (Dunne *et al.*, 1993).

Konstiainen *et al.* (1991) reported the use of frit-FAB LC-high-resolution MS for the confirmation of several type A and B trichothecenes including T-2 toxin, HT-2 toxin, DON, DAS, monoacetoxyscirpenol and triacetoxyscirpenol. The technique was reported to yield abundant glycerol adduct ions, protonated molecules and fragment ions formed by the losses of functional groups.

Both Hill *et al.* (1984) and Kuronen (1989) proposed the use of HPLC retention indices for the identification of numerous mycotoxins. The former method was developed by using an alkylphenone index scale, while the latter was based on a series of 1-[4-(2,3-dihydroxypropoxy)phenyl]-1-alkanones as internal standards. The versatility of DAD was cited by Kuronen (1989); when combined with the retention index data, it increased the specificity of the technique. With the exception of the analysis of an aflatoxin-spiked almond paste sample (Kuronen, 1989), the application of these techniques was not demonstrated in contaminated food or feed matrices.

3.9.4 *Application of GC techniques*

Möller and Gustavsson (1992) reported a method for the co-determination of several type A and B trichothecenes in cereals. The method included extraction with acetonitrile:ethyl acetate:water, followed by partitioning with *n*-hexane and further purification on a Florisil® SPE cartridge. The trichothecenes were then silylated and determined by capillary GC–ECD. Mean recoveries (based on either five or six determinations) ranged between 78 and 100% for DON, FUS-X, NIV, 3-Ac-DON, DAS, NEO, HT-2 toxin, T-2 toxin and trichothecin, spiked at 250 μg/kg into barley, maize, oats, rye, wheat and wheat bran. The method was also applied to other mycotoxins (analysed singly) with recoveries of 77–86% for T-2 tetrol (T-2tol), trichothecolone, scirpentriol (Sctol) and verrucarol (Möller and Gustavsson, 1992). Croteau *et al.* (1994) proposed a multi-trichothecene capillary GC–ECD method for the co-determination of 13 type A and B trichothecenes, as their HFB derivatives. Recoveries at a spiking level of 1000 μg/kg ranged from 43 to 102%, but these tended to decrease at lower spiking levels of 250 and 100 μg/kg. Detection limits for the method were reported to be 50–200 μg/kg (Croteau *et al.*, 1994). A GC–FID multi-trichothecene method was proposed by Furlong and Soares (1995); it incorporated solvent extraction of wheat samples with methanol:aqueous KCl, clarification with ammonium sulphate, partition into chloroform and final purification on a charcoal:alumina:celite clean-up column. The trichothecenes were converted to their HFB derivatives and quantified with a methyl eicosanoate as internal standard. Average recoveries for DAS, NIV, T-2tol and T-2 triol ranged from 83 to 93%, with detection limits of 0.1–0.5 μg/kg (Furlong and Soares, 1995).

A capillary GC method originally developed for the determination of NIV and DON (Scott *et al.*, 1986) was modified to include the additional simultaneous determination of HT-2 toxin, T-2 toxin and DAS in wheat (Scott *et al.*, 1989). The modifications were primarily associated with the sample extract clean-up phase of the method, with the toxins being ultimately separated on a DB-1701 capillary GC column as their TMS (NIV and DON) and HFB (HT-2 toxin, T-2 toxin and DAS) derivatives. The presence of matrix interferences precluded the additional co-determination of the 4- and 16-monoacetoxyscirpenol trichothecene analogues. Recoveries for the five trichothecenes from wheat averaged 71–92% (over the spiking range 800–4000 μg/kg) (Scott *et al.*, 1989). Confirmation of observations was provided by electron ionization GC–MS, which was operated in a selected ion monitoring mode. Scott *et al.* (1993) also proposed a method for the co-determination of DON, NIV, α- and β-ZOL and ZEA in beer by capillary GC–MS, as their HFB derivatives. The method involved initial partitioning of the toxins into a mixture of ethyl acetate and methanol on a hydrophilic matrix, followed by purification on C_{18} SPE cartridges. Recoveries averaging between 90 and 103% were recorded from beers spiked at levels of 5–20 μg/l (with each toxin variant). Detection limits of 0.01–3 μg/l for the toxins could be achieved by using this method (Scott *et al.*, 1993).

3.9.5 *Application of immunochemical techniques*

Pestka (1991) reported a method that combined the sensitivity and selectivity of competitive ELISA with the capacity of HPTLC, to the analysis of structurally related haptens. The immunoblot approach, termed the 'elisagram', involved separation of the haptens by HPTLC followed by blotting of the chromatographic plate with nitrocellulose (NC) that was coated with hapten-specific monoclonal antibody. The NC was then incubated with hapten–enzyme conjugate in order to identify unreacted antibody binding sites. Detection of the bound enzyme conjugate was achieved by using a precipitating substrate, while visual or densitometric assessments of the inhibition bands, corresponding to the cross-reacting haptens, was utilized for quantification purposes (Pestka, 1991). Applied to the analysis of ZEA (and α-ZOL) and the aflatoxins, the detection limits of the method were 300 pg/assay for ZEA and α-ZOL, 380 pg/assay for AFB_1, AFB_2 and AFG_1 and 1500 pg/assay for AFG_2. The technique therefore allowed for the identification and quantitation of the individual toxin variants, with the cross-reactivity of the 'elisagram' being similar to that previously observed for the conventional competitive ELISA (Pestka, 1991).

A test-strip ELISA for the detection of a number of mycotoxins was developed by Schneider *et al.* (1991). Monoclonal and polyclonal antibodies against AFB_1, AFM_1, OA, T-2 toxin, DAS, 3-Ac-DON, roridin A (RA) and ZEA were immobilized on a nylon membrane. By using the corresponding toxin–horseradish peroxidase conjugate in a direct competitive assay, the dot colour developments of toxin-positive test strips were visually and instrumentally compared against negative controls. The technique was applied to the analysis of cereal extracts with visual detection limits for AFB_1, OA, T-2 toxin and ZEA of 10, 100, 20 and 80 μg/kg, respectively. It was concluded that the test strip assay yielded semi-quantitative results and was suitable for screening foods and feeds under field conditions (Schneider *et al.*, 1991). A computer-assisted multi-analyte assay system (CAMAS) for the simultaneous determination of FB_1, AFB_1 and ZEA in cereal extracts, using a line immunoblot system, was proposed by Abouzied and Pestka (1994). Antibodies to the three mycotoxins were bound to segmented sections of an NC membrane. The sectors were exposed to sample extracts together with aliquots of the respective enzyme-labelled 'markers', which then competed for corresponding antibody binding sites. A substrate was added that resulted in colour development, which was inversely proportional to the concentration of the mycotoxin in the original sample. The image analysis system (CAMAS) was adapted to measure the intensity of each line that was exposed to the free mycotoxin(s), and to compare the intensities against controls of known concentrations. Ranges could be estimated for AFB_1 (0.5–10 μg/kg), FB_1 (500–10 000 μg/kg) and ZEA (1–25 μg/kg). Given that the antibodies incorporated into the system also cross-reacted with their corresponding structurally related analogues (i.e. the FB_1 antibody cross-reacted with FB_2 and FB_3; the AFB_1 antibody cross-reacted with AFB_2, AFG_1 and AFG_2; and the ZEA antibody cross-reacted with α-ZOL and β-ZOL), it was concluded that the multianalyte assay could detect total fumonisins, aflatoxins and zearalenones in contaminated samples (Abouzied and Pestka, 1994).

3.9.6 *Application of dual and/or alternative techniques*

Ehrlich and Lee (1984) proposed a multi-mycotoxin method for the analysis of aflatoxins, OA, ZEA, secalonic acid D (SA-D) and DON in grain dust. The mycotoxins were extracted sequentially with dichloromethane (aflatoxins, OA, ZEA and SA-D) followed by acetonitrile:water (DON). Purification was achieved by using selective liquid–liquid partitioning and/or minicolumn chromatography steps. The aflatoxins and OA were determined by silica-gel TLC with fluorescence detection, while the other mycotoxins were quantified by RP-HPLC with UV detection at $\lambda = 214$ nm (for DON), $\lambda = 280$ (for ZEA) and $\lambda = 340$ nm (for SA-D). Recoveries for the toxins were 57–98% (spiked at levels ranging from 0.5 to 1000 μg/kg), with detection levels from 0.5 μg/kg (for the aflatoxins) to 20 μg/kg (for ZEA) (Ehrlich and Lee, 1984).

The simultaneous determination of NIV, DON, Sctol and T-2tol as their parent alcohols has been proposed (Lauren and Agnew, 1991). The toxins were extracted with acetonitrile:methanol:water, purified on a cation-exchange alumina–carbon column, hydrolysed and then purified further on a carbon–celite minicolumn. Analyses were performed by HPLC with UV detection (for NIV and DON) or capillary GC–ECD (NIV, DON, Sctol and T-2tol). Recoveries spiked into wheat and maize extracts at 500 μg/kg ranged between 55 and 103%, with detection limits of 50 μg/kg or less. Following hydrolysis, the purified extracts were reported to be stable, thereby yielding reliable and consistent data. The method also allowed for an additional purification step for the HPLC determination of moniliformin (MON) and ZEA. Recoveries for MON and ZEA from maize were 78 and 72%, respectively, with detection limits for both toxins of 10 μg/kg. However, the presence of matrix co-extractives necessitated peak confirmation at an alternative wavelength for MON determinations (making identification difficult at levels <100 μg/kg) (Lauren and Agnew, 1991).

Young and Games (1992) investigated the use of capillary and packed-column SFC for the separation of *Fusarium* mycotoxins of various structure types, including DON and its acetylated derivatives, T-2 toxin, butenolide, culmorin and ZEA. The degree of resolution was influenced by column temperature, pressure/density of the liquid carbon dioxide eluent and the concentration of the organic modifier (methanol). Retention times on packed columns were significantly shorter than those on capillary columns. In a comparison with conventional HPLC, packed-column SFC offered faster analysis times and resulted in narrow peak widths, attributes normally associated with capillary column GC. Homologous *n*-alkylphenones were applied for the generation of retention-index values using capillary columns, but these were found to be unsuitable for packed-column SFC (Young and Games, 1992). The use of SFC in the analysis of these mycotoxins in naturally contaminated samples was not demonstrated, but the method has been applied (in conjunction with detection by UV and MS) to the analysis of several toxins (including DON, 3-Ac-DON and others) in liquid culture extracts (Young and Games, 1993). Similarly, Roach *et al.* (1989) applied capillary SFC with negative-ion chemical ionization MS for the analysis of T-2 toxin, DON and RA.

3.10   Conclusions

It is clear, from the literature published over the last decade, that methods for the determination of mycotoxins in foods have been the subject of constant revision and improvement. Due to the necessity to determine low levels of mycotoxins in increasingly complex matrices, the development of new clean-up procedures, such as SPE and immunoaffinity columns, has received considerable attention. The general trend has favoured the development of methods based on instrumental techniques such as HPLC and GC, although TLC remains a viable and popular chromatographic procedure (Van Egmond, 1989). Given their inherent high specificity, the development of methods based on immunochemical techniques has advanced dramatically, with a shift in emphasis from RIA to ELISA techniques. The application of antibody technology to affinity chromatography has also been investigated, most notably for the aflatoxins, although the technique has been and will undoubtedly continue to be applied to other mycotoxins. Advanced extraction/separation techniques such as SFC, SFE and CZE appear to have received only limited attention, being used primarily as research tools. The practical applications of these latter techniques to the analysis of foods may be restricted by the chemical nature of some of the mycotoxins. The necessity to confirm observations unequivocally has also become apparent, with MS (on its own or interfaced with GC or HPLC) becoming the analytical technique of choice.

An area of particular concern to the analyst should be the practical limitations of methods. In this regard, a number of collaborative studies published in the last decade have again highlighted the need for such studies to determine the true between-laboratory variability of methods. As analysts strive to determine ever lower levels of environmental contaminants (including mycotoxins), they do so with an increased degree of error. In a study concerning the reliability of mycotoxin assays (based on the evaluation of data generated from several hundred mycotoxin collaborative studies), Horwitz *et al.* (1993) concluded that despite advances in technology, the overall precision patterns for aflatoxin assays had not improved dramatically over a period of some 20 years, and that overall the primary factor affecting RSD_R values was concentration, which was independent of analyte, matrix and/or age of study. These observations directly impact upon the reliability of data generated by the analyst, especially at low analyte concentrations. Together with the problems of the sampling of heterogeneously contaminated materials, these potential sources of error should be considered by analysts and the legislators who act on their results.

References

Abouzied, M.M. and Pestka, J.J. (1994) Simultaneous screening of fumonisin B_1, aflatoxin B_1, and zearalenone by line immunoblot: A computer-assisted multianalyte assay system. *Journal of AOAC International*, **77**, 495–501.

Abramson, D. (1987) Measurement of ochratoxin A in barley extracts by liquid chromatography–mass spectrometry. *Journal of Chromatography*, **391**, 315–320.

Abramson, D., Thorsteinson, T. and Forest, D. (1989) Chromatography of mycotoxins on precoated reverse-phase thin-layer plates. *Archives of Environmental Contamination and Toxicology*, **18**, 327–330.

Albassam, M.A., Young, S.I., Bhatnagar, R., Sharma, A.K. and Prior, M.G. (1987) Histopathologic and electron microscopic studies on the acute toxicity of ochratoxin A in rats. *Veterinary Pathology*, **24**, 427–435.

Alberts, J.F., Gelderblom, W.C.A. and Marasas, W.F.O. (1992) Evaluation of the extraction and purification procedures of the maleyl derivatization HPLC technique for the quantification of the fumonisin B mycotoxins in corn cultures. *Mycotoxin Research*, **8**, 2–12.

Azcona, J.I., Abouzied, M.M. and Pestka, J.J. (1990) Detection of zearalenone by tandem immunoaffinity–enzyme-linked immunosorbent assay and its application to milk. *Journal of Food Protection*, **53**, 577–580.

Azcona-Olivera, J.I., Abouzied, M.M., Plattner, R.D., Norred, W.P. and Pestka, J.J. (1992a) Generation of antibodies reactive with fumonisins B_1, B_2, and B_3 by using cholera toxin as the carrier-adjuvant. *Applied and Environmental Microbiology*, **58**, 169–173.

Azcona-Olivera, J.I., Abouzied, M.M., Plattner, R.D. and Pestka, J.J. (1992b) Production of monoclonal antibodies to the mycotoxins fumonisins B_1, B_2, and B_3. *Journal of Agricultural and Food Chemistry*, **40**, 531–534.

Bagneris, R.W., Gaul, J.A. and Ware, G.M. (1986) Liquid chromatographic determination of zearalenone and zearalenol in animal feeds and grains, using fluorescence detection. *Journal of the Association of Official Analytical Chemists*, **69**, 894–898.

Barmark, A.-L. and Larsson, K. (1994) Immunoaffinity column cleanup/liquid chromatographic determination of aflatoxins: An interlaboratory study. *Journal of AOAC International*, **77**, 46–53.

Barna-Vetró, I., Gyöngyösi, A. and Solti, L. (1994) Monoclonal antibody-based enzyme-linked immunosorbent assay of *Fusarium* T-2 and zearalenone toxins in cereals. *Applied and Environmental Microbiology*, **60**, 729–731.

Bartolomé, B., Bengoechea, M.L., Gálvez, M.C., Pérez-Ilzarbe, F.J., Hernández, T., Estrella, I. and Gómez-Cordovés, C. (1993) Photodiode array detection for elucidation of the structure of phenolic compounds. *Journal of Chromatography A*, **655**, 119–125.

Bartolomé, B. Bengoechea, M.L., Pérez-Ilzarbe, F.J., Hernández, T., Estrella, I. and Gómez-Cordovés, C. (1994) Determination of patulin in apple juice by high-performance liquid chromatography with diode-array detection. *Journal of Chromatography A*, **664**, 39–43.

Beaver, R.W. (1990) Degradation of aflatoxins in common HPLC solvents. *Journal of High Resolution Chromatography*, **13**, 833–835.

Beaver, R.W., James, M.A. and Lin, T.Y. (1991) Comparison of an ELISA-based screening test with liquid chromatography for the determination of aflatoxins in corn. *Journal of the Association of Official Analytical Chemists*, **74**, 827–829.

Bendele, A.M., Carlton, W.W., Krogh, P and Lillehoj, E.B. (1985) Ochratoxin A carcinogenesis in the (C57BL/6J × C3H)F1 mouse. *Journal of the National Cancer Institute*, **75**, 733–742.

Bennett, G.A., Nelsen, T.C. and Miller, B.M. (1994) Enzyme-linked immunosorbent assay for detection of zearalenone in corn, wheat, and pig feed: Collaborative study. *Journal of AOAC International*, **77**, 1500–1509.

Bennett, G.A. and Richard, J.L. (1994) Liquid chromatographic method for analysis of the naphthalene dicarboxaldehyde derivative of fumonisins. *Journal of AOAC International*, **77**, 501–506.

Bennett, G.A. and Shotwell, O.L. (1990) Criteria for determining purity of *Fusarium* mycotoxins. *Journal of the Association of Official Analytical Chemists*, **73**, 270–275.

Bennett, G.A., Shotwell, O.L. and Kwolek, W.F. (1985) Liquid chromatographic determination of α-zearalenol and zearalenone in corn: Collaborative study. *Journal of the Association of Official Analytical Chemists*, **68**, 958–961.

Betina, V. (1993) Thin-layer chromatography of mycotoxins. In *Chromatography of Mycotoxins: Techniques and Applications*, edited by V. Betina, Journal of Chromatography Library, Vol. 54, Elsevier: Amsterdam, pp. 141–251.

Bijl, J.P., Van Peteghem, C.H. and Dekeyser, D.A. (1987) Fluorimetric determination of aflatoxin M_1 in cheese. *Journal of the Association of Official Analytical Chemists*, **70**, 472–475.

Boyacioglu, D. and Gönül, M. (1988) Comparison of four thin-layer chromatographic methods for the determination of aflatoxins in raisins. *Journal of the Association of Official Analytical Chemists*, **71**, 280–282.

Bradburn, N., Coker, R.D., Jewers, K. and Tomlins, K.I. (1990) Evaluation of the ability of different concentrations of aqueous acetone, aqueous methanol and aqueous acetone:methanol (1:1) to extract aflatoxin from naturally contaminated maize. *Chromatographia*, **29**, 435–440.

Branham, B.E. and Plattner, R.D. (1993) Isolation and characterization of a new fumonisin from liquid cultures of *Fusarium moniliforme*. *Journal of Natural Products*, **56**, 1630–1633.

Brause, A.R., Trucksess, M.W., Thomas, F.S. and Page, S.W. (1996) Determination of patulin in apple juice by liquid chromatography: Collaborative study. *Journal of AOAC International*, in press.

Breitholtz-Emanuelsson, A., Olsen, M., Oskarsson, A., Palminger, I. and Hult, K. (1993) Ochratoxin A in cow's milk and in human milk with corresponding human blood samples. *Journal of AOAC International*, **76**, 842–846.

Brumley, W.C., Trucksess, M.W., Adler, S.H., Cohen, C.K., White, K.D. and Sphon, J.A. (1985) Negative ion chemical ionization mass spectrometry of deoxynivalenol (DON): Application to identification of DON in grains and snack foods after quantification/isolation by thin-layer chromatography. *Journal of Agricultural and Food Chemistry*, **33**, 326–330.

Burda, K. (1992) Incidence of patulin in apple, pear, and mixed fruit products marketed in New South Wales. *Journal of Food Protection*, **55**, 796–798.

Candlish, A.A.G., Stimson, W.H. and Smith, J.E. (1988) Determination of ochratoxin A by monoclonal antibody-based enzyme immunoassay. *Journal of the Association of Official Analytical Chemists*, **71**, 961–964.

Canton, J.H., Kroes, R., Van Logten, M.J., Van Schothorst, M., Stavenuiter, J.F.C. and Verhulsdonk, C.A.H. (1975) The carcinogenicity of aflatoxin M_1 in rainbow trout. *Food and Cosmetics Toxicology*, **13**, 441–443.

Carisano, A. and Torre, G.D. (1986) Sensitive reversed-phase high-performance liquid chromatographic determination of aflatoxin M_1 in dry milk. *Journal of Chromatography*, **355**, 340–344.

Casale, W.L., Pestka, J.J. and Hart, L.P. (1988) Enzyme-linked immunosorbent assay employing monoclonal antibody specific for deoxynivalenol (vomitoxin) and several analogues. *Journal of Agricultural and Food Chemistry*, **36**, 663–668.

Cawood, M.E., Gelderblom, W.C.A., Vleggaar, R., Behrend, Y., Thiel, P.G. and Marasas, W.F.O. (1991) Isolation of the fumonisin mycotoxins: A quantitative approach. *Journal of Agricultural and Food Chemistry*, **39**, 1958–1962.

Chamkasem, N., Cobb, W.Y., Latimer, G.W., Salinas, C. and Clement, B.A. (1989) Liquid chromatographic determination of aflatoxins, ochratoxin A, and zearalenone in grains, oilseeds, and animal feeds by post-column derivatization and on-line sample cleanup. *Journal of the Association of Official Analytical Chemists*, **72**, 336–341.

Chang, H.L. and DeVries, J.W. (1984) Short liquid chromatographic method for determination of zearalenone and alpha-zearalenol. *Journal of the Association of Official Analytical Chemists*, **67**, 741–744.

Chiba, J., Kawamura, O., Kajii, H., Ohtani, K., Nagayama, S. and Ueno, Y. (1988) A sensitive enzyme-linked immunosorbent assay for detection of T-2 toxin with monoclonal antibodies. *Food Additives and Contaminants*, **5**, 629–639.

Chu, F.S., Fan, T.S.L., Zhang, G.-S., Xu, Y.-C., Faust, S. and McMahon, P.L. (1987) Improved enzyme-linked immunosorbent assay for aflatoxin B_1 in agricultural commodities. *Journal of the Association of Official Analytical Chemists*, **70**, 854–857.

Chu, F.S. and Lee, R.C. (1989) Immunochromatography of group A trichothecene mycotoxins. *Food and Agricultural Immunology*, **1**, 127–136.

Chu, F.S. and Li, G.Y. (1994) Simultaneous occurrence of fumonisin B_1 and other mycotoxins in moldy corn collected from the People's Republic of China in regions with high incidences of esophageal cancer. *Applied and Environmental Microbiology*, **60**, 847–852.

Cilliers, J.J.L. and Van Niekerk, P.J. (1984) Liquid chromatographic determination of hydroxymethylfurfural in fruit juices and concentrates after separation on two columns. *Journal of the Association of Official Analytical Chemists*, **67**, 1037–1039.

Cohen, H. and Boutin-Muma, B. (1992) Fluorescence detection of trichothecene mycotoxins as coumarin-3-carbonyl chloride derivatives by high-performance liquid chromatography. *Journal of Chromatography*, **595**, 143–148.

Cohen, H. and Lapointe, M. (1984) Capillary gas chromatographic determination of T-2 toxin, HT-2 toxin, and diacetoxyscirpenol in cereal grains. *Journal of the Association of Official Analytical Chemists*, **67**, 1105–1107.

Cohen, H. and Lapointe, M. (1986) Determination of ochratoxin A in animal feed and cereal grains by liquid chromatography with fluorescence detection. *Journal of the Association of Official Analytical Chemists*, **69**, 957–959.

Cole, R.J. and Cox, R.H. (1981) *Fusarium* toxins. In *Handbook of Toxic Fungal Metabolites*, edited by R.J. Cole and R.H. Cox, Academic Press: New York, pp. 893–910.

Cole, R.J. and Dorner, J.W. (1994) Extraction of aflatoxins from naturally contaminated peanuts with different solvent and solvent/peanut ratios. *Journal of AOAC International*, **77**, 1509–1511.

Creppy, E.E., Roschenthaler, R. and Dirheimer, G. (1984) Inhibition of protein synthesis in mice by ochratoxin A and its prevention by phenylalanine. *Food and Chemical Toxicology*, **22**, 883–886.

Croteau, S.M., Prelusky, D.B. and Trenholm, H.L. (1994) Analysis of trichothecene mycotoxins by gas chromatography with electron capture detection. *Journal of Agricultural and Food Chemistry*, **42**, 928–933.

Cullen, J.M., Ruebner, B.H. and Hsieh, D.S.P. (1985) Comparative hepatocarcinogenicity of aflatoxins B_1 and M_1 in the rat. *Food and Chemical Toxicology*, **22**, 1027–1028.

Damoglou, A.P. and Campbell, D.S. (1986) The effect of pH on the production of patulin in apple juice. *Letters in Applied Microbiology*, **2**, 9–11.

Dell, M.P.K., Haswell, S.J., Roch, O.G., Coker, R.D., Medlock, V.F.P. and Tomlins, K. (1990) Analytical methodology for the determination of aflatoxins in peanut butter: Comparison of high-performance thin-layer chromatographic, enzyme-linked immunosorbent assay and high-performance liquid chromatographic methods. *Analyst*, **115**, 1435–1439.

Dixon, D.E., Warner, R.L., Ram, B.P., Hart, L.P. and Pestka, J.J. (1987) Hybridoma cell line production of a specific monoclonal antibody to mycotoxins zearalenone and α-zearalenol. *Journal of Agricultural and Food Chemistry*, **35**, 122–126.

Dixon-Holland, D.E., Pestka, J.J., Bidigare, B.A., Casale, W.L., Warner, R.L., Ram, B.P. and Hart, L.P. (1988) Production of sensitive monoclonal antibodies to aflatoxin B_1 and aflatoxin M_1 and their application to ELISA of naturally contaminated foods. *Journal of Food Protection*, **51**, 201–204.

Doerge, D.R., Howard, P.C., Bajic, S. and Preece, S. (1994) Determination of fumonisins using on-line liquid chromatography coupled to electrospray mass spectrometry. *Rapid Communications in Mass Spectrometry*, **8**, 603–606.

Domínguez, L., Blanco, J.L., Moreno, M.A., Díaz, S.D., Prieta, J., Cámara, J.M., Bayo, J. and Suárez, G. (1992) Diphasic dialysis: A new membrane method for a selective and efficient extraction of low molecular weight organic compounds from aqueous solutions. *Journal of AOAC International*, **75**, 854–857.

Dorner, J.W. and Cole, R.J. (1988) Rapid determination of aflatoxins in raw peanuts by liquid chromatography with postcolumn iodination and modified minicolumn clean-up. *Journal of the Association of Official Analytical Chemists*, **71**, 43–47.

Dorner, J.W. and Cole, R.J. (1989) Comparison of two ELISA screening tests with liquid chromatography for determination of aflatoxins in raw peanuts. *Journal of the Association of Official Analytical Chemists*, **72**, 962–964.

Dragacci, S., Gleizes, E., Fremy, J.M. and Candlish, A.A.G. (1995) Use of immunoaffinity chromatography as a purification step for the determination of aflatoxin M_1 in cheeses. *Food Additives and Contaminants*, **12**, 59–65.

Duhart, B.T., Shaw, S., Wooley, M., Allen, T. and Grimes, C. (1988) Determination of aflatoxins B_1, B_2, G_1, and G_2 by high performance liquid chromatography with electrochemical detection. *Analytica Chimica Acta*, **208**, 343–346.

Dunne, C., Meaney, M., Smyth, M. and Tuinstra, L.G.M.Th. (1993) Multimycotoxin detection and clean-up method for aflatoxins, ochratoxin and zearalenone in animal feed ingredients using high-performance liquid chromatography and gel permeation chromatography. *Journal of Chromatography*, **629**, 229–235.

Dupuy, J., Le Bars, P., Boudra, H. and Le Bars, J. (1993) Thermostability of fumonisin B_1, a mycotoxin from *Fusarium moniliforme*, in corn. *Applied and Environmental Microbiology*, **59**, 2864–2867.

Ehrlich, K.C. and Lee, L.S. (1984) Mycotoxins in grain dust: Method for analysis of aflatoxins, ochratoxin A, zearalenone, vomitoxin, and secalonic acid D. *Journal of the Association of Official Analytical Chemists*, **67**, 963–967.

El-Banna, A.A. and Scott, P.M. (1984) Fate of mycotoxins during processing of foodstuffs III. Ochratoxin A during cooking of faba beans (*Vicia faba*) and polished wheat. *Journal of Food Protection*, **47**, 189–192.

Engelhardt, H. and Haas, P. (1993) Possibilities and limitations of SFE in the extraction of aflatoxin B_1 from food matrices. *Journal of Chromatographic Science*, **31**, 13–19.

Eppley, R.M., Trucksess, M.W., Nesheim, S., Thorpe, C.W. and Pohland, A.E. (1986) Thin layer chromatographic method for determination of deoxynivalenol in wheat: Collaborative study. *Journal of the Association of Official Analytical Chemists*, **69**, 37–40.

Fan, T.S.L., Zhang, G.S. and Chu, F.S. (1984) An indirect enzyme-linked immunosorbent assay for T-2 toxin in biological fluids. *Journal of Food Protection*, **47**, 964–967.

Farjam, A., Van de Merbel, N.C., Nieman, A.A., Lingeman, H. and Brinkman, U.A.Th. (1992) Determination of aflatoxin M_1 using a dialysis-based immunoaffinity sample pretreatment system coupled on-line to liquid chromatography. *Journal of Chromatography*, **589**, 141–149.

Ferguson-Foos, J. and Warren, J.D. (1984) Improved cleanup for liquid chromatographic analysis and fluorescence detection of aflatoxins M_1 and M_2 in fluid milk products. *Journal of the Association of Official Analytical Chemists*, **67**, 1111–1114.

Fernandez, C., Stack, M.E. and Musser, S.M. (1994) Determination of deoxynivalenol in 1991 U.S. winter and spring wheat by high-performance thin-layer chromatography. *Journal of AOAC International*, **77**, 628–630.

Food and Agricultural Organisation, World Health Organisation (1995) Evaluation of certain food additives and contaminants. In *44th Report of the Joint FAO/WHO Expert Committee on Food Additives*. Technical Report Series, WHO, Geneva, Switzerland pp. 36–38.

Forbito, P.R. and Babsky, N.E. (1985) Rapid liquid chromatographic determination of patulin in apple juice. *Journal of the Association of Official Analytical Chemists*, **68**, 950–951.

Francis, O.J. Jr, Kirschenheuter, G.P., Ware, G.M., Carman, A.S. and Kuan, S.S. (1988) β-Cyclodextrin post-column fluorescence enhancement of aflatoxins for reverse-phase liquid chromatography determination in corn. *Journal of the Association of Official Analytical Chemists*, **71**, 725–728.

Fremy, J.M. and Chu, F.S. (1984) Direct enzyme-linked immunosorbent assay for determining aflatoxin M_1 at picogram levels in dairy products. *Journal of the Association of Official Analytical Chemists*, **67**, 1098–1101.

Frisvad, J.C. and Thrane, U. (1993) Liquid column chromatography of mycotoxins. In *Chromatography of Mycotoxins: Techniques and Applications*, edited by V. Betina, Journal of Chromatography Library, Vol. 54, Elsevier: Amsterdam, pp. 143–162.

Frohlich, A.A., Marquardt, R.R. and Bernatsky, A. (1988) Quantitation of ochratoxin A: Use of reversed phase thin-layer chromatography for sample cleanup followed by liquid chromatography or direct fluorescence measurement. *Journal of the Association of Official Analytical Chemists*, **71**, 949–953.

Fukal, L. (1990) A survey of cereals, cereal products, feedstuffs and porcine kidneys for ochratoxin A by radioimmunoassay. *Food Additives and Contaminants*, **7**, 253–258.

Furlong, E.B. and Soares, E.B. (1995) Gas chromatographic method for quantitation and confirmation of trichothecenes in wheat. *Journal of AOAC International*, **78**, 386–390.

Gareis, M., Märtlbauer, E., Bauer, J. and Gedek, B. (1988) Determination of ochratoxin A in human milk. *Proceedings of the Japanese Association of Mycotoxicology*, supplement no. 1, 61–62.

Gelderblom, W.C.A., Jaskiewicz, K., Marasas, W.F.O., Thiel, P.G., Horak, R.M., Vleggaar, R. and Kriek, N.P.J. (1988) Fumonisins—Novel mycotoxins with cancer-promoting activity produced by *Fusarium moniliforme*. *Applied and Environmental Microbiology*, **54**, 1806–1811.

Gelderblom, W.C.A., Kriek, N.P.J., Marasas, W.F.O. and Thiel, P.G. (1991) Toxicity and carcinogenicity of the *Fusarium moniliforme* metabolite, fumonisin B$_1$, in rats. *Carcinogenesis*, **12**, 1247–1251.

Gelderblom, W.C.A., Cawood, M.E., Snyman, S.D., Vleggaar, R. and Marasas, W.F.O. (1993) Structure-activity relationships of fumonisins in short-term carcinogenesis and cytotoxicity assays. *Food and Chemical Toxicology*, **31**, 407–414.

Gifford, L.A., Wright, C. and Gilbert, J. (1990) Robotic analysis of aflatoxin M$_1$ in milk. *Food Additives and Contaminants*, **7**, 829–836.

Gilbert, J., Startin, J.R. and Crews, C. (1985) Optimisation of conditions for the trimethylsilylation of trichothecene mycotoxins. *Journal of Chromatography*, **319**, 376–381.

Goliński, P. and Grabarkiewicz-Szczęsna, J. (1984) Chemical confirmatory tests for ochratoxin A, citrinin, penicillic acid, sterigmatocystin, and zearalenone performed directly on thin layer chromatographic plates. *Journal of the Association of Official Analytical Chemists*, **67**, 1108–1110.

Gordon, W.C. and Gordon, L.J. (1990) Rapid screening method for deoxynivalenol in agricultural commodities by fluorescent minicolumn. *Journal of the Association of Official Analytical Chemists*, **73**, 266–270.

Goto, T., Matsui, M. and Kitsuwa, T. (1988) Determination of aflatoxins by capillary column gas chromatography. *Journal of Chromatography*, **447**, 410–414.

Hagler, W.M. Jr, Mirocha, C.J., Pathre, S.V. and Behrens, J.C. (1979) Identification of the naturally occurring isomer of zearalenol produced by *Fusarium roseum* 'Gibbosum' in rice cultures. *Applied and Environmental Microbiology*, **37**, 849–853.

Hald, B., Wood, G.M., Boenke, A., Schurer, B. and Finglas, P. (1993) Ochratoxin A in wheat: An intercomparison of procedures. *Food Additives and Contaminants*, **10**, 185–207.

Harrison, L.R., Colvin, B.M., Greene, J.T., Newman, L.E. and Cole, J.R. Jr (1990) Pulmonary edema and hydrothorax in swine produced by fumonisin B$_1$, a toxic metabolite of *Fusarium moniliforme*. *Journal of Veterinary Diagnostic Investigation*, **2**, 217–221.

Harrison, M.A. (1989) Presence and stability of patulin in apple products: A review. *Journal of Food Safety*, **9**, 147–153.

Heathcote, J.G. and Hibbert, J.R. (1978) *Aflatoxins: Chemical and Biological Aspects*, Elsevier Applied Science: Amsterdam.

Hetmanski, M.T. and Scudamore, K.A. (1989) A simple quantitative HPLC method for determination of aflatoxins in cereals and animal feedstuffs using gel permeation chromatography clean-up. *Food Additives and Contaminants*, **6**, 35–48.

Hetmanski, M.T. and Scudamore, K.A. (1991) Detection of zearalenone in cereal extracts using high-performance liquid chromatography with post-column derivatization. *Journal of Chromatography*, **588**, 47–52.

Hill, D.W., Kelley, T.R., Langner, K.J. and Miller, K.W. (1984) Determination of mycotoxins by gradient high-performance liquid chromatography using an alkylphenone retention index system. *Analytical Chemistry*, **56**, 2576–2579.

Hisada, K., Terada, H., Yamamoto, K., Tsubouchi, H. and Sakabe, Y. (1984) Reverse phase liquid chromatographic determination and confirmation of aflatoxin M$_1$ in cheese. *Journal of the Association of Official Analytical Chemists*, **67**, 601–606.

Holcomb, M., Wilson, D.M., Trucksess, M.W. and Thompson, H.C. Jr (1992) Determination of aflatoxins in food products by chromatography. *Journal of Chromatography*, **624**, 341–352.

Holcomb, M., Thompson, H.C. Jr and Hankins, L.J. (1993) Analysis of fumonisin B_1 in rodent feed by gradient elution HPLC using precolumn derivatization with FMOC and fluorescence detection. *Journal of Agricultural and Food Chemistry*, **41**, 764–767.

Hongyo, K-I., Itoh, Y., Hifumi, E., Takeyasu, A. and Uda, T. (1992) Comparison of monoclonal antibody-based enzyme-linked immunosorbent assay with thin-layer chromatography and liquid chromatography for aflatoxin B_1 determination in naturally contaminated corn and mixed feed. *Journal of AOAC International*, **75**, 307–312.

Hopkins, J. (1993) The toxicological hazards of patulin. *BIBRA Bulletin*, **32**, 3–4.

Hopmans, E.C. and Murphy, P.A. (1993) Detection of fumonisins B_1, B_2, and B_3 and hydrolyzed fumonisin B_1 in corn-containing foods. *Journal of Agricultural and Food Chemistry*, **41**, 1655–1658.

Horwitz, W., Albert, A. and Nesheim, S. (1993) Reliability of mycotoxin assays—An update. *Journal of AOAC International*, **76**, 461–491.

Hurst, W.J. and Martin, R.A. Jr (1983) High-performance liquid chromatographic determination of ochratoxin A in artificially spiked cocoa beans. *Journal of Chromatography*, **265**, 353–356.

Hurst, W.J., Martin, R.A. Jr and Vestal, C.H. (1991) The use of HPLC/thermospray MS for the confirmation of aflatoxins in peanuts. *Journal of Liquid Chromatography*, **14**, 2541–2550.

Hurst, W.J., Snyder, K.P. and Martin, R.A. Jr (1987a) Determination of aflatoxins in peanut products using disposable bonded-phase columns and post-column reaction detection. *Journal of Chromatography*, **409**, 413–418.

Hurst, W.J., Snyder, K.P. and Martin, R.A. Jr (1987b) High-performance liquid chromatographic determination of the mycotoxins patulin, penicillic acid, zearalenone and sterigmatocystin in artificially contaminated cocoa beans. *Journal of Chromatography*, **392**, 389–396.

International Standards Organisation (1993) Apple juice, apple juice concentrates and drinks containing apple juice—determination of patulin—part 1: Method using high-performance liquid chromatography. *ISO 8128-1:1993(E)*, Geneva, Switzerland.

Jarvis, B.B., Eppley, R.M. and Mazzola, E.P. (1983) Chemistry and bioproduction of macrocyclic trichothecenes. In *Trichothecenes—Chemical, Biological and Toxicological Aspects*, edited by Y. Ueno, Elsevier: Amsterdam, pp. 20–38.

Jiao, Y, Blaas, W., Rühl, C. and Weber, R. (1992) Identification of ochratoxin A in food samples by chemical derivatization and gas chromatography-mass spectrometry. *Journal of Chromatography*, **595**, 364–367.

Jimènez, M., Sanchis, V., Mateo, R. and Hernández, E. (1988) Detection and quantification of patulin and griseofulvin by high pressure liquid chromatography in different strains of *Penicillium griseofulvin* Dierckx. *Mycotoxin Research*, **4**, 59–66.

Jimènez, M., Mateo, R., Mateo, J.J., Huerta, T. and Hernández, E. (1991) Effect of the incubation conditions on the production of patulin by *Penicillium griseofulvin* isolated from wheat. *Mycopathologia*, **115**, 163–168.

Jones, B.D. (1977) Chemistry of mycotoxins—Aflatoxin and related compounds. In *Mycotoxic Fungi, Mycotoxins, Mycotoxicoses—Encyclopedic Handbook—Volume 1*, edited by T.D. Wyllie and L.G. Morehouse, Marcel Dekker: New York, pp. 131–237.

Joshua, H. (1993) Determination of aflatoxins by reversed-phase high-performance liquid chromatography with post-column in-line photochemical derivatization and fluorescence detection. *Journal of Chromatography A*, **654**, 247–254.

Kalinoski, H.T., Udseth, H.R., Wright, B.W. and Smith, R.D. (1986) Supercritical fluid extraction and direct fluid injection mass spectrometry for the determination of trichothecene mycotoxins in wheat samples. *Analytical Chemistry*, **58**, 2421–2425.

Kamimura, H., Nishijima, M., Yasuda, K., Ushiyama, H., Tabata, S., Matsumoto, S. and Nishima, T. (1985) Simple, rapid cleanup method for analysis of aflatoxins and comparison with various methods. *Journal of the Association of Official Analytical Chemists*, **68**, 458–461.

Kane, A., Creppy, E.E., Roth, A., Roschenthaler, R. and Dirheimer, G. (1986) Distribution of the $[^3H]$-label from low doses of radioactive ochratoxin A ingested by rats, and evidence for DNA single-strand breaks caused in liver and kidneys. *Archives of Toxicology*, **58**, 219–224.

Kawamura, O., Nagayama, S., Sato, S., Ohtani, K., Ueno, I. and Ueno, Y. (1988) A monoclonal antibody-based enzyme-linked immunosorbent assay of aflatoxin B_1 in peanut products. *Mycotoxin Research*, **4**, 75–88.

Kawamura, O., Sato, S., Kajii, H., Nagayama, S., Ohtani, K., Chiba, J. and Ueno, Y. (1989)

A sensitive enzyme-linked immunosorbent assay of ochratoxin A based on monoclonal antibodies. *Toxicon*, **27**, 887–897.

Kellerman, T.S., Marasas, W.F.O., Thiel, P.G., Gelderblom, W.C.A., Cawood, M.E. and Coetzer, J.A.W. (1990) Leukoencephalomalacia in two horses induced by oral dosing of fumonisin B_1. *Onderstepoort Journal of Veterinary Research*, **57**, 269–275.

Kientz, C.E. and Verweij, A. (1986) Trimethylsilylation and trifluoroacetylation of a number of trichothecenes followed by gas chromatographic analysis on fused-silica capillary columns. *Journal of Chromatography*, **355**, 229–240.

Kientz, C.E. and Verweij, A. (1987) Decomposition of trifluoroacetyl derivatives of some trichothecenes on fused-silica capillary columns during gas chromatographic analysis. *Journal of Chromatography*, **407**, 340–342.

King, R.R., Greenhalgh, R. and Blackwell, B.A. (1984) Oxidative transformation of deoxynivalenol (vomitoxin) for quantitative and chemical confirmation purposes. *Journal of Agricultural and Food Chemistry*, **32**, 72–75.

Koeltzow, D.E. and Tanner, S.N. (1990) Comparative evaluation of commercially available aflatoxin test methods. *Journal of the Association of Official Analytical Chemists*, **73**, 584–589.

Kok, W.Th. (1994) Derivatization reactions for the determination of aflatoxins by liquid chromatography with fluorescence detection. *Journal of Chromatography B*, **659**, 127–137.

Kok, W.Th., Van Neer, T.C.H., Traag, W.A. and Tuinstra, L.G.M.T. (1986) Determination of aflatoxins in cattle feed by liquid chromatography and post-column derivatization with electrochemically generated bromine. *Journal of Chromatography*, **367**, 231–236.

Korfmacher, W.A., Chiarelli, M.P., Lay, J.O. Jr, Bloom, J., Holcomb, M. and McManus, K.T. (1991) Characterization of the mycotoxin fumonisin B_1: Comparison of thermospray, fast-atom bombardment and electrospray mass spectrometry. *Rapid Communications in Mass Spectrometry*, **5**, 463–468.

Kostiainen, R. (1991) Identification of trichothecenes by thermospray, plasmaspray and dynamic fast-atom bombardment liquid chromatography–mass spectrometry. *Journal of Chromatography*, **562**, 555–562.

Kostiainen, R. and Kuronen, P. (1991) Use of 1-[p-(2,3-dihydroxypropoxy)phenyl]-1-alkanones as retention index standards in the identification of trichothecenes by liquid chromatography–thermospray and dynamic fast atom bombardment mass spectrometry. *Journal of Chromatography*, **543**, 39–47.

Kostiainen, R., Rizzo, A. and Hesso, A. (1989) The analysis of trichothecenes in wheat and human plasma samples by chemical ionization tandem mass spectrometry. *Archives of Environmental Contamination and Toxicology*, **18**, 356–364.

Kostiainen, R., Matsura, K. and Nojima, K. (1991) Identification of trichothecenes by frit-fast atom bombardment liquid chromatography–high-resolution mass spectrometry. *Journal of Chromatography*, **538**, 323–330.

Kubacki, S.J. and Goszcz, H. (1988) A collaborative study of HPLC methods for the determination of patulin in apple juice. *Pure and Applied Chemistry*, **60**, 871–876.

Kuiper-Goodman, T. (1991) Risk assessment to humans of mycotoxins in animal-derived food products. *Veterinary and Human Toxicology*, **33**, 325–333.

Kuiper-Goodman, T., Scott, P.M. and Watanabe, H. (1987) Risk assessment of the mycotoxin zearalenone. *Regulations in Toxicology and Pharmacology*, **7**, 255–306.

Kuronen, P. (1989) High-performance liquid chromatographic screening method for mycotoxins using new retention indexes and diode array detection. *Archives of Environmental Contamination and Toxicology*, **18**, 336–348.

Laamanen, I. and Veijalainen, P. (1992) Factors affecting the results of T-2 mycotoxin ELISA assay. *Food Additives and Contaminants*, **9**, 337–343.

Lamprecht, S.C., Marasas, W.F.O., Alberts, J.F., Cawood, M.E., Gelderblom, W.C.A., Shephard, G.S., Thiel, P.G. and Calitz, F.J. (1994) Phytotoxicity of fumonisins and TA-toxin to corn and tomato. *Phytopathology*, **84**, 383–391.

Langseth, W. and Clasen, P.-E. (1992) Automation of a clean-up procedure for determination of trichothecenes in cereals using a charcoal–alumina column. *Journal of Chromatography*, **603**, 290–293.

Langseth, W., Ellingsen, Y., Nymoen, U. and Okland, E.M. (1989) High-performance liquid chromatographic determination of zearalenone and ochratoxin A in cereals and feed. *Journal of Chromatography*, **478**, 269–274.

Lanin, S.N. and Nikitin, Y.S. (1991) Retention data for five ketotrichothecenes in reversed-phase high-performance liquid chromatography with different eluent systems. *Journal of Chromatography*, **558**, 81–88.

Lauren, D.R. and Agnew, M.P. (1991) Multitoxin screening method for *Fusarium* mycotoxins in grains. *Journal of Agricultural and Food Chemistry*, **39**, 502–507.

Lauren, D.R. and Greenhalgh, R. (1987) Simultaneous analysis of nivalenol and deoxynivalenol in cereals by liquid chromatography. *Journal of the Association of Official Analytical Chemists*, **70**, 479–483.

Lázaro, F., Luque de Castro, M.D. and Valcárcel, M. (1988) Fluorimetric determination of aflatoxins in foodstuffs by high-performance liquid chromatography with flow injection analysis. *Journal of Chromatography*, **448**, 173–181.

Ledoux, D.R., Brown, T.P., Weibking, T.S. and Rottinghaus, G.E. (1992) Fumonisin toxicity in broiler chicks. *Journal of Veterinary Diagnostic Investigation*, **4**, 330–333.

Lee, S.C. and Chu, F.S. (1984) Enzyme-linked immunosorbent assay of ochratoxin A in wheat. *Journal of the Association of Official Analytical Chemists*, **67**, 45–49.

Lepom, P. (1986) Simultaneous determination of the mycotoxins citrinin and ochratoxin A in wheat and barley by high-performance liquid chromatography. *Journal of Chromatography*, **355**, 335–339.

Lin, L.M., Zhang, J., Sui, K. and Sung, W.B. (1993) Simultaneous thin layer chromatographic determination of zearalenone and patulin in maize. *Journal of Planar Chromatography*, **6**, 274–277.

Liu, M.-T., Ram, B.P., Hart, L.P. and Pestka, J.J. (1985) Indirect enzyme-linked immunosorbent assay for the mycotoxin zearalenone. *Applied and Environmental Microbiology*, **50**, 332–336.

Maragos, C.M. (1995) Capillary zone electrophoresis and HPLC for the analysis of fluorescein isothiocyanate-labelled fumonisin B_1. *Journal of Agricultural and Food Chemistry*, **43**, 390–394.

Maragos, C.M. and Richard, J.L. (1994) Quantitation and stability of fumonisins B_1 and B_2 in milk. *Journal of AOAC International*, **77**, 1162–1167.

Maycock, R. and Utley, D. (1985) Analysis of some trichothecene mycotoxins by liquid chromatography. *Journal of Chromatography*, **347**, 429–433.

McElroy, L.J. and Weiss, C.M. (1993) The production of polyclonal antibodies against the mycotoxin derivative patulin hemiglutarate. *Canadian Journal of Microbiology*, **39**, 861–863.

Micco, C., Grossi, M., Miraglia, M. and Brera, C. (1989) A study of the contamination by ochra toxin A of green and roasted coffee beans. *Food Additives and Contaminants*, **6**, 333–339.

Miller, J.D., Savard, M.E., Sibilia, A., Rapior, S., Hocking, A.D. and Pitt, J.I. (1993) Production of fumonisins and fusarins by *Fusarium moniliforme* from southeast Asia. *Mycologia*, **85**, 385–391.

Miller, N.M., Pretorius, H.E. and Trinder, D.W. (1985) Determination of aflatoxins in vegetable oils. *Journal of the Association of Official Analytical Chemists*, **68**, 136–137.

Mills, E.N.C., Alcock, S.M., Lee, H.A. and Morgan, M.R.A. (1990) An enzyme-linked immunosorbent assay for deoxynivalenol in wheat, utilizing novel hapten derivatization procedures. *Food and Agricultural Immunology*, **2**, 109–118.

Mills, E.N.C., Johnston, J.M., Kemp, H.A. and Morgan, M.R.A. (1988) An enzyme-linked immunosorbent assay for diacetoxyscirpenol applied to the analysis of wheat. *Journal of the Science of Food and Agriculture*, **42**, 225–233.

Ministry of Agriculture, Fisheries and Food (1993) *Mycotoxins: Third Report, Food Surveillance Paper no. 36*. Her Majesty's Stationery Office: London, United Kingdom.

Mirocha, C.J. and Christensen, C.M. (1974) Oestrogenic mycotoxins synthesized by *Fusarium*. In *Mycotoxins*, edited by I.F.M. Purchase, Elsevier: Amsterdam, pp. 129–148.

Mirocha, C.J., Schauerhamer, B. and Pathre, S.V. (1974) Isolation, detection, and quantification of zearalenone in maize and barley. *Journal of the Association of Official Analytical Chemists*, **57**, 1104–1110.

Mirocha, C.J., Gilchrist, D.G., Shier, W.T., Abbas, H.K., Wen, Y. and Vesonder, R.F. (1992) AAL toxins, fumonisins (biology and chemistry) and host-specificity concepts. *Mycopathologia*, **117**, 47–56.

Möller, T.E. and Gustavsson, H.F. (1992) Determination of type A and B trichothecenes in cereals by gas chromatography with electron capture detection. *Journal of AOAC International*, **75**, 1049–1053.

Möller, T.E. and Josefsson, E. (1980) Rapid high pressure liquid chromatography of patulin in apple juice. *Journal of the Association of Official Analytical Chemists*, **63**, 1055–1056.

Morgan, M.R.A., McNerny, R. and Chan, H.W.-S. (1983) Enzyme-linked immunosorbent assay of ochratoxin A in barley. *Journal of the Association of Official Analytical Chemists*, **66**, 1481–1484.

Morgan, M.R.A., Kang, A.S. and Chan, H.W.-S. (1986) Aflatoxin determination in peanut butter by enzyme-linked immunosorbent assay. *Journal of the Science of Food and Agriculture*, **37**, 908–914.

Mortimer, D.N., Gilbert, J. and Shepherd, M.J. (1987a) Rapid and highly sensitive analysis of aflatoxin M_1 in liquid and powdered milks using an affinity column cleanup. *Journal of Chromatography*, **407**, 393–398.

Mortimer, D.N., Shepherd, M.J., Gilbert, J. and Morgan, M.R.A. (1987b) A survey of the occurrence of aflatoxin B_1 in peanut butters by enzyme-linked immunosorbent assay. *Food Additives and Contaminants*, **5**, 127–132.

Mortimer, D.N., Shepherd, M.J., Gilbert, J. and Clark, C. (1988) Enzyme-linked immunosorbent (ELISA) determination of aflatoxin B_1 in peanut butter: Collaborative trial. *Food Additives and Contaminants*, **5**, 601–608.

Murphy, P.A., Rice, L.G. and Ross, P.F. (1993) Fumonisins B_1, B_2, and B_3 content of Iowa, Wisconsin, and Illinois corn and corn screenings. *Journal of Agricultural and Food Chemistry*, **41**, 263–266.

Nakajima, M., Terada, H., Hisada, K., Tsubouchi, H., Yamamoto, K., Uda, T., Itoh, Y., Kawamura, O. and Ueno, Y. (1990) Determination of ochratoxin A in coffee beans and coffee products by monoclonal antibody affinity chromatography. *Food and Agricultural Immunology*, **2**, 189–195.

Nawaz, S., Coker, R.D. and Haswell, S.J. (1992) Development and evaluation of analytical methodology for the determination of aflatoxins in palm kernels. *Analyst*, **117**, 67–74.

Nesheim, S. and Trucksess, M.W. (1986) Thin-layer chromatography for mycotoxin determination. In *Modern Methods in the Analysis and Structural Elucidation of Mycotoxins*, edited by R.J. Cole, Academic Press: New York, pp. 239–264.

Nesheim, S., Stack, M.E., Trucksess, M.W., Eppley, R.M. and Krogh, P. (1992) Rapid solvent-efficient method for liquid chromatographic determination of ochratoxin A in corn, barley, and kidney: Collaborative study. *Journal of AOAC International*, **75**, 481–487.

Niessen, L., Böhm-Schraml, M., Vogel, H. and Donhauser, S. (1993) Deoxynivalenol in commercial beer—screening for the toxin with an indirect competitive ELISA. *Mycotoxin Research*, **9**, 99–109.

Nieuwenhof, F.F.J., Hoolwerf, J.D. and Van den Bedem, J.W. (1990) Evaluation of an enzyme immunoassay for the determination of aflatoxin M_1 in milk using antibody-coated polystyrene beads. *Milchwissenschaft*, **45**, 584–588.

Novak, T.J. and Quinn-Doggett, K. (1991) 2-(Diphenylacetyl)-1,3-indanedione-1-hydrazone (DIPAIN) derivatives for detection of trichothecene mycotoxins. *Analytical Letters*, **24**, 913–924.

Panalaks, T. and Scott, P.M. (1977) Sensitive silica gel-packed flowcell for fluorimetric detection of aflatoxins by high pressure liquid chromatography. *Journal of the Association of Official Analytical Chemists*, **60**, 583–589.

Park, D.L., DiProssimo, V., Abdel-Malek, E., Trucksess, M.W., Nesheim, S., Brumley, W.C., Sphon, J.A., Barry, T.L. and Petzinger, G. (1985) Negative ion chemical ionization mass spectrometric method for confirmation of identity of aflatoxin B_1: Collaborative study. *Journal of the Association of Official Analytical Chemists*, **68**, 636–640.

Park, D.L., Miller, B.M., Nesheim, S., Trucksess, M.W., Vekich, A., Bidigare, B., McVey, J.L. and Brown, L.H. (1989) Visual and semiquantitative spectrophotometric ELISA screening method for aflatoxin B_1 in corn and peanut products: Follow-up collaborative study. *Journal of the Association of Official Analytical Chemists*, **72**, 638–643.

Park, D.L., Trucksess, M.W., Nesheim, S., Stack, M. and Newell, R.F. (1994) Solvent-efficient thin-layer chromatographic method for the determination of aflatoxins B_1, B_2, G_1, and G_2 in corn and peanut products: Collaborative study. *Journal of AOAC International*, **77**, 637–646.

Paterson, R.M.A. and Kemmelmeier, C. (1990) Neutral, alkaline and difference ultraviolet spectra of secondary metabolites from *Penicillium* and other fungi, and comparisons to published maxima from gradient high-performance liquid chromatography with diode-array detection. *Journal of Chromatography*, **511**, 195–221.

Patey, A.L., Sharman, M. and Gilbert, J. (1991) Liquid chromatographic determination of aflatoxin levels in peanut butters using an immunoaffinity column cleanup method: International collaborative trial. *Journal of the Association of Official Analytical Chemists*, **74**, 76–81.

Patey, A.L., Sharman, M. and Gilbert, J. (1992) Determination of total aflatoxin levels in peanut butter by enzyme-linked immunosorbent assay: Collaborative study. *Journal of AOAC International*, **75**, 693–697.

Pestka, J.J. (1991) High performance thin layer chromatography ELISAGRAM. Application of a multi-hapten immunoassay to analysis of the zearalenone and aflatoxin mycotoxin families. *Journal of Immunological Methods*, **136**, 177–183.

Pestka, J.J., Azcona-Olivera, J.I., Plattner, R.D., Minervini, F., Doko, M.B. and Visconti, A. (1994) Comparative assessment of fumonisin in grain-based foods by ELISA, GC–MS, and HPLC. *Journal of Food Protection*, **57**, 169–172.

Phillips, T.D., Clement, B.A., Sarr, A.B., Huang, Z.-G. and Williams, S.M. (1992) Chemiselective immobilization and detection of fumonisin B_1. Proceedings of the VIII International IUPAC Symposium on Mycotoxins and Phycotoxins, November 6–13, 1992, Mexico City, Mexico, p. 42.

Pittet, A., Parisod, V. and Schellenberg, M. (1992) Occurrence of fumonisins B_1 and B_2 in corn-based products from the Swiss market. *Journal of Agricultural and Food Chemistry*, **40**, 1352–1354.

Plattner, R.D. and Branham, B.E. (1994). Labeled fumonisins: Production and use of fumonisin B_1 containing stable isotopes. *Journal of AOAC International*, **77**, 525–532.

Plattner, R.D., Norred, W.P., Bacon, C.W., Voss, K.A., Peterson, R., Shackelford, D.D. and Weisleder, D. (1990) A method of detection of fumonisins in corn samples associated with field cases of equine leukoencephalomalacia. *Mycologia*, **82**, 698–702.

Plattner, R.D., Ross, P.F., Reagor, J., Stedelin, J. and Rice, L.G. (1991) Analysis of corn and cultured corn for fumonisin B_1 by HPLC and GC–MS by four laboratories. *Journal of Veterinary Diagnostic Investigation*, **3**, 357–358.

Plattner, R.D., Weisleder, D., Shackelford, D.D., Peterson, R. and Powell, R.G. (1992) A new fumonisin from solid cultures of *Fusarium moniliforme*. *Mycopathologia*, **117**, 23–28.

Pohland, A.E., Nesheim, S. and Friedman, L. (1992) Ochratoxin A: A review. *Pure and Applied Chemistry*, **64**, 1029–1046.

Prelusky, D.B., Warner, R.M. and Trenholm, H.L. (1989) Sensitive analysis of the mycotoxin zearalenone and its metabolites in biological fluids by high-performance liquid chromatography. *Journal of Chromatography*, **494**, 267–277.

Prelusky, D.B., Scott, P.M., Trenholm, H.L. and Lawrence, G.A. (1990) Minimal transmission for zearalenone to milk of dairy cows. *Journal of Environmental Science and Health*, **B25**, 87–103.

Prieta, J., Moreno, M.A., Blanco, J.L., Suárez, G. and Domínguez, L. (1992) Determination of patulin by diphasic dialysis extraction and thin-layer chromatography. *Journal of Food Protection*, **55**, 1001–1002.

Prieta, J., Moreno, M.A., Bayo, J., Díaz, S., Suárez, G., Domínguez, L., Canela, R. and Sanchis, V. (1993) Determination of patulin by reversed-phase high-performance liquid chromatography with extraction by diphasic dialysis. *Analyst*, **118**, 171–173.

Qian, G.S., Yasei, P. and Yang, G.C. (1984) Rapid extraction and detection of aflatoxin M_1 in cow's milk by high-performance liquid chromatography and radioimmunoassay. *Analytical Chemistry*, **56**, 2079–2080.

Quiroga, N.M., Sola, I. and Varavsky, E. (1994) Selection of a simple and sensitive method for detecting zearalenone in corn. *Journal of AOAC International*, **77**, 939–941.

Rajakylä, E., Laasasenaho, K. and Sakkers, P.J.D. (1987) Determination of mycotoxins in grain by high-performance liquid chromatography and thermospray liquid chromatography–mass spectrometry. *Journal of Chromatography*, **384**, 391–402.

Ram, B.P., Hart, L.P., Shotwell, O.L. and Pestka, J.J. (1986) Enzyme-linked immunosorbent assay of aflatoxin B_1 in naturally contaminated corn and cottonseed. *Journal of the Association of Official Analytical Chemists*, **69**, 904–907.

Ramakrishna, N., Lacey, J., Candlish, A.A.G., Smith, J.E. and Goodbrand, I.A. (1990) Monoclonal antibody-based enzyme linked immunosorbent assay of aflatoxin B_1, T-2 toxin, and ochratoxin A in barley. *Journal of the Association of Official Analytical Chemists*, **73**, 71–76.

Ramakrishna, Y., Sashidhar, R.B. and Bhat, R.V. (1989) Minicolumn technique for the detection of deoxynivalenol in agricultural commodities. *Bulletin of Environmental Contamination and Toxicology*, **42**, 167–171.

Rheeder, J.P., Marasas, W.F.O., Thiel, P.G., Sydenham, E.W., Shephard, G.S. and Van Schalkwyk, D.J. (1992) *Fusarium moniliforme* and fumonisins in corn in relation to human esophageal cancer in Transkei. *Phytopathology*, **82**, 353–357.

Richardson, K.E., Hagler, W.M., Jr. and Mirocha, C.J. (1985) Production of zearalenone, α- and β-zearalenol, and α- and β-zearalanol by *Fusarium* spp. in rice culture. *Journal of Agricultural and Food Chemistry*, **33**, 862–866.

Roach, J.A.G., Sphon, J.A., Easterling, J.A. and Calvey, E.M. (1989) Capillary supercritical fluid chromatography/negative ion chemical ionization mass spectrometry of trichothecenes. *Biomedical and Environmental Mass Spectrometry*, **18**, 64–70.

Roch, O.G., Blunden, G., Coker, R.D. and Nawaz, S. (1992) The development and validation of a solid phase extraction/HPLC method for the determination of aflatoxins in groundnut meal. *Chromatographia*, **33**, 208–212.

Romer, T.R. (1986) Use of small charcoal/alumina cleanup columns in determination of trichothecene mycotoxins in foods and feeds. *Journal of the Association of Official Analytical Chemists*, **69**, 699–703.

Rosen, R.T., Rosen, J.D. and DiProssimo, V.P. (1984) Confirmation of aflatoxins B_1 and B_2 by gas chromatography/mass spectrometry/selected ion monitoring. *Journal of Agricultural and Food Chemistry*, **32**, 276–278.

Ross, P.F., Rice, L.G., Plattner, R.D., Osweiler, G.D., Wilson, T.M., Owens, D.L., Nelson, H.A. and Richard, J.L. (1991) Concentrations of fumonisin B_1 in feeds associated with animal health problems. *Mycopathologia*, **114**, 129–135.

Rottinghaus, G.E., Coatney, C.E. and Minor, H.C. (1992) A rapid, sensitive thin layer chromatographic procedure for the detection of fumonisin B_1 and B_2. *Journal of Veterinary Diagnostic Investigation*, **4**, 326–329.

Rousseau, D.M., Slegers, G.A. and Van Peteghem, C.H. (1985) Radioimmunoassay of ochratoxin A in barley. *Applied and Environmental Microbiology*, **50**, 529–531.

Rovira, R., Ribera, F., Sanchis, V. and Canela, R. (1993) Improvements in the quantitation of patulin in apple juice by high-performance liquid chromatography. *Journal of Agricultural and Food Chemistry*, **41**, 214–216.

Roybal, J.E., Munns, R.K., Morris, W.J., Hurlbut, J.A. and Shimoda, W. (1988) Determination of zeranol/zearalenone and their metabolites in edible animal tissue by liquid chromatography with electrochemical detection and confirmation by gas chromatography/mass spectrometry. *Journal of the Association of Official Analytical Chemists*, **71**, 263–271.

Sashidar, R.B. (1993) Dip-strip method for monitoring environmental contamination of aflatoxin in food and feed—Use of a portable aflatoxin detection kit. *Environmental Health Perspectives*, **101** (suppl. 3), 43–46.

Sano, A., Matsutani, S., Suzuki, M. and Takitani, S. (1987) High-performance liquid chromatographic method for determining trichothecene mycotoxins by post-column fluorescence derivatization. *Journal of Chromatography*, **410**, 427–436.

Schneider, E., Dietrich, R., Märtlbauer, E., Usleber, E. and Terplan, G. (1991) Detection of aflatoxins, trichothecenes, ochratoxin A and zearalenone by test strip enzyme immunoassay: A rapid method for screening cereals for mycotoxins. *Food and Agricultural Immunology*, **3**, 185–193.

Schwadorf, K. and Müller, H.-M. (1991) Determination of trichothecenes in cereals by gas chromatography with ion trap detection. *Chromatographia*, **32**, 137–142.

Schwadorf, K. and Müller, H.-M. (1992) Determination of α- and β-zearalenol and zearalenone in cereals by gas chromatography with ion-trap detection. *Journal of Chromatography*, **595**, 259–267.

Scott, P.M. (1989) Methods for determination of aflatoxin M_1 in milk and milk products—a review of performance characteristics. *Food Additives and Contaminants*, **6**, 283–305.

Scott, P.M. (1993) Fumonisins. *International Journal of Food Microbiology*, **18**, 257–270.

Scott, P.M. and Lawrence, G.A. (1988) Liquid chromatographic determination of zearalenone and α- and β-zearalenols in milk. *Journal of the Association of Official Analytical Chemists*, **71**, 1176–1179.

Scott, P.M. and Lawrence, G.A. (1992) Liquid chromatographic determination of fumonisins with 4-fluoro-7-nitrobenzofurazan. *Journal of AOAC International*, **75**, 829–834.

Scott, P.M. and Lawrence, G.A. (1994) Stability and problems in recovery of fumonisins added to corn-based foods. *Journal of AOAC International*, **77**, 541–545.

Scott, P.M., Kanhere, S.R. and Tarter, E.J. (1986) Determination of nivalenol and deoxynivalenol in cereals by electron-capture gas chromatography. *Journal of the Association of Official Analytical Chemists*, **69**, 889–893.

Scott, P.M., Lombaert, G.A., Pellaers, P., Bacler, S., Kanhere, S.R., Sun, W.F., Lau, P.-Y. and Weber, D. (1989) Application of capillary gas chromatography to a survey of wheat for five trichothecenes. *Food Additives and Contaminants*, **6**, 489–500.

Scott, P.M., Kanhere, S.R. and Weber, D. (1993) Analysis of Canadian and imported beers for *Fusarium* mycotoxins by gas chromatography–mass spectrometry. *Food Additives and Contaminants*, **10**, 381–389.

Scott, P.M., Delgado, T., Prelusky, D.B., Trenholm, H.L. and Miller, J.D. (1994) Determination of fumonisins in milk. *Journal of Environmental Science and Health*, **B29**, 989–998.

Scott, P.M., Kanhere, S.R., Lawrence, G.A., Daley, E.F. and Farber, J.M. (1995) Fermentation of wort containing added ochratoxin A and fumonisins B_1 and B_2. *Food Additives and Contaminants*, **12**, 31–40.

Shannon, G.M., Peterson, R.E. and Shotwell, O.L. (1985) Rapid screening method for detection of deoxynivalenol. *Journal of the Association of Official Analytical Chemists*, **68**, 1126–1128.

Sharman, M., Patey, A.L. and Gilbert, J. (1989) Application of an immunoaffinity column sample clean-up to the determination of aflatoxin M_1 in cheese. *Journal of Chromatography*, **474**, 457–461.

Sharman, M., MacDonald, S. and Gilbert, J. (1992) Automated liquid chromatographic determination of ochratoxin A in cereals and animal products using immunoaffinity column clean-up. *Journal of Chromatography*, **603**, 285–289.

Shephard, G.S., Sydenham, E.W., Thiel, P.G. and Gelderblom, W.C.A. (1990) Quantitative determination of fumonisins B_1 and B_2 by high-performance liquid chromatography with fluorescence detection. *Journal of Liquid Chromatography*, **13**, 2077–2087.

Shepherd, M.J. and Gilbert, J. (1984) An investigation of HPLC post-column iodination conditions for the enhancement of aflatoxin B_1 fluorescence. *Food Additives and Contaminants*, **4**, 325–335.

Shepherd, M.J., Holmes, M. and Gilbert, J. (1986) Comparison and critical evaluation of six published extraction and clean-up procedures for aflatoxin M_1 in liquid milk. *Journal of Chromatography*, **354**, 305–315.

Shier, W.T., Abbas, H.K. and Mirocha, C.J. (1991) Toxicity of the mycotoxins fumonisins B_1 and B_2 and *Alternaria alternata* f. sp. *lycopersici* toxin (AAL) in cultured mammalian cells. *Mycopathologia*, **116**, 97–104.

Shrivastava, A.K. and Ansari, A.A. (1992) Isolation and determination of zearalenone in rice cultures using liquid chromatography and diode array detection. *Food Additives and Contaminants*, **9**, 331–336.

Sidwell, W.J., Chan, H.W.-S. and Morgan, M.R.A. (1989) Reductions in assay time for a double antibody, indirect enzyme-linked immunosorbent assay applied to ochratoxin A. *Food and Agricultural Immunology*, **1**, 111–118.

Smith, J.E., Lewis, C.W., Anderson, J.G. and Solomons, G.L. (1994) *Mycotoxins in Human Nutrition and Health*, European Commission Directorate General XII, Brussels Report No. EUR16048EN.

Soares, L.M.V. and Rodriguez-Amaya, D.B. (1985) Screening and quantitation of ochratoxin A in corn, peanuts, beans, rice and cassava. *Journal of the Association of Official Analytical Chemists*, **68**, 1128–1130.

Soares, L.M.V. and Rodriguez-Amaya, D.B. (1989) Survey of aflatoxins, ochratoxin A, zearalenone, and sterigmatocystin in some Brazilian foods by using multi-toxin thin-layer chromatographic method. *Journal of the Association of Official Analytical Chemists*, **72**, 22–26.

Steyn, P.S. (1984) Ochratoxins and related dihydroisocoumarins. In *Mycotoxins—Production, Isolation, Separation and Purification*, edited by V. Betina, Elsevier: Amsterdam, pp. 183–216.

Stockenström, S., Sydenham, E.W. and Thiel, P.G. (1994) Determination of fumonisins in corn: Evaluation of two purification procedures. *Mycotoxin Research*, **10**, 9–14.

Stoloff, L.S. and Scott, P.M. (1984) Natural poisons. In *Official Methods of Analysis of the Association of Official Analytical Chemists, 14th Edition*, edited by S. Williams, AOAC: Arlington, Virginia, pp. 477–500.

Stoloff, L., Van Egmond, H.P. and Park, D.L. (1991) Rationales for the establishment of limits and regulations for mycotoxins. *Food Additives and Contaminants*, **8**, 213–222.

Stratton, G.W., Robinson, A.R., Smith, H.C., Kittilsen, L. and Barbour, M. (1993) Levels of five mycotoxins in grains harvested in Atlantic Canada as measured by high performance liquid chromatography. *Archives of Environmental Contamination and Toxicology*, **24**, 399–409.

Stray, H. (1978) High pressure liquid chromatographic determination of patulin in apple juice. *Journal of the Association of Official Analytical Chemists*, **61**, 1359–1362.

Stubblefield, R.D. (1987) Optimum conditions for formation of aflatoxin M_1-trifluoroacetic acid derivative. *Journal of the Association of Official Analytical Chemists*, **70**, 1047–1049.

Stubblefield, R.D. and Kwolek, W.F. (1986) Rapid liquid chromatographic determination of aflatoxins M_1 and M_2 in artificially contaminated fluid milks: Collaborative study. *Journal of the Association of Official Analytical Chemists*, **69**, 880–885.

Sudershan, R.V., Prasad, G.S., Krishna, T.P. and Bhat, R.V. (1992) Field level evaluation of aflatoxin detection kit. *Journal of Food Protection*, **55**, 392–394.

Swanson, S.P., Corley, R.A., White, D.G. and Buck, W.B. (1984) Rapid thin layer chromatographic method for determination of zearalenone and zearalenol in grains and animal feeds. *Journal of the Association of Official Analytical Chemists*, **67**, 580–582.

Sydenham, E.W. (1989) The chromatographic determination of *Fusarium* toxins in maize associated with human oesophageal cancer. M.Sc. thesis, University of Cape Town, South Africa.

Sydenham, E.W. and Thiel, P.G. (1987) The simultaneous determination of diacetoxyscirpenol and T-2 toxin in fungal cultures and grain samples by capillary gas chromatography. *Food Additives and Contaminants*, **4**, 277–284.

Sydenham, E.W., Thiel, P.G. and Marasas, W.F.O. (1988) Occurrence and chemical determination of zearalenone and alternariol monomethyl ether in sorghum-based mixed feeds associated with an outbreak of suspected hyperestrogenism in swine. *Journal of Agricultural and Food Chemistry*, **36**, 621–625.

Sydenham, E.W., Gelderblom, W.C.A., Thiel, P.G. and Marasas, W.F.O. (1990a) Evidence for the natural occurrence of fumonisin B_1, a mycotoxin produced by *Fusarium moniliforme*, in corn. *Journal of Agricultural and Food Chemistry*, **38**, 285–290.

Sydenham, E.W., Thiel, P.G., Marasas, W.F.O., Shephard, G.S., Van Schalkwyk, D.J. and Koch, K.R. (1990b) Natural occurrence of some *Fusarium* mycotoxins in corn from low and high esophageal cancer prevalence areas of the Transkei, southern Africa. *Journal of Agricultural and Food Chemistry*, **38**, 285–290.

Sydenham, E.W., Shephard, G.S., Thiel, P.G., Marasas, W.F.O. and Stockenström, S. (1991) Fumonisin contamination of commercial corn-based human foodstuffs. *Journal of Agricultural and Food Chemistry*, **39**, 2014–2018.

Sydenham, E.W., Shephard, G.S. and Thiel, P.G. (1992) Liquid chromatographic determination of fumonisins B_1, B_2, and B_3 in foods and feeds. *Journal of AOAC International*, **75**, 313–318.

Sydenham, E.W., Stockenström, S., Thiel, P.G., Shephard, G.S., Koch, K.R. and Marasas, W.F.O. (1995a). The potential of alkaline hydrolysis for the removal of fumonisins from contaminated corn. *Journal of Agricultural and Food Chemistry*, **43**, 1198–1201.

Sydenham, E.W., Vismer, H.F., Marasas, W.F.O., Brown, N., Schlechter, M., Van der Westhuizen, L. and Rheeder, J.P. (1995b) Reduction of patulin in apple juice samples—influence of initial processing. *Food Control*, **6**, 195–200.

Sydenham, E.W., Shephard, G.S., Thiel, P.G., Bird, C. and Miller, B.M. (1996a) Determination of fumonisins in corn: Evaluation of competitive immunoassay and HPLC techniques. *Journal of Agricultural and Food Chemistry*, **44**, 159–164.

Sydenham, E.W., Shephard, G.S., Thiel, P.G., Stockenström, S. and Van Schalkwyk, D.J. (1996b) Liquid chromatographic determination of fumonisins B_1, B_2, and B_3 in corn: IUPAC/AOAC interlaboratory collaborative study. *Journal of AOAC International* (In press).

Sydenham, E.W., Stockenström, S., Thiel, P.G., Doko, M.B. and Miller, B.M. (1996c) Polyclonal antibody-based ELISA and HPLC methods for the determination of fumonisins in corn: Comparative study. *Journal of Food Protection* (In press).

Takeda, N. (1984) Determination of aflatoxin M_1 in milk by reversed-phase high-performance liquid chromatography. *Journal of Chromatography*, **288**, 484–488.

Takeda, N., Akiyama, Y. and Shibasaki, S. (1991) Solid-phase extraction and cleanup for liquid chromatographic analysis of ochratoxin A in pig serum. *Bulletin of Environmental Contamination and Toxicology*, **47**, 198–203.

Tanaka, T., Hasegawa, A., Matsuki, Y., Ishii, K. and Ueno, Y. (1985a) Improved methodology for the simultaneous detection of the trichothecene mycotoxins deoxynivalenol and nivalenol in cereals. *Food Additives and Contaminants*, **2**, 125–137.

Tanaka, T., Hasegawa, A., Matsuki, Y., Lee, U.-S. and Ueno, Y. (1985b) Rapid and sensitive determination of zearalenone in cereals by high-performance liquid chromatography with fluorescence detection. *Journal of Chromatography*, **328**, 271–278.

Tanaka, T., Teshima, R., Ikebuchi, H., Sawada, J., Terao, T. and Ichinoe, M. (1993) Sensitive determination of zearalenone and α-zearalenol in barley and job's-tears by liquid chromatography with fluorescence detection. *Journal of AOAC International*, **76**, 1006–1009.

Tanaka, T., Teshima, R., Ikebuchi, H., Sawada, J. and Ichinoe, M. (1995) Sensitive enzyme-linked immunosorbent assay for mycotoxin zearalenone in barley and job's-tears. *Journal of Agricultural and Food Chemistry*, **43**, 946–950.

Tanner, H. and Zanier, C. (1976) Ueber eine neue Patulinbestimmung in Fruchtsäften und Konzentraten. *Schweizerische Zeitschrift für Obst- und Weinbau*, **112**, 656–662.

Tarter, E.J. and Scott, P.M. (1991) Determination of patulin by capillary gas chromatography of the heptafluorobutyrate derivative. *Journal of Chromatography*, **538**, 441–446.

Taylor, S.L., King, J.W., Richard, J.L. and Greer, J.I. (1993) Analytical-scale supercritical fluid extraction of aflatoxin B_1 from field-inoculated corn. *Journal of Agricultural and Food Chemistry*, **41**, 901–913.

Terada, H., Tsubouchi, H., Yamamoto, K., Hisada, K. and Sakabe, Y. (1986) Liquid chromatographic determination of ochratoxin A in coffee beans and coffee products. *Journal of the Association of Official Analytical Chemists*, **69**, 960–964.

Teshima, R., Hirai, K., Sato, M., Ikebuchi, H., Ichinoe, M. and Terao, T. (1990a) Radio-immunoassay of nivalenol in barley. *Applied and Environmental Microbiology*, **56**, 764–768.

Teshima, R., Kawase, M., Tanaka, T., Hirai, K., Sato, M., Sawada, J., Ikebuchi, H., Ichinoe, M. and Terao, T. (1990b) Production and characterization of a specific monoclonal antibody against mycotoxin zearalenone. *Journal of Agricultural and Food Chemistry*, **38**, 1618–1622.

Thakur, R.A. and Smith, J.S. (1994) Analysis of fumonisin B_1 by negative-ion thermospray mass spectrometry. *Rapid Communications in Mass Spectrometry*, **8**, 82–88.

Thiel, P.G. (1986) HPLC determination of aflatoxins and mammalian aflatoxin metabolites. In *Mycotoxins and Phycotoxins: A collection of invited papers presented at the Sixth International IUPAC Symposium on Mycotoxins and Phycotoxins*, edited by P.S. Steyn and R. Vleggaar, Elsevier: Amsterdam, pp. 329–340.

Thiel, P.G., Stockenström, S. and Gathercole, P.S. (1986) Aflatoxin analysis by reverse phase HPLC using post-column derivatization for enhancement of fluorescence. *Journal of Liquid Chromatography*, **9**, 103–112.

Thiel, P.G., Sydenham, E.W., Shephard, G.S. and Van Schalkwyk, D.J. (1993) Study of the reproducibility characteristics of a liquid chromatographic method for the determination of fumonisins B_1 and B_2 in corn: IUPAC collaborative study. *Journal of AOAC International*, **76**, 361–366.

Tomlins, K.I., Jewers, K. and Coker, R.D. (1989a) Evaluation of non-polar bonded-phases for the clean-up of maize extracts prior to aflatoxin assay by HPTLC. *Chromatographia*, **27**, 67–70.

Tomlins, K.I., Jewers, K., Coker, R.D. and Nagler, M.J. (1989b) A bi-directional HPTLC development method for the detection of low levels of aflatoxin in maize extracts. *Chromatographia*, **27**, 49–52.

Tosch, D., Waltking, A.E. and Schlesier, J.F. (1984) Comparison of liquid chromatography and high performance thin layer chromatography for determination of aflatoxin in peanut products. *Journal of the Association of Official Analytical Chemists*, **67**, 337–339.

Toyoda, M., Saisho, K., Aoki, G., Kobayashi, A., Saito, Y. and Martinez, M.U. (1994) Repeated use of a single immunoaffinity column for sample clean-up in the HPLC determination of aflatoxin M_1 in powdered milk. *International Dairy Journal*, **4**, 369–375.

Traag, W.A., Van Trijp, J.M.P., Tuinstra, L.G.M.T. and Kok, W.Th. (1987) Sample clean-up and post-column derivatization for the determination of aflatoxin B_1 in feedstuffs by liquid chromatography. *Journal of Chromatography*, **396**, 389–394.

Trenholm, H.L., Warner, R.M. and Fitzpatrick, D.W. (1984) Rapid, sensitive liquid chromatographic method for determination of zearalenone and α- and β-zearalenol in wheat. *Journal of the Association of Official Analytical Chemists*, **67**, 968–972.

Trenholm, H.L., Warner, R.M. and Prelusky, D.B. (1985) Assessment of extraction procedures in the analysis of naturally contaminated grain products for deoxynivalenol (vomitoxin). *Journal of the Association of Official Analytical Chemists*, **68**, 645–649.

Trucksess, M.W. and Stack, M.E. (1994) Enzyme-linked immunosorbent assay of total aflatoxins B_1, B_2, and G_1 in corn: Follow-up collaborative study. *Journal of AOAC International*, **77**, 655–658.

Trucksess, M.W., Brumley, W.C. and Nesheim, S. (1984) Rapid quantitation and confirmation of aflatoxins in corn and peanut butter, using a disposable silica gel column, thin layer chromatography, and gas chromatography/mass spectrometry. *Journal of the Association of Official Analytical Chemists*, **67**, 973–975.

Trucksess, M.W., Flood, M.T. and Page, S.W. (1986) Thin layer chromatographic determination of deoxynivalenol in processed grain products. *Journal of the Association of Official Analytical Chemists*, **69**, 35–36.

Trucksess, M.W., Flood, M.T., Mossoba, M.M. and Page, S.W. (1987) High-performance thin-layer chromatographic determination of deoxynivalenol, fusarenon-x, and nivalenol in barley, corn, and wheat. *Journal of Agricultural and Food Chemistry*, **35**, 445–448.

Trucksess, M.W., Stack, M.E., Nesheim, S., Park, D.L. and Pohland, A.E. (1989) Enzyme-linked immunosorbent assay of aflatoxins B_1, B_2, and G_1 in corn, cottonseed, peanuts, peanut butter, and poultry feed: Collaborative study. *Journal of the Association of Official Analytical Chemists*, **72**, 957–962.

Trucksess, M.W., Young, K., Donahue, K.F., Morris, D.K. and Lewis, E. (1990) Comparison of two immunochemical methods with thin-layer chromatographic methods for determination of aflatoxins. *Journal of the Association of Official Analytical Chemists*, **73**, 425–428.

Trucksess, M.W., Stack, M.E., Nesheim, S., Page, S.W., Albert, R.H., Hansen, T.J. and Donahue, K.F. (1991) Immunoaffinity column coupled with solution fluorometry or liquid chromatography postcolumn derivatization for determination of aflatoxins in corn, peanuts, and peanut butter: Collaborative study. *Journal of the Association of Official Analytical Chemists*, **74**, 81–88.

Trucksess, M.W., Stack, M.E., Nesheim, S., Albert, R.H. and Romer, T.R. (1994) Multifunctional column coupled with liquid chromatography for determination of aflatoxins B_1, B_2, G_1, and G_2 in corn, almonds, Brazil nuts, peanuts, and pistachio nuts: Collaborative study. *Journal of AOAC International*, **77**, 1512–1521.

Tsubouchi, H., Terada, H., Yamamoto, K., Hisada, K. and Sakabe, Y. (1988) Ochratoxin A found in commercial roast coffee. *Journal of Agricultural and Food Chemistry*, **36**, 540–542.

Tuinstra, L.G.M.Th., Kienhuis, P.G.M. and Dols, P. (1990) Automated liquid chromatographic determination of aflatoxin M_1 in milk using on-line dialysis for sample preparation. *Journal of the Association of Official Analytical Chemists*, **73**, 969–973.

Tuinstra, L.G.M.Th., Roos, A.H. and Van Trijp, J.M.P. (1993) Liquid chromatographic determination of aflatoxin M_1 in milk powder using immunoaffinity columns for clean-up: Interlaboratory study. *Journal of AOAC International*, **76**, 1248–1254.

Ueno, Y. (1983) General toxicology. In *Trichothecenes—Chemical, Biological and Toxicological Aspects*, edited by Y. Ueno, Elsevier: Amsterdam, pp. 135–146.

Ueno, Y. (1984) Toxicological features of T-2 toxin and related trichothecenes. *Fundamentals in Applied Toxicology*, **4**, S124–S132.

Usleber, E., Renz, V. Märtlbauer, E. and Terplan, G. (1992) Studies on the application of enzyme immunoassays for *Fusarium* mycotoxins deoxynivalenol, 3-acetyldeoxynivalenol, and zearalenone. *Journal of Veterinary Medicine B*, **39**, 617–627.

Usleber, E., Straka, M. and Terplan, G. (1994) Enzyme immunoassay for fumonisin B_1 applied to corn-based food. *Journal of Agricultural and Food Chemistry*, **42**, 1392–1396.

Van Egmond, H.P. (1989) Current situation on regulations for mycotoxins. Overview of tolerances and status of standard methods of sampling and analysis. *Food Additives and Contaminants*, **6**, 139–188.

Van Egmond, H.P. and Wagstaffe, P.J. (1987) Development of milk powder reference materials certified for aflatoxin M_1 content (part I). *Journal of the Association of Official Analytical Chemists*, **70**, 605–610.

Van Egmond, H.P. and Wagstaffe, P.J. (1990) Aflatoxin B_1 in compound-feed reference materials: An intercomparison of methods. *Food Additives and Contaminants*, **7**, 239–251.

Van Egmond, H.P., Heisterkamp, S.H. and Paulsch, W.E. (1991) EC-collaborative study on the determination of aflatoxin B_1 in animal feeding stuffs. *Food Additives and Contaminants*, **8**, 17–29.

Voyksner, R.D., Hagler, W.M. Jr. and Swanson, S.P. (1987) Analysis of some metabolites of T-2 toxin, diacetoxyscirpenol and deoxynivalenol by thermospray high-performance liquid chromatography–mass spectrometry. *Journal of Chromatography*, **394**, 183–199.

Wang, C.-R. and Chu, F.S. (1991) Production and characterization of antibodies against nivalenol tetraacetate. *Applied and Environmental Microbiology*, **57**, 1026–1030.

Wannemacher, R.W. Jr., Bunner, D.L. and Neufeld, H.A. (1991) Toxicity of trichothecenes and other related mycotoxins in laboratory animals. In *Mycotoxins and Animal Foods*, edited by J.E. Smith and R.S. Henderson, CRC Press: Boca Raton, Florida, pp. 499–552.

Ware, G.M., Francis, O.J., Kuan, S.S. and Carmen, A.S. (1989) Determination of zearalenol and zearalenone using electrochemical detection. *Analytical Letters*, **22**, 2335–2352.

Ware, G.M., Francis, O., Kuan, S.S., Umrigar, P., Carman, A., Carter, L. and Bennett, G.A. (1993) Determination of fumonisin B_1 in corn by high performance liquid chromatography with fluorescence detection. *Analytical Letters*, **26**, 1751–1770.

Ware, G.M., Umrigar, P.P., Carman, A.S. Jr. and Kuan, S.S. (1994) Evaluation of fumonitest immunoaffinity columns. *Analytical Letters*, **27**, 693–715.

Warner, R., Ram, B.P., Hart, L.P. and Pestka, J.J. (1986) Screening for zearalenone in corn by competitive direct enzyme-linked immunosorbent assay. *Journal of Agricultural and Food Chemistry*, **34**, 714–717.

Wei, R.-D. and Chu, F.S. (1987) Production and characterization of a generic antibody against group A trichothecenes. *Analytical Biochemistry*, **160**, 399–408.

Weibking, T.S., Ledoux, D.R., Brown, T.P. and Rottinghaus, G.E. (1993) Fumonisin toxicity in turkey poults. *Journal of Veterinary Diagnostic Investigation*, **5**, 75–83.

Whitaker, T.B., Dickens, J.W. and Giesbrecht, F.G. (1986) Optimum methanol concentration and solvent/peanut ratio for extraction of aflatoxin from raw peanuts by modified AOAC method II. *Journal of the Association of Official Analytical Chemists*, **69**, 508–510.

Wilson, T.J. and Romer, T.R. (1991) Use of the Mycosep multifunctional cleanup column for liquid chromatographic determination of aflatoxins in agricultural products. *Journal of the Association of Official Analytical Chemists*, **74**, 951–956.

Wilson, T.M., Ross, P.F., Rice, L.G., Osweiler, G.D., Nelson, H.A., Owens, D.L., Plattner, R.D., Reggiardo, C., Noon, T.H. and Pickrell, J.W. (1990) Fumonisin B_1 levels associated with an epizootic of equine leukoencephalomalacia. *Journal of Veterinary Diagnostic Investigation*, **2**, 213–216.

Wreford, B.J. and Shaw, K.J. (1987) Analysis of deoxynivalenol as its trifluoroacetyl ester by gas chromatography–electron ionization mass spectrometry. *Food Additives and Contaminants*, **5**, 141–147.

Xu, Y.-C., Zhang, G.-S. and Chu, F.S. (1986) Radioimmunoassay of deoxynivalenol in wheat and corn. *Journal of the Association of Official Analytical Chemists*, **69**, 967–969.

Xu, Y.-C., Zhang, G.-S. and Chu, F.S. (1988) Enzyme-linked immunosorbent assay for deoxynivalenol in corn and wheat. *Journal of the Association of Official Analytical Chemists*, **71**, 945–949.

Yagen, B., Sintov, A. and Bialer, M. (1986) New, sensitive thin-layer chromatographic–high-performance liquid chromatographic method for detection of trichothecene mycotoxins. *Journal of Chromatography*, **356**, 195–201.

Young, J.C. and Games, D.E. (1992) Supercritical fluid chromatography of *Fusarium* mycotoxins. *Journal of Chromatography*, **627**, 247–254.

Young, J.C. and Games, D.E. (1993) Analysis of *Fusarium* mycotoxins by supercritical fluid chromatography with ultraviolet or mass spectrometric detection. *Journal of Chromatography*, **653**, 374–379.

Young, J.C. and Lafontaine, P. (1993) Detection and characterization of fumonisin mycotoxins as their methyl esters by liquid chromatography/particle-beam mass spectrometry. *Rapid Communications in Mass Spectrometry*, **7**, 352–359.

Zhang, G.-S., Li, S.W. and Chu, F.S. (1986) Production and characterization of antibody against deoxynivalenol triacetate. *Journal of Food Protection*, **49**, 336–339.

4 Inductively coupled plasma-mass spectrometry (ICP-MS) for the analysis of trace element contaminants in foods

HELEN M. CREWS

Summary

This chapter describes the inductively coupled plasma, its interface to the mass spectrometer and the advantages and disadvantages of inductively coupled plasma-mass spectrometry (ICP-MS). Examples are given of applications in food analysis for single- and multi-element determinations, isotopic measurements and speciation studies. The applications are not restricted to contaminating elements such as arsenic, cadmium, mercury and lead, since the multi-element nature of the technique enables the analyst to measure a range of elements in a sample, including both toxic and nutritional elements, which may or may not be present as contaminants. Future prospects and requirements are considered.

Abbreviations

AC	alternating current
AES	atomic emission spectrometry
CE	capillary electrophoresis
cps	counts per second
DC	direct current
DIN	direct injection nebulisation
ES-MS	electrospray mass spectrometry
ETAAS	electrothermal vaporisation atomic absorption spectrometry
FAAS	flame atomic absorption spectrometry
FAB	fast atom bombardment
FI	flow injection
GC	gas chromatography
HG	hydride generation
IC	ion chromatography
ICP-MS	inductively coupled plasma-mass spectrometry
IDA	isotope dilution analysis
IP	ionisation potential

IQC   internal quality control
LA   laser ablation
LODs   limits of detection
NAA   neutron activation analysis
NIST   National Institute for Standards and Testing
OES   optical emission spectrometry
R resolution
RF   radio frequency
RPC   reverse-phased chromatography
RSD   relative standard deviation
SEC   size exclusion chromatography
SIMS   secondary ion mass spectrometry
SRM   standard reference material
SSMS   spark source mass spectrometry
TIMS   thermal ionisation mass spectrometry
UV   ultraviolet

4.1 Introduction

> Of the commercially available inorganic mass spectrometric techniques, inductively
> coupled plasma-mass spectrometry appears to hold the greatest potential for food
> analysis but it is evident that the transition from the analysis of simple solutions to
> that of complex real samples is not straightforward and the problems of matrix
> effects and interferences need to be further investigated (Ure and Bacon, 1987).

The first analytical mass spectra from an inductively coupled plasma were
produced at the Ames Laboratory, Iowa, by Houk and co-workers in 1978
and published in 1980 (Houk *et al.*, 1980). Prior to this, Gray had demonstrated the feasibility of plasma sampling using a capillary arc (DC)
plasma for metal samples (Gray, 1975, 1978). In 1981, papers appeared
detailing the use of both microwave-induced (Douglas and French, 1981)
and inductively coupled plasmas (Houk *et al.*, 1981a,b) as sources for mass
spectrometry. Commercial inductively coupled plasma-mass spectrometers
appeared in 1983. As the above quote implies, inductively coupled plasma-
mass spectrometry (ICP-MS) was not without problems some years later
but its potential as a powerful analytical tool had been recognised.

The problems and potential for exploitation applied to all types of
analyses—geological, clinical, biological, nuclear and industrial. For food
samples in particular, the potential offered by ICP-MS was that of rapid
multi-element analysis both at endogenous levels and at higher concentrations indicative of contamination, providing a powerful tool for surveillance,
legislative and emergency work. Published work dealing with the analysis of
food and related matrices by ICP-MS is still in the minority compared with
environmental and geological applications, for example. Branch *et al.* (1991b)

noted that although ICP-MS was hailed originally as a multi-element technique, it had been used more frequently for the determination of one or two elements, for speciation studies and to measure isotopic composition for nutritional studies and for identifying sources of environmental exposure.

The major problem with the multi-element analysis of unknown samples is that a thorough evaluation of the accuracy and precision of the determination can be required for each element. In food analysis, the matrices involved can differ widely in composition. It therefore takes considerable effort to evolve methods which account for all the possible matrices and any associated interferences and include good quality assurance, while still being able to take advantage of the speed and sensitivity that ICP-MS offers. Additionally, the large amount of data produced requires careful processing and this can also be time consuming. Therefore, efficient data handling routines need to be developed in tandem with the analytical procedures.

4.2 Principles of ICP-MS

There are several textbooks and reviews which describe the principles of ICP-MS and only a relatively brief overview of some of this work is given here. Chapters by Williams (1992a,b) on instrument options and sample introduction for liquids and gases, to be found in an excellent handbook of ICP-MS (Jarvis *et al.*, 1992), are particularly recommended to the reader.

4.2.1 *The ICP ion source*

Figure 4.1 shows a schematic diagram of an ICP-mass spectrometer based on the PlasmaQuad (VG Elemental), and Figure 4.2 is a simplified version of the interface region. An induction coil surrounds a series of concentric quartz tubes (the torch) and argon gas is introduced in a tangential flow (Douglas and Houk, 1985). A high-temperature plasma is produced by the interaction of an induced magnetic field and the flow of argon gas. The plasma operates at powers of between 1 and 2 kW, usually at a frequency of 27 MHz, and is at atmospheric pressure. Temperatures in the plasma range from 5000 to 10 000K. Samples are sprayed into the plasma with a nebuliser. After the liquid sample has been aspirated, its state changes as it travels from the nebuliser to the torch exit (Hieftje and Vickers, 1989). In the nebuliser, aerosol droplets are formed and those greater than $4\,\mu$m diameter are lost in the spray chamber (normally water cooled to approximately 10°C; much lower, -15°C, if organics are being analysed). The remaining droplets are transported to the plasma via the torch capillary. In the central channel of the plasma a sequence of desolvation, volatilisation, and dissociation forms

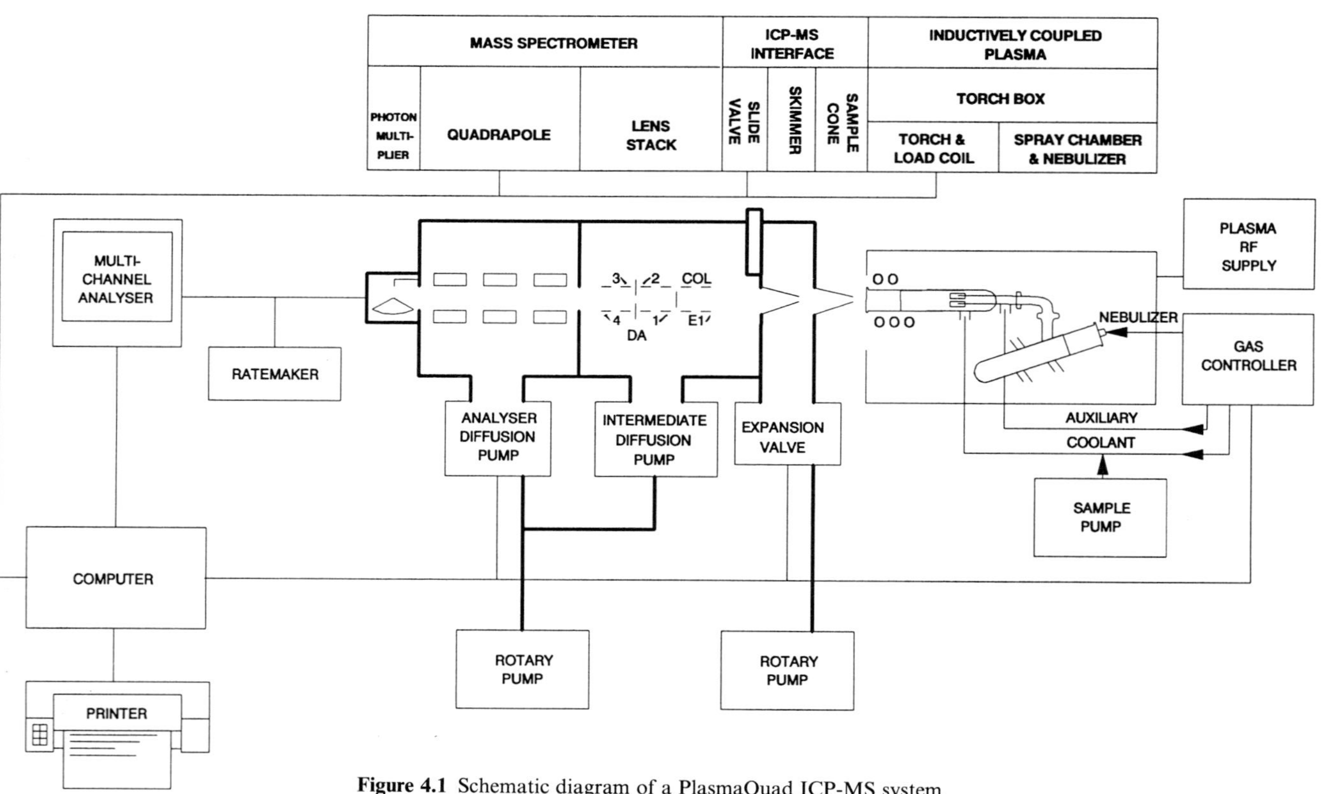

Figure 4.1 Schematic diagram of a PlasmaQuad ICP-MS system.

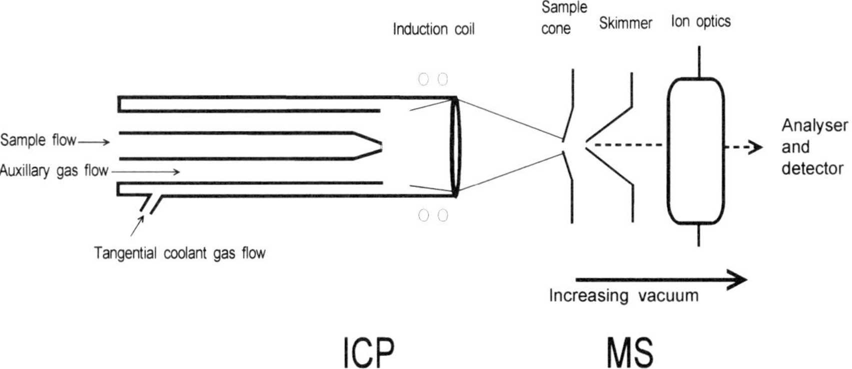

Figure 4.2 Schematic diagram of ICP-MS interface.

molecular vapour from the droplets, which in turn undergoes excitation and ionisation so that a mixture of ions and atoms leaves the torch.

The degree of ionisation of an element can be estimated from the Saha Equation:

$$\frac{M^+}{M^0} = \frac{1}{n_e} \frac{2\pi m_e kT}{h^2} \frac{3}{2} \frac{Q^+}{Q^0} e^{-IP/kT}$$

where n_e is the electron density, m_e is the mass of the electron, k is Boltzmann's constant, T is the temperature, h is Planck's constant, Q^+ and Q^0 are partition functions of the ion and the neutral atom and IP is the ionisation potential of the element. This equation assumes that the ICP is in thermal equilibrium, but there is evidence that this is not the case (Houk *et al.*, 1981a; Douglas and Houk, 1985) and that the degree of ionisation is higher than that calculated. The mean energy of an argon plasma is 13.6 eV, which is higher than the first ionisation potential of most elements but below the second ionisation potential of the majority of elements. Therefore, singly charged positive ions are formed for almost all of the elements, which greatly simplifies the complexity of the mass spectrum when compared to an atomic emission spectrum (Taylor, 1986).

4.2.2   *The ICP-MS interface*

The successful development of ICP-MS as a viable analytical system lay in the interface between the argon plasma and the quadrupole mass spectrometer. The former operates at atmospheric pressure whilst the latter requires an operating pressure of no more than 5×10^{-5} torr. The ions must be transported from the high-temperature gas source to the low-temperature quadrupole without cooling or perturbing the plasma. This has been achieved by the use of a water-cooled cone with a central extraction aperture of about 1 mm diameter (see Figure 4.2). A reduced pressure (approximately 1 torr) is maintained behind the sample cone by a mechanical (rotary) pump, and this ensures a flow of gas containing ions from the centre of the tail flame of the plasma into the vacuum system (Douglas and French, 1981). Gas is sampled from an area approximately three times the sample cone orifice diameter, which represents a large fraction of the total gas in the axial channel (Houk, 1986). A second cone, the skimmer, is located directly behind the sample cone and is positioned to ensure that as much of the sampled beam as possible is transmitted into the second reduced-pressure stage (see Figures 4.1 and 4.2). Also, to ensure that ions are extracted from the resulting supersonic expansion behind the sampling cone, the skimmer orifice is normally of similar size to that of the sampler orifice (Hieftje and Vickers, 1989). The area between the sampler and skimmer cones is referred to as the expansion chamber, and the region behind the skimmer, which contains the ion lenses to collect and focus the beam, is known as the intermediate stage. The vacuum behind the skimmer is maintained when the ICP is not operating by closing a slide valve between the expansion chamber and intermediate stage. The operating pressure of the intermediate stage is about 10^{-4} torr. The quadrupole is housed in a further stage which operates at 1×10^{-7} to 5×10^{-8} torr. The vacuum in these two stages is achieved by use of diffusion or turbomolecular pumps.

To be observed by the mass spectrometer, an ion must be present in the plasma and survive the extraction process or be formed during the extraction process (Houk, 1986). The interface is the crucial region in ICP-MS and its complexities are not fully understood. Much debate (Fulford and Douglas, 1986; Gray, 1986a; Zhu and Browner, 1987; Douglas and Kerr, 1988) continues as to its role in the stability of analyte signals and in particular in the production of doubly charged and oxide species. The first sample cone orifices were smaller than 1 mm and the plasma expanded into a region of 10^{-3} torr or less (Houk *et al.*, 1980; Date and Gray, 1981) causing the sampling orifice to be eroded rapidly, and clogging occurred even with low salt content solutions. Douglas and French (1988) investigated the gas dynamics of the 1 mm sampler orifice with the 1 torr expansion chamber and found that the ICP can be sampled with little or no ionic recombination. They reported that the interface region offers little chance for ion–molecule reactions to alter the sampled ion distribution. Thus, to a very good approximation, a representative sample of the ions in the plasma can be obtained. The use of the larger sampling orifice has a number of advantages from the analytical point of view. For example, it allows a greater flow rate of ions to the vacuum system and it is more resistant to plugging by solids (Houk, 1986). Sample cones are commonly made of nickel because of its reasonably good thermal conductivity, stability and machinability. Platinum-tipped cones are also available, and offer a greater degree of inertness.

4.2.3 The quadrupole

The mass analyser in ICP-MS systems to date has been a quadrupole. There are several reasons for this choice (Miller and Denton, 1986). The relatively short distance between the ion source and the detector, combined with the strong focusing properties of such devices, make the quadrupole mass analyser useful at comparatively high pressures (5×10^{-5} torr). Quadrupoles resolve ions on the basis of judicious selection of DC and RF voltages (V). By controlling the ratio V_{dc}/V_{rf}, the quadrupole field can be established so that only ions of a particular m/z^* ratio will have stable paths through the quadrupole rods. Therefore, the separation achieved results from the ion's intrinsic instability or stability within the device. This contrasts with magnetic or sector instruments, which use momentum or kinetic energy to disperse ions with different m/z ratios. Thus, for quadrupoles, unit mass resolution may be preserved even when ion populations which have wide velocity distributions are being sampled. Unlike other types of mass analysers, quadrupole devices do not rely on the use of magnetic fields for their mass discriminating properties, so that the size, weight, cost and slow scan speeds often associated with magnets are avoided.

Detailed descriptions of the principles and operating concepts of quadrupoles are given in publications by Miller and Denton (1986) and by Dawson (1986). Physically, the quadrupole mass analyser consists of a set of four electrodes (rods), ideally of hyperbolic but usually of circular cross section with opposite pairs connected together. The filtering action is obtained by application of a combination of a time-independent (DC) and a time-dependent (AC or RF) potential. If the amplitude of DC and RF voltages is properly chosen, ions of one mass are transmitted and ions of all other masses have unstable trajectories and are not transmitted. A mass scan is accomplished by varying the amplitude of the voltages while keeping the ratio of RF to DC constant (Douglas and Houk, 1985; Miller and Denton, 1986). All scanning is electronic, and the mass transmitted increases linearly with the magnitude of the voltages, resulting in a linear mass scale. Since only ions of a given m/z value have a stable path through the mass analyser at any time, ICP-MS should be described as a rapid sequential multi-element technique as opposed to a simultaneous method. However, the time spent switching between elements is minimal, of the order of 30 ms for m/z 1 to 300. The entire mass range or selected range may be scanned, or a selected number of masses may be measured by peak jumping or hopping.

The mass resolution is set electronically by the ratio of the RF to DC voltages. There is a limiting resolution R, given by $R = n^2/c$, where n is the number of RF cycles spent by ions in the RF field and c is a constant (Douglas and Houk, 1985). If ions of too high energy enter the quadrupole, the number of cycles is reduced, the resolution degrades and, in practice, the

---

$^*m/z$ = mass/charge.

peaks appear unsymmetrical and tail towards lower mass. The energy of ions at the quadrupole entrance is mainly due to the difference in DC potential between the source of the ions, the plasma, and the rod system, and this difference can be set as required to give optimum ion energy (Gray, 1989).

In addition to resolution, the abundance sensitivity is an important characteristic of quadrupole mass spectra. This is a measure of the degree of overlap between adjacent peaks, and can be expressed as the ratio of the response at the correct peak mass to that at adjacent unit mass positions (Douglas and Houk, 1985; Gray, 1989). A large numerical value for abundance sensitivity is desirable. Quadrupole peaks are rarely completely symmetrical, and the abundance sensitivity is usually worse on the low mass side. Values are given for the low mass side of 10^6 and on the high mass side of 10^8, although values one or two orders of magnitude smaller can be more common in routine use (Gray, 1989; Jarvis *et al.*, 1992).

4.2.4 *The detector*

The detector commonly used in ICP-MS instruments is a continuous dynode channel electron multiplier. These have a fast enough response, can count ion pulse rates up to 10^6 counts per second (cps), and have low background levels of less than 1 cps (Gray, 1989). However, despite good analytical performances they have significant disadvantages. The lifetime is limited to about one year under normal use, and the use of pulse-counting techniques, in which a high bias voltage is applied to the multiplier so that its gain is saturated, probably accelerates the rate of gain loss (Huang *et al.*, 1987). There is some evidence that the detector gain can be degraded temporarily by scanning the mass analyser across an intense peak, for example one exceeding 10^7 cps (Huang *et al.*, 1987; H.M. Crews and M.J. Baxter, unpublished data). This type of counting fatigue will affect the accuracy and precision of isotope ratio data, for example, if isotopes of very different concentrations are involved, and emphasises the need to acquire preliminary data to ensure optimum procedures for sample preparation and measurement. The use of channel multipliers at lower gains in a mean current mode, in conjunction with current amplifiers to give linear responses for higher ion densities (Gray, 1986b), has been studied. In this way a linear response may be obtained up to the analyte concentration at which the plasma equilibrium becomes significantly disturbed and the degree of ionisation of the analyte decreases, usually at about 10 mg/ml (Gray, 1989). With a pulse-counting detector, a linear response may be obtained over 5–6 decades of concentration, but, if mean current detection is also used at the top end, this may be extended to about 8 decades. This is the basis for the extended dynamic range facility on newer inductively coupled plasma-mass spectrometers which in 'dual' mode use both pulse-counting and analogue detection. Other types of detector are under investigation, such as scintillation devices

(Huang *et al.*, 1987), which should not deteriorate with time, discrete-dynode electron multipliers (Hieftje and Vickers, 1989), which are capable of higher counting rates and are not subject to fatigue caused by high ion flux, and Faraday cup ion current collectors which may be used to extend the upper limit of the dynamic range (Jarvis *et al.*, 1992). Data from the detector is stored initially in a multi-channel analyser and from there the data is dumped to a computer for processing. A good overview of calibration and data handling is given by Jarvis *et al.* (1992).

4.3 The advantages and disadvantages of ICP-MS

The high initial cost of ICP-MS instruments (although prices are becoming much more competitive as the number of instrument suppliers increases), and their running costs in terms of consumables and skilled staff, are prohibitive for many establishments. The longer established multi-element technique, ICP-optical emission spectrometry (OES), uses simpler and less expensive instrumentation with lower running costs, and displays better tolerance to total salt content and matrix effects when compared with ICP-MS (Barnes, 1993; Chang *et al.*, 1993). However, the advantages of ICP-MS, when compared with ICP-OES and flame and electrothermal vaporisation atomic absorption spectrometry (FAAS and ETAAS), include the selectivity and simplicity of spectra with fewer spectral interferences than ICP-OES, sensitivity and wide dynamic range, and the ability to acquire multi-element and isotopic information rapidly. The most common criticism levelled at ICP-MS, apart from its cost, is that it suffers from spectral and non-spectral interferences.

Spectral interferences take the form of (i) isobaric interferences, i.e. overlap of isotopes of the same unit mass, (ii) polyatomic species due to overlap of the analyte peak by molecular fragments present in the plasma or formed from plasma and matrix components, and (iii) doubly charged species, which appear at mass positions $m/2$.

Non-spectral interferences can be the most difficult to deal with. They can be broadly categorised as (iv) transport effects, (v) ionisation interferences, and (vi) ion sampling effects.

Brief descriptions of these types of interferences as well as the advantages of ICP-MS are given below.

4.3.1 Isobaric overlap

Indium is the only element which does not have one isotope which is free of isobaric overlap; ^{113}In, relative abundance 4.28%, is overlapped by ^{113}Cd, relative abundance 12.26%; ^{115}In, relative abundance 95.72%, is overlapped by ^{115}Sn, relative abundance 0.35%. The severity of this particular isobaric problem is not great, since ^{115}In is frequently employed as an internal standard. The contribution from ^{115}Sn is negligible unless Sn levels are excessive. Isobaric interferences and their abundances are predictable from isotope tables and can be avoided in most cases. Computer programs used for isotope selection contain the necessary information to allow selection of appropriate isotopes for individual applications. The use of an argon plasma does preclude the use of the main isotope of calcium, ^{40}Ca, but the levels of this element are usually such that the next most abundant isotope, ^{44}Ca, can be used.

4.3.2 Polyatomic interferences

When samples are introduced into the plasma by nebulisation of aqueous solutions the major plasma species are H, O and Ar. The presence of solvent acids can also produce high concentrations of N, Cl, S or P ions. These species can form polyatomic ions which are usually binary, but some may include the addition of a hydrogen atom. Tan and Horlick (1986) characterised basic background spectra for water and 5% solutions of nitric, sulphuric and hydrochloric acids. Prior to that, Douglas and Houk (1985) had presented background peaks from a deionised water blank. In both cases, only some of the isotopes in the lower mass range, 1–84 m/z, are affected. Table 4.1 summarises some of the polyatomic species that are important in the analysis of food samples. Figures 4.3 and 4.4 are examples of ICP-MS spectra obtained from the measurement of nitric acid and sulphuric acid reagent blanks (Fordham et al., 1995).

The understanding of the processes by which polyatomic species are formed is incomplete, but they appear to be greatly dependent upon system geometry (Gray, 1986b). The contribution from water can be reduced

Table 4.1 Examples of polyatomic ions associated with the presence of argon, chloride and nitrogen, which can interfere with the determination of some low mass elements

Element	Mass	Ar	Cl	N
V	51		$^{35}Cl^{16}O$ $^{37}Cl^{14}N$	$^{36}Ar^{14}NH$
Cr	52		$^{35}Cl^{16}OH$	$^{38}Ar^{14}N$
	53		$^{37}Cl^{16}O$	
Mn	55			$^{40}Ar^{15}N$ $^{40}Ar^{14}NH$
Zn	64			$^{36}Ar^{14}N^{14}N$
	67		$^{35}Cl^{16}O^{16}O$	
	68		$^{35}Cl^{16}O^{17}O$	$^{40}Ar^{14}N^{14}N$
As	75		$^{40}Ar^{35}Cl$	
Se	74		$^{37}Cl^{37}Cl$	
	76	$^{40}Ar^{36}Ar$		
	77		$^{40}Ar^{37}Cl$	
	78	$^{40}Ar^{38}Ar$		

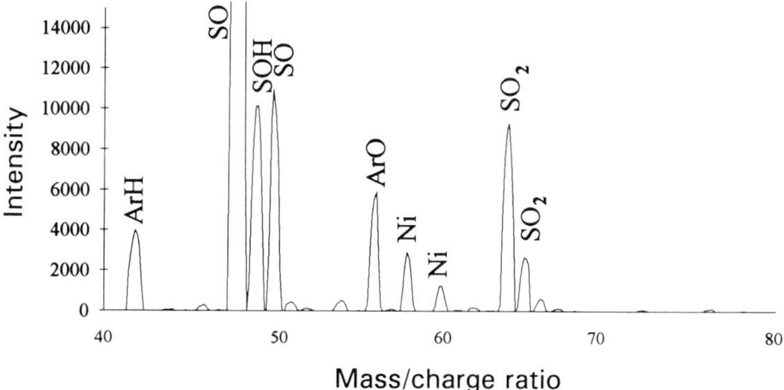

Figure 4.3 Spectrum of nitric acid/sulphuric acid reagent blank with some of the major interferences shown.

by using a cooled spray chamber, thus reducing the water loading of the plasma. When the polyatomic species are generated by reagent solutions (see Figures 4.3 and 4.4) or the plasma, and do not involve sample components, then blank subtraction may be sufficient to correct for their contribution. Alternatively, the water loading on the plasma may be reduced by, for example, multiple cryogenic desolvation (Alves *et al.*, 1992).

In other cases, for example that of $^{40}Ar^{35}Cl$ on ^{75}As, the chloride contribution may be from the sample. By judicious use of the responses on masses 77 and 82, as well as that on 75, the $^{35}Cl/^{37}Cl$ ratio may be used to calculate the chloride contribution on ^{75}As. The $^{40}Ar^{37}Cl$ contribution on ^{77}Se may be calculated from the $^{77}Se/^{82}Se$ ratio. However, the usefulness of this approach is limited by the counting statistics of the instrument, so that at low levels of arsenic this method of correction does not work. Other approaches can be used to overcome the problem. Chromatographic techniques have been

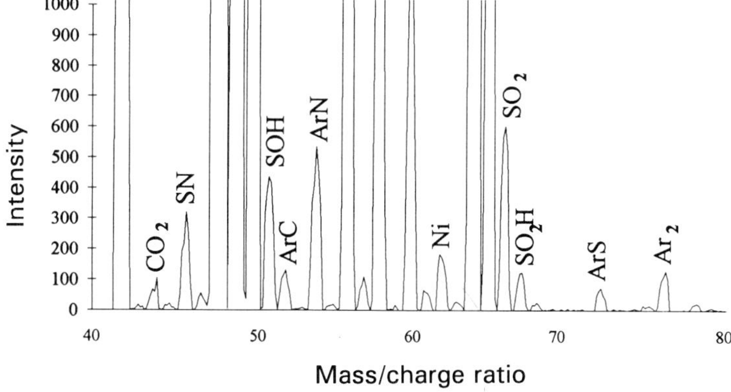

Figure 4.4 As Figure 4.3, but with the intensity scale expanded ×14, to show some of the minor interferences.

used either to separate the analyte from chloride or to remove the chloride from the analyte (Ebdon *et al.*, 1993; Dean *et al.*, 1994). Ion-exchange columns have also been used for concentration and specific elution of trace elements (Boomer *et al.*, 1990). Both arsenic- and selenium-containing species were separated by Thompson and Houk (1986), using a reverse-phase column coupled to ICP-MS. Lyon *et al.* (1988) eliminated the chloride interference from the determination of selenium in serum by de-salting samples with a Sephadex G-25 gel-filtration column. They reported that the de-salting eliminated OCl^+, $HOCl^+$ and $ArCl^+$ from the spectra. Several elements were determined in the sera, but copper and zinc were partially retained by the Sephadex gel. The use of nitrogen added to an argon plasma has been reported as reducing the polyatomic interferences in arsenic determinations, and for other elements such as vanadium and selenium which can suffer from chloride interferences (Evans and Ebdon, 1990; Branch *et al.*, 1991a; Laborda *et al.*, 1994). Cryogenic desolvation has also been used (Alves *et al.*, 1992) to reduce the chloride load in the plasma.

The levels of oxide species can be significant, particularly for the lighter elements. Elements whose bond strengths are above about 500 kJ/mol (e.g. Ba, P, Mo, Si and some of the rare earths) can show ratios of MO^+/M^+ from about 0.1% up to a few percent for the most refractory species (Gray and Williams, 1987). Their levels are, however, stable enough to be corrected for with the use of standard solutions. A preliminary scan of unknown samples will provide much information with regard to the likely polyatomic species which may be encountered and when possible should be carried out routinely.

4.3.3 *Doubly charged species*

The effect of doubly charged species is generally much less serious than that of oxides and other polyatomic species. In mineral matrices, for example, $^{138}Ba^{2+}$ can cause a problem, since barium can be present at high levels and the doubly charged species coincides with $^{69}Ga^+$. Both doubly charged ions and oxides vary in response with plasma parameters, particularly injector flow rate, but different instruments show different behaviour (Gray, 1989).

4.3.4 *Transport effects*

These effects are associated with the sample and the sample introduction system. They include viscosity effects, orifice clogging due to high dissolved solids, and nebuliser problems. The design of nebulisers and the inherent inefficiency of those in common use are areas in which there is much scope for development (Fassel, 1986; Michaud-Poussel and Mermet, 1986;

Sharp, 1988a,b; Denoyer, 1991; Smith *et al.*, 1991; Shum *et al.*, 1992a,b). Other strategies include flow injection to cope with high levels of dissolved solids or organics (Dean *et al.*, 1988; Hutton and Eaton, 1988), and electrothermal vaporisation as a means of sample introduction (Nisamaneepong *et al.*, 1985; Park and Hall, 1986; Park *et al.*, 1987a,b, Grégoire *et al.*, 1992; Williams, 1992b; Marawi *et al.*, 1994).

4.3.5 *Ionisation interferences*

Signal enhancement or suppression can result from the presence of concomitant elements in the matrix. Beauchemin *et al.* (1987) found that some elements (Na, K, Cs, Mg, Ca) induced an enhancement of the analyte signal (several elements were measured across the mass range), and others (B, Al, U) caused a suppression. Houk and co-workers (Olivares and Houk, 1986; Thompson and Houk, 1987) reported that ICP-MS was more susceptible to ionisation suppression effects than was ICP-OES, but that the effects could be compensated for by stable isotope dilution or internal standardisation. The validity of the correction by internal standardisation relies on the internal standard being suppressed to the same extent as the analyte. Therefore, care is needed in choosing the appropriate standard(s).

4.3.6 *Ion-sampling effect*

This effect has been described as the dependence of ion signal strength on sample matrix element concentration at dilution levels not normally associated with transport or ionisation interferences (Riddle *et al.*, 1988). However, ion sampling and ion optical conditions are thought to influence ionisation effects (Thompson and Houk, 1987). Work has also been reported suggesting that beam defocusing downstream from the skimmer and the effect of space charge repulsion on ion transmission through the skimmer contribute to ion sampling and matrix effect (Gillson *et al.*, 1988).

4.3.7 *Sensitivity and dynamic range*

The sensitivity and dynamic range achievable using ICP-MS are frequently cited as primary advantages of ICP-MS. More than 50 elements can be detected below the 10 pg/ml level (Barnes, 1993). In a comparison of neutron activation analysis (NAA) with ICP-MS for the analysis of 50 different foods and diets for trace elements, ICP-MS had superior sensitivity for a wide range of elements (Fardy and Warner, 1992). However, whilst NAA was found to be time-consuming and inconvenient for many applications, it did not suffer from the blank problems encountered with ICP-MS when low limits of detection (LODs) were required for trace concentrations in solid samples.

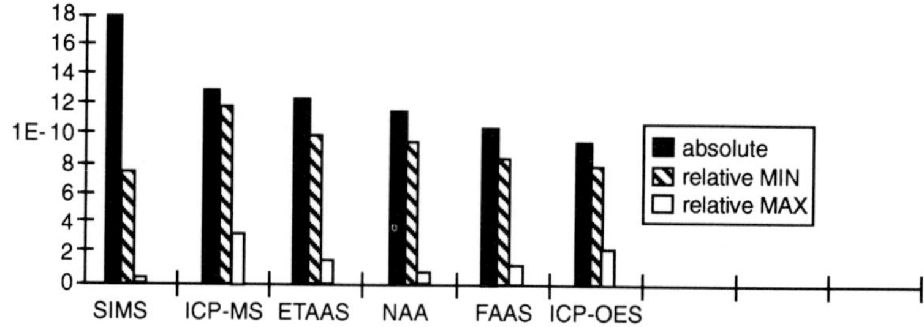

Figure 4.5 Rough estimates of absolute $(1E - g)$ limits of detection and relative concentrations $(1E - g/ml)$, adapted from Tölg (1993).

A good discussion of limits of detection (LODs) and accuracy in trace element analysis in general has been published by Kirchmer (1994). Figure 4.5 (adapted from Tölg, 1993) shows a comparison of estimated absolute LODs and relative concentrations in solution for ICP-MS with those of NAA, FAAS and ETAAS, secondary ion MS (SIMS) and ICP-OES. It is apparent that, with the exception of SIMS (absolute LOD 10^{-18} g), which cannot compete in terms of sample throughput, ICP-MS offers the best absolute LOD (estimated at 10^{-14} g). For ICP-MS, the estimated relative concentrations in solution which can be detected range from 10^{-3} to 10^{-12} g/ml. This compares with, for example, 10^{-2} to 10^{-10} g/ml for ETAAS and 10^{-3} to 10^{-8} g/ml for ICP-OES (Tölg, 1993).

In practice, the LOD obtained is strongly dependent upon the sample being analysed, the dissolution procedure and the background levels of the analytes in reagents. The LODs given in Table 4.2 were obtained in the author's laboratory after multi-element analysis of total diet samples by ICP-MS. These are complex samples representing a wide variety of food groups. Thirty-six elements were determined after fresh weight (0.5 g) sub-samples (i.e. not oven- or freeze-dried, with two exceptions discussed below) were digested with nitric acid (5 ml) in Teflon-lined stainless steel pressure bombs. The resultant digests were diluted 50-fold for sodium, magnesium, phosphorus, potassium and calcium, which were determined in one measurement run, and tenfold for the remaining elements, which were determined in a separate run. The LOD was defined as three times the standard deviation of the reagent blank corrected for sample weight and any digest dilution. These values can be reduced by an order of magnitude for those matrices and analytes which are amenable to freeze-drying. For example, the LODs quoted in Table 4.2 for cadmium and mercury in milk samples and mixed beverages were obtained by analysis of the samples after freeze-drying. The actual concentrations of analytes in the sample solutions measured by ICP-MS ranged from approximately 10^{-1} mg/kg (i.e. 10^{-7} g/ml) to 10^{-6} mg/kg (i.e. 10^{-12} g/ml), and fall within the lower range of concentrations in solution that can be detected by ICP-MS as estimated by Tölg (1993).

Table 4.2 Limits of detection, mg/kg fresh weight of food sample, following pressure-bomb digestion with nitric acid and multi-element analysis by ICP-MS

Element	Limit of detection
Lithium	0.02
Boron	0.05
Aluminium	0.05
Chromium	0.01
Manganese	0.01
Iron	1.0
Cobalt	0.005
Nickel	0.05
Copper	0.2
Zinc	0.3
Arsenic	0.01
Selenium*	0.002
Molybdenum	0.01
Tin	0.05
Lanthanum	0.001
Cerium	0.001
Praseodymium	0.001
Neodymium	0.001
Samarium	0.001
Europium	0.001
Gadolinium	0.001
Terbium	0.001
Dysprosium	0.001
Holmium	0.001
Erbium	0.001
Thulium	0.001
Ytterbium	0.001
Lutetium	0.001
Lead	0.01
Sodium	60
Magnesium	1
Phosphorus	40
Potassium	250
Calcium	300
Cadmium	0.01 #0.001 Milk #0.001 Beverages
Mercury*	0.001 #0.0005 Milk #0.0005 Beverages

*Using HG-ICP-MS; # samples freeze dried.

4.3.8   Multi-element and multi-isotopic capability

The ability to acquire rapidly both multi-element and multi-isotopic data has revolutionised trace element analysis. It is the major advantage of ICP-MS and is directly responsible for the wide range of applications for which the technique is used. In March 1971, the design requirements for a new analytical mass spectrometer included a speed of analysis of four to six samples per

Table 4.3 Analysis of selected elements in certified reference material mussel tissue MAM2/TM by Ar plasma and N_2–Ar plasma. Reference chlorine content: 8.71%

| Nebuliser flow rate (1/min) | | 0.755 | 0.755 | Certified (μg/g) |
| Nebuliser nitrogen (%) | | 0 | 8 | |
Element	Mass	Measured (μg/g)*		
B	11	40.3 ± 0.8	40.3 ± 0.6	
Al	27	155 ± 2	165 ± 1	
V	51	11.0 ± 0.6	1.61 ± 0.00	
Cr	52	1.62 ± 0.03	1.35 ± 0.05	1.25 (0.95–1.62)
	53	49.5 ± 2.9	1.60 ± 0.01	
Mn	55	64.8 ± 0.2	67.8 ± 0.1	67.1 (60.7–75.3)
Fe	57	312 ± 7	327 ± 0.4	256.2 (229.2–268.2)
Co	59	0.905 ± 0.024	0.898 ± 0.005	0.88 (0.75–1.07)
Ni	60	1.32 ± 0.03	1.29 ± 0.05	
Cu	65	10.0 ± 0.1	7.54 ± 0.15	7.96 (7.53–8.44)
Zn	67	156 ± 1	159 ± 1	156.5 (152.8–166.7)
	68	149 ± 2	154 ± 0	
As	75	32.4 ± 1.6	13.1 ± 0.1	12.8 (11.8–14.4)
Se	77	172 ± 12	2.63 ± 0.10	2.27 (1.70–2.56)
	78	30.7 ± 0.4	3.59 ± 0.19	
	82	2.56 ± 0.05	<LOD	
Y	89	0.181 ± 0.011	0.189 ± 0.003	
Mo	95	0.968 ± 0.036	0.698 ± 0.015	
Cd	111	1.44 ± 0.03	1.40 ± 0.03	1.32 (1.16–1.54)
Sn	120	0.677 ± 0.086	0.582 ± 0.014	
La	139	0.145 ± 0.001	0.146 ± 0.007	
Ce	140	0.204 ± 0.003	0.196 ± 0.008	
Pr	141	0.027 ± 0.001	0.031 ± 0.003	
Nd	146	0.107 ± 0.002	0.093 ± 0.019	
Sm	152	0.021 ± 0.004	<LOD	
Eu	153	0.007 ± 0.001	<LOD	
Gd	157	0.022 ± 0.004	<LOD	
Tb	159	0.004 ± 0.000	<LOD	
Dy	162	0.027 ± 0.003	0.029 ± 0.006	
Ho	165	0.005 ± 0.001	0.005 ± 0.001	
Er	166	0.014 ± 0.001	0.022 ± 0.006	
Tm	169	0.002 ± 0.000	<LOD	
Yb	174	0.007 ± 0.003	<LOD	
Lu	175	0.002 ± 0.001	0.004 ± 0.000	
Hg	202	0.934 ± 0.056	0.862 ± 0.149	0.95 (0.85–1.06)
Pb	208	2.16 ± 0.02	2.13 ± 0.08	1.92 (1.53–2.50)#
Th	232	0.030 ± 0.000	0.035 ± 0.001	
U	238	0.188 ± 0.002	0.190 ± 0.007	

*Average ± standard deviation of duplicates; # reference value.

hour with a recorded m/z range of 6–238 (Jarvis *et al.*, 1992). Almost 25 years later and some 15 years after its introduction, ICP-MS permits the analyst to measure tens of samples per hour (depending upon sample concentration and data acquisition times) and, with some exceptions due to interferences, to determine elements and isotopes across the entire mass range.

Although ICP-OES allows the analyst to carry out multi-element determinations, compared with ICP-MS it has limited sensitivity for some elements.

Table 4.4 Analysis of selected elements in certified reference material lobster hepatopancreas TORT-1 by Ar plasma and N_2–Ar plasma. Certified chlorine content: 5.58%

| Nebuliser flow rate (1/min) | | 0.755 | 0.755 | Certified (μg/g) |
Nebuliser nitrogen (%)		0	8	
Element	Mass	Measured (μg/g)[*]		
B	11	5.41 ± 0.15	5.82 ± 0.06	
Al	27	27.2 ± 0.5	29.3 ± 0.6	
V	51	7.17 ± 1.74	1.44 ± 0.00	1.4 ± 0.3
Cr	52	2.81 ± 0.25	2.66 ± 0.35	2.4 ± 0.6
	53	46.0 ± 12.3	2.64 ± 0.43	
Mn	55	20.9 ± 0.3	21.7 ± 0.4	23.4 ± 1.0
Fe	57	212 ± 1	222 ± 2	186 ± 11
Co	59	0.461 ± 0.002	0.480 ± 0.014	0.42 ± 0.05
Ni	60	3.49 ± 0.54	3.55 ± 0.66	2.3 ± 0.3
Cu	65	409 ± 3	409 ± 1	439 ± 22
Zn	67	155 ± 1	160 ± 1	177 ± 10
	68	149 ± 0	157 ± 1	
As	75	38.3 ± 1.9	27.8 ± 0.1	24.6 ± 2.2
Se	77	86.0 ± 13.5	7.40 ± 0.04	6.88 ± 0.47
	78	22.2 ± 12.6	7.44 ± 0.35	
	82	6.96 ± 0.02	<LOD	
Y	89	1.57 ± 0.03	1.66 ± 0.01	
Mo	95	1.33 ± 0.06	1.29 ± 0.06	1.5 ± 0.3
Cd	111	26.0 ± 0.1	26.3 ± 0.4	26.3 ± 2.1
Sn	120	0.108 ± 0.008	0.135 ± 0.013	0.139 ± 0.011
La	139	5.15 ± 0.01	5.11 ± 0.08	
Ce	140	4.39 ± 0.04	4.32 ± 0.01	
Pr	141	0.627 ± 0.002	0.637 ± 0.010	
Nd	146	2.54 ± 0.03	2.51 ± 0.01	
Sm	152	0.333 ± 0.003	0.344 ± 0.009	
Eu	153	0.069 ± 0.005	0.079 ± 0.008	
Gd	157	0.371 ± 0.011	0.423 ± 0.016	
Tb	159	0.041 ± 0.002	0.040 ± 0.004	
Dy	162	0.190 ± 0.002	0.179 ± 0.009	
Ho	165	0.036 ± 0.001	0.030 ± 0.002	
Er	166	0.086 ± 0.001	0.097 ± 0.006	
Tm	169	0.008 ± 0.001	0.011 ± 0.002	
Yb	174	0.040 ± 0.003	0.042 ± 0.002	
Lu	175	0.005 ± 0.000	0.005 ± 0.001	
Hg	202	0.247 ± 0.021	0.196 ± 0.023	0.330 ± 0.060
Pb	208	9.25 ± 0.41	9.58 ± 0.31	10.4 ± 2.0
Th	232	0.007 ± 0.000	0.009 ± 0.002	
U	238	0.098 ± 0.006	0.091 ± 0.002	

[*]Average ± standard deviation of duplicates.

Several publications have reported studies assessing the use of ICP-MS for multi-element determinations in foods, using reference materials as test matrices (Boorn *et al.*, 1985; Hutton, 1986; Munro *et al.*, 1986; Satzger, 1988; Dalgarno *et al.*, 1989; Dean *et al.*, 1989; Vanhoe, 1993). In general, fair agreement was reported between measured and certified levels. The advantages of internal standardisation for compensating for the effects of changes in viscosity, surface tension and aerosol characteristics of standards

Table 4.5 Analysis of selected elements in certified reference material oyster tissue NIST SRM 1566a by Ar plasma and N_2–Ar plasma. Certified chlorine content: 0.829%

| Nebuliser flow rate (1/min) | | 0.755 | 0.755 | Certified (μg/g) |
Nebuliser nitrogen (%)		0	8	
Element	Mass	Measured (μg/g)[*]		
B	11	9.01 ± 0.43	9.06 ± 0.21	
Al	27	119 ± 1	127 ± 2	202.5 ± 12.5
V	51	5.51 ± 0.50	4.67 ± 0.06	4.68 ± 0.15
Cr	52	1.39 ± 0.19	1.33 ± 0.00	1.43 ± 0.46
	53	4.46 ± 1.53	1.40 ± 0.12	
Mn	55	11.4 ± 0.1	11.2 ± 0.1	12.3 ± 1.5
Fe	57	477 ± 3	477 ± 8	539 ± 15
Co	59	0.538 ± 0.014	0.547 ± 0.000	0.57 ± 0.11
Ni	60	1.85 ± 0.02	1.94 ± 0.10	2.25 ± 0.44
Cu	65	63.2 ± 0.0	60.0 ± 0.7	66.3 ± 4.3
Zn	67	798 ± 6	755 ± 9	830 ± 57
	68	760 ± 5	742 ± 9	
As	75	15.3 ± 0.5	13.3 ± 0.4	14.0 ± 1.2
Se	77	12.9 ± 3.6	1.84 ± 0.07	2.21 ± 0.24
	78	9.21 ± 11.69	2.33 ± 0.60	
	82	2.49 ± 0.30	<LOD	
Y	89	0.370 ± 0.001	0.392 ± 0.001	
Mo	95	0.411 ± 0.183	0.219 ± 0.030	
Cd	111	4.15 ± 0.03	4.12 ± 0.15	4.15 ± 0.38
Sn	120	2.25 ± 0.12	2.20 ± 0.09	
La	139	0.234 ± 0.007	0.223 ± 0.013	0.3#
Ce	140	0.312 ± 0.002	0.325 ± 0.032	0.4#
Pr	141	0.54 ± 0.002	0.060 ± 0.001	
Nd	146	0.236 ± 0.001	0.224 ± 0.021	
Sm	152	0.045 ± 0.003	0.044 ± 0.001	0.06#
Eu	153	0.011 ± 0.001	<LOD	0.01#
Gd	157	0.052 ± 0.001	0.069 ± 0.003	
Tb	159	0.009 ± 0.001	0.008 ± 0.001	0.007#
Dy	162	0.053 ± 0.000	0.063 ± 0.015	
Ho	165	0.011 ± 0.002	0.012 ± 0.003	
Er	166	0.033 ± 0.003	0.032 ± 0.015	
Tm	169	0.005 ± 0.001	0.006 ± 0.001	
Yb	174	0.029 ± 0.001	0.034 ± 0.002	
Lu	175	0.006 ± 0.002	0.007 ± 0.001	
Hg	202	<LOD	<LOD	0.0642 ± 0.0067
Pb	208	0.350 ± 0.004	0.372 ± 0.045	0.371 ± 0.014
Th	232	0.033 ± 0.002	0.035 ± 0.002	0.04#
U	238	0.122 ± 0.004	0.114 ± 0.006	0.132 ± 0.012

[*]Average $\pm$ standard deviation of duplicates; # reference value.

and samples were also reported. A preliminary comparison of ICP-MS and ICP-OES for the analysis of four biological reference materials was made by Pickford and Brown (1986). They reported that the sensitivity of ICP-MS for elements above mass 80 was clearly superior to that obtainable by ICP-OES. However, at that time the precision of ICP-MS for multi-element analysis was poor compared to ICP-OES, and it was anticipated that this would improve as better quadrupoles became available and more stable instruments

Table 4.6 Analysis of selected elements in certified reference material peach leaves NIST SRM 1547 by Ar plasma and N_2–Ar plasma. Certified chlorine content: 0.036%

Nebuliser flow rate (1/min) Nebuliser nitrogen (%)		0.755 0	0.755 8	Certified (μg/g)
Element	Mass	Measured (μg/g)[*]		
B	11	26.4 ± 1.3	28.0 ± 0.2	29 ± 2
Al	27	199 ± 3	219 ± 1	249 ± 8
V	51	0.425 ± 0.060	0.354 ± 0.012	0.37 ± 0.03
Cr	52	0.854 ± 0.004	0.885 ± 0.005	1#
	53	1.11 ± 0.16	0.865 ± 0.025	
Mn	55	89.4 ± 1.3	96.3 ± 0.9	98 ± 3
Fe	57	264 ± 2	280 ± 3	220#
Co	59	0.082 ± 0.003	0.078 ± 0.008	0.07#
Ni	60	0.733 ± 0.008	0.707 ± 0.057	0.69 ± 0.09
Cu	65	3.92 ± 0.12	3.64 ± 0.17	3.7 ± 0.4
Zn	67	19.6 ± 0.1	32.6 ± 0.6	17.9 ± 0.4
	68	18.3 ± 0.4	29.1 ± 0.3	
As	75	<LOD	<LOD	0.060 ± 0.018
Se	77	<LOD	<LOD	0.120 ± 0.009
	78	<LOD	<LOD	
	82	<LOD	<LOD	
Y	89	2.99 ± 0.03	3.20 ± 0.01	0.2#
Mo	95	0.141 ± 0.034	0.051 ± 0.003	0.060 ± 0.008
Cd	111	0.032 ± 0.014	0.020 ± 0.005	0.03
Sn	120	0.066 ± 0.004	0.071 ± 0.000	< 0.2#
La	139	9.12 ± 0.02	10.4 ± 0.2	9#
Ce	140	11.6 ± 0.1	12.7 ± 0.1	10#
Pr	141	1.81 ± 0.15	1.91 ± 0.00	
Nd	146	6.94 ± 0.03	7.83 ± 0.14	7#
Sm	152	1.15 ± 0.00	1.11 ± 0.00	1#
Eu	153	0.220 ± 0.003	0.206 ± 0.011	0.17#
Gd	157	1.11 ± 0.00	1.03 ± 0.3	
Tb	159	0.121 ± 0.000	0.108 ± 0.006	0.1#
Dy	162	0.566 ± 0.008	0.493 ± 0.023	
Ho	165	0.087 ± 0.002	0.093 ± 0.006	
Er	166	0.226 ± 0.002	0.245 ± 0.005	
Tm	169	0.025 ± 0.001	0.028 ± 0.002	
Yb	174	0.133 ± 0.001	0.132 ± 0.002	
Lu	175	0.019 ± 0.002	0.020 ± 0.001	
Hg	202	<LOD	<LOD	0.031 ± 0.007
Pb	208	0.829 ± 0.003	0.855 ± 0.026	0.87 ± 0.03
Th	232	0.045 ± 0.002	0.044 ± 0.003	0.05#
U	238	0.010 ± 0.000	0.011 ± 0.001	0.015#

[*]Average $\pm$ standard deviation of duplicates; # reference value.

evolved. Since then, instruments have become more stable, and accurate and precise multi-element analysis of certified reference materials is possible.

For example, the data in Tables 4.3 to 4.7 were obtained using standard ICP-MS operating conditions following nitric acid digestion in pressure bombs of five certified reference materials (Laborda *et al.*, 1994). The results were obtained from one analytical batch of reference materials digested in duplicate. Diluted digests (5% nitric acid) were measured. For information,

Table 4.7 Analysis of selected elements in standard reference material mixed diet NIST SRM 8431a by Ar plasma and N_2–Ar plasma. Chlorine content unknown

| Nebuliser flow rate (1/min) | | 0.755 | 0.755 | Recommended (μg/g) |
Nebuliser nitrogen (%)		0	8	value
Element	Mass	Measured (μg/g)[*]		
B	11	3.94 ± 0.52	4.14 ± 0.38	
Al	27	4.82 ± 0.05	5.18 ± 0.09	4.39 ± 1.07
V	51	0.333 ± 0.192	0.053 ± 0.001	
Cr	52	0.161 ± 0.041	0.091 ± 0.002	0.102 ± 0.006
	53	1.07 ± 0.65	0.090 ± 0.023	
Mn	55	7.86 ± 0.02	7.86 ± 0.14	8.12 ± 0.31
Fe	57	41.4 ± 3.2	41.8 ± 0.6	37.0 ± 2.6
Co	59	0.043 ± 0.001	0.046 ± 0.002	0.038 ± 0.008
Ni	60	0.659 ± 0.060	0.716 ± 0.020	0.644 ± 0.151
Cu	65	3.43 ± 0.57	3.23 ± 0.02	3.36 ± 0.33
Zn	67	15.1 ± 0.8	15.1 ± 0.2	17.0 ± 0.6
	68	14.8 ± 0.1	15.0 ± 0.3	
As	75	1.85 ± 0.31	0.992 ± 0.028	0.924 ± 0.344
Se	77	4.52 ± 1.87	0.231 ± 0.042	0.242 ± 0.030
	78	1.22 ± 1.56	0.908 ± 0.089	
	82	0.432 ± 0.182	<LOD	
Y	89	0.005 ± 0.001	0.004 ± 0.001	
Mo	95	0.305 ± 0.028	0.298 ± 0.040	0.288 ± 0.029
Cd	111	0.034 ± 0.018	0.041 ± 0.004	0.042 ± 0.011
La	139	0.010 ± 0.000	0.013 ± 0.002	
Ce	140	0.009 ± 0.002	0.010 ± 0.001	
Pr	141	<LOD	<LOD	
Nd	146	<LOD	<LOD	
Sm	152	<LOD	<LOD	
Eu	153	<LOD	<LOD	
Gd	157	<LOD	<LOD	
Tb	159	<LOD	<LOD	
Dy	162	<LOD	<LOD	
Ho	165	0.002 ± 0.001	<LOD	
Er	166	<LOD	<LOD	
Tm	169	<LOD	<LOD	
Yb	174	<LOD	<LOD	
Lu	175	<LOD	<LOD	
Hg	202	<LOD	<LOD	
Pb	208	0.136 ± 0.044	0.170 ± 0.024	
Th	232	<LOD	<LOD	
U	238	0.006 ± 0.001	<LOD	

[*]Average ± standard deviation of duplicates.

values are also given for non-certified elements and for all elements when 8% nitrogen was added to the aerosol carrier (nebuliser) gas flow to reduce polyatomic interferences. This improved the measurement of vanadium ([51]V), chromium ([53]Cr), zinc ([67]Zn and [68]Zn, except for NIST SRM Peach leaves), arsenic ([75]As) and selenium ([77]Se). The LODs for this work are given in Table 4.8 and standard instrument conditions in Table 4.9. The increase in LODs when nitrogen was added was not so great as to cause any problems with the multi-element determinations.

Table 4.8 Limits of detection (3σ) for selected elements in an Ar plasma and N_2–Ar plasma; matrix: 5% HHO_3

| Nebuliser flow rate (1/min) | | 0.755 | 0.755 | 0.755 | 0.755 |
| Nubuliser nitrogen (%) | | 0 | 8 | 0 | 8 |
Element	Mass	Solution LOD (ng/ml)[*]		Sample LOD (μg/g)*#	
Li	7	0.8	20	0.2	4
Be	9	0.1	2	0.02	0.4
B	11	0.6	6	0.1	1.2
Mg	24	0.3	0.6	0.06	0.1
Al	27	0.2	1.1	0.04	0.2
V	51	0.1	0.2	0.02	0.04
Cr	52	0.3	0.2	0.05	0.04
	53	0.3	0.5	0.05	0.1
Mn	55	0.07	0.7	0.01	0.1
Fe	57	7	23	1	4
Co	59	0.05	0.1	0.01	0.03
Ni	60	0.2	3	0.04	0.6
Cu	63	0.1	0.4	0.02	0.08
	65	0.3	0.6	0.04	0.1
Zn	64	0.2	0.5	0.06	0.06
	66	0.2	0.3	0.05	0.8
	68	0.6	5	0.1	1
As	75	0.4	0.3	0.07	0.06
Se	74	37	49	7	10
	76	120	66	24	13
	77	3	4	0.7	0.9
	78	19	9	4	2
	82	2	32	0.5	6
Br	81	0.3	1.5	0.07	0.3
Y	89	0.02	0.04	0.005	0.007
Mo	95	0.1	0.1	0.03	0.02
Cd	111	0.2	0.2	0.04	0.03
Sn	120	0.2	0.2	0.04	0.06
La	139	0.02	0.03	0.005	0.006
Ce	140	0.04	0.04	0.008	0.008
Pr	141	0.01	0.03	0.003	0.005
Nd	146	0.1	0.1	0.02	0.03
Sm	152	0.06	0.1	0.01	0.02
Eu	153	0.02	0.06	0.004	0.01
Gd	157	0.09	0.2	0.02	0.04
Tb	159	0.01	0.02	0.002	0.004
Dy	162	0.05	0.05	0.01	0.01
Ho	165	0.01	0.03	0.002	0.005
Er	166	0.03	0.05	0.006	0.01
Tm	169	0.01	0.03	0.002	0.006
Yb	174	0.02	0.1	0.005	0.02
Lu	175	0.01	0.01	0.002	0.004
Hg	200	0.3	1	0.05	0.2
	202	0.3	1	0.06	0.2
Pb	206	0.1	0.3	0.02	0.05
	207	0.07	0.5	0.01	0.09
	208	0.03	0.2	0.007	0.04
Th	232	0.02	0.02	0.003	0.004
U	238	0.01	0.02	0.002	0.006

[*]$n = 10$; # dilution 1 : 200.

Table 4.9 Default instrumental ICP-MS conditions

Instrumental parameters	
VG Elemental PlasmaQuad (PQ1)	
R.f. power	1350 W
Reflected power	<7 W
Ar gas flows	
outer	14 l/min
intermediate	0.5 l/min
nebuliser	0.755 l/min
Sample uptake rate	0.70 ml/min
Spray chamber temperature	10°C
Sample cone	1.00 mm Nicone
Skimmer	0.75 mm Nicone

Measurement parameters	
Measuring mode	Scanning
Mass range	6.02 to 239.54 u
Number of channels	2048
Number of scan sweeps	100
Dwell time	320 μs
Points per peak	5
Collector type	Pulse

Isotopic determinations for isotope ratio measurements and isotope dilution analysis, prior to ICP-MS, were confined to mass spectrometric methods involving, for example, thermal ionisation (TIMS), secondary ionisation (SIMS), fast atom bombardment (FAB-MS) and spark sources (SSMS). However, all of these techniques require specialist sample pretreatment and sample throughput is slow. In contrast, sample throughput for ICP-MS is rapid and several isotopes can be determined after simple acid dissolution. The major criticism of ICP-MS for isotope ratio determinations is that it is less precise than the conventional MS methods. This is due to a combination of factors including sensitivity and counting statistics, dead time (the interval during which the detector and its associated counting electronics are unable to resolve successive pulses, resulting in a non-linear response above count rates typically greater than about 1×10^6 counts per second), mass bias (when the observed or measured isotope ratio differs from the true value as a function of the difference in mass between the two isotopes), resolution (directly related to the quadrupole), and abundance sensitivity (Jarvis *et al.*, 1992).

Other factors such as noise from the plasma and electronics can lead to precisions which are poorer than those which would be predicted from the counting statistics alone, and this subject has recently been studied (Jarvis *et al.*, 1992; Begley and Sharp, 1994). Begley and Sharp (1994) found that whilst an educated selection of acquisition parameters can give rise to a practical precision of about 0.05% for ^{107}Ag/^{109}Ag, for example, the precision is limited by inaccuracies associated with the operation of the quadrupole mass analyser and the statistical error arising from the random arrival of ions at the detector.

A comparison of ICP-MS with TIMS and FAB-MS was made (Eagles *et al.*, 1989) and good agreement was found for zinc isotope ratios (^{64}Zn/^{67}Zn), in both accuracy and precision, between the three techniques. Mean %RSDs were 0.6 (FAB-MS), 0.5 (TIMS), and 0.3 for ICP-MS with continuous nebulisation. If flow injected samples were used for ICP-MS, the precision was only 5.9% RSD, although the accuracy of the zinc ratio was very similar by all methods.

4.4 Sample preparation

For all trace analytical techniques, the prevention of contamination of samples, reagents and apparatus with the analytes of interest or potential interferents is paramount. This is particularly important for determinations using a sensitive technique like ICP-MS and where the analytes can be ubiquitous in the environment, e.g. aluminium, lead and zinc. Therefore the use of clean-room facilities is strongly recommended, and wherever possible the purest reagents available should be used. In addition, all glass and plastic ware should be appropriately cleaned in, for example, dilute nitric acid followed by rinsing with purified water. The instrument itself can contribute to contamination either by analyte build-up or retention in transfer tubes, nebulisers and spray chambers. Whenever new elements are to be measured, the efficiency of washout systems should be tested in advance of the analyses. Some very good, recent accounts of methods for sampling and sample preparation (Woittiez and Sloof, 1994), of systematic errors in trace analysis (Tölg and Tschöpel, 1994) and sample preparation for ICP-MS (Jarvis, 1992) are available, and the general aspects of sample preparation will not be detailed here. However, the following sections will give examples of food analysis by ICP-MS, and where appropriate special precautions that were taken will be noted.

4.5 Total analyte determinations

4.5.1 Minimum sample preparation and isotope dilution analysis

In an ideal world, the best LODs and fastest sample throughputs could be obtained if all samples could be analysed directly with little or no sample preparation. In practice this is rarely possible because few samples consist of the analyte only and no matrix! In this laboratory we have recently performed multi-element determinations on wine samples by ICP-MS following simple dilution with water and flow injection into a stream of 5% ethanol and 0.5% nitric acid. Standards were matrix matched by adding ethanol. One of the earliest papers reporting the use of ICP-MS for food analysis was that by Al-Swaiden (1988). Iron, tin and lead were determined by direct aspiration of homogenised orange, apple and pineapple juice samples after simple dilution with water. Calibration was by standard additions. Recoveries of added analyte were around 100%. Concentrations in the samples were 4.49 to 8.19 mg/kg for iron, 5.16 to 200 mg/kg for tin and 0.203 to 0.332 mg/kg for lead.

Dean *et al.* (1987a) used slurries of dried milk powders when measuring lead isotope ratios. Milk powder (5.0 g) was mixed with 0.1% Triton-X as dispersant and 200 ml water. The pH was adjusted to 7.4 with ammonia solution to prevent precipitation of the protein. The samples were spiked with enriched ^{202}Pb (5 ng/ml) and the total lead content was determined by isotope dilution analysis (IDA). Correction for mass bias was achieved by using NIST SRM 981 lead standard, and all results were blank corrected. The use of IDA permitted accurate data to be obtained with an RSD of 4.2% for milk powder containing less than 30 ng/g lead. The same method for slurries was used by Baxter *et al.* (1991) to measure aluminium in infant formulae by ICP-MS. This provided an independent check on aluminium data obtained by pressure digestion with nitric acid and measurement by ETAAS.

Isotope dilution analysis was also used to establish the concentration of lead in 'in-house' reference materials (Baxter *et al.*, 1989). Sub-samples of acid digests of NIST SRM 1573 Tomato leaves, of NIST SRM 1568 Rice flour, and of 'in-house' reference materials cabbage, Brussels sprouts, lettuce and curly kale were fortified with enriched ^{206}Pb (NIST SRM 983) and the ^{208}Pb/^{206}Pb ratio was measured. The results obtained by wet ashing and IDA were compared with those obtained by dry ashing and measurement by ETAAS, and are given in Table 4.10.

The data in Table 4.10 tell us two things. Firstly, they tell us how good ETAAS is for measuring total lead concentrations. If the only application

Table 4.10 Comparison of lead data obtained by dry ashing with measurement by ETAAS with that obtained by ICP-MS and IDA after wet ashing

Sample	ETAAS (mean mg/kg)	ICP-MS + IDA (mean mg/kg)	Certified mean (mg/kg) (RM = ref. material)
NIST Tomato leaves	6.2 ± 0.5	5.9 ± 0.5	6.3 ± 0.3
NIST Rice flour	0.043 ± 0.004	0.038 ± 0.005	0.045 ± 0.010
Cabbage	0.0031 ± 0.0006	0.0027 ± 0.0002	In-house RM
Brussels sprouts	0.0187 ± 0.0017	0.0222 ± 0.0015	In-house RM
Lettuce	0.0596 ± 0.0093	0.0586 ± 0.0007	In-house RM
Curly kale	0.477 ± 0.027	0.487 ± 0.014	In-house RM

required is the determination of total lead then ETAAS is an excellent technique. Secondly, they tell us how useful it is to be able to measure isotope ratios for IDA as a confirmatory method, or as a way of measuring very low concentrations of multi-isotopic elements.

4.5.2 Measurement by ICP-MS after digestion of foods and related samples with acid(s)

Negretti de Brätter *et al.* (1995) have compared microwave-based digestion and pressurised ashing systems using different acid mixtures for the determination of mineral and trace elements in total diet samples. They were measuring samples by ICP-OES, but the results are of equal interest for ICP-MS users. With regard to the influence of the acid mixture on sensitivity, they concluded that the best results were obtained from digestion with nitric acid alone or with nitric acid and hydrogen peroxide (2:1) for both pressurised ashing and microwave-based digestion.

Microwave dissolution (for preliminary breakdown of samples) combined with an open-beaker digestion using nitric acid and hydrogen peroxide has been reported (Friel *et al.*, 1990) for the determination of 24 elements in some reference materials (oyster, liver and animal muscle) by ICP-MS. Total sample preparation time was 2 h per sample. An on-line five-element internal standard was used for correction of matrix and instrument drifts. Linear extrapolation between the internal standards provided effective correction.

Compounded animal feed and imported gluten pellets, as well as milk, some dairy products and various animal tissues and fluids were subject to microwave dissolution with nitric acid following an incident in which animal feed became contaminated with lead (Crews *et al.*, 1992). The use of microwave digestion (12 digests per batch) permitted rapid sample throughput using a microwave power programme which lasted 20 min. Lead, cadmium, magnesium, zinc, copper, iron, cobalt and manganese were determined in dilute nitric acid digests. (Arsenic was measured by hydride generation–AAS and nickel and chromium by ETAAS.) Good agreement with the certified values for these elements in the reference material TORT-1 (lobster hepatopancreas) was obtained.

Figure 4.6 compares some multi-element data for contaminated and non-contaminated feed. It is apparent that lead was the major contaminant, with zinc present at elevated concentrations and the other elements at slightly above normal levels (Crews *et al.*, 1992). In Figure 4.7, lead and cadmium data for a control and test cow fed contaminated feed are given. Whilst elevated levels are found in the offal of the test animal, by far the greatest sink for lead was bone. The lead value for the humerus bone was actually off scale and was approximately 20 mg/kg, well above the statutory maximum of 1 mg/kg. This had implications for meat on the bone and resulted

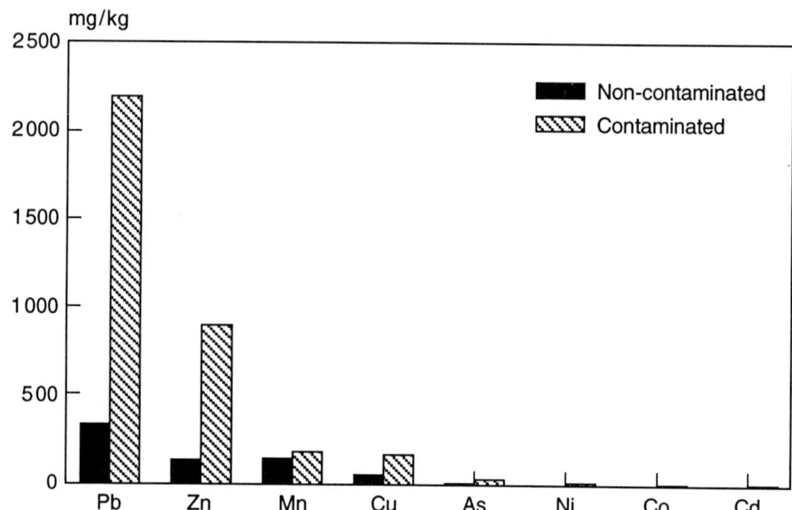

Figure 4.6 Multi-element data for contaminated and non-contaminated feed.

in further work on the leaching of lead from bone during cooking (Baxter *et al.*, 1992).

The multi-element analysis of certified reference materials (Tables 4.3–4.7) and the LODs (Table 4.2) obtained during the analysis of total diet samples for 36 elements were discussed earlier. This information was obtained from samples which were digested with nitric acid only, in stainless steel, Teflon-lined pressure bombs with indium as internal standard. The use of a single pure acid (in this case nitric acid) for sample digestion is currently one of the best ways of achieving good sample solutions for measurement by ICP-MS across the mass range with the minimum of interferences. However, for the heavier elements, where polyatomic interferences are not such a problem, mixed acids have been used to good effect.

Shiraishi *et al.* (1990, 1991) have analysed duplicate diet samples for trace and ultratrace levels of the elements barium, molybdenum, nickel, cobalt, cadmium, caesium, thallium, lead, bismuth, thorium and uranium. They dried and then dry-ashed samples in a muffle furnace prior to digesting the ash with nitric acid (prepared in a sub-boiling still) and ultra-pure perchloric acid in a microwave oven. The sample solutions were then taken to dryness on a hot plate before being dissolved in nitric acid. Diluted nitric acid digests were measured by ICP-MS with rhenium as internal standard. Good agreement with reference materials was reported and the LODs (defined as 3× the square of the error counts obtained for the blank solution) for thorium and uranium were 1.3 and 2.9 pg/ml respectively (Shiraishi *et al.*, 1991).

Wet ashing with nitric acid and hydrogen peroxide was used by Muto *et al.* (1994) to prepare freeze-dried food samples for measurement by ICP-MS.

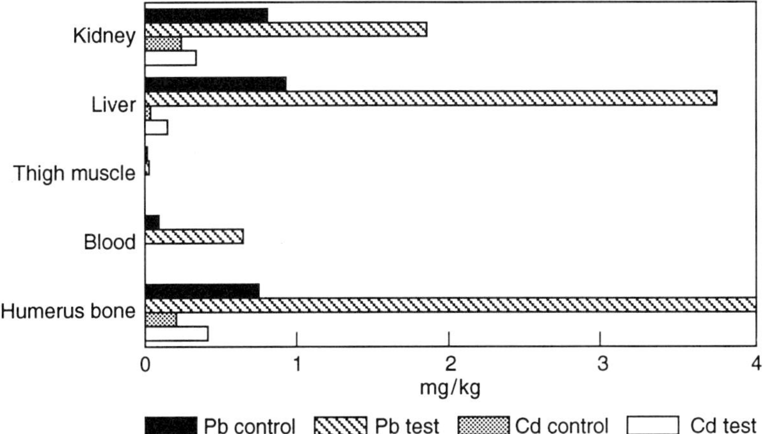

Figure 4.7 Pb and Cd data for a control and a test cow after ingestion of contaminated feed.

Thirteen elements were determined with indium as internal standard. No LODs were reported, but the investigators stated that the simplicity, adequate sensitivity and accuracy of ICP-MS made it suitable for this type of work.

Some elements are particularly difficult to measure in biological matrices. Cox *et al.* (1992) developed a method for measuring iodine as ^{129}I in vegetable samples by ICP-MS. Previous methods of analysis included NAA, which includes lengthy pre- and post-irradiation procedures with labour-intensive separation chemistry. Prior to measurement by ICP-MS, cabbage samples were homogenised, dried and then ashed for 6 h. The ash was refluxed with sulphuric acid for 30 min and iodine trapped as iodide in a U-tube containing $NaHSO_3$. The ^{129}I was introduced into the plasma as iodine, using $NaNO_2$ as oxidant. The LOD for the cabbage on a fresh weight basis was 1.4 ng/kg. Although ICP-MS cannot match the sensitivity achievable by NAA (an LOD of $8.6 \times 10^{-8} \mu g$ of ^{129}I was given as an example by Cox *et al.*, 1992), ICP-MS offers speed and simplicity, and thus substantially less effort is required.

Of direct relevance to the issue of contaminants in food is the use of food packaging materials which may contain trace elements. Polymers may contain low levels of the chemicals used in their manufacture or substances derived from them, such as catalyst residues. For polymers intended for food contact use, such as packaging, these substances may migrate (transfer) from the polymer into food, so that there is a need to identify possible food contaminants. The greater proportion of proposed limits for residual elements in food contact polymers are in the form of migration limits (EEC, 1992), but there is clearly a need to identify and determine the levels of

Table 4.11 Comparison of LA-ICP-MS data for antimony and cobalt in PET I with that for ICP-MS and NAA, obtained using PET polymer reference material as calibrant

	Semi-quantitative ICP-MS	Quantitative ICP-MS	NAA	LA-ICP-MS
Co	58	42	36	33
Sb	158	168	136	166

trace elements present in the polymer in order to target substances for later migration work. Using microwave digestion with nitric and sulphuric acids, trace elements were determined by ICP-MS in a wide range of polymers (Fordham *et al.*, 1995). The results were compared to those obtained by laser ablation LA-ICP-MS and NAA. Whilst the use of sulphuric acid means that additional interferences are present (see Figures 4.3 and 4.4), the use of specially prepared reference materials enabled a comparison of the three techniques for some elements to be made. For example, Table 4.11 compares the results for antimony and cobalt in an unknown polymer sample PET 1, using an in-house polymer reference material as calibrant for LA-ICP-MS. Although generally accuracy in the semi-quantitative mode of ICP-MS was not exceptional, combined with microwave digestion it did allow rapid multi-element screening of polymers prior to full quantitative analysis. The polymers require vigorous digestion procedures, but ideally an alternative is needed to sulphuric acid if all the elements of interest are to be measured with adequate LODs.

4.6   Speciation studies of foods

During the past fifteen years, there have been many publications reporting speciation studies in biological, environmental and clinical samples. The reader is referred to two texts, edited by Batley and Krull respectively, which provide excellent overviews of methods for trace metal speciation and some of the problems associated with the techniques (Batley, 1989; Krull, 1991).

Arsenic, particularly in fish, is a good example of the requirement to determine both total levels of the element and the chemical form. Total arsenic levels have been determined in foods by ICP-MS (Lásztity *et al.*, 1995). The inorganic forms of arsenic (As III and As V) are much more toxic than the organic forms monomethylarsonic acid and dimethylarsinic acid, which are in turn more toxic than arsenocholine and arsenobetaine. The last-named is frequently reported as the main organic arsenic compound in fish and has been isolated, for example, from seafood products by using anion exchange chromatography coupled to ICP-MS (Ybáñez *et al.*, 1995) and by reverse-phase chromatography (RPC)-ICP-MS from blue mussels (Gailer *et al.*, 1995). Hansen *et al.* (1992) isolated seven arsenic species by using anion and cation exchange systems with ICP-MS. Arsenic has also been monitored following HPLC–ICP-MS to determine the presence of the growth promotor, 4-hydroxy-3-phenylarsonic acid, in chicken tissue (Dean *et al.*, 1994).

Other authors have reported the use of HPLC with direct injection nebulisation-ICP-MS to determine arsenic, tin, lead and mercury species (Shum *et al.*, 1992a,b), the role of chemical speciation on the determination of ICP-MS of mercury in cod (Campbell *et al.*, 1992), and the use of ion-pairing- and RPC-ICP-MS for the measurement of organo-lead compounds (Al-Rashdan *et al.*, 1991). Aluminium species have been measured by SEC–ICP-MS in tea (Owen *et al.*, 1992a) and in intestinal digests (Owen *et al.*, 1994). Inorganic halogen species have been determined in water and soup using LC–ICP-MS (Salov *et al.*, 1992). Bromate has been measured in bread by GC–MS (Himata *et al.*, 1994), by GC with electron capture detection and the results compared with data from ion-exchange chromatography(IC)–ICP-MS (Dennis *et al.*, 1994) and by IC–ICP-MS only (Heit-kemper *et al.*, 1994).

However, the papers which perhaps illustrate the way in which HPLC–ICP-MS has progressed over the last decade or so are those which deal with the separation and characterisation of some metal-binding proteins. Morita *et al.* (1980) reported using gel-permeation chromatography coupled to the nebuliser of an argon plasma for AES. Using this system, cobalt, copper, iron, manganese, phosphorus, zinc and carbon were determined simultaneously in a mixture of proteins. Some problems were encountered with adsorption of protein on to stainless steel tubing, making absolute quantification of the elements impossible.

The potential of ICP-MS for speciation studies was reported by Dean *et al.* (1987b), who also used protein standards to characterise HPLC coupled to ICP-MS. The system used is shown schematically in Figure 4.8. A scavenger column containing Chelex 100 was used to ensure that buffers were free of

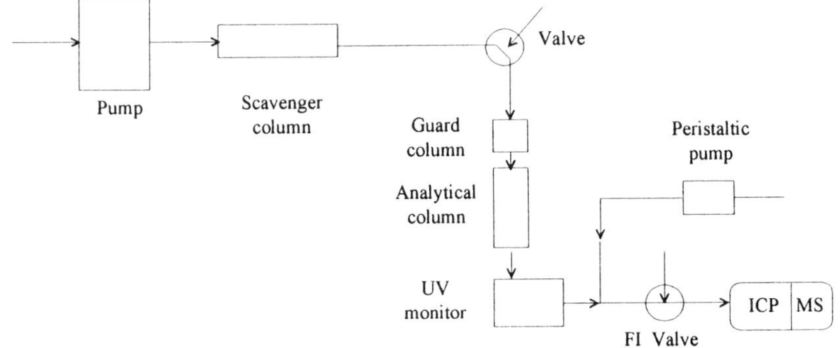

Figure 4.8 Schematic diagram of an HPLC–ICP-MS system.

the elements of interest. A mixed protein standard, which included vitamin B12, was separated by using size exclusion chromatography (SEC) and the [59]Co signal monitored by ICP-MS. A comparison of the UV response with ICP-MS response clearly showed the cobalt associated with the vitamin B12 (Dean *et al.*, 1987b). In a separate experiment reported in the same publication, SEC was used to analyse a mixture of equine metallothionein and ferritin and the [114]Cd signal was monitored. The specificity of ICP-MS enabled detection of cadmium in the metallothionein standard. The methodology was used by the same group to study the speciation of cadmium in retail samples of pig kidney (Crews *et al.*, 1989). They were able to show that the majority of soluble cadmium in the pig kidney was associated with a metallothionein-like protein that survived both cooking and *in vitro* gastrointestinal digestion. In addition, the work demonstrated that ICP-MS was sensitive enough to permit the study of cadmium at background levels. For example, the raw kidney contained 0.22 mg cadmium/kg fresh weight and cadmium concentrations of 2 ng/ml could be measured on-line, in fractions separated by SEC.

Work by Mason *et al.* (1990) reported the use of multi-element monitoring for HPLC–ICP-MS studies. They described simultaneous data acquisition on up to eight elements following separation of metalloproteins in cytosolic extracts by SEC. They encountered the following problems, when optimising the HPLC and ICP-MS systems. Firstly, although a higher ionic strength of the mobile phase (0.25 M NaCl, 0.06 M tris-HCl, pH 7.5 in 0.05% sodium azide) gave better chromatographic separation of the protein standard, the direct aspiration of the high salt buffer compromised the elemental analysis of the protein by ICP-MS. The presence of sodium caused severe ion suppression resulting in loss of attenuation and reduced sensitivity. In addition, the background counts for copper, zinc and, to a lesser extent, cadmium were elevated, giving poorer LODs. The high salt content caused deposition of solids on the torch and nebuliser (standard Meinhard nebuliser), which the authors postulated as being the cause of both the instability in the analyte signal and the irreproducibility of analyses that they experienced. The use of a lower ionic-strength buffer (0.06 M tris-HCl, 0.05% sodium azide, pH 7.5) gave fewer theoretical plates than the high ionic-strength buffer but similar values for peak tailing and symmetry, and was used for the remainder of the study. They reported relative LODs for [114]Cd, for example, of approximately 500–2000 pg/ml depending upon the metallothionein concentration. The necessity to 'optimise compromise conditions' for HPLC–ICP-MS is a common feature of speciation studies using this technique.

Owen *et al.* (1992b) used multi-isotopic analysis to investigate the feasibility of on-line measurement of elements associated with protein mixtures and to determine isotope ratios of zinc following labelling of chicken meat with stable zinc. Figure 4.9 and 4.10 show the chromatograms and element profiles following SEC–ICP-MS measurement of mixtures of thyroglobulin,

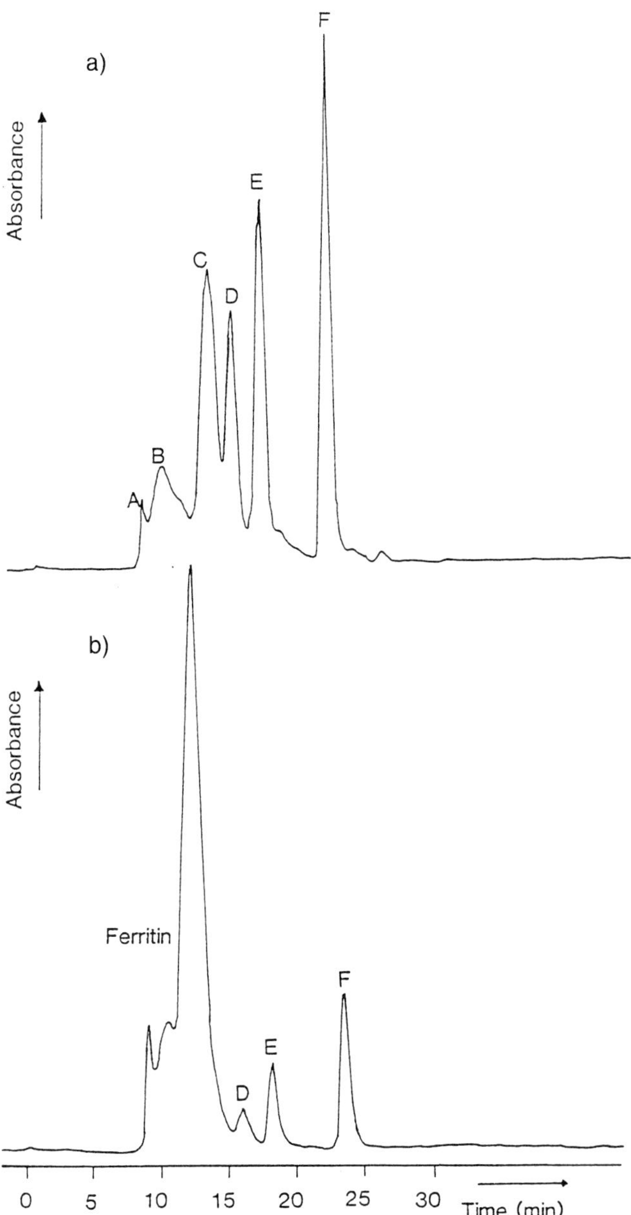

Figure 4.9 Chromatograms obtained with UV detection (280 nm) for (a) Bio-Rad standard and (b) a mixture of Bio-Rad standard, ferritin and metallothionein. A, protein aggregates; B, thyroglobulin; C, gamma-globulin; D, ovalbumin; E, myoglobulin; and F, vitamin B12.

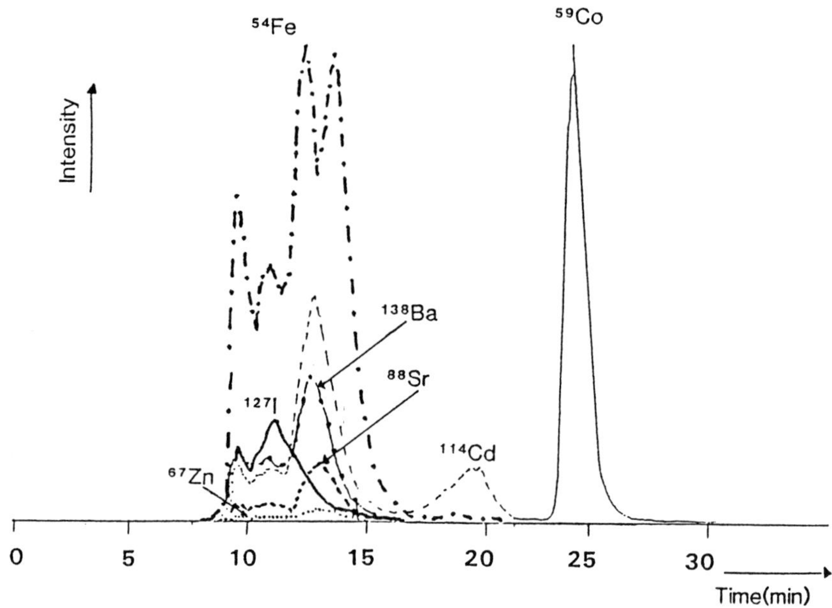

Figure 4.10 Isotope profiles following SEC–ICP-MS of a mixture of Bio-Rad standard, ferritin and metallothionein. The ^{59}Co, ^{67}Zn, ^{114}Cd, ^{127}I and ^{138}Ba are all on the same scale; ^{54}Fe is scaled down by a factor of 4; ^{88}Sr is scaled up by a factor of 8.

gamma-globulin, ovalbumin, myoglobulin, vitamin B12, metallothionein and ferritin. Seven elements are shown in Figure 4.10. Lead was also measured and is not shown for clarity, but it followed a very similar profile to strontium and was approximately eight times more intense. Similarly, copper is not shown, but its behaviour was similar to that of cadmium but was approximately eight times less intense. The buffer used was 0.12 M tris, pH 7.3, and the RSDs for peak retention times ($n = 3$) ranged from 0 to 3.6%. Whilst no quantitative data were reported, the results indicate that SEC–ICP-MS should be useful for assessing the associations between proteins and elements. However, as stated above, the optimisation of such systems is necessarily a compromise. For example, Owen *et al.* (1992b) comment that, in addition to the nine isotopes reported, molybdenum was also present, either associated with the proteins or as a contaminant. However, the data from repeated runs indicated that reproducibility was poor, for retention times and for a qualitative assessment of the amount of molybdenum present. Thus, in addition to optimising protein separations and ICP conditions, the need to consider the effect of these conditions on different isotopes is also apparent.

The results obtained for zinc isotope ratios measured by direct nebulisation-ICP-MS, by flow injection(FI)-ICP-MS, and following RPC-ICP-MS were generally in good agreement (Owen *et al.*, 1992b). Measurements by FI-ICP-MS of the ^{64}Zn:^{67}Zn ratio in enzyme digests of chicken meat which had been intrinsically labelled with ^{67}Zn were compared with those obtained using thermal ionisation mass spectrometry (TIMS). The value obtained by TIMS was 3.05 which compared well with the value, 3.09, obtained by ICP-MS.

Most recently, gel filtration coupled to ICP-MS has been used to investigate the copper, zinc and cadmium contents of freshwater mussels and their relationship to metallothionein content as part of an evaluation of potential reference materials (High *et al.*, 1995a,b). An electrospray mass spectrometry (ES-MS) protocol was developed to measure accurately the molecular weight of commercially available rabbit liver metallothioneins (High *et al.*, 1995a). For a comparison of mechanisms and performance of electrospray and ion spray see Ikonomou *et al.* (1991) and for an assessment of ES-MS for elemental speciation, see Agnes *et al.* (1994). However, the importance of knowing the degree of metal saturation of metallothionein and metallothionein-like proteins was emphasised in both publications and, in analytical terms, the authors suggested that the coupling of metal saturation/reduction treatments with HPLC–ES-MS (for molecular-weight determinations), HPLC–ICP-MS (for metal profiles) and polarography or HPLC–AES-ICP (for sulphur content) will provide valuable tools for certifying potential reference materials (High *et al.*, 1995a).

4.7 Future developments

For speciation studies, the combination of capillary electrophoresis (CE) with its high separation efficiency and ICP-MS with its isotopic specific detection, low LODs and rapid analysis times is currently an area of great interest. Olesik *et al.* (1995) have reported initial results for rapid elemental speciation using solutions of standard compounds. They report LODs for CE–ICP-MS as low as 0.06 ng/ml which corresponded to, for example, 8 fg of strontium in the injected sample. If a direct injection nebuliser (DIN) is used to interface CE with the ICP (Liu *et al.*, 1995), detection limits (pg/ml) of, for example, 7 for Tl$^+$, 60 for Cd^{2+}, 100 and 20, for As III and V respectively, 600 for Pb^{2+}, and 1000 pg/ml for K$^+$ were obtained. Further work on the interface with real samples is required as well as studies to understand why the ratio of LODs obtained by continuous nebulisation ICP-MS with those from CE/DIN–ICP-MS differ considerably between species (Liu *et al.*, 1995).

For multi-element determinations by ICP-MS, the use of appropriate quality control procedures needs addressing. Currently, the best analyses of food samples are judged by using criteria adopted from single-element analyses. A recent publication by Thompson and Wood (1995) discusses harmonised guidelines for internal quality control (IQC) in analytical chemistry laboratories. They state that, for multi-analyte data, IQC methods are still the subject of research and were not regarded as sufficiently established for inclusion in their guidelines. They caution that, because multi-analyte data may require a series of univariate IQC tests, care is needed to avoid inappropriate frequent rejection of data. As the approaches for data handling and assessment are developed, the instrument and software manufacturers will also need to accommodate the new systems.

Finally, ICP-MS is a rapid, multi-element technique. Currently, the sample throughputs achievable are limited by the sample preparation methods available. Whilst liquid samples can be dealt with relatively efficiently, any form of treatment for solid samples, from slurry analysis combined with on-line microwave digestion (which requires significant input to achieve homogeneous slurries of suitable particle size) to wet or dry ashing methods, is the rate-limiting step. It is likely, therefore, that, as has happened during the last twenty years, the future will see more work on sample-preparation methods for fast sample throughput. For ICP-MS in particular, novel approaches both to straightforward sample preparation and to analytical and instrumental methods for addressing the problems of complex matrix effects and interferences will still be required. The advent of high-resolution ICP-MS will enable isobaric overlaps, for example, to be resolved from the isotope of interest. Rottman and Heumann (1994) have already reported the use of HPLC with high resolution ICP-MS to allow interference-free detection of iron species.

References

Agnes, G.R., Stewart, I.I. and Horlick, G. (1994) Elemental speciation measurements with electrospray mass spectrometry: An assessment. *Applied Spectroscopy*, **48**, 1347–1359.

Al-Rashdan, A., Heitkemper, D. and Caruso, J.A. (1991) Lead speciation by HPLC–ICP-AES and HPLC–ICP-MS. *Journal of Chromatographic Science*, **29**, 98–102.

Al-Swaidan, H.M. (1988) Analysis of fruit juices by inductively coupled plasma-mass spectroscopy. *Analytical Letters*, **21**, 1469–1475.

Alves, L.C., Wiederin, D.R. and Houk, R.S. (1992) Reduction of polyatomic ion interferences in inductively coupled plasma mass spectrometry by cryogenic desolvation. *Analytical Chemistry*, **64**, 1164–1169.

Barnes, R.M. (1993) Advances in inductively coupled plasma mass spectrometry: human nutrition and toxicology. *Analytica Chimica Acta*, **283**, 115–130.

Batley, G.E. (1989) *Trace Element Speciation: Analytical Methods and Problems*, CRC Press Inc., Boca Raton, Florida.

Baxter, M.J., Burrell, J.A., Crews, H.M., Massey, R.C. and McWeeny, D.J. (1989) A procedure for the determination of lead in green vegetables at concentrations down to 1 μg/kg. *Food Additives and Contaminants*, **6**, 341–349.

Baxter, M.J., Burrell, J.A., Crews, H.M. and Massey, R.C. (1991) Aluminium levels in milk and infant formulae. *Food Additives and Contaminants*, **8**, 653–660.

Baxter, M.J., Burrell, J.A., Crews, H.M., Smith, A. and Massey, R.C. (1992) Lead contamination during domestic preparation and cooking of potatoes and leaching of bone-derived lead on roasting, marinating and boiling beef. *Food Additives and Contaminants*, **9**, 225–235.

Beauchemin, D., McLaren, J.W. and Berman, S.S. (1987) Study of the effects of concomitant elements in inductively coupled plasma mass spectrometry. *Spectrochimica Acta*, **42B**(3), 467–490.

Begley, I.S. and Sharp, B.L. (1994) Occurrence and reduction of noise in inductively coupled plasma-mass spectrometry for enhanced precision in isotope ratio measurements. *Journal of Analytical Atomic Spectrometry*, **9**, 171–176.

Boomer, D.W., Powell, M.J. and Hipfner, J. (1990) Characterization and optimisation of HPIC for on-line preconcentration of trace metals with detection by ICP-mass spectrometry. *Talanta*, **37**, 127–134.

Boorn, A., Fulford, J.E. and Wegscheder, W. (1985) Determination of trace elements in organic material by inductively coupled plasma mass spectrometry. *Mikrochim. Acta*, **11**, 171–178.

Branch, S., Ebdon, L., Ford, M., Foulkes, M. and O'Neill, P. (1991a) Determination of arsenic in samples with high chloride content by inductively coupled plasma mass spectrometry. *Journal of Analytical Atomic Spectrometry*, **6**, 151–154.

Branch, S., Crews, H.M., Halls, D.J. and Taylor, A. (1991b) Atomic spectrometry update— clinical and biological materials, foods and beverages. *Journal of Analytical Atomic Spectrometry*, **6**, 69R–108R.

Campbell, M.J., Vermeir, G., Dams, R. and Quevauviller, P. (1992) Influence of chemical species on the determination of mercury in a biological matrix (cod muscle) using inductively coupled plasma mass spectrometry. *Journal of Analytical Atomic Spectrometry*, **7**, 617–621.

Chang, S.K.C., Rayas-Duarte, P., Holm, E. and McDonald, C. (1993) Food. *Analytical Chemistry*, **65**, 334R–363R.

Cox, R.J., Pickford, C.J. and Thompson, M. (1992) Determination of iodine 129 in vegetable samples by inductively coupled plasma mass spectrometry. *Journal of Analytical Atomic Spectrometry*, **7**, 635–640.

Crews, H.M., Dean, J.R., Ebdon, L. and Massey, R.C. (1989) Application of high-performance liquid chromatography–inductively coupled plasma mass spectrometry to the investigation of cadmium speciation in pig kidney following cooking and *in vitro* gastro-intestinal digestion. *Analyst*, **114**, 895–899.

Crews, H.M., Baxter, M.J., Bigwood, T., Burrell, J.A., Owen, L.M., Robinson, C., Wright, C. and Massey, R.C. (1992) Lead in feed incident—multi-element analysis of cattle feed and tissues by inductively coupled plasma-mass spectrometry and co-operative quality assurance scheme for lead analysis of milk. *Food Additives and Contaminants*, **9**, 365–378.

Dalgarno, B.G., Brown, R.M. and Pickford, C.J. (1989) Use of ICP-MS to assess the bioavailability of trace elements in the human diet. *Nutrient Availability: Chemical and Biological Aspects*, **72**, 31–33.

Date, A.R. and Gray, A.L. (1981) Plasma source mass spectrometry using an inductively coupled plasma and a high resolution quadrupole mass filter. *Analyst*, **106**, 1255–1267.

Dawson, P.H. (1986) Quadrupole mass analysers: Performance, design and some recent applications. *Mass Spectrometry Reviews*, **5**, 1–37.

Dean, J.R., Ebdon, L. and Massey, R.C. (1987a) Selection of mode for the measurement of lead isotope ratios by inductively coupled plasma mass spectrometry and its application to milk powder analysis. *Journal of Analytical Atomic Spectrometry*, **2**, 369–374.

Dean, J.R., Munro, S., Ebdon, L., Crews, H.M. and Massey, R.C. (1987b) Studies of metalloprotein species by directly coupled high-performance liquid chromatography inductively coupled plasma mass spectrometry. *Journal of Analytical Atomic Spectrometry*, **2**, 607–610.

Dean, J.R., Ebdon, L., Crews, H.M. and Massey, R.C. (1988) Characteristics of flow injection-ICP-MS for trace metal analysis. *Journal of Analytical Atomic Spectrometry*, **3**, 349–354.

Dean, J.R., Crews, H.M. and Ebdon, L. (1989) Applications in food science. In *Applications of Inductively Coupled Plasma Mass Spectrometry*, eds Date, A.R. and Gray, A.L., Blackie, Glasgow, pp. 141–168.

Dean, J.R., Ebdon, L., Foulkes, M.E., Crews, H.M. and Massey, R.C. (1994) Determination of the growth promotor, 4-hydroxy-3-nitrophenyl-arsonic acid in chicken tissue by coupled high-performance liquid chromatography–inductively coupled plasma mass spectrometry. *Journal of Analytical Atomic Spectrometry*, **9**, 615–618.

Dennis, M.J., Burrell, J.A., Mathieson, K., Willetts, P. and Massey, R.C. (1994) The determination of the flour improver potassium bromate in bread by gas chromatographic and ICP-MS methods. *Food Additives and Contaminants*, **11**, 633–639.

Denoyer, E.R. (1991) Current trends in ICP-mass spectrometry. *Atomic Spectroscopy*, **12**, 215–224.

Douglas, D.J. and French, J.B. (1981) Elemental analysis with a microwave induced plasma quadrupole mass spectrometer system. *Analytical Chemistry*, **53**, 37–41.

Douglas, D.J. and French, J.B. (1988) Gas dynamics of the inductively coupled plasma mass spectrometry interface. *Journal of Analytical Atomic Spectrometry*, **3**, 743–747.

Douglas, D.J. and Houk, R.S. (1985) Inductively-coupled plasma mass spectrometry (ICP-MS). *Progress in Analytical Atomic Spectroscopy*, **8**, 1–18.

Douglas, D.J. and Kerr, L.A. (1988) Study of solids deposition on inductively coupled plasma mass spectrometry samplers and skimmers. *Journal of Analytical Atomic Spectrometry*, **3**, 749–752.

Eagles, J., Fairweather-Tait, S.J., Mellon, F.A., Portwood, D.E., Self, R., Gotz, A., Heumann, K.G. and Crews, H.M. (1989) Comparison of fast-atom bombardment, thermal ionisation and inductively coupled plasma-mass spectrometry for the measurement of $^{64}Zn/^{67}Zn$ stable isotopes in human nutrition studies. *Rapid Communications in Mass Spectrometry*, **3**(6), 203–205.

Ebdon, L., Fisher, A.S., Worsfold, P.J., Crews, H.M. and Baxter, M.J. (1993) On-line removal of interferences in the analysis of biological materials by flow injection inductively coupled plasma mass spectrometry. *Journal of Analytical Atomic Spectrometry*, **8**, 691–695.

EEC (1992) Draft resolution on control of aids to polymerisation (technical adjuvants) for plastic materials intended to come into contact with foodstuffs. Adopted by the Committee of Ministers at the 482nd meeting on 19/10/92.

Evans, E.H. and Ebdon, L. (1990) Effect of organic solvents and molecular gases on polyatomic ion interferences in inductively coupled plasma mass spectrometry. *Journal of Analytical Atomic Spectrometry*, **5**, 425–430.

Fardy, J.J. and Warner, I.M. (1992) A comparison of NAA with ICP-MS for the determination of trace elements in foods and diets. *Journal of Radioanalysis and Nuclear Chemistry*, **157**(2), 239–244.

Fassel, V.A. (1986) Analytical inductively coupled plasma mass spectroscopies—Past, present and future. *Fresenius Zeitschrift für Analytische Chemie*, **324**, 511–518.

Fordham, P.J., Gramshaw, J.W., Castle, L., Crews, H.M., Thompson, D., Parry, S.J. and McCurdy, E. (1995) Determination of trace elements in food contact polymers by semi-quantitative inductively coupled plasma mass spectrometry. Performance evaluation using alternative multi-element techniques and in-house polymer reference materials. *Journal of Analytical Atomic Spectrometry*, **10**, 303–309.

Friel, J.K., Skinner, C.S., Jackson, S.E. and Longerich, H.P. (1990) Analysis of biological reference materials, prepared by microwave dissolution, using inductively coupled plasma mass spectrometry. *Analyst*, **115**, 269–273.

Fulford, J.E. and Douglas, D.J. (1986) Ion kinetic energies in inductively coupled plasma/mass spectrometry (ICP-MS). *Applied Spectroscopy*, **40**(7), 971–975.

Gailer, J., Francesconi, K.A., Edmonds, J.S. and Irgolic, K.J. (1995) Metabolism of arsenic compounds by the blue mussel *Mytilus edulis* after accumulation from seawater spiked with arsenic compounds. *Applied Organometallic Chemistry*, **9**, 341–355.

Gillson, G.R., Douglas, D.J., Fulford, J.E., Halligan, K.W. and Tanner, S.D. (1988) Non-spectroscopic inter-element interferences in inductively coupled plasma mass spectrometry. *Analytical Chemistry*, **60**, 1472–1474.

Gray, A.L. (1975) Mass spectrometric analysis of solutions using an atmospheric pressure ion source. *Analyst*, **100**, 289–299.

Gray, A.L. (1978) Isotope ratio determination on solutions with a plasma ion source. In *Dynamic Mass Spectrometry*, vol. 5, ed. Todd, J.F.J., Heyden, London, pp. 106–113.

Gray, A.L. (1986a) Mass spectrometry with an inductively coupled plasma as an ion source: The influence on ultratrace analysis of background and matrix response. *Spectrochimica Acta*, **41B**, 151–167.

Gray, A.L. (1986b) Ions or photons—An assessment of the relationship between emission and mass spectrometry with the ICP. *Fresenius Zeitschrift für Analytische Chemie*, **324**, 561–570.

Gray, A.L. (1989) The origins, nebulization and performance of ICP-MS systems. In *Applications of ICP-MS*, eds Date, A.R. and Gray, A.L., Blackie, Glasgow, pp. 1–42.

Gray, A.L. and Williams, J.G. (1987) Oxide and doubly charged ion response of a commercial inductively coupled plasma mass spectrometry instrument. *Journal of Analytical Atomic Spectrometry*, **2**, 81–82.

Grégoire, D.C., Lamoureux, M., Chakrabarti, C.L., Al-Maawali, S. and Byrne, J.P. (1992) Electrothermal vaporisation for inductively coupled plasma mass spectrometry and atomic absorption spectrometry: Symbiotic analytical techniques. *Journal of Analytical Atomic Spectrometry*, **7**, 579–585.

Hansen, S.H., Larsen, E.H., Pritzl, G. and Cornett, C. (1992) Separation of seven arsenic compounds by high-performance liquid chromatography with on-line detection by hydrogen–argon flame atomic absorption spectrometry and inductively coupled plasma mass spectrometry. *Journal of Analytical Atomic Spectrometry*, **7**, 629–634.

Heitkemper, D.T., Kaine, L.A., Jackson, D.S. and Wolnik, K.A. (1994) Practical applications of element-specific detection by inductively coupled plasma atomic emission spectroscopy and inductively coupled plasma mass spectrometry to ion chromatography of foods. *Journal of Chromatography A*, **671**, 101–108.

Hieftje, G.M. and Vickers, G.H. (1989) Developments in plasma source/mass spectrometry. *Analytica Chimica Acta*, **216**, 1–24.

High, K.A., Methven, B.A.J., McLaren, J.W., Siu, K.W.M., Wang, J., Klaverkamp, J.F. and Blais, J.S. (1995a) Physico-chemical characterisation of metal binding proteins using HPLC–ICP-MS, HPLC–MA-AAS and electrospray-MS. *Fresenius Journal of Analytical Chemistry*, **315**, 393–402.

High, K.A., Blais, J.S., Methven, B.A.J. and McLaren, J.W. (1995b) Probing the characteristics of metal-binding proteins using high-performance liquid chromatography–atomic absorption and inductively coupled plasma mass spectrometry. *Analyst*, **120**, 629–634.

Himata, K., Kuwahara, T., Ando, S. and Maruoka, H. (1994) Determination of bromate in bread by capillary gas chromatography with a mass detector (GC/MS). *Food Additives and Contaminants*, **11**, 559–569.

Houk, R.S. (1986) Mass spectrometry of inductively coupled plasmas. *Analytical Chemistry*, **58**, 97–105.

Houk, R.S., Fassel, V.A., Flesch, G.D., Svec, H.J., Gray, A.L. and Taylor, C.E. (1980) Inductively coupled argon plasma as an ion source for mass spectrometric determination of trace elements. *Analytical Chemistry*, **52**, 2283–2289.

Houk, R.S., Svec, H.J. and Fassel, V.A. (1981a) Mass spectrometric evidence for suprathermal ionisation in an ICP. *Applied Spectroscopy*, **35**, 380–384.

Houk, R.S., Fassel, V.A. and Svec, H.J. (1981b) Inductively coupled plasma-mass spectrometry: Sample introduction, ionisation, ion extraction and analytical results. In *Dynamic Mass Spectrometry*, vol. 6, eds. Price, D. and Todd, J.F.T., Heyden and Son, London, pp. 234–251.

Huang, L.Q., Jiang, S.J. and Houk, R.S. (1987) Scintillation-type ion detection for inductively coupled plasma mass spectrometry. *Analytical Chemistry*, **59**, 2316–2320.

Hutton, R.C. (1986) Application of inductively coupled plasma source mass spectrometry (ICP-MS) to the determination of trace metals in organics. *Journal of Analytical Atomic Spectrometry*, **1**, 259–263.

Hutton, R.C. and Eaton, A.N. (1988) Analysis of solutions containing high levels of dissolved solids by inductively coupled plasma mass spectrometry. *Journal of Analytical Atomic Spectrometry*, **3**, 547–550.

Ikonomou, M.G., Blades, A.T. and Kebarle, P. (1991) Electrospray-ion spray: A comparison of mechanisms and performance. *Analytical Chemistry*, **63**, 1989–1998.

Jarvis, I. (1992) Sample preparation for ICP-MS. In *Handbook for ICP-MS*, eds Jarvis, K.E., Gray, A.L. and Houk, R.S., Blackie, Glasgow, pp. 173–224.

Jarvis, K.E., Gray, A.L. and Houk, R.S. (eds) (1992) *Handbook for ICP-MS*, Blackie, Glasgow.

Kirchmer, C.J. (1994) Limits of detection and accuracy in trace elements analysis. In *Determination of Trace Elements*, ed. Alfassi, Z.B., VCH Publishers, Weinheim, pp. 39–58.

Krull, I.S. (1991) *Trace Metals Analysis and Speciation*, Elsevier, Amsterdam.

Laborda, F., Baxter, M.J., Crews, H.M. and Dennis, J. (1994) Reduction of polyatomic interferences in inductively coupled plasma mass spectrometry by selection of instrumental parameters and using an argon–nitrogen plasma: Effect on multi-element analyses. *Journal of Analytical Atomic Spectrometry*, **9**, 727–736.

Lásztity, A., Krushevska, A., Kotrebai, M., Barnes, R.M. and Amarasiriwardena, D. (1995) Arsenic determination in environmental, biological and food samples by inductively coupled plasma mass spectrometry. *Journal of Analytical Atomic Spectrometry*, **10**, 505–510.

Liu, Y., Lopez-Avila, V., Zhu, J.J., Wiederin, D.R. and Beckert, W.F. (1995) Capillary electrophoresis coupled on-line with inductively coupled plasma mass spectrometry for elemental speciation. *Analytical Chemistry*, **67**, 2020–2025.

Lyon, T.D.B., Fell, G.S., Hutton, R.C. and Eaton, A.N. (1988) Elimination of chloride interference on the determination of selenium in serum by inductively coupled plasma mass spectrometry. *Journal of Analytical Atomic Spectrometry*, **3**, 601–604.

Marawi, I., Wang, J. and Caruso, J.A. (1994) Graphite furnace hydride preconcentration and subsequent detection by inductively coupled plasma mass spectrometry. *Analytica Chimica Acta*, **291**, 127–136.

Mason, A.Z., Storms, S.D. and Jenkins, K.D. (1990) Metalloprotein separation and analysis by directly coupled size exclusion high-performance liquid chromatography inductively coupled plasma mass spectroscopy. *Analytical Biochemistry*, **186**, 187–201.

Michaud-Poussel, E. and Mermet, J.M. (1986) Comparison of nebulizers working below $0.81\,min^{-1}$ in inductively coupled plasma atomic emission spectrometry. *Spectrochimica Acta*, **41B**, 49–61.

Miller, P.E. and Denton, M.B. (1986) The quadrupole mass filter: Basic operating concepts. *Journal of Chemical Education*, **63**, 617–622.

Morita, M., Uehiro, T. and Fuwa, K. (1980) Speciation and elemental analysis of mixtures by high performance liquid chromatography with inductively coupled argon plasma emission spectrometric detection. *Analytical Chemistry*, **52**, 349–351.

Munro, S., Ebdon, L. and McWeeny, D.J. (1986) Application of inductively coupled plasma mass spectrometry (ICP-MS) for trace metal determination in foods. *Journal of Analytical Atomic Spectrometry*, **1**, 211–219.

Muto, H., Abe, T., Takizawa, Y., Kawabata, K., Yamaguchi, K. and Saitoh, K. (1994) Simultaneous multi elemental analysis of daily food samples by inductively coupled plasma mass spectrometry. *The Science of the Total Environment*, **144**, 231–239.

Negretti de Brätter, V.E., Brätter, P., Reinicke, A., Schulze, G., Alvarez, W.A.L. and Alvarez, N. (1995) Determination of mineral trace elements in total diet by inductively coupled plasma mass emission spectrometry: Comparison of microwave-based digestion and pressurised ashing systems using different acid mixtures. *Journal of Analytical Atomic Spectrometry*, **10**, 487–491.

Nisamaneepong, W., Carnso, J.A. and Ng, K.C. (1985) Electrothermal vaporisation as an interface for HPLC introduction to the inductively coupled plasma. *J. Chromatographic Science*, **23**, 465–470.

Olesik, O.W., Kinzer, J.A. and Olesik, S.V. (1995) Capillary electrophoresis inductively coupled plasma spectrometry for rapid elemental speciation. *Analytical Chemistry*, **67**, 1–12.

Olivares, J.A. and Houk, R.S. (1986) Suppression of analyte signal by various concomitant salts in inductively coupled plasma mass spectrometry. *Analytical Chemistry*, **58**, 20–25.

Owen, L.M.W., Crews, H.M. and Massey, R.C. (1992a) Aluminium in tea: SEC–ICP-MS speciation studies of infusions and simulated gastrointestinal digests. *Chemical Speciation and Bioavailability*, **4**, 89–96.

Owen, L.M.W., Crews, H.M., Hutton, R.C. and Walsh, A. (1992b) Preliminary study of metals in proteins by HPLC–ICP-MS using multi-element time-resolved analysis. *Analyst*, **117**, 649–655.

Owen, L.M.W., Crews, H.M., Bishop, N.J. and Massey, R.C. (1994) Aluminium uptake from some foods by guinea pigs and the characterisation of aluminium in *in vivo* intestinal digesta by SEC–ICP-MS. *Food Chemistry and Toxicology*, **32**, 697–705.

Park, C.J. and Hall, G.E.M. (1986) Electrothermal vaporisation as a means of sample introduction into an inductively coupled plasma mass spectrometer: A preliminary report of a new analytical technique. *Current Research, Part B, Geological Survey of Canada*, **86-1B**, 767–773.

Park, C.J., Van Loon, J.C., Arrowsmith, P. and French, J.B. (1987a) Design and optimisation of an electrothermal vaporiser for use in plasma source mass spectrometry. *Canadian Journal of Spectroscopy*, **32**, 29–36.

Park, C.J., Van Loon, J.C., Arrowsmith, P. and French, J.B. (1987b) Sample analysis using plasma source mass spectrometry with electrothermal sample introduction. *Analytical Chemistry*, **59**, 2191–2196.

Pickford, C.J. and Brown, R.M. (1986) Comparison of ICP-MS with ICP-ES: detection power and interference experienced with complex matrices. *Spectrochimica Acta*, **41B**, 183–187.

Riddle, C., Vander Voet, A. and Doherty, W. (1988) Rock analysis using inductively coupled plasma mass spectrometry: A review. *Geostandards Newsletter*, **12**(1), 203–234.

Rottman, L. and Heumann, K.G. (1994) Determination of heavy metal interactions with dissolved organic materials in natural aquatic systems by coupling a high-performance liquid chromatography system with an inductively coupled plasma mass spectrometer. *Analytical Chemistry*, **66**, 3709–3715.

Salov, V.V., Yoshinaga, J., Shibata, Y. and Morita, M. (1992) Determination of inorganic halogen species by liquid chromatography with inductively coupled argon plasma mass spectrometry. *Analytical Chemistry*, **64**, 2425–2428.

Satzger, R.D. (1988) Evaluation of inductively coupled plasma mass spectrometry for the determination of trace elements in food. *Analytical Chemistry*, **60**, 2500–2504.

Sharp, B.L. (1988a) Pneumatic nebulisers and spray chambers for inductively coupled plasma spectrometry. A review. Part I. Nebulisers. *Journal of Analytical Atomic Spectrometry*, **3**, 613–652.

Sharp, B.L. (1988b) Pneumatic nebulisers and spray chambers for inductively coupled plasma spectrometry. A review. Part 2. Spray chambers. *Journal of Analytical Atomic Spectrometry*, **3**, 939–963.

Shiraishi, K., McInroy, J.F. and Igarashi, Y. (1990) Simultaneous multi element analysis of diet samples by inductively coupled plasma mass spectrometry and inductively coupled plasma atomic emission spectrometry. *Journal of Nutritional Science and Vitaminology*, **36**, 81–86.

Shiraishi, K., Takaku, Y., Yoshimizu, K., Igarashi, Y., Masuda, K., McInroy, J.F. and Tanaka, G. (1991) Determination of thorium and uranium in total diet samples by inductively coupled plasma mass spectrometry. *Journal of Analytical Atomic Spectrometry*, **6**, 335–338.

Shum, S.C.K., Neddersen, R. and Houk, R.S. (1992a) Elemental speciation by liquid chromatography–inductively coupled plasma mass spectrometry with direct injection nebulization. *Analyst*, **117**, 577–582.

Shum, S.C.K., Pang, H.-M. and Houk, R.S. (1992b) Speciation of mercury and lead compounds by microbore column liquid chromatography–inductively coupled plasma mass spectrometry with direct injection nebulization. *Analytical Chemistry*, **64**, 2444–2450.

Smith, F.G., Wiederin, D.R., Houk, R.S., Egan, C.B. and Serfass, R.E. (1991) Measurement of boron concentration and isotope ratios in biological samples by inductively coupled plasma mass spectrometry with direct injection nebulization. *Analytica Chimica Acta*, **248**, 229–234.

Tan, S.H. and Horlick, G. (1986) Background spectral features in inductively coupled plasma mass spectrometry. *Applied Spectroscopy*, **40**, 445–460.

Taylor, H.E. (1986) Inductively coupled plasma-mass spectrometry: An introduction. *Spectroscopy*, **1**, 20–22.

Thompson, J.J. and Houk, R.S. (1986) Inductively coupled plasma mass spectrometric detection for multi-element flow injection analysis and elemental speciation by reversed-phase liquid chromatography. *Analytical Chemistry*, **58**, 2541–2548.

Thompson, J.J. and Houk, R.S. (1987) A study of internal standardisation of inductively coupled plasma-mass spectrometry. *Applied Spectroscopy*, **41**, 801–806.

Thompson, M. and Wood, R. (1995) Harmonised guidelines for internal quality control in analytical chemistry laboratories. *Pure and Applied Chemistry*, **67**, 649–666.

Tölg, G. (1993) Problems and trends in extreme trace analysis for the elements. *Analytica Chimica Acta*, **283**, 3–18.

Tölg, G. and Tschöpel, P. (1994) Systematic errors in trace analysis. In *Determination of Trace Elements*, ed. Alfassi, B.Z., VCH Publishers, Weinheim, pp. 1–38.

Ure, A.M. and Bacon, J.R. (1987) Inorganic trace analysis by mass spectroscopy. In *Applications of Mass Spectrometry in Food Science*, ed. Gilbert, J.G., Elsevier Applied Science, London, pp. 343–376.

Vanhoe, H. (1993) A review of the capabilities of ICP-MS for trace element analysis in body fluids and tissues. *Journal of Trace Elements, Electrolytes, Health and Disease*, **7**, 131–139.

Williams, J.G. (1992a) Instrument options. In *Handbook for ICP-MS*, eds Jarvis, K.E., Gray, A.L. and Houk, R.S., Blackie, Glasgow, pp. 58–80.
Williams, J.G. (1992b) Sample introduction for liquids and gases. In *Handbook for ICP-MS*, eds Jarvis, K.E., Gray, A.L. and Houk, R.S., Blackie, Glasgow, pp. 81–124.
Woittiez, J.R.W. and Sloof, J.E. (1994) Sampling and sample preparation. In *Determination of Trace Elements*, ed. Alfassi, Z.B., VCH Publishers, Weinheim, pp. 59–107.
Ybáñez, N., Velez, D., Tejedor, W. and Montoro, R. (1995) Optimisation of the extraction, clean-up and determination of arsenobetaine in manufactured seafood products by coupling liquid chromatography with inductively coupled plasma atomic emission spectrometry. *Journal of Analytical Atomic Spectrometry*, **10**, 459–465.
Zhu, G. and Browner, R.F. (1987) Investigation of experimental parameters with a quadrupole ICP/MS. *Applied Spectroscopy*, **41**, 349–359.

5 Applications of immunoassay to pesticide analysis

BENNETT M. KAUFMAN

Summary

Determination of the presence and levels of pesticides and their residues in food is the basis of monitoring and regulatory programmes. Most pesticide measurement techniques generally require extensive sample clean-up and most often utilise instrumental analysis, such as gas chromatography. There is a need for rapid, easily performed, and inexpensive tests, such as immunoassays, which could be used for screening or quantitation of pesticides. A number of immunoassay measurement techniques for pesticides have been developed which may often be used directly on the test portion, or require only minimal sample clean-up. Assays have been developed for the full range of currently used pesticides, such as organophosphorus, carbamate, triazine, pyrethroid, and other compounds, and these assays have been applied to the determination of such compounds in water, crops, and processed foodstuffs. Many of these assays have been made commercially available as test kits in a variety of formats. The US government agencies responsible for monitoring and regulating pesticide use and the presence of pesticide residues in water, crops, foods, and livestock, the Environmental Protection Agency, the Food and Drug Administration, and the Department of Agriculture, are working with the Association of Official Analytical Chemists (AOAC International) and other concerned groups in evaluating these test kits and developing guidelines and specifications, as part of the process of officially accepting test kits as accurate and practical methods of screening and analysis.

5.1 Introduction

As the use of pesticides has increased, and the concern about their residues in soil, water, crops, and foods has become an environmental and health issue (Kutz *et al.*, 1992; Yess *et al.*, 1993; US Food and Drug Administration, 1994), greater interest has developed in the methods for the detection and quantitation of pesticides and their residues. Approved methods for the determination of pesticides and their residues in the USA are compiled in the Pesticide Analytical Manual, Volume I (multiple residue methods) and Volume II (individual residue methods). These methods describe the extraction, clean-up, and analysis of samples by instrumental techniques such as gas chromatography (GC), mass spectrometry (MS), high performance liquid chromatography, and others. To increase the number of samples analysed, and to broaden the scope of analyses, rapid and simple tests are needed. In this regard, the area of immunoassay of pesticides has become one of increasing interest and has been the subject of several recent extensive reviews (Kaufman and Clower, 1991, 1995; Sherry, 1992). Traditional instrumental analyses of pesticides are usually very sensitive and selective, but are limited by their low throughput, requirement for complex equipment, lack of portability, high cost per analysis, need for extensive sample clean-up, etc. Immunoassays, on the other hand, are relatively simple, inexpensive, portable, robust, and can often be conducted on samples after minimal clean-up. However, immunoassays may not be very selective (identifying not only the parent compound, but its breakdown products, isomers, etc.), and may not be as sensitive as instrumental techniques.

5.1.1 Immunoassay technology

Immunoassays depend on the recognition of an antigen (e.g. a protein, a virus, or a pesticide) by an antibody which was elicited by the introduction of the antigen into a responding animal. The antibodies are present in the serum of the immunised animal, and this serum serves as the reactive component in the immunoassay. The use of monoclonal antibody technology (Köhler and Milstein, 1975) allows the production of potentially unlimited amounts of pure, highly specific antibodies. The theory and basic methodologies of immunoassay have been the subject of previous reviews (Voller *et al.*, 1979; Monroe, 1984; Kurstak, 1986; Harrison and Hammock, 1988), and the advantages and disadvantages, and problems, of applying immunoassays to the analysis of pesticides and foods have been discussed (Newsome, 1986a; Schwalbe-Fehl, 1986; Aherne, 1987; Allen and Smith, 1987; Vanderlaan *et al.*, 1988; Ellis, 1989; Fukal and Kás, 1989; Jung *et al.*, 1989a; Kaufman and Clower, 1991; Lee and Richman, 1991; Sherry, 1992; Van Emon and Lopez-Avila, 1992). Some of the important advantages and disadvantages of immunoassay are also described in section 5.1.2 and Table 5.1.

5.1.1.1 Types of immunoassays.

In one type of direct assay, an (unlabelled) antibody specific for the analyte is bound to a surface (beads, plastic tubes, or the wells of a plastic microtitre plate). The sample containing the analyte is added and allowed to react with the antibody. The unbound material is washed off, and labelled antibody directed against the analyte is added. After further washing, the amount of bound labelled antibody is determined. The signal produced is proportional to the amount of analyte present in the sample. The use of known amounts of analyte allow the

Table 5.1 Advantages and disadvantages of immunoassay for the determination of pesticides

Advantages	Disadvantages
Rapid	Tolerant of only low amounts of organic solvents
Inexpensive	May have low sensitivity
Large throughput	Matrix effects may reduce sensitivity
Usually minimal sample clean-up	May detect only some members of a pesticide group
May be specific for a single compound	
Colour readers simple and inexpensive; samples may be screened by eye.	

construction of a 'dose–response' curve. In a competitive direct assay, a known amount of tracer-labelled hapten is added along with the sample to the well or tube containing the bound antibody; the amount of tracer detected is inversely proportional to the amount of analyte present in the sample. Alternatively, an indirect competitive assay may be performed; this type is commonly used for pesticide determination. Usually, a known amount of standard analyte is bound to a solid surface as described above (often coupled to a carrier protein to facilitate binding). The test sample is then mixed with the specific antibody, and the mixture incubated with the bound analyte. If the analyte–antibody mixture is preincubated before addition to the well or tube containing the bound hapten, this is often referred to as a two-step indirect assay.

The amount of antibody available to bind to the fixed analyte is reduced by any analyte present in the test sample. This antibody either is detected directly, if labelled, or may be detected by the use of a second, labelled, anti-antibody. By using known amounts of analyte as competitor, a dose–response curve may also be generated. There are, of course, several variations on all these immunoassays.

Although there are many ways of detecting the signal produced in an immunoassay, the two most common are either by measuring radioactivity (radioimmunoassay, RIA) or by the development of a colour (enzyme immunoassay, EIA, or enzyme-linked immunosorbent assay, ELISA), which is measured spectrophotometrically. Other methods exist, and are under development, and are discussed in section 5.1.1.2 below. In most cases, the dose–response curve is non-linear. One standard method for linearising the standard curve is the logit transformation (Oellerich *et al.*, 1982; Kurstak, 1986) in which the transformed response (cpm, in RIA, or absorbance, in the case of EIA/ELISA) is plotted as a function of [log(concentration)]

$$\operatorname{logit} p = \ln\left[p/1(1-p)\right]$$

$$p = \frac{(A_x - \mathrm{NSB})}{(B_0 - \mathrm{NSB})}$$

where A_x is the signal produced in the presence of the analyte, B_0 is the signal obtained in the absence of analyte (i.e. the maximum response), and NSB is the non-specific, or reagent, blank.

Solvent selection may be one of the complicating factors in developing immunoassays for pesticides. Antibodies generally are functional in an aqueous environment; however, many pesticides are soluble only in organic solvents, or are sparingly soluble in water. Likewise, most analytes are extracted from food or plant tissue by non-polar solvents which may not be miscible with the aqueous buffers used for immunoassay. The effects of non-aqueous solvents on immunoassay performance have not been systematically studied. Russell *et al.* (1989) studied the interaction of a monoclonal antibody (bound to glass beads) against the antigen 4-aminobiphenyl in the presence of water, dioxane, acetonitrile, 1-propanol, 1-butanol, and 1-pentanol. Although the antibody retained its specificity for the antigen in all the systems tested, the binding affinities decreased several hundredfold as the hydrophobicities of the solvents increased. Additionally, the time required for the system to reach equilibrium increased similarly. Weetall (1991) studied the binding of monoclonal antiprogesterone antibodies (coupled to magnetic beads) with progesterone both in phosphate buffer and in hexane. Although the time course of the reaction was delayed in hexane, eventually the same degree of binding was obtained in the hexane system as in the buffer system. These results were attributed to a thin shell of water surrounding the immobilised antibodies, so that the antigen diffused slowly through the hexane phase into the water shell, wherein the reaction occurred. A similar mechanism would explain the results of Russell *et al.* (1989). The effects of some organic solvents on assay performance have been investigated in several of the specific immunoassays described below, and will be discussed in the appropriate sections. In general, it was found that many assays are tolerant of solvents such as methanol, with up to 20% concentrations resulting in no degradation of assay performance. In many cases, the test sample is actually a solvent extract of the pesticide-containing material, so one must have a method for determining the percent recovery of the analyte from the analytical portion (Provost and Elder, 1983).

One of the purported advantages of immunoassay over instrumental analysis is the reduced or absent requirement for sample clean-up, so that the need to concentrate the analyte present in the sample may be an important factor in assay performance (Fukal and Kás, 1989). In this regard, the use of solid-phase extraction (SPE) methods, utilising, for instance, silica particles bonded with an organic substrate, is valuable for concentrating materials present in extracts or water samples. Crepeau *et al.* (1991) reported the use of C_8 filter disks to extract and concentrate the pesticides carbofuran, 2,4-D, dinoseb, prometryn, prometon, and fenamiphos from (artificial) soil leachates. The simplicity and speed of use of these units make them valuable assets for pesticide assay.

5.1.1.2 New technology/instrumentation. In addition to the common RIA and EIA/ELISA techniques described above, several new technologies utilising biosensors (a bioactive material coupled to a transducer that reacts to the biochemical event) have been developed for the determination of very small amounts of specific analytes, and have been applied to the determination of pesticides. Many of these technologies currently require sensitive and/or complex equipment, and do not yet represent robust, field-portable techniques for pesticide analysis. These will also be mentioned again where they have been used to determine specific pesticides.

One technology that has not been widely applied to the immunoassay of pesticides is piezoelectric or electroconductive sensors (see Mathewson and Finley, 1992). The first of these techniques (Guilbault and Schmid, 1991) employs piezoelectric crystals: materials with the ability to oscillate when a voltage is applied across them. The frequency of oscillation is a function of the crystal's mass and dimensions. If the crystal is coated with a substance (e.g. an antibody) that adsorbs the compound of interest (e.g. a pesticide), when the crystal is exposed to the ligand, absorption of the ligand changes the mass, and thus, the frequency, of the crystal's oscillation. The change in frequency can then be correlated with the amount of material adsorbed on the crystal, and represents a quantitative measure of the ligand's concentration. This technique has been applied to the determination of parathion (Ngeh-Ngwainbi *et al.*, 1986) and atrazine in drinking water (Guilbault *et al.*, 1992). In the second technique, a series of coupled reactions initiated by the antigen–antibody reaction of a substrate-bound antibody (or other bioactive material) is used to generate a reactive molecule which changes the conductive properties of an underlying film. The change in conductive properties is proportional to the amount of reactive molecule produced, itself a function of the original antigen–antibody reaction (Sandberg *et al.*, 1992). This technique has been applied to the determination of the organophosphate pesticides carbaryl and heptenophos, using cholinesterases as the bound bioactive materials (Skládal *et al.*, 1994). These techniques suffer from some disadvantages, including the difficulty of fabrication of the sensor elements, the small number of samples that can be run at the same time [although a 32-sensor array of electroconductive cells is under development (Sandberg *et al.*, 1992)], and the deterioration of the sensors with time.

One commercially available biosensor device (BIAcore™, Pharmacia Biosensor, Sweden) has been used for the measurement of pesticides. A BIAcore (BIA = biospecific interaction analysis) device uses a glass slide coated on one side with a thin gold film fixed with a flexible polymer capable of binding biomolecules. When plane-polarised light is shone on to the film, a dip in reflectance intensity is observed at a specific angle of reflection (the resonance angle). Any interaction of the attached biomolecule with its ligand results in a measurable change in the resonance angle ('surface plasmon resonance') (Jönsson *et al.*, 1991; Fägerstam *et al.*, 1992). This technique has been applied to the measurement of atrazine in drinking water (Minunni and Mascini, 1993).

Another relatively new technique with potential pesticide monitoring applications is flow injection immunoanalysis (FIIA). Anti-pesticide antibodies are immobilised on a membrane contained within a replaceable cartridge. A set of pumps forces the test portion (containing the pesticide) over the membrane, then washes it. Enzyme-labelled hapten is pumped over the membrane, the membrane is washed, and a fluorogenic substrate is then pumped through the cartridge. The fluorescent product is measured by a downstream detector; the resultant fluorescence signal is inversely proportional to the pesticide concentration (Schmid and Künnecke, 1990). This technique has been applied to the determination of triazines in water (Krämer and Schmid, 1991). Although FIIA reduces analysis time per sample from the 1–3 h of an ELISA to approximately 6–7 min, there are some disadvantages: only one sample can be analysed at a time, unlike microwell ELISAs in which dozens of samples can be assayed at once; single-use cartridges and large amounts of antibody and enzyme–hapten are employed; and generally higher CVs are found than in ELISAs. However, FIIA might be employed, for instance, for monitoring water supply stations or effluent streams in pesticide production plants.

Anis *et al.* (1992) described a technique in which quartz fibres were coated with parathion conjugates, then incubated with antiserum in the presence of parathion-containing samples. After washing, the fibres were mounted in a flow cell, and perfused with fluorescein-labelled anti-antibody, giving rise to a measurable fluorescence signal proportional to the amount of parathion in the original sample. Although the sensitivity of the method appeared to be quite high, the technique did not appear usable for multiple samples, and the problems of regenerating the fibres for re-use, and the robustness of the system, remain unaddressed.

5.1.2 *Advantages and disadvantages of immunoassay*

As noted above in section 5.1.1, the general area of immunoassay has been thoroughly reviewed, and the limitations as well as the advantages of the technology have been discussed. All of the positive and negative aspects of immunoassay apply to those assays used for pesticide analysis. These are again summarised in Table 5.1. Although many of the first immunoassays for pesticides were radioimmunoassays, these have been superseded by enzyme-based assays because of the difficulties of radioactively labelling the reagents, and the environmental concerns related to exposure of personnel and disposal of the reagents.

5.2   Pesticide determination by immunoassay

The major classes of pesticides are listed in Table 5.2, and the applications of immunoassay to the detection and determination of these compounds are discussed below. Throughout the following sections, the term 'pesticide' will be used to describe insecticides, fungicides, rodenticides, herbicides, and other agents that are used against pests causing damage to food crops and food-producing animals.

5.2.1   Paraquat

Paraquat and diquat are bipyridinium herbicides widely used as contact weed killers. Mortality from paraquat poisoning is high (Haley, 1979); therefore, the ability to monitor the presence of paraquat in clinical specimens, as well as to detect paraquat residues on food crops, is quite important. However, the majority of immunoassays developed for paraquat have been applied to clinical use, and few have been used for the detection of paraquat in soils, crops, or foods.

Quite probably due to the clinical importance of paraquat poisoning, one of the first immunoassays for the determination of any pesticide was a radio-immunoassay developed for the determination of paraquat in the blood of people who had ingested commercial formulations of paraquat solutions (Levitt, 1977, 1979). This original assay was able to detect paraquat in plasma or urine at levels as low as 0.6 μg/ml. The radioassay was applied to the measurement of paraquat in poisoned patients in order to develop a correlation between paraquat levels and prognosis (Proudfoot *et al.*, 1979) to help differentiate the patients who required minimal treatment (levels below 1 mg/ml) from those for whom treatment would not be beneficial

Table 5.2 Major types of chemical pesticides

Type of compound	Examples
Organophosphorus	Parathion, malathion, paraoxon, fenamiphos
Carbamate	Benomyl, carbaryl, carbofuran, aldicarb
Organohalogen	DDT, 2,4-D, alachlor
Cyclodiene	Endrin, dieldrin, chlordane
Hexachlorocyclohexane	Lindane, BHC
Rotenoid	Rotenone
Pyrethroid	Allethrin
Triazine	Atrazine, prometryne, ametryne, simazine
Substituted urea	Diflubenzuron, chlorsulfuron, diuron
Bipyridyliums	Paraquat, diquat
Arsenicals	Ammonium and sodium salts of methanearsonate (MAMA, MSMA, DSMA)
Fluorine rodenticides	Fluoroacetate, fluoroacetamide
Nicotinoid	Nicotine
Warfarin anticoagulants	Warfarin

(above 1 mg/ml) and those for whom aggressive treatment could be efficacious (Braithwaite, 1987; Scherrmann *et al.*, 1987). Fatori and Hunter (1980) developed two competitive RIAs for serum paraquat using either a ^{3}H tracer (able to determine paraquat in the 10–2500 ng/ml range) or an ^{125}I tracer (usable down to 0.1 ng/ml). Additional RIAs for the determination of paraquat in serum, urine, or acid extracts of tissue have been reported by Nagao *et al.* (1989).

Coxon *et al.* (1988) developed a fluorimetric immunoassay for serum paraquat: fluorescein-labelled paraquat was mixed with test serum containing unlabelled paraquat, sheep anti-paraquat antibody was added, and antibody-bound paraquat was precipitated by the addition of donkey anti-sheep antibodies. This assay was able to detect serum paraquat only in the range of 20–2000 ng/ml.

An enzyme immunoassay developed to detect paraquat in serum samples at levels of 0.3–10 ng/ml (Niewola *et al.*, 1983) was modified using monoclonal antibodies and applied to the detection of paraquat in acid-extracted soil samples (Niewola *et al.*, 1985, 1986).

Van Emon *et al.* (1985, 1986) developed an enzyme immunoassay that was applied to material extracted by sonication in 6M HCl, which was applicable to the determination of paraquat in milk, beef, and potatoes at levels of 2.5 μg/kg in these matrices (Van Emon *et al.*, 1987).

Mouse monoclonal antibodies directed against paraquat have been reported (Niewola *et al.*, 1985; Hogg *et al.*, 1987; Bowles *et al.*, 1988; Johnston *et al.*, 1988; Nagao *et al.*, 1989; Bowles and Pond, 1990, 1992).

5.2.2 *Organophosphorus compounds*

The organophosphorus (OP) pesticides are anti-cholinesterase (anti-AChE) agents operating at cholinergic neuromuscular junctions; these compounds can cause both acute and chronic toxicity in humans (Koelle, 1981). OPs are widely used in agriculture around the world, and yet the development of immunoassays for determination of these compounds has been slow. Also, many assays for OPs can be based on colorimetric reactions with immobilised cholinesterases (Soong and Lo, 1989; Skládal *et al.*, 1994). Rabbit antisera against malathion were produced over 25 years ago (Centeno and Johnson, 1967; Haas and Guardia, 1968). However, much of the work on assays (and immunoassays) for organophosphorus compounds subsequently has been spurred by the military interest in nerve agents (Koelle, 1981).

Rabbit antibodies to the insecticide paraoxon were used in animal protection studies (Sternberger *et al.*, 1974), and Hunter and Lenz (1982) developed a competition ELISA for paraoxon able to detect the compound in buffer at levels as low as 28 pg/ml (and tenfold higher in serum). No assays for this compound in crops or foods have been reported.

Both polyclonal and monoclonal antibodies have been produced against the pesticide fenitrothion (FN) (Hill *et al.*, 1992; McAdam *et al.*, 1992), and were used to develop an assay for the determination of this pesticide. The optimised assay employed a monoclonal antibody immobilised in the wells of a microtitre plate. FN-containing samples were mixed with a known amount of enzyme-labelled FN and applied to the wells. Thus, the amount of enzyme-labelled FN ultimately bound to the plate was inversely proportional to the amount of unlabelled FN in the sample. FN-treated grain was extracted with methanol for analysis; the assay was able to tolerate up to 10% methanol with no degradation of performance, detecting FN at levels of 0.1–20 mg/kg in the grain, with no matrix effect. This type of assay was extended by using additional antibodies (Skerritt *et al.*, 1992a) for the determination of chlorpyrifos-methyl (CPM) using a monoclonal antibody, and pirimiphos-methyl (PIRM) using a polyclonal antibody, in wheat grain and flour-milling fractions. These assays were able to detect the pesticides at levels of 0.08 mg/kg FN, 0.2 mg/kg CPM, and 0.03 mg/kg PIRM in the grain fractions.

Ercegovich *et al.* (1981) developed a competitive RIA for parathion that was able to detect this pesticide in fortified extracts of lettuce at levels of 0.1 mg/kg. Both the lettuce extract and extracts of tissue from apple, carrots, celery, corn, cabbage, endive, and parsley showed some matrix effects, as did fortified human serum samples. An extensive study of the immunogenicity of several parathion–bovine serum albumin conjugates, in which bridging groups of differing length and structure were inserted between the carrier protein and pesticide, reported the failure to obtain any parathion-specific antibodies (Vallejo *et al.*, 1982). Although the authors observe that parathion, due to its small size (MW = 291), may present special problems with respect to its immunogenicity, the successful development of other immunoassays for this pesticide would seem to belie this apparent difficulty. Anis *et al.* (1992) reported the development of a fibre-optic immunosensor for the determination of parathion (see also section 5.1.1.2 above). This assay format was able to detect parathion at levels of 0.3 μg/ml in buffer. However, as described above, this immunosensor technique does not yet represent a practical, robust method for pesticide determination.

Several monoclonal antibody-based enzyme immunoassays for the chemical nerve agent soman have been described (Hunter *et al.*, 1982; Erhard *et al.*, 1990; Glikson *et al.*, 1992); the best of these is able to detect soman at levels of 80 ng/ml (Erhard *et al.*, 1990). Enzyme immunoassays (Schmidt *et al.*, 1988; Erhard *et al.*, 1989) and a chemiluminescent immunoassay (Erhard *et al.*, 1992) have been reported for the model organophosphorus compound methyl phosphonic acid, *p*-aminophenyl 1,2,2-trimethylpropyl diester (MATP). An optimised ELISA employing a monoclonal antibody was able to detect MATP at levels as low as 40 ng/ml in water and animal sera (Erhard *et al.*, 1989). These assays are of relevance to the detection of chemical nerve agents in the environment (or in exposed personnel), but have less importance in the area of pesticide determination.

5.2.3   Carbamates

The benzimidazole (carbamate) fungicides benomyl and thiabendazole are widely used pre- and post-harvest on fruits and vegetables to prevent storage rot (Newsome and Collins, 1987; Bushway *et al.*, 1992). Residue limits range from 0.4 to 10 mg/kg in Canada for thiabendazole and benomyl, and its decomposition product, methyl 2-benzimidazole carbamate (MBC; carbendazim) (Newsome and Collins, 1987), and from 0.12 to 35 mg/kg in the USA (Bushway *et al.*, 1992), and the World Health Organization has established an acceptable daily intake of 0.3 mg/kg of body weight/day (Brandon *et al.*, 1993).

One of the first immunoassays for benomyl was based on the change in fluorescence polarisation which occurred upon the binding of an anti-benomyl antibody to fluorescent benomyl (Lukens *et al.*, 1977). Although able to measure benomyl in the sub-nanogram/ml range, the assay could be used only in a neutral aqueous environment, and required sophisticated equipment with which to measure the fluorescence polarisation.

Rabbit antisera were used to develop both an RIA (Newsome and Shields, 1981) and, later, an EIA (Newsome and Collins, 1987) for benomyl and thiabendazole in commodities such as oranges, lemons, apples, grapes, peaches, tomatoes, and cucumber. The samples were extracted with ethyl acetate in sodium carbonate, and the residue was re-dissolved in methanol after the original solvent was evaporated. Good recovery was obtained, and the assay was able to detect benomyl at levels of 0.35 mg/kg and thiabendazole at levels of 0.03 mg/kg (Newsome and Collins, 1987).

Bushway *et al.* (1992, 1993) utilised commercial benomyl assay kits (Envirogard[TM] EIA kits, available from Millipore Corp., Bedford, MA, USA) to assay for MBC in blueberries (Bushway *et al.*, 1992) and wine (Bushway *et al.*, 1993). Methanol extracts of blueberries were dried, then re-dissolved in water; for analysis, a format consisting of plastic tubes coated with anti-MBC antibodies was used. The assay was able to measure MBC in fortified blueberry extracts at levels of 18 μg/kg (Bushway *et al.*, 1992). Both the tube format and microtitre-plate format were employed in measuring MBC in samples of red and white wines (Bushway *et al.*, 1993). For best results in the tube assay format, the wine samples were dried, then re-dissolved in a diluted acetonitrile–methanol–water–monoethanolamine assay buffer containing skim/non-fat dry milk. Wine diluted 10–100-fold in water could be assayed directly in the microtitre plate-based assay. The limits of quantitation for the two assays were 5 μg/kg and 50 μg/mg for the tube and the plate formats, respectively (Bushway *et al.*, 1993).

Brandon and co-workers have developed a monoclonal antibody-based ELISA for thiabendazole in bovine urine and liver (Brandon *et al.*, 1992) and in potatoes and apples (Brandon *et al.*, 1993). Thiabendazole could be extracted from liver tissue with water, and urine diluted eight-fold with water, for assay. The assay of urine was usable for screening purposes, and could determine the difference between control and near-tolerance samples; thiabendazole could be detected in extracts of 0.1 gm liver at levels of <20 μg/kg (Brandon *et al.*, 1992). When this assay was applied to potatoes and apples, extracts of the peels (where the pesticide would be presumed to be concentrated, due to the use of this pesticide as a post-harvest dip or spray) from the crops were prepared in water, clarified, and used directly. Under these conditions, the limit of detection in apple peel was 160 μg/kg and the limits for potatoes were 360 μg/kg in the whole potato and 120 μg/kg in the peel, well below the USA tolerance of 10 mg/kg.

Itak *et al.* (1993a) coupled rabbit antiserum against carbendazim to magnetic particles (Advanced Magnetics, Cambridge, MA, USA; see also Rubio *et al.*, 1991, Itak *et al.*, 1992 and Lawruk *et al.*, 1993a,b for further discussions of this technology) to develop an assay for this fungicide. Although this technique substitutes a magnetic particle for the usual microtitre well as the solid phase to which the antibody is bound, the competitive assay is conducted in an otherwise similar fashion: enzyme-labelled carbendazim competes with unlabelled material present in the sample, and the final colour development is inversely proportional to the amount of carbendazim in the sample. A variety of water samples and samples of orange, apple, grapefruit, and pineapple juices were analysed, as well as methanol and methanol/sodium hydroxide extracts of soil. The limits of detection for these matrices were 1 μg/kg in water, 300 μg/kg in the juices, and <40 μg/kg in 75% methanol/25% 0.5 N NaOH extracts of soil. Magnetic particle-based ELISAs have also been used for the detection of triazine and aldicarb (see below).

Molinate (Ordram, *S*-ethyl hexahydroazepine-1-carbothioate) is used as a herbicide to control grasses in rice fields (Gee *et al.*, 1988; Harrison *et al.*, 1989a). Rabbit antisera were developed against molinate derivatives conjugated to carrier proteins. When ELISA standard curves were produced using the best combination of antiserum and coating antigen, the lowest part of the standard curve represented a concentration of 15 ng/ml, whereas the lowest amount of molinate that could actually be measured in buffer was 3 ng/ml (Gee *et al.*, 1988). The ELISA was tested in comparison to a GC method using fortified and field samples of air, water, and soil. Liquid–liquid extraction into propylene glycol and acetonitrile and solid-phase extraction using C_{18} cartridges were also applicable to the ELISA, because molinate could be eluted from the cartridge with methanol or acetonitrile. The combination of SPE plus ELISA yielded a detection limit of 1 μg/kg (Li *et al.*, 1989) and excellent correlation with the results obtained by GC was found (Harrison *et al.*, 1989a; Li *et al.*, 1989).

Carbaryl was one of the first carbamate insecticides, introduced in 1958. It is currently used on more than 100 crops, on lawns, and on domestic animals for controlling over 100 major pests (Baron, 1991). Marco *et al.* (1993) described a competitive ELISA for carbaryl utilising rabbit antisera bound in microtitre wells and enzyme-labelled tracer to compete with analyte in the sample being assayed. Carbaryl-fortified samples of water, milk, honey, urine, and extracts of soil were assayed. Although water could be assayed directly, the other samples had to be diluted with buffer (1/25 for soil extract, 1/50 for urine, and 1/1000 for honey) to eliminate matrix effects; milk had to undergo sample clean-up procedures, such as SPE. The assay was able to detect carbaryl in these matrices at levels as low as 0.05 ng/ml. This assay was sensitive to organic solvents, with acetone, acetonitrile, and polypropylene glycol having negative effects on assay sensitivity. However, this assay system was able to tolerate up to 10% methanol, as has been seen with many other assay systems examined.

Itak *et al.* (1993b) applied the magnetic particle-based ELISA system (Carbaryl RaPID Assay®, Ohmicron Corporation, Newtown, PA, USA) to the determination of carbaryl in water. The assay, carried out in disposable plastic tubes, had a limit of detection of 0.25 μg/kg of carbaryl in water, and correlated well with carbaryl values determined by an HPLC method over a 1–100 μg/kg range.

Aldicarb is a widely used insecticide applied to the soil of many crops, and may be found in the tissues of animals fed aldicarb-treated feed; tolerance levels in meat and meat by-products of domestic animals is 0.05 mg/kg, and milk tolerance is 0.01 mg/kg. Brady *et al.* (1989) described an EIA for aldicarb, specific for the parent compound, and not the sulphoxide nor the sulphone. The antibody was absorbed to the wells of a microtitre plate, and the standard type of competitive assay using enzyme-labelled tracer was employed. Matrices such as stream water, lemonade, limeade, orange juice, liquid from watermelon, urine, saliva, and rabbit and sheep sera were simply filtered, fortified with aldicarb and assayed, although the pH needed to be adjusted to above 6.0 (e.g. in the citrus juice material) to avoid antibody inactivation. Very similar standard curves were obtained with these different matrices; the usual linear response range was 15.6–2000 ng aldicarb/100 μl volume in the well; the least detectable level of aldicarb was 0.3 mg/kg.

Itak *et al.* (1992) used the magnetic particle-based format to develop an immunoassay for aldicarb in water, which was also applied to the detection of aldicarb in fortified lemonade samples. The assay was usable over the pH range of 3–10, showed no matrix effect in lemonade samples if they were diluted 1:50 or greater or 1:20 if filtered. This assay, in contrast to the one reported by Brady *et al.* (1989), detected the parent compound and the sulphone equally well, but was less sensitive to the sulphoxide. The limit of detection of this assay in water samples was 0.25 μg/kg aldicarb, and it could detect aldicarb at 25–50 μg/kg in lemonade (aldicarb tolerance for lemons, 300 μg/kg; see Title 40 US Code of Federal Regulations, 150–159).

Lehotay and Argauer (1993) used the EnviroGard™ carbofuran test kit (tube format) and the EnviroGard™ aldicarb test kit (microtitre plate format) (both from Millipore Corporation, Bedford, MA, USA) to assay for the presence of these compounds in meat (ground chuck, liver, and cat food) and in meat and liquids (milk, blood, and urine), respectively. For carbofuran determinations in meat, the samples were extracted with water or acetonitrile (the acetonitrile extract was dried, then re-dissolved in water for assay), both methods giving comparable results. Aldicarb sulphone determinations were done either on water extracts of meat (although the assay was able to tolerate up to 1% acetonitrile) or directly on the milk, blood, and urine samples. These assays were able to detect the pesticides of interest at levels at or below the tolerance for all the matrices tested, except for pig liver, which appears to contain enzymes capable of decomposing carbofuran. In general, the authors concluded that these assays, due to their sensitivity to matrix effects, and lack of ability to determine the presence of the pesticides quantitatively, were useful only as screening assays (Lehotay and Argauer, 1993).

5.2.4 Organohalogens

The premier example of the organohalogen group of pesticides, 2,2-bis-(p-chlorophenyl)-1,1,1-trichloroethane (DDT), although no longer in use, is a highly stable compound, insoluble in water and soluble in organic solvents (including depot fat, which allows it to accumulate in body tissues). It is metabolised in the body to a number of compounds, chiefly 2,2-bis-(p-chlorophenyl)-acetic acid, DDA. Centeno and Johnson (1970) reported the production of rabbit antisera to DDA (and also to malathion, an organophosphorus pesticide), although the sera did not react with DDT, nor was an immunoassay developed from these materials. Banerjee (1987) also developed an ELISA employing rabbit antisera for the detection of DDA in urine with a detection limit of 5–10 ng/ml; this assay was not applied to crops or water samples.

Langone and Van Vanukis (1975) developed a competitive RIA for dieldrin and aldrin, utilising serum from rabbits immunised with a human serum-conjugated derivative of dieldrin. The detection limits for these compounds were 50 pg/ml and 2 ng/ml, respectively. Again, this assay was not employed in the determination of residues in foods or crops. Interestingly, they reported that in order for the assay to work, the buffers had to contain 10% horse serum; dieldrin is far more soluble in serum-containing buffers than in non-protein-containing buffers. Dreher and Podratzki (1988) reported a capture-type EIA for the insecticide endosulfan capable of detecting this compound at levels of 3 μg/kg (or a measuring range between 3 and 400 ng/ml) in aqueous solutions (soil extracts and water). However, the assay had a high cross-reactivity with endrin. Wigfield and Grant (1992) evaluated the use of the Res-I-Mune[TM] immunoassay kit (available from Immuno-Systems, Inc., Scarborough, ME, USA). This kit utilises a plastic tube coated with antibodies to the pesticide; it can detect endrin, endosulfan, and dieldrin. Crop materials (apples, tomato, and lettuce) were extracted with organic solvents, and the extracts were dried and re-dissolved in aqueous buffers for analysis. Although the kit was able to detect the pesticides studied at levels of 10–30 μg/kg in crops, the authors felt the kit was useful mostly as a quick 'yes/no' screening tool.

Several groups have been developing immunoassays for the herbicides 2,4-dichlorophenoxyacetic acid (2,4-D) and 2,4,5-trichlorophenoxyacetic acid (2,4,5-T) in water, in which the maximum limit is 70 μg/kg. Canadian regulations limit 2,4-D to 0.1 mg/kg in corn, strawberries, and cereals, 2 mg/kg on citrus, and 1 mg/kg on apples (Newsome and Collins, 1989). Most of the assay systems described below employed methanol extraction of the compounds from the matrix, clean-up/concentration on C_8 cartridges, removal of the alcohol, and re-dissolution in water or buffer prior to analysis. Rinder and Fleeker (1981) described an RIA using rabbit sera for these compounds with limits of detection of 0.1 ng/ml and 1 ng/ml, respectively. However, the assay was unable to distinguish between the two compounds if both were present. A further development of this assay system (Fleeker, 1987) into an EIA employed two different enzyme-hapten labels and antibodies raised against two different immunogens in an attempt to reduce cross-reactivity and increase sensitivity. However, the detection limits were no better than the previous RIA, and the assay, run in tubes, required an overnight (18 h) incubation period. Knop *et al.* (1985) likewise used an RIA to detect 2,4-D in water, plasma, and urine. This assay also had a detection limit of approximately 5 ng/ml, but required a 24 h incubation period. Hall *et al.* (1989) reported both an enzyme immunoassay for 2,4-D and RIAs for 2,4-D and picloram (4-amino-3,5,6-trichloro-2-pyridinecarboxylic acid) in urine and water. The 2,4-D ELISA could detect the compound at levels below 100 ng/ml, and the RIAs were able to detect 2,4-D and picloram at levels under 50 ng/ml. Total incubation times for all assays were under 4 h. The authors indicated that the RIA was preferable to the EIA because it was simpler, and more efficient and rapid. However, the disadvantages of RIA have been previously discussed (section 5.1.2). Deschamps *et al.* (1990) compared a rabbit polyclonal antiserum-based immunoassay for picloram with one based on a mouse monoclonal antibody. The polyclonal assay had a working range of 5–5000 ng/ml, with a detection limit of 5 ng/ml, whereas the monoclonal assay had a working range of 1–200 ng/ml, with a detection limit of 1 ng/ml. The assays were applied to the analysis of fortified water, soil extracts, urine, and plant tissue (grass clippings). In this case, the authors favoured the ELISA format for its sensitivity and precision. In order for their polyclonal serum-based ELISA to detect 2,4-D in food and crop materials, Newsome and Collins (1989) used a preparation procedure employing acid–methanol extraction followed by partitioning with dichloromethane. The assay was used to determine 2,4-D in fortified corn, strawberry, oats, barley, apple, and orange. With the extraction procedure, the assay was able to detect the herbicide in the 50–200 μg/kg range in corn, strawberry, and cereal samples, and in the 0.5–2 mg/kg range in apple and orange. Lawruk *et al.* (1994) used a magnetic particle-based immunoassay (see section 5.1.1.2 above) to determine 2,4-D in groundwater samples. They reported a sensitivity of 0.7 μg/kg, approximately 1/100 of the allowable USA level in drinking water.

Alachlor, the active ingredient of Lasso[®], is the most widely used of the chloroacetanilide herbicides. The maximum contaminant level in drinking water is zero (USEPA). Other herbicides in this class of compounds include metolachlor, amidochlor, butachlor, and the related chloroacetamide herbicide metazachlor (see Table 5.1). Feng *et al.* (1990) described an ELISA for alachlor in environmental water samples. The samples (1 ml) could be analysed without any pretreatment, and could detect alachlor in water in the range of 0.2–8 μg/l. However, this ELISA method was generally less accurate and less precise than GC–MS analysis, and had much larger CVs than the instrumental method. This assay was later modified (Feng *et al.*, 1994) using an enzyme-labelled hapten to produce an assay system potentially able to monitor urine levels of alachlor and its metabolites to assess human exposure to alachlor. This assay could detect alachlor residues in the 0.25–16 μg/kg range in the urine of monkeys given alachlor intravenously. France and King (1991) used supercritical fluid extraction (SFE) of meat products followed by enzyme immunoassay (a commercial kit, the Res-I-Quant Alachlor Immunoassay Kit, ImmunoSystems, Scarborough, ME, USA) to detect alachlor residues (SPE was also used to extract carbofuran—see sections 5.1.1.2 and 5.2.3 above). Alachlor could be confirmed in lard spiked at levels of 10 μg/kg or greater. Lawruk *et al.* (1992) have also applied the magnetic particle-based assay (Ohmicron Corporation, Newtown, PA, USA) to the detection of alachlor in water. This assay can detect alachlor in water samples down to a level of 50 ng/l.

Metolachlor is the active ingredient of Dual[®], a herbicide used on corn, soybeans, cotton, potato, peanuts, and sugar cane (Schlaeppi *et al.*, 1991; Lawruk *et al.*, 1992). A monoclonal antibody-based assay for this compound, developed in both indirect (competitive; enzyme-labelled second antibody) and direct (enzyme-labelled antigen) ELISA formats, was able to detect metolachlor in fortified soil samples which had been extracted with methanol/water, dried, and redissolved in PBS–Tween buffer (Schlaeppi *et al.*, 1991). The minimum amounts of metolachlor detectable by the two formats were 0.05 and 0.1 μg/kg by indirect and direct ELISA, respectively. Although the direct ELISA was slightly less sensitive, it was faster (4 h versus 5.5 h). The magnetic particle-based assay of Lawruk *et al.*

(1993b) showed a detection range of 3–300 µg/kg metolachlor in methanol/ water-extracted fortified soils, and 0.05 µg/l in water samples. In addition to an ELISA for alachlor (Feng *et al.*, 1990), Feng *et al.* (1992) developed specific ELISAs for the related chloroacetanilide herbicides metolachlor, amidochlor, and butachlor. Hapten–protein conjugates for immunisation were easily made by reacting the native compound with carrier protein (human serum albumin or sheep IgG) by using *S*-acetylmercaptosuccinic anhydride (AMSA) or *N*-acetylhomocysteine thiolactone (AHT). Each of the (competitive) ELISAs was capable of detecting its specific analyte at levels of 6–10 µg/kg.

A direct competitive ELISA for metazachlor, a related α-chloroacetamide herbicide and the active ingredient of Butisan S® was developed by Scholz and Hock (1991). Because this group of pesticides reacts well with sulphydryl groups, BSA with activated –SH groups was used as the carrier protein to produce the immunising conjugates. The optimised assay was capable of detecting metazachlor in the range of 0.01–1 µg/ml, well within the European Community (EC) guidelines for drinking water of 0.1 µg/ml for a single pesticide, or 0.5 µg/ml for the sum of pesticides.

Additional immunoassays for several other organohalogen pesticides have been developed by several groups. These assays are summarised in Table 5.3.

5.2.5 Triazines

Derivatives of the *s*-triazines have gained wide acceptance as herbicides; their persistence has become a problem in soil and groundwater contamination. The most commonly used compound is atrazine, often used in crop weed and grass management, and found in corn and fruit. Because of the potential for contamination of water and foods, a great deal of effort has gone into the development of immunoassays for the members of the *s*-triazine group of herbicides.

Several different groups have developed assays for *s*-triazines that detect either one or several members of this class, and have used polyclonal rabbit and sheep (Wüst and Hock, 1992) antibodies and monoclonal mouse antibodies (Schlaeppi *et al.*, 1989; Giersch and Hock, 1990; Schneider and Hammock, 1992; Giersch, 1993; Muldoon *et al.*, 1993) in a variety of assay formats. Likewise, there has been a great deal of interest in the development of commercial immunoassay kits for these compounds (see Table 5.6 below). Additionally, several different assay formats and technologies have been exploited during the development of all these assays, and the effects of solvents (Stöcklein *et al.*, 1990; Harrison *et al.*, 1991), matrices (Stearman and Wells, 1993), extraction procedures (Stearman and Adams, 1992) and other assay parameters have been extensively explored.

Guilbault *et al.* (1992) developed a piezoelectric atrazine sensor (i.e. a crystal coated with antibodies to atrazine; see section 5.1.1.2 above) to

Table 5.3 Immunoassays for organohalogen pesticides

Pesticide/herbicide	Assay format	Matrix	Detection range	Reference(s)
Clomazone (Command[R])	Competitive ELISA	Soil (acetonitrile extract)	$0.5-500\,\mu$g/kg	Koppatschek *et al.* (1990)
Norflurazon	Competitive ELISA	Water	$1-1000\,\mu$g/l	Riggle and Dunbar (1990)
Trifluralin	Competitive ELISA	Water	$0.1-1.0\,$mg/l	Riggle (1991)
Pentachlorophenol	Competitive ELISA	Water	$30-400\,\mu$g/l	Van Emon and Gerlach (1992)
Bromacil	Competitive ELISA	Buffer; water; basic extracts of soil	$0.1-160\,\mu$g/kg	Szurdoki *et al.* (1992), Bekheit *et al.* (1993)
Triazole fungicides (tetraconazole, etc.)	Competitive ELISA	Buffer	$12-850\,\mu$g/l	Forlani *et al.* (1992)
Dichlorprop	Fluorescence polarisation	Apple extract	$0.01-100\,$mg/l	Sánchez *et al.* (1993a,b)
Diclofop-methyl (diclofop)	Fluoroimmunoassay	Soil, sugar beets, wheat, soybeans, milk, urine, serum	$45\,\mu$g/l	Schwalbe *et al.* (1984)
	Competitive EIA		$23\,\mu$g/l	

measure this compound in water. The sensor could detect atrazine in water at levels of 0.03–100 μg/kg, but could only be re-used eight or nine times. Antibody-coated magnetic beads have also been used to develop assays for atrazine and relate triazines with a sensitivity of 50 ng/l (Rubio *et al.*, 1991), and for cyanazine in water and soil extracts with detection limits of 35 ng/l and 3.5 μg/kg, respectively (Lawruk *et al.*, 1993a).

Giersch (1993) developed a 'dipstick' assay format for atrazine in which antibody-coated membranes in plastic carriers were dipped into tubes containing samples plus enzyme-labelled tracer, washed, and developed in substrate. Although semi-quantitative, with a detection limit of 0.5 μg/ml, this type of system most readily lends itself to use as a qualitative screening technique.

In a series of papers (Goh *et al.*, 1992a,b, 1993), Goh and co-workers described the use of an immunoassay (Lucas *et al.*, 1991) for atrazine and simazine for compliance monitoring of these pesticides in different Californian soils.

A number of groups have evaluated commercially available pesticide immunoassay kits (see below and Table 5.6) for the measurement of atrazine exposure of field workers (Reed *et al.*, 1990), atrazine in water and soil (Bushway *et al.*, 1988, 1991; Thurman *et al.*, 1990; Stearman and Wells, 1993), and atrazine in cornmeal and corn products (Wigfield and Grant, 1993). Additionally, a multi-laboratory comparison of several commercial and research assays for *s*-triazines in water has been reported (Hock *et al.*, 1991).

Although the majority of assays for triazines have been developed and used for the detection of these compounds in water and soil, they have also been used for the detection of atrazine in milk, fruit juices, corn and potato products, and similar matrices (Bushway *et al.*, 1989; Wigfield and Grant, 1993; Wittmann and Hock, 1993a).

Additional information on immunoassays developed for *s*-triazine pesticides is presented in summary form in Table 5.4.

5.2.6 *Substituted urea compounds*

Researchers at the University of California have been mainly responsible for the development of immunoassays for the substituted urea group of pesticides. Diuron is used as a herbicide for soil sterilisation at high concentrations, and at lower concentration on potatoes, cotton, corn, beans, cereals, and orchards and vineyards; tolerance for meat and most commodities is 1–2 mg/kg, but can be as low as 100 μg/kg in bananas and peaches (Karu *et al.*, 1994a). The first assays developed for diuron employed rabbit antisera in an indirect competitive format (Newsome and Collins, 1990). Later assays (Karu *et al.*, 1994a,b) were either direct competitive, employing murine monoclonal antibodies bound to the microtitre plate and an enzyme-labelled diuron conjugate, or two-step indirect competitive, with hapten

Table 5.4 Immunoassays for *s*-triazine herbicides

Compounds	Assay format	Matrices	Sensitivity	Reference(s)
Atrazine	Direct (tube format)	Water, soil	0.5–10 μg/l	Bushway *et al.* (1988)
Atrazine	Indirect	Buffer	0.1–1 μg/l	Dunbar *et al.* (1990)
Atrazine	Direct (plate)	Water	1.1–2200 μg/l	Huber (1985)
	Direct (microspheres)	Water	0.1–55 μg/l	
Atrazine	Direct	Soil extracts	<5 mg/l	Schewes *et al.* (1994)
Atrazine	Direct	Water	1 μg/l	Wittmann and Hock (1989)
Propazine			0.1 μg/l	
Atrazine	Indirect	Buffer	1–100 μg/l	Goodrow *et al.* (1990)
Simazine			10–200 μg/l	Harrison *et al.* (1991)
Diethylatrazine	Direct	Water, soil	0.01 μg/l (ppb)	Wittmann and Hock (1991)
Diisopropylatrazine				Wittmann and Hock (1993b)
Hydroxyatrazine,	Direct (double antibody coat)	Water, soil, manure, urine	0.1–1 μM	Lucas *et al.* (1993)
Hydroxysimazine				
Terbutryn	Direct	Water	25–1500 μg/l	Huber and Hock (1985)
Terbutryn	Indirect	Buffer	0.1–1 μg/l	Böcher *et al.* (1992)

conjugate bound to the plate, and the sample plus standard preincubated before being applied to the plate; detection was via enzyme-labelled second antibody. The indirect assay was able to detect diuron in the 2–20 µg/kg range in groundwater. The indirect assay was also able to tolerate up to 15% methanol, an advantage due to the increased solubility of diuron in methanol. This assay was further validated (Karu *et al.*, 1994b) using groundwater samples. Water samples (200 ml) were filtered and applied to a C_{18} column, which was eluted with methanol. Samples were diluted with assay buffer to 12.5% methanol before analysis. The final optimised assay had a limit of detection of 0.07 µg/kg. A set of direct competitive assays utilising different immobilized rabbit anti-diuron antibodies and enzyme-labelled haptens was developed for the determination of diuron, monuron, and linuron in water, milk, and orange juice matrices (Schneider *et al.*, 1994). The appropriate combination of antiserum with enzyme–hapten gave assay sensitivities for monuron, diuron, and linuron of 0.05 µg/l, 0.04 µg/l, and 0.08 µg/l, respectively. These assays could tolerate 10% methanol (some, up to 50% methanol), and could be used directly on the milk and juice materials after simple dilution.

Two-step indirect ELISAs for the benzoylphenylurea insect growth regulator diflubenzuron have been described (Wei *et al.*, 1982; Wei and Hammock, 1982). The anti-pesticide rabbit antibodies used in the assays were either detected with enzyme-labelled second antibody, or were themselves labelled with enzyme (modified indirect assay). The indirect assay could detect diflubenzuron in milk at a level of 40 µg/kg, the modified assay could detect this pesticide in water at a level of 1 µg/kg. Later developments of this assay system (Wei and Hammock, 1984) allowed the detection of analyte in milk at the level of 2 µg/l.

Immunoassays have also been reported for chlorsulfuron (Kelley *et al.*, 1985) and for methabenzthiazuron (Kreissig *et al.*, 1991).

5.2.7 Pyrethroids

Pyrethroids (pyrethrins), derived from plants such as the chrysanthemum, have become important insecticides due to their effectiveness, low mammalian toxicity, and rapid decomposition. A radioimmunoassay has been reported for *S*-bioallethrin, the marketed insecticidal isomer of allethrin (Wing *et al.*, 1978; Wing and Hammock, 1979). The competitive RIA was able to detect the presence of *S*-bioallethrin at concentrations in the 0.01 nM range.

Stanker *et al.* (1989) developed a monoclonal antibody-based competitive immunoassay for permethrin in meat in which 3-phenoxybenzoic acid (Pba) coupled to BSA was used to coat microtitre plate wells. Meat was extracted with acetonitrile–water and partitioned into hexane, the hexane was run through an alumina column which was eluted with benzene, and the eluate was dried, then re-dissolved and assayed in 6% acetonitrile/buffer. Samples were added to wells, followed by the monoclonal antibody. After incubation, followed by washing, enzyme-labelled second antibody was used to develop the plates. The response of the assay was linear in the 50–500 μg/kg range, well within the regulatory limit of 150 μg/kg. The assay was able to tolerate up to 5% acetonitrile.

This assay was improved (Skerritt *et al.*, 1992b) by coupling enzyme directly to competitor Pba, and coating the microtitre plate wells with the monoclonal antibody. This assay was then applied to the detection of permethrin and 1(*R*)-phenothrin in wheat grain and flour milling fractions. Samples were extracted with methanol, acetonitrile, hexane or acetone, depending on the matrix (methanol and acetonitrile extracts of ground grain were cleaned up on alumina columns), then dried and re-dissolved in methanol before analysis. The optimised assay system could detect methanol-extracted permethrin at 1.5 μg/l (75 μg/l in the original sample) and phenothrin at 2 μg/l.

Hill *et al.* (1993) reported a polyclonal antibody-based immunoassay for bioresmethrin (BRM). Enzyme-conjugated BRM was added to wells containing bound anti-BRM antibody, along with methanol-extracted whole or ground wheat or barley samples. This assay system could detect 2 μg/kg BRM in buffer, and 50 μg/kg in whole wheat and barley grain samples.

5.2.8 Miscellaneous pesticide compounds

A variety of immunoassays have been reported for other pesticide compounds. These are briefly described in Table 5.5, which lists the pesticide, the type of assay (see above), the assay's sensitivity or detection range, and references to the literature.

5.3 Commercial pesticide immunoassay kits

Several companies have either developed assay kits or licensed kits developed by academic research groups, and are marketing these for use in a variety of matrices. These kits come in a variety of formats, but are generally enzyme-based (EIAs). A brief listing of manufacturers and commercially available kits and related equipment is presented in Table 5.6.

5.4 Official evaluation/acceptance of pesticide immunoassay kits

In the USA, three government agencies are mainly responsible for monitoring the presence of pesticides in the environment and foodstuffs. The Environmental Protection Agency (USEPA) is the agency that registers commercial marketed pesticides and monitors their presence in the general

Table 5.5 Immunoassays for miscellaneous pesticides

Pesticide	Assay format	Matrices	Sensitivity	Reference(s)
Warfarin	Direct RIA	Plasma (treated rats)	$0.025\,\mu$g/l	Cook *et al.* (1977, 1979)
Metalaxyl	2-step competitive ELISA	Cucumber, avocado, potato, squash, tomato (methanol extract)	0.10–2.0 mg/kg	Newsome (1985)
Triadimefon	2-step competitive ELISA	Apple, pear, grape, pineapple (methanol or ethyl acetate extract)	0.5 mg/kg	Newsome (1986b)
Bentazon	2-step competitive ELISA	Water (buffer)	1–$100\,\mu$g/l (only *N*-derivatives)	Li *et al.* (1991)
Fenpropimorph	Competitive/capture ELISA	Water	$0.1\,\mu$g/l	Jung *et al.* (1989b)
Imazaquin	Direct competitive ELISA	Water (buffer)	$0.40\,\mu$g/l	Wong and Ahmed (1992)
Captan	2-step competitive ELISA	Raspberry, nectarine, apple, apricot, grape (methanol extract)	$1\,\mu$g/l (as tetrahydrophthalimide)	Newsome *et al.* (1993)
Maleic hydrazide	2-step competitive ELISA (monoclonal)	Buffer	<1 mg/l	Harrison *et al.* (1989b)
Methoprene	Indirect ELISA	Tobacco (acetonitrile extract)	$5\,\mu$g/kg	Mei *et al.* (1991)
	Competitive EIA	Tobacco	$1\,\mu$g/kg	Hill *et al.* (1991)
		Grain; ground wheat	$0.25\,\mu$g/kg	
	EIA (EnvironGard kit)	Tobacco	0.64 mg/kg	Heckman *et al.* (1992)

Table 5.6 Manufacturers of pesticide detection kits and equipment

Manufacturer/location	Assay kit(s)	Comments
ACTIO, Inc. Oakland, CA	Pesticide Detection Kit (parathion, phosmet, etc)	Colorimetric cholinesterase inhibition; water, crop extracts
Agri-Diagnostics Cinnaminson, NJ	Pesticide ELISA kits	Soil testing, also fungal toxin tests
Environmental Diagnostics Burlington, NC	EZ-Screen (paraquat)	Urine, serum, crop and food extracts
EnzyTech, Inc. Lenexa, KS	ENZYTEC Pesticide Detector Ticket (organophosphorus/ carbamate pesticides)	Cholinesterase inhibition; water testing
ImmunoSystems, Inc. Scarborough, ME	Res-I-Mune (aldicarb, chlordane, carbofuran, 2,4-D)	Tube format; water, food extracts
	Res-I-Quant (alachlor, benomyl, triazines)	Plate format; water, food extracts
Millipore, Inc. Bedford, MA	EnviroGard (aldicarb, carbofuran, benomyl)	Soil, water; also make portable ELISA reader
Ohmicron Corp. Newtown, PA	RaPID assay (aldicarb, carbaryl, alachlor)	Magnetic-particle based; water
J T Baker Phillipsburg, NJ	N/A	Distributor of Ohmicron Corp RaPID assays

environment, such as soil and groundwater. The Food and Drug Administration (USFDA) has the responsibility for measuring the presence of pesticide residues in both USA-produced and imported commodities, crops and food-stuffs, and helps set tolerance levels for these materials, and has enforcement authority for seizing violative materials. The Department of Agriculture (USDA), through its Food Safety and Inspection Service (FSIS), monitors pesticide residue levels in meat, poultry and similar products. Although some of their monitoring activities overlap, each agency is developing its own guidelines and validation procedures for the acceptance and use of laboratory immunoassays and prepackaged commercial kits. Additionally, the Association of Official Analytical Chemists (AOAC International) has also established a task force and guidelines on test kits [Mastrorocco and Brodsky, 1990; see also *Journal of the Association of Official Analytical Chemists*, **72**, 694–704 (1989)].

The Analytical Environmental Immunochemical Consortium (AEIC), whose members include kit manufacturers, agriculture, industry, and government and private representatives, has been established to provide a joint forum for those involved in environmental immunochemistry, to establish test performance standards, and to furnish expertise to regulatory agencies.

5.4.1 USEPA

The USEPA has issued draft guidelines applicable to the submission of a test kit by a manufacturer to the EPA for evaluation (Van Emon, 1989).

Generally, a test must be well documented, including detailed protocols; the manufacturer must be able to demonstrate adequate stocks of important reagents, especially antibodies used in the assay; and the test should be compared to other established methodologies, such as instrumental analysis. After evaluation by the EPA of any submitted test (including a 3–5 laboratory collaborative study), the agency would formulate recommendations as to the acceptability of the test. An additional guidance memorandum has also been issued by the Office of Pesticide Programs (OPP) which describes the performance requirements for environmental chemistry methods, which includes immunoassay methodology.

5.4.2   USFDA

Although the USFDA has not yet published any formal recommendations or guidelines for the evaluation or adoption of pesticide test kits, it maintains a program of immunoassay evaluation utilising in-house and extramurally developed assays (Pohland *et al.*,1989).

5.4.3   USDA

The FSIS of USDA has published a proposed review and approval process [see *Federal Register*, **54**(158), August 17, 1989] similar to that of the USEPA, including a collaborative laboratory study as part of the validation process. As in all other cases, the processes and total costs of all development and validation are borne by the applicant.

5.5   Conclusions

Immunoassays represent a cost- and time-effective approach to the determination of pesticides and their residues in water, soil and food crops. Assays have been developed for the majority of pesticide classes used in agriculture. Many of these assays are quantitative, and may supplement or replace current instrumental assays. Others may be useful for screening, with the presumptive violative materials being quantitated by instrumental methods. Sensitivity can be increased at the expense of added analysis time and reagent costs. For appropriate analytes the requisite sensitivity can be obtained at much lower cost than with classical (instrumental) techniques. The other major advantage offered by immunochemical analysis is speed. Much of the time and expense in an analysis is involved in sample clean-up. Immunoassays minimise the need for extensive clean-up. Immunoassays can also be readily automated. Several immunoassay kits are being marketed. The USA government agencies responsible for regulating pesticides are evaluating kits submitted to them, as well as drafting and implementing guidelines for kit performance and acceptance. Immunoassays will become an increasingly important adjunct to the monitoring of pesticides in the environment and in foods.

References

Aherne, G.W. (1987) Immunoassays in environmental analysis. *Analytical Proceedings*, **24**, 140–141.

Allen, J.C. and Smith, C.J. (1987) Enzyme-linked immunoassay kits for routine food analysis. *Trends in Biotechnology*, **5**, 193–199.

Anis, N.A. *et al.* (1992) A fiber-optic immunosensor for detecting parathion. *Analytical Letters*, **25**, 627–635.

Banerjee, B.D. (1987) Development of an enzyme-linked immunosorbent assay for the quantification of DDA (2,2-bis-(*p*-chlorophenyl)-acetic acid) in urine. *Bulletin of Environmental Contamination and Toxicology*, **38**, 798–804.

Baron, R.L. (1991) Carbamate insecticides. In *Handbook of Pesticide Toxicology*, Vol. 3, Hays, Jnr, W.H. and Laws, Jnr, E.R., eds, Academic Press, San Diego, 1125–1189.

Bekheit, H.K.M. *et al.* (1993) An enzyme immunoassay for the environmental monitoring of the herbicide bromacil. *Journal of Agricultural and Food Chemistry*, **41**, 2220–2227.

Böcher, M. *et al.* (1992) Dextran, a hapten carrier in immunoassays for *s*-triazines. *Journal of Immunological Methods*, **151**, 1–8.

Bowles, M.R. and Pond, S.M. (1990) The importance of electrostatic interactions in the binding of paraquat to its elicited monoclonal antibody. *Molecular Immunology*, **27**, 847–852.

Bowles, M.R. *et al.* (1988) Large scale production and purification of paraquat and desipramine monoclonal antibodies and their Fab fragments. *International Journal of Immunopharmacology*, **10**, 537–545.

Bowles, M.R. *et al.* (1992) Quantitation of paraquat in biological samples by radioimmunoassay using a monoclonal antibody. *Fundamental and Applied Toxicology*, **19**, 375–379.

Braithwaite, R.A. (1987) Emergency analysis of paraquat in biological fluids. *Human Toxicology*, **6**, 83–86.

Brady, J.F. *et al.* (1989) Enzyme immunoassay for aldicarb. In: *Biological Monitoring for Pesticide Exposure*, ACS Symposium no. 382. American Chemical Society, Washington, DC, pp. 263–284.

Brandon, D.L. *et al.* (1992) Monoclonal antibody-based ELISA for thiabendazole in liver. *Journal of Agricultural and Food Chemistry*, **40**, 1722–1726.

Brandon, D.L. *et al.* (1993) Analysis of thiabendazole in potatoes and apples by ELISA using monoclonal antibodies. *Journal of Agricultural and Food Chemistry*, **41**, 996–999.

Bushway, R.J. *et al.* (1988) Determination of atrazine residues in water and soil by enzyme immunoassay. *Bulletin of Environmental Contamination and Toxicology*, **40**, 647–654.

Bushway, R.J. *et al.* (1989) Determination of atrazine residues in food by enzyme immunoassay. *Bulletin of Environmental Contamination and Toxicology*, **42**, 899–904.

Bushway, R.J. *et al.* (1991) Comparison of enzyme-linked immunosorbent assay and high-performance liquid chromatography for the analysis of atrazine in water from Czechoslovakia. *Archives of Environmental Contamination and Toxicology*, **21**, 365–370.

Bushway, R.J. *et al.* (1992) Determination of methyl 2-benzimidazolecarbamate in blueberries by competitive inhibition enzyme immunoassay. *Journal of the Association of Official Analytical Chemists International*, **75**, 323–327.

Bushway, R.J. *et al.* (1993) Determination of methyl 2-benzimidazolecarbamate in wine by competitive inhibition enzyme immunoassay. *Journal of the Association of Official Analytical Chemists International*, **76**, 851–856.

Centeno, E.R. and Johnson, W.J. (1967) Antibodies to two common pesticides: DDT and malathion. *Federation Proceedings*, **26**, 704.

Centeno, E.R. and Johnson, W.J. (1970) Antibodies to two common pesticides: DDT and malathion. *International Archives of Allergy*, **37**, 1–13.

Cook C.E. *et al.* (1977) Immunoassay as a tool for the analysis of enantiomers: Stereoselective radioimmunoassay for *R*- and *S*-warfarin. *The Pharmacologist*, **19**, 128 (abstract 006).

Cook, C.E. *et al.* (1979) Warfarin enantiomer disposition: Determination by stereoselective radioimmunoassay. *Journal of Pharmacology and Experimental Therapeutics*, **210**, 391–398.

Coxon, R.E. *et al.* (1988) Development of a simple, fluoroimmunoassay for paraquat. *Clinica Chimica Acta*, **175**, 297–306.

Crepeau, K.L. *et al.* (1991) Extraction of pesticides from soil leachate using sorbent disks. *Bulletin of Environmental Contamination and Toxicology*, **46**, 512–518.

Deschamps, R.J.A. *et al.* (1990) Polyclonal and monoclonal enzyme immunoassays for picloram detection in water, soil, plants, and urine. *Journal of Agricultural and Food Chemistry*, **38**, 1881–1886.

Dreher, R.M. and Podratzki, B. (1988) Development of an enzyme immunoassay for endosulfan and its degradation products. *Journal of Agricultural and Food Chemistry*, **36**, 1072–1075.

Dunbar, B. *et al.* (1990) Development of enzyme immunoassay for the detection of triazine herbicides. *Journal of Agricultural and Food Chemistry*, **38**, 433–437.

Ellis, R.L. (1989) Changing pesticide technology in meat and poultry products. *Journal of the Association of Official Analytical Chemists*, **72**, 521–524.

Ercegovich, C.D. *et al.* (1981) Development of a radioimmunoassay for parathion. *Journal of Agricultural and Food Chemistry*, **29**, 559–563.

Erhard, M.H. *et al.* (1989) Development of an ELISA for detection of an organophosphorus compound using monoclonal antibodies. *Archives of Toxicology*, **63**, 462–468.

Erhard, M.H. *et al.* (1990) Detection of the organophosphorus nerve agent soman by an ELISA using monoclonal antibodies. *Archives of Toxicology*, **63**, 580–585.

Erhard, M.H. *et al.* (1992) Development of a direct and indirect chemiluminescence immunoassay for the detection of an organophosphorus compound. *Journal of Immunoassay*, **13**, 273–287.

Fägerstam, L.G. *et al.* (1992) Biospecific interaction analysis using surface plasmon resonance detection applied to kinetic, binding site and concentration analysis. *Journal of Chromatography*, **597**, 397–410.

Fatori, D. and Hunter, W.M. (1980) Radioimmunoassay for serum paraquat. *Clinica Chimica Acta*, **100**, 81–90.

Feng, P.C.C. *et al.* (1990) Development of an enzyme-linked immunosorbent assay for alachlor and its application to the analysis of environmental water samples. *Journal of Agricultural and Food Chemistry*, **38**, 159–163.

Feng, P.C.C. *et al.* (1992) A general method for developing immunoassays to chloroacetanilide herbicides. *Journal of Agricultural and Food Chemistry*, **40**, 211–214.

Feng, P.C.C. *et al.* (1994) Quantitation of alachlor residues in monkey urine by ELISA. *Journal of Agricultural and Food Chemistry*, **42**, 316–319.

Fleeker, J. (1987) Two enzyme immunoassays to screen for 2,4-dichlorophenoxyacetic acid in water. *Journal of the Association of Official Analytical Chemists*, **70**, 874–878.

Forlani, F. *et al.* (1992) Development of an enzyme-linked immunosorbent assay for triazole fungicides. *Journal of Agricultural and Food Chemistry*, **40**, 328–331.

France, J.E. and King, J.W. (1991) Supercritical fluid extraction/enzyme assay: A novel technique to screen for pesticide residues in meat products. *Journal of the Association of Official Analytical Chemists*, **74**, 1013–1016.

Fukal, L. and Kás, J. (1989) The advantages of immunoassay in food analysis. *Trends in Analytical Chemistry*, **8**, 112–116.

Gee, S.J. *et al.* (1988) Development of an enzyme-linked immunosorbent assay for the analysis of the thiocarbamate herbicide molinate. *Journal of Agricultural and Food Chemistry*, **36**, 863–870.

Giersch, T. (1993) A new monoclonal antibody for the sensitive detection of atrazine with immunoassay in microtiter plate and dipstick format. *Journal of Agricultural and Food Chemistry*, **41**, 1006–1011.

Giersch, T. and Hock, B. (1990) Production of monoclonal antibodies for the determination of *s*-triazines with enzyme immunoassay. *Food and Agricultural Immunology*, **2**, 85–97.

Glikson, M. *et al.* (1992) Characterization of soman-binding antibodies raised against soman analogs. *Molecular Immunology*, **29**, 903–910.

Goh, K.S. *et al.* (1992a) ELISA of simazine in soil: Applications for a field leaching study. *Bulletin of Environmental Contamination and Toxicology*, **48**, 554–560.

Goh, K.S. *et al.* (1992b) Enzyme-linked immunosorbent assay (ELISA) of simazine for Delhi and Yolo soils in California. *Bulletin of Environmental Contamination and Toxicology*, **49**, 348–353.

Goh, K.S. *et al.* (1993) ELISA regulatory application: Compliance monitoring of simazine and atrazine in California soils. *Bulletin of Environmental Contamination and Toxicology*, **51**, 333–340.

Goodrow, M.H. *et al.* (1990) Hapten synthesis, antibody development, and competitive inhibition enzyme immunoassay for *s*-triazine herbicides. *Journal of Agricultural and Food Chemistry*, **38**, 990–996.

Guilbault, G.G. and Schmid, R.D. (1991) Biosensors for the determination of drug substances. *Biotechnology and Applied Biochemistry*, **14**, 133–145.

Guilbault, G.G. *et al.* (1992) A piezoelectric immunobiosensor for atrazine in drinking water. *Biosensors and Bioelectronics*, **7**, 411–419.

Haas, G.J. and Guardia, E.J. (1968) Production of antibodies against insecticide–protein conjugates. *Proceedings of the Society for Experimental Biology and Medicine*, **129**, 546–551.

Haley, T.J. (1979) Review of the toxicology of paraquat (1,1'-dimethyl-4,4'-bipyridinium chloride). *Clinical Toxicology*, **14**, 1–46.

Hall, J.C. *et al.* (1989) Immunoassays for the detection of 2,4-D and picloram in river water and urine. *Journal of Agricultural and Food Chemistry*, **37**, 981–984.

Harrison, R.O. and Hammock, B.D. (1988) Location dependent biases in automatic 96-well microplate readers. *Journal of the Association of Official Analytical Chemists*, **71**, 981–987.

Harrison, R.O. *et al.* (1989a) Evaluation of an enzyme-linked immunosorbent assay (ELISA) for the direct analysis of molinate (Ordram®) in rice field water. *Food and Agricultural Immunology*, **1**, 37–51.

Harrison, R.O. *et al.* (1989b) Development of a monoclonal antibody based enzyme immunoassay for analysis of maleic hydrazide. *Journal of Agricultural and Food Chemistry*, **37**, 958–964.

Harrison, R.O. *et al.* (1991) Competitive inhibition ELISA for the *s*-triazine herbicides: Assay optimization and antibody characterization. *Journal of Agricultural and Food Chemistry*, **39**, 122–128.

Heckman, R.A. *et al.* (1992) Validation of an enzyme immunoassay for analysis of methoprene residues on tobacco. *Journal of Agricultural and Food Chemistry*, **40**, 2530–2532.

Hill, A.S. *et al.* (1991) Determination of the insect growth regulator methoprene in wheat grain and milling fractions using an enzyme immunoassay. *Journal of Agricultural and Food Chemistry*, **39**, 1882–1886.

Hill, A.S. *et al.* (1992) Mono- and poly-clonal antibodies to the organophosphate fenitrothion. 2. Antibody specificity and assay performance. *Journal of Agricultural and Food Chemistry*, **40**, 1471–1474.

Hill, A.S. *et al.* (1993) Quantitation of bioresmethrin, a synthetic pyrethroid grain protectant, by enzyme immunoassay. *Journal of Agricultural and Food Chemistry*, **41**, 2011–2018.

Hock, B. *et al.* (1991) Enzyme immunoassays for the determination of *s*-triazines in water samples: Two interlaboratory tests. *Analytical Letters*, **4**, 529–549.

Hogg, P.J. *et al.* (1987) Evaluation of equilibrium constants for antigen–antibody interactions by solid-phase immunoassay: The binding of paraquat to its elicited mouse monoclonal antibody. *Molecular Immunology*, **24**, 797–801.

Huber, S.J. (1985) Improved solid-phase enzyme immunoassay systems in the ppt range for atrazin in fresh water. *Chemosphere*, **14**, 1795–1803.

Huber, S.J. and Hock, B. (1985) A solid-phase enzyme immunoassay for quantitative determination of the herbicide terbutryn. *Journal of Plant Diseases and Protection*, **92**, 147–156.

Hunter, K.W. *et al.* (1982) Quantification of the organophosphorus nerve agent soman by competitive inhibition enzyme immunoassay using monoclonal antibody. *Federation of European Biochemical Societies Letters*, **149**, 147–151.

Hunter, K.W. and Lenz, D.E. (1982) Detection and quantification of the organophosphate insecticide paraoxon by competitive inhibition enzyme immunoassay. *Life Sciences*, **30**, 355–361.

Itak, J.A. *et al.* (1992) Development and evaluation of a magnetic particle based enzyme immunoassay for aldicarb, aldicarb sulfone and aldicarb sulfoxide. *Chemosphere*, **24**, 11–21.

Itak, J.A. *et al.* (1993a) Determination of benomyl (as carbendazim) and carbendazim in water, soil, and fruit juice by a magnetic particle-based immunoassay. *Journal of Agricultural and Food Chemistry*, **41**, 2329–2332.

Itak, J.A. *et al.* (1993b) Validation of a paramagnetic particle-based ELISA for the quantitative determination of carbaryl in water. *Bulletin of Environmental Contamination and Toxicology*, **51**, 260–267.

Johnston, S.C. *et al.* (1988) Comparison of paraquat-specific murine monoclonal antibodies produced by in vitro and in vivo immunization. *Fundamental and Applied Toxicology*, **11**, 261–267.

Jönsson, U. *et al.* (1991) Real-time biospecific interaction analysis using surface plasmon resonance and a sensor chip technology. *BioTechniques*, **5**, 620–624.

Jung, F. *et al.* (1989a) Use of immunochemical techniques for the analysis of pesticides. *Pesticide Science*, **26**, 303–317.

Jung, F. *et al.* (1989b) Development of a sensitive enzyme-linked immunosorbent assay for the fungicide fenpropimorph. *Journal of Agricultural and Food Chemistry*, **37**, 1183–1187.

Karu, A.E. *et al.* (1994a) Synthesis of haptens and derivation of monoclonal antibodies for immunoassay of the phenylurea herbicide diuron. *Journal of Agricultural and Food Chemistry*, **42**, 301–309.

Karu, A.E. *et al.* (1994b) Validation of a monoclonal immunoassay for diuron in groundwater. *Journal of Agricultural and Food Chemistry*, **42**, 310–315.

Kaufman, B.M. and Clower, M. (1991) Immunoassay of pesticides. *Journal of the Association of Official Analytical Chemists*, **74**, 239–247.

Kaufman, B.M. and Clower, M. (1995) Immunoassay of pesticides: An update. *Journal of the Association of Official Analytical Chemists International*, **78**, 1079–1090.

Kelley, M.M. *et al.* (1985) Chlorsulfuron determination in soil extracts by enzyme immunoassay. *Journal of Agricultural and Food Chemistry*, **33**, 962–965.

Knopp, D. *et al.* (1985) Radioimmunoassay for 2,4-dichlorophenoxyacetic acid. *Archives of Toxicology*, **58**, 27–32.

Koelle, G.B. (1981) Organophosphate poisoning—An overview. *Fundamental and Applied Toxicology*, **1**, 129–134.

Köhler, G. and Milstein, C. (1975) Continuous cultures of fused cells secreting antibody of predefined specificity. *Nature*, **256**, 495–497.

Koppatschek, F.K. *et al.* (1990) Development of an enzyme-linked immunosorbent assay for the detection of the herbicide clomazone. *Journal of Agricultural and Food Chemistry*, **38**, 1519–1522.

Krämer, P. and Schmid, R. (1991) Flow injection immunoanalysis (FIIA)—A new immunoassay format for the determination of pesticides in water. *Biosensors and Bioelectronics*, **6**, 239–243.

Kreissig, S. and Hock, B. (1991) An enzyme immunoassay for the determination of methabenzthiazuron. *Analytical Letters*, **24**, 1729–1739.

Kurstak, E. (1986) *Enzyme Immunodiagnosis*. Academic Press, New York.

Kutz, F.W. *et al.* (1992) Selected pesticide residues and metabolites in urine from a survey of the U.S. general population. *Journal of Toxicology and Environmental Health*, **37**, 277–291.

Langone, J.J. and Van Vanukis, H. (1975) Radioimmunoassay for dieldrin and aldrin. *Research Communications in Chemical Pathology and Pharmacology*, **10**, 163–171.

Lawruk, T.S. *et al.* (1992) Quantification of alachlor in water by a novel magnetic particle-based ELISA. *Bulletin of Environmental Contamination and Toxicology*, **48**, 643–650.

Lawruk, T.S. *et al.* (1993a) Quantification of cyanazine in water and soil by a magnetic particle-based ELISA. *Journal of Agricultural and Food Chemistry*, **41**, 747–752.

Lawruk, T.S. *et al.* (1993b) Determination of metolachlor in water and soil by a rapid magnetic particle-based ELISA. *Journal of Agricultural and Food Chemistry*, **41**, 1426–1431.

Lawruk, T.S. *et al.* (1994) Quantification of 2,4-D and related chlorophenoxy herbicides by a magnetic particle-based ELISA. *Bulletin of Environmental Contamination and Toxicology*, **52**, 538–545.

Lee, S.M. and Richman, S. (1991) Thoughts on the use of immunoassay techniques for pesticide residue analysis. *Journal of the Association of Official Analytical Chemists*, **74**, 893.

Lehotay, S.J. and Argauer, R.J. (1993) Detection of aldicarb sulfone and carbofuran in fortified meat and liver with commercial ELISA kits after rapid extraction. *Journal of Agricultural and Food Chemistry*, **41**, 2006–2010.

Levitt, T. (1977) Radioimmunoassay for paraquat. *Lancet*, **ii**, 358.

Levitt, T. (1979) Determinations of paraquat in clinical practice using radioimmunoassay. *Proceedings of the Analytical Division of the Chemical Society*, **16**, 72–76.

Li, Q.X. *et al.* (1989) Comparison of an enzyme-linked immunosorbent assay and a gas chromatographic procedure for the determination of molinate residues. *Analytical Chemistry*, **61**, 819–823.

Li, Q.X. *et al.* (1991) Development of an enzyme-linked immunosorbent assay for the herbicide bentazon. *Journal of Agricultural and Food Chemistry*, **39**, 1537–1544.

Lucas, A.D. *et al.* (1991) Determination of atrazine and simazine in water and soil using polyclonal and monoclonal antibodies in enzyme linked immunosorbent assays. *Food and Agricultural Immunology*, **3**, 155–167.

Lucas, A.D. *et al.* (1993) Development of antibodies against hydroxyatrazine and hydroxysimazine: Application to environmental samples. *Journal of Agricultural and Food Chemistry*, **41**, 1523–1529.

Lukens, H.R. *et al.* (1977) Fluorescence immunoassay technique for detecting organic environmental contaminants. *Environmental Science and Technology*, **11**, 292–297.

Marco, M.-P. *et al.* (1993) Development of an enzyme-linked immunosorbent assay for carbaryl. *Journal of Agricultural and Food Chemistry*, **41**, 423–430.

Mastrorocco, D. and Brodsky, M. (1990) Recommendations on test kit methods: Task force report. *Journal of the Association of Official Analytical Chemists*, **73**, 331–332.

Mathewson, P.R. and Finley, J.W. (eds) (1992) *Biosensor Design and Application*, ACS Symposium no. 511. American Chemical Society, Washington, DC.

McAdam, D.P. *et al.* (1992) Mono and polyclonal antibodies to the organophosphate fenitrothion. 1. Approaches to hapten–protein conjugation. *Journal of Agricultural and Food Chemistry*, **40**, 1466–1470.

Mei, J.V. *et al.* (1991) Hapten synthesis and development of immunoassays for methoprene. *Journal of Agricultural and Food Chemistry*, **39**, 2083–2090.

Minunni, M. and Mascini, M. (1993) Detection of pesticide in drinking water using real-time biospecific interaction analysis (BIA). *Analytical Letters*, **26**, 1441–1460.

Monroe, D. (1984) Enzyme immunoassay. *Analytical Chemistry*, **56**, 920A–931A.

Muldoon, M.T. *et al.* (1993) Evaluation of ELISA for the multianalyte analysis of *s*-triazines in pesticide waste and rinsate. *Journal of Agricultural and Food Chemistry*, **41**, 322–328.

Nagao, M. *et al.* (1989a) Development and application of immunoassay for paraquat: Radioimmunoassay. *Journal of Forensic Sciences*, **34**, 547–552.

Nagao, M. *et al.* (1989b) Development and characterization of monoclonal antibodies reactive with paraquat. *Journal of Immunoassay*, **10**, 1–17.

Newsome, W.H. (1985) An enzyme-linked immunosorbent assay for metalaxyl in foods. *Journal of Agricultural and Food Chemistry*, **33**, 528–530.

Newsome, W.H. (1986a) Potential and advantages of immunochemical methods for analysis of foods. *Journal of the Association of Official Analytical Chemists*, **69**, 919–922.

Newsome, W.H. (1986b) Development of an enzyme-linked immunosorbent assay for triadimefon in foods. *Bulletin of Environmental Contamination and Toxicology*, **36**, 9–14.

Newsome, W.H. and Collins, P.G. (1987) Enzyme-linked immunosorbent assay of benomyl and thiabendazole in some foods. *Journal of the Association of Official Analytical Chemists*, **70**, 1025–1027.

Newsome, W.H. and Collins, P.G. (1989) Determination of 2,4-D in foods by enzyme-linked immunosorbent assay. *Food and Agricultural Immunology*, **1**, 203–210.

Newsome, W.H. and Collins, P.G. (1990) Development of an ELISA for urea herbicides in foods. *Food and Agricultural Immunology*, **2**, 75–84.

Newsome, W.H. and Shields, J.B. (1981) A radioimmunoassay for benomyl and 2-benzimidazolecarbamate on food crops. *Journal of Agricultural and Food Chemistry*, **29**, 220–222.

Newsome, W.H. *et al.* (1993) Development of enzyme immunoassay for captan and its degradation product tetrahydrophthalimide in foods. *Journal of the Association of Official Analytical Chemists International*, **76**, 381–386.

Ngeh-Ngwainbi, J. *et al.* (1986) Parathion antibodies on piezoelectric crystals. *Journal of the American Chemical Society*, **108**, 5444–5447.

Niewola, Z. *et al.* (1983) Enzyme linked immunosorbent assay (ELISA) for paraquat. *International Journal of Immunopharmacology*, **5**, 211–218.

Niewola, Z. *et al.* (1985) Quantitative estimation of paraquat by an enzyme linked immunosorbent assay using a monoclonal antibody. *Clinica Chimica Acta*, **148**, 149–156.

Niewola, Z. *et al.* (1986) Determination of paraquat residues in soil by an enzyme linked immunosorbent assay. *Analyst*, **111**, 399–403.

Oellerich, M. *et al.* (1982) Evaluation of EMIT adapted to the 'Cobas' biocentrifugal analyzer. *Journal of Clinical Chemistry and Clinical Biochemistry*, **20**, 765–772.

Pohland, A. *et al.* (1989) Immunoassays in food safety applications: Developments and perspectives. 198th National Meeting of the American Chemical Society, Abstract no. 45.

Proudfoot, A.T. *et al.* (1979) Paraquat poisoning: Significance of plasma-paraquat concentration. *Lancet*, **ii**, 330–332.

Provost, L.P. and Elder, R.S. (1983) Interpretation of percent recovery data. *American Laboratory*, **15**(19), 57–63.

Reed, J.P. *et al.* (1990) Measurement of ATV applicator exposure to atrazine using an ELISA method. *Bulletin of Environmental Contamination and Toxicology*, **44**, 8–12.

Riggle, B. (1991) Development of a preliminary enzyme-linked immunosorbent assay for the herbicide trifluralin. *Bulletin of Environmental Contamination and Toxicology*, **46**, 404–409.

Riggle, B. and Dunbar, B. (1990) Development of enzyme immunoassay for the detection of the herbicide norflurazon. *Journal of Agricultural and Food Chemistry*, **38**, 1922–1925.

Rinder, D.F. and Fleeker, J.R. (1981) A radioimmunoassay to screen for 2,4-dichlorophenoxyacetic acid and 2,4,5-trichlorophenoxyacetic acid in surface water. *Bulletin of Environmental Contamination and Toxicology*, **26**, 375–380.

Rubio, F.M. *et al.* (1991) Performance characteristics of a novel magnetic-particle-based enzyme-linked immunosorbent assay for the quantitative analysis of atrazine and related triazines in water samples. *Food and Agricultural Immunology*, **3**, 113–125.

Russell, A.J. *et al.* (1989) Antibody–antigen binding in organic solvents. *Biochemical and Biophysical Research Communications*, **158**, 80–85.

Sánchez, F.G. *et al.* (1993a) Polarization fluoroimmunoassay of the herbicide dichlorprop. *Journal of Agricultural and Food Chemistry*, **41**, 2215–2219.

Sánchez, F.G. *et al.* (1993b) Phase-modulation fluorescence lifetime immunoassay of dichlorprop. *Analytical Biochemistry*, **214**, 359–365.

Sandberg, R.G. *et al.* (1992) A conductive polymer-based immunosensor for the analysis of pesticide residues. In: *Biosensor Design and Application*, eds Mathewson and Finley, ACS Symposium no. 511. American Chemical Society, Washington, DC, pp. 81–88.

Scherrmann, J.M. *et al.* (1987) Prognostic value of plasma and urine paraquat concentration. *Human Toxicology*, **6**, 91–93.

Schewes, R. *et al.* (1994) Determination of weathered atrazine residues in soil by enzyme immunoassay and HPLC: An evaluation study. *Analytical Letters*, **27**, 487–494.

Schlaeppi, J.-M. *et al.* (1989) Hydroxyatrazine and atrazine determination in soil and water by enzyme-linked immunosorbent assay using specific monoclonal antibodies. *Journal of Agricultural and Food Chemistry*, **37**, 1532–1538.

Schlaeppi, J.-M. *et al.* (1991) Determination of metolachlor by competitive enzyme immunoassay using a specific monoclonal antibody. *Journal of Agricultural and Food Chemistry*, **39**, 1533–1536.

Schmid, R.D. and Künnecke, W. (1990) Flow injection analysis (FIA) based on enzymes or antibodies—Applications in the life sciences. *Journal of Biotechnology*, **14**, 3–31.

Schmidt, P. *et al.* (1988) A competitive inhibition enzyme immunoassay for detection and quantification of organophosphorus compounds. *Zeitschrift für Naturforschung*, **43c**, 167–172.

Schneider, P. and Hammock, B.D. (1992) Influence of the ELISA format and the hapten–enzyme conjugate on the sensitivity of an immunoassay for *s*-triazine herbicides using monoclonal antibodies. *Journal of Agricultural and Food Chemistry*, **40**, 525–530.

Schneider, P. *et al.* (1994) A highly sensitive and rapid ELISA for the arylurea herbicides diuron, monuron, and linuron. *Journal of Agricultural and Food Chemistry*, **42**, 413–422.

Scholz, H.M. and Hock, B. (1991) Development of an enzyme immunoassay for the determination of metazachlor. *Analytical Letters*, **24**, 413–427.

Schwalbe, M. *et al.* (1984) Enzyme immunoassay and fluoroimmunoassay for the herbicide diclofop-methyl. *Journal of Agricultural and Food Chemistry*, **32**, 734–741.

Schwalbe-Fehl, M. (1986) Immunoassays in environmental analytical chemistry. *International Journal of Environmental Analytical Chemistry*, **26**, 295–304.

Sherry, J.P. (1992) Environmental chemistry: The immunoassay option. *Critical Reviews in Analytical Chemistry*, **23**, 217–300.

Skerritt, J.H. *et al.* (1992a) Enzyme-linked immunosorbent assay for quantitation of organophosphate pesticides: Fenitrothion, chlorpyrifos-methyl, and pirimiphos-methyl in wheat grain and flour-milling fractions. *Journal of the Association of Official Analytical Chemists International*, **75**, 519–528.

Skerritt, J.H. *et al.* (1992b) Analysis of the synthetic pyrothroids, permethrin and 1(R)-phenothrin, in grain using a monoclonal antibody-based test. *Journal of Agricultural and Food Chemistry*, **40**, 1287–1292.

Skládal, P. *et al.* (1994) Pesticide biosensor based on coimmobilized acetylcholinesterase and butyrylcholinesterase. *Analytical Letters*, **27**, 29–40.

Soong, T.-S.T. and Lo, M.-C. (1989) Pesticide diagnostic kit development and its use in massive screening for pesticide residues in vegetables. *Food Laboratory News*, **15**, 36–39.

Stanker, L.H. *et al.* (1989) An immunoassay for pyrethroids: Detection of permethrin in meat. *Journal of Agricultural and Food Chemistry*, **37**, 834–839.

Stearman, G.K. and Adams, V.D. (1992) Atrazine soil extraction techniques for enzyme immunoassay microtiter plate analysis. *Bulletin of Environmental Contamination and Toxicology*, **48**, 144–151.

Stearman, G.K. and Wells, J.M. (1993) Enzyme immunoassay microtiter plate response to atrazine and metalochlor in potentially interfering matrices. *Bulletin of Environmental Contamination and Toxicology*, **51**, 588–595.

Sternberger, L.A. *et al.* (1974) Antibodies to organophosphorus haptens: Immunity to paraoxon poisoning. *Federation Proceedings*, **33**, 728.

Stöcklein, W. *et al.* (1990) Binding of triazine herbicides to antibodies in anhydrous organic solvents. *Analytical Letters*, **23**, 1465–1476.

Szurdoki, F. *et al.* (1992) Synthesis of haptens and conjugates for an enzyme immunoassay for analysis of the herbicide bromacil. *Journal of Agricultural and Food Chemistry*, **40**, 1459–1465.

Thurman, E.M. *et al.* (1990) Enzyme-linked immunosorbent assay compared with gas chromatography/mass spectrometry for the determination of triazine herbicides in water. *Analytical Chemistry*, **62**, 2043–2048.

United States Food and Drug Administration (1994) Food and Drug Administration pesticide program—Residue Monitoring—1993. *Journal of the Association of Official Analytical Chemists International*, **78**, 1179–1436.

Vallejo, R.P. *et al.* (1982) Effects of hapten structure and bridging groups on antisera specificity in parathion immunoassay development. *Journal of Agricultural and Food Chemistry*, **30**, 572–580.

Vanderlaan, M. *et al.* (1988) Environmental monitoring by immunoassay. *Environmental Science and Technology*, **22**, 247–254.

Van Emon, J.M. *et al.* (1985) Applications of immunoassay to paraquat and other pesticides. In: *Bioregulators for Pest Control*, ACS Symposium no. 276. American Chemical Society, Washington, DC, pp. 307–316.

Van Emon, J.M. *et al.* (1986) Enzyme-linked immunosorbent assay for paraquat and its application to exposure analysis. *Analytical Chemistry*, **58**, 1866–1873.

Van Emon, J.M. *et al.* (1987) Application of an enzyme-linked immunosorbent assay (ELISA) to determine paraquat residues in milk, beef, and potatoes. *Bulletin of Environmental Contamination and Toxicology*, **39**, 490–497.

Van Emon, J.M. (1989) EPA evaluation of immunochemical methods. 198th National Meeting of the American Chemical Society, Abstract no. 46.

Van Emon, J.M. and Gerlach, R.W. (1992) Evaluation of a pentachlorophenol immunoassay for environmental water samples. *Bulletin of Environmental Contamination and Toxicology*, **48**, 635–642.

Van Emon, J.M. and Lopez-Avila, V. (1992) Immunochemical methods for environmental analysis. *Analytical Chemistry*, **64**, 79A–88A.

Voller, A. *et al.* (1979) *The Enzyme Linked Immunosorbent Assay*. Dynatech Laboratories, Alexandria, VA.

Weetall, H.H. (1991) Antibodies in water immiscible solvents. Immobilized antibodies in hexane. *Journal of Immunological Methods*, **136**, 139–142.

Wei, S.I. and Hammock, B.D. (1982) Development of enzyme-linked immunosorbent assays for residue analysis of diflubenzuron and BAY SIR 8514. *Journal of Agricultural and Food Chemistry*, **30**, 949–957.

Wei, S.I. and Hammock, B.D. (1984) Comparison of coating and immunizing antigen structure on the sensitivity and specificity of immunoassays for benzoylphenylurea insecticides. *Journal of Agricultural and Food Chemistry*, **32**, 1294–1301.

Wei, S.I. *et al.* (1982) Synthesis of haptens and potential radioligands and development of antibodies to insect growth regulators diflubenzuron and BAY SIR 8514. *Journal of Agricultural and Food Chemistry*, **30**, 943–948.

Wigfield, Y.Y. and Grant, R. (1992) Evaluation of an immunoassay kit for the detection of certain organochlorine (cyclodiene) pesticide residues in apple, tomato, and lettuce. *Bulletin of Environmental Contamination and Toxicology*, **49**, 342–347.

Wigfield, Y.Y. and Grant, R. (1993) Analysis of atrazine in fortified cornmeal and corns using a commercially available enzyme immunoassay microtiter plate. *Bulletin of Environmental Contamination and Toxicology*, **51**, 171–177.

Wing, K.D. *et al.* (1978) Development of an *S*-bioallethrin specific antibody. *Journal of Agricultural and Food Chemistry*, **26**, 1328–1333.

Wing, K.D. and Hammock, B.D. (1979) Stereoselectivity of a radioimmunoassay for the insecticide *S*-bioallethrin. *Experientia*, **35**, 1619–1620.

Wittmann, C. and Hock, B. (1989) Improved enzyme immunoassay for the analysis of *s*-triazines in water samples. *Food and Agricultural Immunology*, **1**, 211–224.

Wittmann, C. and Hock, B. (1991) Development of an ELISA for the analysis of atrazine metabolites deethylatrazine and deisopropylatrazine. *Journal of Agricultural and Food Chemistry*, **39**, 1194–1200.

Wittmann, C. and Hock, B. (1993a) Analysis of atrazine residues in food by an enzyme immunoassay. *Journal of Agricultural and Food Chemistry*, **41**, 1421–1425.

Wittmann, C. and Hock, B. (1993b) Application and performance characteristics of a novel ELISA for the quantitative analysis of the atrazine metabolite deethylatrazine. *Journal of Agricultural and Food Chemistry*, **41**, 1795–1799.

Wong, R.B. and Ahmed, Z.H. (1992) Development of an enzyme-linked immunosorbent assay for imazaquin herbicide. *Journal of Agricultural and Food Chemistry*, **40**, 811–816.

Wüst, S. and Hock, B. (1992) A sensitive enzyme immunoassay for the detection of atrazine based upon sheep antibodies. *Analytical Letters*, **25**, 1025–1037.

Yess, N. *et al.* (1993) Food and Drug Administration monitoring of pesticide residues in infant foods and adult foods eaten by infants/children. *Journal of the Association of Official Analytical Chemists International*, **76**, 492–507.

6 Bioassay and chemical methods for analysis of paralytic shellfish poison

PEDRO A. BURDASPAL

Summary

Paralytic shellfish poison (PSP) is the common name assigned to a group of potent neurotoxins that may be produced by some algal species. Their accumulation in filter feeding shellfish may cause these marine products to become toxic for human consumption. As a consequence, a number of countries have regulations trying to limit the exposure of the population to these toxins. Control, monitoring and research tasks on this subject demand suitable analytical methods.

Nowadays, there exists a range of different techniques and methods that can be used efficiently for the analysis of PSP. The analyst can choose from the traditional mouse bioassay, which has been widely used in many laboratories for more than fifty years, to chemical methods such as the fluorimetric assays or those based on high performance liquid chromatography (HPLC), which at the moment possess probably the greatest potential as alternative methods. Other recent and very promising methods should be *in vitro* cell-culture assays based on the sodium channel-blocking activity of PSP toxins. All these methods and others which are also described in this chapter, such as the fly bioassay or the enzyme-linked immunosorbent assays (ELISA), offer a broad scope of possibilities in order to solve what can be considered one of the most complex analytical problems in the field of natural toxicants. Unfortunately, all of these methods have some disadvantages that handicap their generalised acceptance and application. Thus, the need for reliable and validated analytical methods, the lack of pure PSP standards and the availability of certified reference materials still constitute an open area for further development.

6.1 Introduction

'Any month with an r in its name is safe for shellfish harvesting.' This sentence reproduces a popular saying which is well known in some parts of the world. Although the message is not totally true, it demonstrates a widespread empirical observation of some seasonal coincidence in the appearance of episodes of molluscs becoming unsuitable for human consumption. Some of those episodes have been related historically to the simultaneous occurrence of massive phytoplankton blooms commonly known as 'red tides'. Proliferation of toxic species of phytoplankton, visible or not, may cause filter-feeding molluscs to accumulate toxins, those shellfish acting further as vectors in some food poisoning syndromes.

Paralytic shellfish poisoning (PSP) is one of the most severe forms of intoxication mediated by shellfish and may cause potentially deadly illness. The PSP toxins group comprises more than 20 structurally related compounds; saxitoxin being the first one to be chemically characterised. The PSP complex can be divided into four subgroups commonly labelled as carbamates, sulphocarbamoyl, decarbamoyl and deoxydecarbamoyl toxins (Figure 6.1). Chemical and enzymatic transformations are possible both *in vivo* and *in vitro* between analogues from different subgroups (Oshima, 1995a). Comparative studies between toxin profiles from causative dinoflagellates and contaminated shellfish indicate that transformations can occur actually inside the shellfish organism (Oshima *et al.*, 1990). Other important issues which still require further investigation are the proposed bacterial involvement in the production of PSP toxins and the interactions between bacteria and harmful algae (Tamplin, 1990; Doucette, 1995).

Although specific toxic potencies are not well established for all the PSP toxins, it is recognised that carbamates are the most toxic compounds, their sulphocarbamoyl analogues exhibiting a low toxic potency and decarbamoyl toxins placed in between. Toxicity has been normally expressed in terms of saxitoxin equivalency.

The reported PSP levels at which intoxication can be produced vary greatly ranging from a few hundreds to thousands of micrograms of saxitoxin equivalents per one hundred grams of shellfish edible meat (Van Egmond *et al.*, 1993). It is quite probable that a part of the published data may be not very reliable due to epidemiological uncertainties, analytical difficulties, and other reasons derived from the impossibility of analysing the actual meal portion responsible for the poisoning. The minimal lethal dose for children in an outbreak which occurred in Guatemala in 1987 was estimated at approximately $500\,\mu g$ saxitoxin equivalent/$100\,g$ meat, while in an intoxication case in Massachusetts in 1990, levels of $4280\,\mu g/100\,g$ were reported in the remaining uneaten cooked mussels, with six affected people developing symptoms but no fatalities. Additionally, in a well-documented fatal accident, the patient's gastric contents were found to contain $370\,\mu g/100\,g$ of PSP toxin, and a sample of the clam broth from the meal contained $2650\,\mu g/100\,g$ (US Department of Health and Human Services, 1991). Similarly, in a case of a woman's death presumably promoted by consumption of contaminated spontaneously growing (uncultured) mussels, the gastric contents were found to contain $122\,\mu g$ saxitoxin equivalent/$100\,g$ by mouse bioassay, $98.3\,\mu g$ saxitoxin equivalent/$100\,g$ by fluorometric assay and

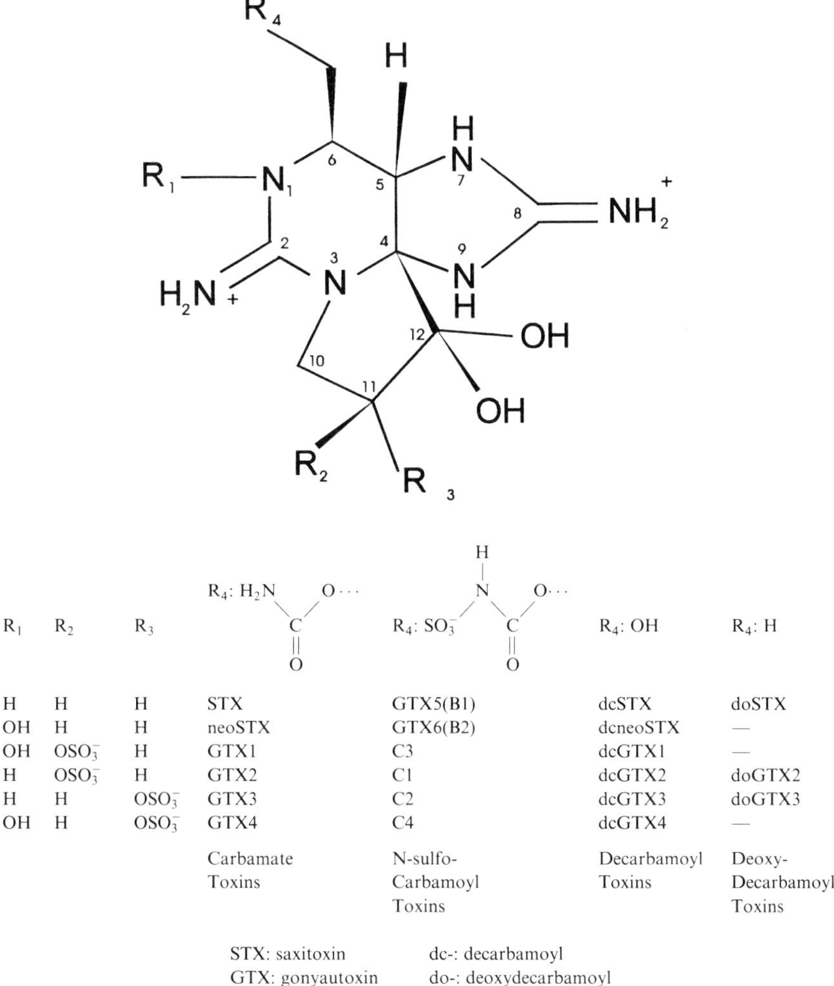

R₁	R₂	R₃	R₄: H₂N-CO-O··· (Carbamate Toxins)	R₄: SO₃⁻-NH-CO-O··· (N-sulfo-Carbamoyl Toxins)	R₄: OH (Decarbamoyl Toxins)	R₄: H (Deoxy-Decarbamoyl Toxins)
H	H	H	STX	GTX5(B1)	dcSTX	doSTX
OH	H	H	neoSTX	GTX6(B2)	dcneoSTX	—
OH	OSO₃⁻	H	GTX1	C3	dcGTX1	—
H	OSO₃⁻	H	GTX2	C1	dcGTX2	doGTX2
H	H	OSO₃⁻	GTX3	C2	dcGTX3	doGTX3
OH	H	OSO₃⁻	GTX4	C4	dcGTX4	—

STX: saxitoxin   dc-: decarbamoyl
GTX: gonyautoxin   do-: deoxydecarbamoyl

Figure 6.1 Chemical structures of PSP toxins.

24 μg decarbamoylsaxitoxin + saxitoxin/100 g by HPLC, the presence of decarbamoylsaxitoxin, gonyautoxins 2 and 3, C1 + C2 and B1 toxins being also detected. Mussels collected from the same area gave levels of 3035 μg saxitoxin equivalent/100 g by bioassay, 4579 μg saxitoxin equivalent/100 g by fluorimetric assay and 32 μg saxitoxin + 444 μg decarbamoylsaxitoxin/100 g by HPLC; the PSP toxins identified were saxitoxin, decarbamoylsaxitoxin, gonyautoxins 1–4, neosaxitoxin, C1 + C2 toxins, decarbamoylgonyautoxin 2 and B1 toxin (Burdaspal, 1994).

The human health concern has prompted for years an enormous effort on research and development of surveillance and monitoring programmes.

Many scientific articles, books and bulletins have been published and specific conferences have been held on this subject. A number of countries have in force or have proposed regulatory limits, with surveillance plans and alerting systems being currently established in many parts of the world. Most of these activities rely on analytical results obtained from marine plankton or shellfish sample testing. The following pages try to offer a comprehensive view of the plethora of available methods to determine the presence and amount of PSP toxins.

6.2   Biological methods

6.2.1   Mouse bioassay

The mouse bioassay determines the net toxicity of a test solution obtained after aqueous hydrochloric acid extraction from a shellfish sample. The solution is injected intraperitoneally into laboratory mice, toxicity being established in terms of a single response by correlating the time taken from the injection to death of each mouse with a table of the dose and time to death relationship.

This bioassay with mice was originally devised by Sommer and Meyer (1937), and was afterwards standardised and adopted as an official method for paralytic shellfish poison by the Association of Official Analytical Chemists (AOAC International, as it is known today), and currently used worldwide (AOAC, 1995).

The test involves several different steps, which can be summarised as follows.

Preparation of sample.   After thorough cleaning of the external surface of shellfish (clams, oysters and mussels) with fresh water, the inside is also rinsed and meat is obtained by cutting the adductor muscle. Other shellfish are prepared by separating just the edible portions. Meat weighing 100–150 g is collected, drained and ground. In the case of canned shellfish, the entire contents of the can are blended until homogeneous.

Extraction.   100 g of homogenised material is mixed with 100 ml of 0.1 M HCl and heated for 5 min at a gentle boil. The cooled mixture is adjusted to pH 2.0–4.0 and diluted to 200 ml. The extract is allowed to settle until there is separation of enough translucent supernatant to perform the mouse test. In some cases it may be necessary to centrifuge or filter the extract.

Mouse test.   No less than three mice weighing preferably between 19 and 21 g are inoculated intraperitoneally with 1 ml of acid extract, and the time to death is exactly recorded. If death time is <5 min, the extract must be diluted in order to obtain death times of 5–7 min, which represent the portion of the death time response curve where the dose is most accurately determined from the time of death.

Calculation of toxicity. The median death time of mice is converted into mouse units by comparison with values from the so-called Sommer's Table which correlates both parameters for paralytic shellfish poison (saxitoxin). Death time of survivors is considered as >60 min. The official method indicates that if test animals weigh <19 g or >21 g, the mouse unit value must be multiplied by a factor obtained from a correction table for weight of mice between 10 and 23 g. A mouse unit, as it was initially defined by Sommer and Meyer (1937) is the minimum amount of poison that is required to kill a 20 g mouse in 15 min when 1 ml of the extract is injected intraperitoneally. The assay is completed by converting mouse units into an equivalent amount of poison. That figure for final toxicity is achieved by multiplying the number of mouse units by a conversion factor (CF) which is obtained by previous standardisation of the stock mouse colony. The CF values seem to be normally around 0.2.

Standardisation. This step includes the inoculation of a number of mice with serial dilutions of a standard of pure saxitoxin in water following a clearly defined protocol. The CF value must be checked periodically in order to detect any significant change in response of mice to saxitoxin standard or in the technique of assay, such as mouse colony differences, laboratory personnel changes or other technical details. The method has an accepted variability of 20%. Trials carried out to determine laboratory performance of this assay have found that a reproducibility coefficient of variation approximating 20% can be normally achievable (Adams and Furfari, 1984).

As mentioned above, the detailed operating procedure of the mouse bioassay is described in Official Methods of Analysis of AOAC (AOAC, 1995) and it has remained with no variations since 1959. Nevertheless, laboratories that use this method on a routine basis frequently have introduced a number of apparently minor modifications in order to facilitate it to daily practice.

Apart from its merits, the mouse bioassay has a number of drawbacks and uncertainties which are worth briefly reviewing.

Firstly, one of the main disadvantages of mouse bioassay is a limit of detection of 35–40 μg PSP/100 g meat, which is not far (only one-half) from the most world-wide generalised regulatory limit of 80 μg PSP/100 g meat (Van Egmond *et al.*, 1992).

Using mice as an experimental model, their number, size, sex and strain can influence the precision of the test; even the time of the day in which the assay is performed may affect the sensitivity of the animals against the toxins. Mouse bioassay requires the continuous maintenance of a mouse colony. Mice must be controlled in their weight-gain process in order to be suitable for performing the assay (19–21 g). Overweight mice (>23 g) must be rejected for this purpose and devoted to other objectives or sacrificed. Mice obtained from different commercial sources must be carefully checked in order to detect any differences in their response to the poisons.

Determination of the exact death time relies on a subjective estimation, although experienced personnel may not find any problem in recognising and recording the last gasping breath as the death time. It is important to point out that PSP syndrome includes typical symptoms such as mouth opening, shaking and violent jumping. Any lack or significant alteration of typical symptoms may be due to a different causative factor, as can happen in extracts with high concentration of some metals such as zinc (McCulloch *et al.*, 1989). It is known that some species of oysters can accumulate high levels of metals such as zinc and cadmium.

The relationship between toxin level (dose) and death time is not linear. This fact makes it advisable to dilute extracts to give 5–7 min death time. Used in this manner, the assay is essentially a titration of the sample to a point close to 6 min death time, so that eventual divergences in the slope of the dose–response curves for different toxins or mixtures of them are of little consequence.

One of the most common modifications introduced by laboratories working on a routine basis is adjusting the pH of the final extracts to a fixed point optionally chosen between 2 and 4, in order to better standardise the protocol of the assay. Potential for foaming and loss by 'spluttering' of particles during the boiling step requires constant and careful surveillance of this critical point. Different anti-foaming agents were tested at three distinct concentrations and all of them were found to reduce foaming substantially (Park *et al.*, 1986). The same authors also suggested that the AOAC method should be revised to allow the use of that kind of product, and clearly outlined the pH adjustment procedure and the need to confirm the proper pH.

The assay is affected by the presence of salts. For example, the presence of 1% of sodium chloride in the assay solution reduces the toxicity by 50%. In the case of 0.5% of sodium chloride, the recovery of toxin was about 69%. Addition of ashes (estimated to contain 0.56% chloride) obtained from an acid extract of clams to a purified toxin solution reduced the recovery to 63.5%, which is almost the same effect as that noted by direct addition of an equal amount of sodium chloride to a standard solution of saxitoxin. This property may be of great importance, because it means that toxicity of extracts from toxic clams with low levels of contamination (near the 80 μg/100 g level), which do not need further dilutions in the mouse test, may underestimate the results as much as 60% (Schantz *et al.*, 1958). This fact also has potential consequences in the analysis of commercial shellfish products with salt as an additive reflected in the list of ingredients; this can be the case with some canned products.

Park *et al.* (1986) showed that dilution factors required to obtain a 5–7 min death time with standard solutions of saxitoxin at different concentrations were approximately double those of shellfish extracts at the same concentration. The authors suggested that these results could be explained if other factors influence the toxicity to mice of shellfish extracts. The salt effect may be responsible for a decrease in toxicity at lower levels when low dilutions are required, but its effect should be minimal at higher dilutions. The observed partial inhibition of toxicity of PSP in extracts compared with the toxic potency exhibited by the same amount of PSP in aqueous solution could explain epidemiological findings that indicate that eating the soup of toxic clams is a stronger risk factor than eating just the toxic clams.

The mouse bioassay in fact may not reflect exactly the effect of the actual toxin composition in the shellfish. As previously described, the test is usually performed by extraction of 100 g samples with 100 ml 0.1 M HCl and further boiling for 5 min. Under those conditions, some hydrolysis of sulphamate toxins to their more toxic carbamate analogues may take place. This effect is known as the 'Proctor enhancement' due to Proctor's observations about increased toxicity of cell free extracts of *Gonyaulax* following heating at low pH. It was reported that heating at 100°C for 5 min in aqueous HCl with a free acid concentration of 0.1 M was necessary to ensure complete hydrolysis (Hall and Reichardt, 1984).

Notwithstanding all that has been already discussed concerning the mouse bioassay, there still remain a number of aspects in the official AOAC protocol that should be clarified. One problem is that the mouse test does not mention which solvent must be used for diluting the sample extracts in case of death times <5 min. Some analysts understand that water should be used, in logical correspondence with the way the solutions are made in the standardisation of the method. Other analysts could think it more convenient to use 0.05 M HCl, as it is the solvent of the extract obtained after mixing the sample with the 0.1 M HCl. Another possible question could be: why not perform the standardisation of the bioassay by diluting aliquots of the saxitoxin working standard solution with 0.05 M HCl instead of water?

Another aspect to be considered could be the need to give some additional instructions in the case of analysing canned shellfish or other commercial processed products such as frozen cooked shellfish. In this particular case, it is very important from a health and regulatory point of view to consider the moisture content of the samples. As an example, we can imagine a batch of fresh mussels with a typical moisture content of 86%, containing

42 μg PSP/100 g, as estimated by mouse bioassay; after industrial processing by steaming and freezing, the mussel meat now has a moisture content of 73%. If an inspector takes a 100 g sample of that commercial frozen product to be analysed by mouse bioassay, the result would be 81 μg PSP/100 g and the product would be probably considered illegal or unfit for human consumption. This representative case does not take into account any eventual conversion or loss of PSP toxins.

Although the official protocol calculates the toxicity in terms of μg poison/ 100 g meat, many laboratories have adopted a more logical expression which is μg of saxitoxin equivalents/100 g edible meat. The mouse bioassay cannot estimate individual PSP toxins. This lack of specificity also could be interpreted as an advantage, because it allows one to determine the total toxicity. Last but not least, it must be pointed out that in some countries animal welfare opinion plays a role against the use of mouse bioassay for routine monitoring.

The AOAC official protocol ends with a final statement as follows. 'Consider any value $>80\ \mu$g/100 g as hazardous and unsafe for human consumption.' In this respect, it seems that none of the countries with regulations for PSP at that level or even lower have quantified the risks considered acceptable, which should be scientific rationale(s) for regulatory limits (Van Egmond *et al.*, 1992, 1993).

6.2.2 *Fly bioassay*

An alternative animal assay was proposed by using the common house fly (*Musca domestica*) as an experimental model able to give a unique response to all PSP toxins and to counter the disadvantages of the mouse bioassay (Siger *et al.*, 1984). The protocol of the assay establishes the use of cold treatment in order to immobilise the flies before injection by microsyringe with 1.5 μl of the test solution obtained according to the mouse bioassay procedure (Ross *et al.*, 1985). After 10 min, each fly is scored as paralysed or active. Samples were preliminarily diluted into the range of the LD_{50} and injected into at least ten flies. The number of paralysed flies at each dilution was compared in a previously built standard curve, thus allowing an estimation of the toxin content. The practical limit of detection was fixed at about 20 μg/100 g of meat. The fly bioassay was claimed to be simple, sensitive and inexpensive. The PSP content was not underestimated, as happened with the mouse bioassay at levels close to 80 μg/100 g, because flies were free of the 'salt effect'. Despite its claimed advantages, this bioassay has not been widely used so far. In an intercomparison of various assay methods for the detection of shellfish toxins, it was stated that this method may serve for early warning of toxin levels, although all the observed toxicity may not be PSP-related given that other substances may interfere (Hurst *et al.*, 1985).

6.3 Biochemical assays

6.3.1 In vitro *tissue-culture assay*

A series of papers has been published in recent years on a new and promising approach for the determination of PSP toxins. A tissue-culture assay was initially developed by taking advantage of the ability of saxitoxin to protect neuroblastoma cells in culture against the combined action of two biologically active compounds—veratridine and ouabain (Kogure *et al.*, 1988).

Saxitoxin and other related toxins are neurotoxins which alter the normal function of sodium channels in nerve axons by inhibiting the ion transport (Catterall, 1985). Sodium channels exhibit a high affinity for saxitoxin. Despite little direct experimental support so far, it has been proposed that the blocking of ion transport by saxitoxin could be the consequence of its binding within and physically occluding the ion-conducting pore. In contrast, veratridine is a lipid-soluble compound which is known to stimulate uncontrolled sodium influx into nerve cells by acting on the sodium channel at receptor sites different than those of saxitoxin. Intracellular Na^+ concentration can be kept low by membrane Na^+/K^+ ATPase, which is able to exchange intracellular Na^+ ions for extracellular K^+ ions, but this enzyme in turn can be specifically inhibited by ouabain. Combined addition of both veratridine and ouabain to mouse neuroblastoma cells results in elevated intracellular sodium levels, leading to pronounced morphological changes and their eventual death. The combined effect of veratridine and ouabain was known to be prevented or reduced by the addition of sodium-channel blockers such as saxitoxin, cells remaining alive and morphologically normal (Catterall and Nirenberg, 1973). Mouse neuroblastoma cell line Neuro-2A was used in the assay, a combination of 0.05 mM veratridine and 1 mM ouabain being recommended for routine analysis (Kogure *et al.*, 1988). Morphological changes in the absence of saxitoxin were observed under the assay conditions, less than 1 h after the addition of both compounds, cells becoming swollen, round and granular. After 3 h, morphological differences with and without saxitoxin were reported to be evident. The fraction of the cells which remained alive, keeping clear and distinct cell membranes, was in direct proportion to the concentration of saxitoxin present.

Further attempts were made to replace the tedious and subjective microscopic counting of survivor cells by detecting and quantifying the live cells with a microtitre plate reader at 540 nm by staining with the dye Neutral Red (Gallacher and Birkbeck, 1992). This principle was adopted in the description of a practical method which was readily used to assay PSP toxins in mussel extracts, initially processed according to the standard mouse bioassay (Gallacher *et al.*, 1993). In that study, it was reported to be advantageous to test a range of dilutions of the extracts in order to circumvent or minimise the influence of some heat-labile PSP inhibitors which may be present in some bivalve extracts. High dilution of extracts, or heating at 100°C for 60 min and concomitant destruction of inhibitors, allowed the detection of PSP toxicity in extracts otherwise absent of sodium channel blocking activity. It was also suggested that an acidic protein, which has been found to be induced in some species of shore crab by exposure to saxitoxin and conferred resistance to them (Barber *et al.*, 1988), could play a similar role in mussels.

A similar assay had already been described by another group of investigators (Jellet *et al.*, 1992). This version of the tissue culture bioassay used crystal violet as a staining agent for live neuroblastoma cells and absorbance measurement at 595 nm with a microplate reader. Cells of neuroblastoma cell line Neuro-2A were grown at 37°C in RPMI 1640 medium supplemented with 10% fetal calf serum and 1% antibiotic antimycotic solution containing penicillin G, streptomycin sulphate and amphotericin B. Flat-bottomed wells of microtitre plates were inoculated with 200 ml of a suspension of rapidly growing cells and incubated for 24 h at 37°C. Aliquots (10 μl) of 0.05 mM veritridine and 0.5 mM ouabain were added to wells which had first received 10 μl aliquots of saxitoxin standard or sample extract, using complete growth medium in control wells. After further incubation for 24 h at 37°C, dead cells were removed by taking advantage of their diminished adherence, by emptying and rinsing the wells with phosphate buffered saline (PBS). PSP-treated cells retain adherence, and were fixed with 10% formalin for 15 min. Fixative was removed and the wells were filled with Gram's crystal violet for 5 min. After rinsing with tap water and drying at room temperature, the plates could be kept for several days before absorbance measurement. Stained cells remaining in each well were digested with 100 μl of 33% acetic acid for 1 h and then read at 595 nm. Shellfish extracts were prepared according to the standard mouse bioassay, sub-samples being sterilised by filtration through a cellulose acetate membrane filter. These extracts were found to necessitate an additional solid-phase extraction clean-up through C_{18} cartridges in cases of samples with very low toxicity ($<$74 μg/100 g). The reason seems to be the elimination of interfering co-extracted unknown materials which otherwise could affect toxicity values in undiluted extracts. Other studies (Sato *et al.*, 1988) served to suggest that co-extracted proteins could exhibit some adsorbing capacity for toxins added to non-toxic extracts (Jellet *et al.*, 1992). The assay response in terms of purple colour absorbance was directly related to the added amount of saxitoxin standard and was linear between 0 and about 600 pg/10 μl. The limit of detection of this assay under optimal conditions was about 2.0 μg saxitoxin equivalent/100 g shellfish meat. Samples assayed by both the cell and mouse bioassays gave almost identical results ($r>0.96$), precision of both bioassays being comparable.

Further attempts were made to improve the cell assay by simplifying the end-point estimation of PSP-treated cells and introducing other modifications

(Manger *et al.*, 1993). A colorimetric test based upon the ability of metaboli-cally active cells to reduce the 3-(4,5-dimethylthiazol-2-yl)-2,5-diphenyl tetrazolium salt was incorporated. Conversion of the tetrazolium salt into a blue formazan product takes place only in living cells by action of the mitochondrial enzyme succinate dehydrogenase. The biochemical reaction is performed directly in the wells, making the washing and rinsing steps unnecessary. The method maintained similar limits of detection to those mentioned above.

Cell-based detection methods appear to offer a number of advantages over the mouse bioassay and other proposed alternative methods. The cell assay is relatively inexpensive, simple and sensitive, its precision seems to be similar to mouse assay and it does not require highly sophisticated equipment nor specialised technical skills. Additionally, it requires only a saxitoxin standard and can estimate the total toxicity of PSP toxins present, it needs low main-tenance, and one person can assay 100 samples or more in one day (Jellet *et al.*, 1992). Although a limitation could be the response time of 24–48 h, the cell bioassay provides the possibility of daily screening of a large number of samples and is worth consideration as a real candidate to be checked as an alternative to live animal testing in monitoring programs.

6.3.2 Enzyme-linked immunosorbent assay (ELISA)

A competitive indirect enzyme-linked immunosorbent assay (ELISA) was developed by using antibodies raised from rabbits immunised with saxi-toxin–bovine serum albumin (Chu and Fan, 1985). Antiserum solution is incubated together with sample extract or saxitoxin standard in a microtitre plate coated with saxitoxin antigen. The fraction of total antibodies which is bound to saxitoxin from sample extracts is eliminated. The amount of anti-bodies which remains bound to saxitoxin coated to the plate is determined by colorimetric measurement of the product of the enzymatic reaction mediated by goat antirabbit IgG–peroxidase conjugate, which is added and bound to immobilised saxitoxin antibodies from rabbit. The detection limit appears to be in the range of 5–10 μg/100 g; however, the assay was not totally specific to saxitoxin and some interferences from mussel meat were observed, making it necessary to start the analysis from mussel dark glands instead of edible tissue.

Other polyclonal antisaxitoxin antibody was prepared by immunisation of rabbits with saxitoxin covalently coupled to poly(alanine–lysine) (Cembella and Lamoureux, 1993). This antibody was incorporated into an ELISA kit and evaluated with PSP-producing cultures of marine phytoplankton. The antibody exhibited high affinity and sensitivity for purified saxitoxin and cross-reacted with neoStx (neosaxitoxin), GTX-2 (gonyautoxin 2) and GTX-3 (gonyautoxin 3), being apparently unreactive towards C toxins. The observed underestimation of the total PSP content was due to the lack of response to the whole amount of possible PSP toxins present in tested extracts.

A direct enzyme immunoassay (Usleber *et al.*, 1991) was found to over-estimate the saxitoxin content in some samples tested in an intercomparison study of methods for PSP toxins (Van Egmond *et al.*, 1994). The reason, once again, seemed to be the lack of specificity for saxitoxin and cross-reaction of antibodies towards other PSP toxins present.

The antibody cross-reaction capacity towards some PSP toxins and the lack of response towards some others seem to be the main limitations at present preventing ELISA methods from being widely accepted. Neverthe-less, their high sensitivity may make them useful as screening tests in some cases.

6.4 Chemical methods

As already indicated, considerable efforts have been made to develop alter-native chemical methods for detection of PSP which could fulfil analysts' requirements of suitability either for research or routine monitoring tasks.

In spite of the progress achieved in this field, that objective still remains a major challenge. The 'best' method of analysis should be capable of extracting, separating, detecting, identifying and quantifying as many as possible (ideally each and every one) of up to twenty different components, which can be present mixed in unpredictable proportions in complex bio-logical matrices such as shellfish meat. Although all those compounds are structurally related, they exhibit different physicochemical properties and toxicological potencies. Besides that, there is the possibility of decomposition and transformation between them depending on storage or experimental conditions.

The candidate method should if possible meet some other requirements besides those normally demanded with respect to acceptable accuracy, pre-cision, sensitivity and recoveries for the different toxins. These additional performance characteristics should include wide applicability, relative sim-plicity, and availability for normally equipped laboratories, and the method should not be excessively time-consuming.

Some chemical assays have been developed so far that try to offer an alternative to the mouse bioassay. Saxitoxin is known to give colorimetric reactions with a number of reagents such as Benedict–Behre, Weber, Jaffé (Mold *et al.*, 1957), Folin–Ciocalteau (Price and Lee, 1971) or biacetyl, in the last-named case after its hydrogen peroxide oxidation to guanidine (Reichardt *et al.*, 1978). Some of these detection methods were intended for PSP assay, although for a number of reasons they have not been widely used. The method based on the Jaffé test (McFarren *et al.*, 1959) was less sensitive than the bioassay and showed a lack of precision, derived mainly from its lack of specificity for saxitoxin. The colorimetric assay based upon the reaction of biacetyl with guanidine or other compounds containing the guanidino group (Gershey *et al.*, 1977), was demonstrated to be a relatively simple, rapid and quantitative method of analysis when applied to purified saxitoxin. However, necessary further refinement of the technique for application to shellfish extracts, in order to reduce the analysis time and increase the precision, displaced the interest of researchers to other more promising chemical procedures.

6.4.1   Fluorimetric assay

In the course of the studies that were conducted to elucidate the structure of saxitoxin (Wong *et al.*, 1971), a compound was obtained by alkaline hydrogen peroxide oxidation of saxitoxin, which was identified as 8-amino-6-hydroxymethyl-2-iminopurine-3(2H)-propionic acid. The reaction undergone was a Baeyer–Villiger oxidation of the C-12 hydrated ketone in the saxitoxin ring molecule; the carbamoyl moiety was hydrolysed, and cleavage of the bond between carbons 4 and 12 had occurred with an aromatisation of the remaining rings. The fluorescent properties of this compound made possible the design of a chemical method (Bates and Rapoport, 1975), which in spite of its limitations has subsequently been widely used either in the original version or after further modifications. The same principle has served as the basis for other useful methods which are discussed later in this section.

The initial procedure (Bates and Rapoport, 1975) was improved and modified by the same authors (Bates *et al.*, 1978) for better accuracy and reproducibility. The main steps of that improved version are as follows.

Extraction.   Ground shellfish meat (2 g) is extracted with 2 ml of 0.5 M trichloroacetic acid. The mixture is heated to an internal temperature of 85–90°C for 10 min. After cooling to 20°C, it is neutralised by adding about 0.2 ml of 10% NaOH to a pH of 5.0–5.5. The supernatant obtained after centrifugation is purified by ion-exchange chromatography.

Ion-exchange chromatography.   Bio-Rex 70 resin is prepared by rinsing 200 ml of resin (wet volume) with 3×600 ml of water, 3×600 ml of 0.5 M H_2SO_4, 600 ml of water, 3×600 ml of 1 M NaOH and 3×600 ml of water. The resin is suspended in 600 ml of 0.2 M acetic acid and pH is adjusted to 5.0 with H_2SO_4. Finally, the resin is rinsed with 2×600 ml of 0.2 M sodium acetate buffer, pH 5.0. The extract obtained after centrifugation is applied to a column filled with 2 ml of resin. The column is washed with 30 ml of 0.2 M sodium acetate buffer, pH 5, 25 ml of water and 1 ml of 0.25 M H_2SO_4. Saxitoxin is eluted by passing through the column 3.9 ml of 0.25 M H_2SO_4.

Oxidation reaction. The eluent is divided into two portions of equal volume. Alkaline oxidation is performed on one portion by adding 2 ml of 1.3 M NaOH and 0.05 ml of 10% hydrogen peroxide. The other portion acts as the blank by adding water instead of hydrogen peroxide. Both solutions are centrifuged and, 40 min after the addition of hydrogen peroxide, they are adjusted to pH 5 with about 0.16 ml of glacial acetic acid.

Determination. Fluorescence is measured in a fluorimeter using excitation at 330 nm and emission at 380 nm. Calibration of the fluorimeter is carried out by preparing a number of dilutions of oxidised saxitoxin standard. The Raman peak of water (excitation 330 nm, emission 371 nm) can be used as a standard, and corresponds to approximately $0.024\,\mu g$ of saxitoxin per gram of shellfish meat. The method was reported to be 100 times more sensitive than the mouse bioassay.

An abbreviated assay could be performed by omitting the ion-exchange chromatography. Its application to the analysis of New England shellfish gave results only slightly higher than those of the bioassay, which could be consistent with the presence of other PSP toxins that were not detected by the normal column method (Shoptaugh *et al.*, 1981). Very soon after the first description of the fluorimetric method, it was found (Buckley *et al.*, 1976) that other PSP toxins were able to form fluorescent derivatives after fractionation of partially purified mixtures of toxins by thin-layer chromatography and heating of the developed plates in the presence of hydrogen peroxide. Gonyautoxin 2 was one of the first PSP toxins identified apart from saxitoxin known to afford similar fluorescent products after alkaline oxidation (Shimizu *et al.*, 1976). This behaviour was subsequently similarly shown by other gonyautoxins (Buckley *et al.*, 1978). However, today it is well known that peroxide oxidation only yields fluorescent products with the non-N-1-hydroxy containing toxins, except for a weak response to gonyautoxin 1 (Lawrence *et al.*, 1991).

Given that the fluorimetric method is based upon the strong adsorption of saxitoxin on Bio-Rex 70, which is a weak cation-exchange resin, the assay probably cannot detect other PSP toxins, which are only weakly bound on the resin (Boyer *et al.*, 1979; Hall *et al.*, 1979). This limitation prompted the introduction of some modifications (Childress, 1979) so that all PSP toxins occurring in shellfish and able to give fluorescent derivatives after alkaline peroxide oxidation could be detected as a batch.

The main changes from the original method were as follows.

Extraction. Shellfish samples were extracted by following the same procedure as for mouse bioassay.

Ion-exchange chromatography. Preparation of Bio-Rex 70 was performed in a similar way to the above, except by using ammonium acetate buffer, pH 5, for final rinsing and storage. Acid extract (4 ml) was adjusted to pH 5–5.5, centrifuged and transferred to the column. After washing with 10 ml of water and 1 ml of 0.5 N H_2SO_4, toxins were eluted with 4 ml of 0.5 N H_2SO_4.

Oxidation reaction.   Oxidation was achieved by addition of two drops of 10% hydrogen peroxide and 2 ml of 1 N NaOH to a 2 ml portion of eluate. Addition of water instead of hydrogen peroxide to the remaining 2 ml portion formed a blank. Centrifugation, time and stopping of reaction were as formerly described.

Determination.   Fluorescence was measured with excitation at 335 nm and maximum emission wavelength at about 392 nm. Net fluorescence of samples was compared with that of a 2 ml aliquot of saxitoxin standard solution treated in the same manner.

This method was intended to be collaboratively studied by IUPAC, but initial results revealed several shortcomings, most participants being unable to achieve reproducible quantitative recoveries (Ragelis, 1982). Nevertheless, the modified fluorimetric method, implemented by using the initially pre-scribed protocol for resin preparation (Bates *et al.*, 1978) and other minor refinements, has been reported to be extensively used as part of monitoring activities on fresh and canned shellfish (Burdaspal *et al.*, 1991). Results obtained by this method are expressed in micrograms of saxitoxin equivalent per 100 g of edible shellfish meat, and they have been systematically higher than those obtained by mouse bioassay for the same samples, with a set of variable proportionality factors ranging normally from 2 to 3.8 times.

The apparent discrepancies observed frequently between results from bioassay and fluorimetric methods, whichever version is used (Bose and Reid, 1979; Berenguer *et al.*, 1993), may be logical, bearing in mind the different type of information that each method supplies. Thus, the significant fluorescence differences shown after the alkaline oxidation reaction (Jonas-Davies *et al.*, 1984) and the different degrees of toxicity shown by the various PSP toxins in the mouse assay (Sullivan *et al.*, 1985a) will give a coincidental value only in the case that saxitoxin were the only toxin present in the extract. Discrepancies indicate the probable presence of a mixture of toxins, which could together give an absolute toxicity different from that of the saxitoxin and also with a different absolute fluorescence. Although the N-1-hydroxy toxins (i.e. neosaxitoxin, B-2, GTX-1, GTX-4, C-3 or C-4) give very low yield, if any, in fluorescence under the experimental conditions of the fluori-metric assay, this may be considered a valuable indicator for PSP levels, especially in cases where non-N-1-hydroxy toxins are predominant. Its better sensitivity and easier operation, and the fact that it can be used with the same extraction procedure as in the mouse test, make it a practical tool to detect the presence of PSPs before they can be detected by the mouse test, to detect any trend in PSP levels or simply to complement the mouse bioassay. In fact, a modified version of the original fluorimetric method, which included as an additional step the precipitation of protein with ethanol, has been reported to be used in monitoring programs in Norway. The fluorimetric method did not give any false negative result, samples found PSP positive being also tested in the mouse assay, and this approach permitted a reduction in the number of mice used by about 80% (Yndestad and Underdal, 1985).

A further version of the fluorimetric method was developed for the determination of saxitoxin in canned shellfish products (Hellwig and Petuely, 1980). The extraction was performed with a solution of trichloroacetic acid and Bio-Rex 70 was used once again for the purification of the extract by ion-exchange chromatography. The elution of saxitoxin from the column was achieved by passing $2\,M$ acetic acid/$1\,M$ HCl (1:1), oxidation being performed with alkaline hydrogen peroxide. This procedure was claimed to offer the advantage that, parallel with the fluorimetric determination, a bioassay with mice could be carried out with the same extract eluted from the column. A modification of this method was included in the Collection of Official Methods under article 35 of the German Federal Foods Act (Bundesgesundheitsamt, 1989) as a screening method for determining the content of water-soluble algal toxins of the PSP type in molluscs and mollusc products at levels ranging from not less than $5\,\mu g$ to $300\,\mu g/100\,g$. The main changes introduced in this version were adoption of $0.5\,N\,H_2SO_4$ as elution solvent in the ion-exchange chromatography and oxidation with periodic acid in an alkaline medium.

Fluorimetry and colorimetry also could be combined in a single procedure, allowing detection from crude extracts of saxitoxin, GTX-2, GTX-3, neosaxitoxin and GTX-1/GTX-4. The method combined the fluorimetric and the Folin–Ciocalteau reagent assays, coupled with linear gradient elution from a strong cation-exchange minicolumn, being superior to the mouse bioassay for research purposes (Mosley *et al.*, 1985). Folin–Ciocalteau reagent spray has also been used as a sensitive method for detection of PSP toxins on TLC plates and paper electrophoretograms (Ikawa *et al.*, 1985).

6.4.2 HPLC methods

High performance liquid chromatography (HPLC) is undoubtedly one of the most powerful techniques available today for the separation and determination of PSP toxins. Most of the HPLC methods for PSP determination are based on the extraction of samples with hydrochloric acid, pre- or post-column oxidation of the PSP toxins, liquid chromatographic separation of either the oxidised derivatives or the natural toxins, and fluorescence detection of the PSP toxins' oxidation products.

One of the first methods using high pressure pumps and columns was a toxin analyser developed as a conventional continuous-flow liquid chromatograph equipped with a Bio-Gel column and a postcolumn derivatisation system involving alkaline oxidation of PSP toxins with hydrogen peroxide (Buckley *et al.*, 1978). This procedure was unable to detect all the toxins and quantify them individually. Another attempt was made using *tert*-butyl hydroperoxide as an oxidant and a weak cation-exchange resin in a similar continuous liquid chromatography system (Oshima *et al.*, 1984). The new oxidising reagent was able to convert all the PSP toxins, including those non-fluorescent by hydrogen peroxide oxidation, into fluorescent derivatives. However, the analyser failed to separate GTX-1 from GTX-4 and GTX-3 from GTX-5, despite a complex pretreatment of extracts. A new attempt using HPLC with postcolumn derivatisation was based on the application of *o*-phthalaldehyde as fluorogenic reagent (Onoue *et al.*, 1983). Its lack of enough specificity and the necessity for prior purification of crude extracts limit its practical value.

Sullivan method.   In 1983 the first of a series of papers describing an HPLC procedure was published which, after a number of changes, reached a certain degree of popularity and has been used effectively in monitoring programs. The pioneering HPLC method (Sullivan and Iwaoka, 1983) used a bonded phase cyano column with detection by fluorescence following postcolumn alkaline oxidation with periodic acid. Alkalinity was achieved in the postcolumn reaction by pumping ammonium hydroxide in the reaction manifold, temperature was maintained at 40–50°C, and acetic acid was used to stop the reaction. The ionic strength of the mobile phase was the most important parameter affecting toxin separation on polar bonded-phase columns. Complete separation of toxins in a reasonable amount of time was attempted by using a step gradient with ammonium phosphate to control both the ionic strength and the pH. The main drawbacks of this procedure were, among others, the instability of the cyano columns and the low sensitivity shown to N-1-hydroxy toxins. Good correlation between this technique and the mouse bioassay was obtained in the analysis of shellfish at or below the limit of 80 μg toxin/100 g, while at higher levels of toxicity the HPLC technique tended to underestimate total toxicity slightly (Sullivan *et al.*, 1983). The method was semi-automatic, using an auto-sampler and auto-analyser continuous flow reaction system (Jonas-Davies *et al.*, 1984). In spite of non-correlation of certain samples with the mouse test, it was proposed as a screening method to eliminate samples found to be very high (250 μg/ 100 g) and very low (60 μg/100 g) in toxin content.

An improved version of this HPLC method was developed involving separation of the toxins on a polystyrene divinylbenzene resin column (PRP-1) in the reverse-phase mode by ion-interaction chromatography, using a mobile phase containing hexane and heptanesulphonic acid as ion-pair reagents, ammonium as the co-cation and methanol as the organic modifier. Alkaline oxidation was performed as above with the exception that nitric acid was used in place of acetic acid. The temperature of the reaction was changed to 75°C, and the periodate concentration was lowered as a compromise to produce adequate sensitivity for all the toxins. Incorporation of these changes enabled complete separation of all the carbamate and N-sulphocarbamoyl toxins with the exception of C-1 and C-2 to be achieved. The sensitivity was increased by at least a factor of four for each of the individual toxins detected (Sullivan and Wekell, 1984). Application of this HPLC procedure resulted in a correlation coefficient of 0.92 for 40 samples of different shellfish species exhibiting toxicity in the bioassay (Sullivan *et al.*, 1985a).

In a further development of the previously reported methods, several improvements were incorporated by introducing a number of alterations covering almost all the stages of the technique (Sullivan *et al.*, 1985b). In earlier works, shellfish samples were extracted by the standard bioassay procedure followed by protein precipitation with trichloroacetic acid (Sullivan and Wekell, 1984). This technique was replaced by utilising ultrafiltration membranes for protein removal. In addition, the oxidant and base were pre-mixed instead of being separately pumped into the system, and a sodium phosphate buffer solution was used for pH control during oxidation to alleviate precipitation problems resulting from the use of ammonium hydroxide. The pH of the oxidant was 7.8 in order to obtain a near optimum response for neosaxitoxin and GTX-1–4 while still maintaining acceptable sensitivity to the other toxins. Reaction temperature was increased to 90°C. Change of the organic modifier to acetonitrile and other modifications in the chromatographic conditions improved toxin peak shapes and column re-equilibration time.

The last version of the so-called 'Sullivan method' did not change much from the former description, but nevertheless a number of further modifications and improvements were introduced for its use in shellfish monitoring programmes (Sullivan and Wekell, 1987). To maintain chromatographic conditions it was imperative to undertake a rigorous cleaning of the postcolumn reaction system. An important addition was continuous purging of postcolumn reagents and mobile phases with helium. Some data were also reported showing that the boiling step could be eliminated with no significant effect on HPLC results, at least in the case of those samples analysed. It was apparent that boiling did not increase the efficiency of removal of toxins from the shellfish tissue. However, given that it is possible that some conversion of B-11-hydroxy sulphate toxins to their corresponding A-counterpart (for example, GTX-3 to GTX-2) and also a certain degree of hydrolysis of sulphamate toxins to their carbamate counterparts took place (Sullivan and Wekell, 1987), it was established that the boiling step could cause a significant increase in toxicity values, especially in those samples containing sulphamate toxins as the major portion of total PSP toxins.

The Sullivan method has been used by many laboratories in research or for routine analysis. In practice, there were found a number of disadvantages with such a complicated set-up and operation of the equipment, considerable daily maintenance and the degree of skill demanded. The proposed PRP-1 column was found to be unable to separate carbamate and decarbamoyl toxins (Luckas, 1990). Although the alternative use of a reversed-phase C_{18} column improved the resolution, today it is well established that neither the PRP-1 nor the C_{18} column is able to separate saxitoxin and decarbamoylsaxitoxin under the above mentioned chromatographic conditions (see Figure 6.5). Inadequate separation of saxitoxin and decarbamoylsaxitoxin was considered a major reason for the overestimation of saxitoxin content found by all laboratories that used this method in an intercomparison exercise undertaken within the framework of the European Commission's Standards, Measurements and Testing Programme activities (Van Egmond *et al.*, 1994). At this point, one can only speculate about scientific and even regulatory consequences derived from erroneous interpretation of HPLC chromatograms. Correlation studies between mouse bioassay and this HPLC method may have been seriously affected by identifying as saxitoxin what was really decarbamoylsaxitoxin or a mixture of both. Additionally, a failed precollaborative trial among nine laboratories has been reported using the Sullivan method (Van Egmond *et al.*, 1993). All these facts have contributed to this method having fallen nowadays into some disfavour, supported by the availability at this moment of a number of more promising, advantageous and attractive HPLC methods.

Oshima method. At the time of writing this chapter, three main HPLC methods are finding favour with laboratory users. These methods can be referred as Oshima, Lawrence and Thielert for short. The first one was initially developed especially for the determination of the *N*-sulphocarbamoyl and decarbamoyl toxins and permitted the identification of fifteen PSP naturally occurring toxins in mussels contaminated by *G. catenatum* (Oshima *et al.*, 1989). The method was based on ion-pair chromatography; three separate chromatographic runs on a C_8 bonded silica gel column under isocratic conditions were required to determine all the toxins in every sample. In each run a different mobile phase was adopted for separating toxins grouped by their basicity. Extracts obtained from shellfish according to the standard mouse bioassay had to be purified through a Sep-Pak C_{18} cartridge to remove possible interfering compounds. Postcolumn derivatisation was performed once again by using periodic acid in sodium phosphate buffer as oxidising reagent at 65°C and with acetic acid as acidifying reagent. In a recent version of this method, further improvements of the system have been described (Oshima, 1995b). The following mobile phases were recommended:

(a) 1 mM Tetrabutylammonium phosphate adjusted to pH 5.8 with acetic acid for C1–C4 toxins.
(b) 2 mM Sodium 1-heptanesulphonate in 10 mM ammonium phosphate buffer, pH 7.1, for the gonyautoxin group (GTX-1–CTX-6), enabling the separation of dcGTX-2 and dcGTX-3 also to be achieved.
(c) 2 mM Sodium 1-heptanesulphonate in 30 mM ammonium phosphate buffer, pH 7.1–acetonitrile (100:5) for separation of STX, dcSTX and neoSTX. It seems that this mobile phase allows also the identification of some of the 13-deoxydecarbamoyl toxins (Oshima *et al.*, 1993).

Lawrence method.   The Lawrence method is based on previous studies of the oxidation of a number of toxins associated with PSP (Lawrence *et al.*, 1991). The new method (Lawrence and Ménard, 1991) involved two prechromatographic oxidations of the toxins with hydrogen peroxide and periodic acid respectively, followed by reverse-phase HPLC separation and fluorescence detection of the resulting compounds. This approach combined the advantages of both types of oxidation in order to achieve a better identification of the toxins, one reaction acting also as a confirmation of the other. Reactions carried out before the chromatographic analysis were easier to perform and control, eliminating additionally the postcolumn reaction system and its maintenance requirements. The addition of ammonium formate to the periodate oxidation reaction was shown to be an important factor in enhancing the yield of fluorescent derivatives for neoSTX, GTX-1, B-2 and C-3 toxins. The basic steps of the procedure were as follows. Clean-up of an aliquot of shellfish extract obtained according to the standard mouse bioassay method was achieved through a previously conditioned SPE-C_{18} cartridge and adjustment of the final eluate to pH 8. Peroxide reaction was performed by adding 100 µl of purified sample extract to a mixture of 25 µl of 10% hydrogen peroxide and 250 µl of 1 N NaOH, the reaction being stopped after 2 min at room temperature by adding 20 µl of acetic acid. Periodate oxidation was carried out by adding 500 µl of 0.03 M periodic acid–0.3 M Na_2HPO_4–0.3 M ammonium formate (1:1:1) at pH 8 to 100 µl purified sample extract, reaction being stopped by adding 10 µl acetic acid after 3 min at room temperature. The HPLC system used a silica-based C_{18} column and elution of the oxidation products was achieved with a linear gradient of 0–5% acetonitrile in 0.1 M ammonium formate adjusted to pH 6.

The periodate oxidation reaction which yields fluorescent derivatives from all the PSP toxins has been reported to have been successfully automated using an auto-sampler system and the procedure applied to samples of shellfish and phytoplankton by means of a programmable auto-injector (Janecek and Quilliam, 1993).

After further investigations and refinements (Lawrence *et al.*, 1995), it was found that no single set of oxidation conditions was the optimum for all the toxins. Peroxide oxidation was particularly useful for all the non-N-hydroxy-lated PSP toxins, given that they produce single oxidation products; however, some of them yield the same oxidation product being quantified together, as is the case of GTX-2, GTX-3 and C-1, C-2. Periodate oxidation gave good results with the N-hydroxylated toxins, although they each gave a mixture of mainly three compounds. NeoSTX and B-2 yield the same major products, as do GTX-1, GTX-4 and C-3, C-4. This problem can be overcome by removal of the less toxic sulphamate toxins B-2, C-3 and C-4. The selective clean-up accomplished previously, by introducing in the protocol if desired an additional step of cation-exchange solid-phase extraction (Lawrence and Ménard, 1991), was further developed by being replaced by a strong anion-exchange solid-phase extraction clean-up (Lawrence *et al.*, 1995). Sometimes, an unidentified fluorescent peak could appear in the chromatograms. Injection of an aliquot of the extract without oxidation could serve to confirm that it was not a PSP toxin but a naturally fluorescent compound. On the other hand, some studies on the extraction of the PSP toxins were conducted by comparison of results from extracts obtained with and without the boiling step (Lawrence *et al.*, 1995). Results for some toxins in boiled extracts were double those found in extraction omitting that step. It appeared that boiling of extracts according to the standard mouse bioassay procedure could remove or destroy some matrix components that can inhibit the oxidation reactions for part of the PSP toxins in the prechromatographic oxidation approach. Elimination of those components during purification of extracts, or even differences arising in retention times and their consequent separation from PSP toxins in HPLC, could justify the lack of significant differences observed earlier between boiled and unboiled extracts using the postcolumn oxidation approach (Sullivan and Wekell, 1987).

Thielert method. Studies performed to separate toxins that eluted together in the analysis of naturally contaminated mussels by Sullivan's method (Luckas *et al.*, 1990), served to prove the utility of using an RP-C_{18} column and *n*-octanesulphonic acid as ion-pair reagent in the achievement of good separation performances. A new HPLC method was developed on this basis (Thielert *et al.*, 1991). Samples of shellfish or algae were extracted with 0.03 N acetic acid, and aliquots were directly injected into the HPLC system without prior purification. Elution of the toxins from an RP-C_{18} column was performed isocratically in two steps, switching from first to second eluent 12.5 min after extract injection. The first eluent was formed by 98.5% 50 mM ammonium phosphate buffer at pH 6.6 with 7 mM 1-octanesulphonic acid and 1.5% tetrahydrofuran. The second eluent was composed of 86% 60 mM ammonium phosphate buffer at pH 7 with 8 mM 1-octanesulphonic acid, 12.5% acetonitrile and 1.5% tetrahydrofuran. Postcolumn oxidation of the PSP toxins was carried out at ambient temperature by adding 10 mM periodic acid and 1 M ammonium hydroxide, the reaction mixture being brought down to pH 5 with 1.2 M acetic acid. The method was slightly refined with respect to reagent concentrations (Luckas *et al.*, 1994), and good separations for C-1–C-4, GTX-1, dcGTX-2, dcGTX-3, GTX-2, GTX-3, neoSTX, dcSTX and STX (the latter very close to dcSTX) were obtained. This method provided an interesting alternative to other postcolumn derivatisation HPLC procedures by achievement of a practical compromise, of good separation of a number of relevant PSP toxins in a single chromatographic run.

More recently, a postcolumn reaction method was reported which combined the three separate runs approach from the Oshima method and adoption of octanesulphonic acid as ion-pair reagent which was formerly used in the Thielert method. In this procedure (Franco and Fernandez-Vila, 1993), mollusc samples were extracted three times with 0.1 M acetic acid and purified by passing through a Sep-Pak C_{18} cartridge. Chromatographic operations, including a postcolumn oxidation reaction, were basically the same as described in the Oshima method (Oshima *et al.*, 1989), except for the use of octanesulphonic acid in two of the isocratic

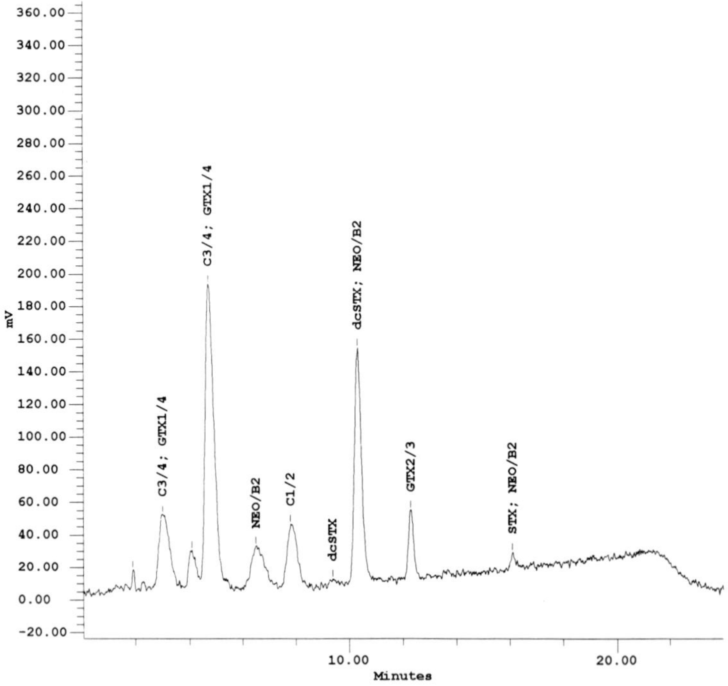

Figure 6.2 Chromatogram obtained in the analysis of contaminated mussel sample, using the Lawrence method, with periodate as oxidant (Lawrence and Ménard, 1991b). The sample was found to contain 3035 µg saxitoxin equivalents/100 g by mouse bioassay and 4579 µg saxitoxin equivalent/100 g by fluorimetric assay (Burdaspal, 1994).

eluents and other minor changes. The method has promising features and its application to research studies on toxin profiles of phytoplankton has been reported (Franco *et al.*, 1994, 1995).

Disadvantages of HPLC procedures. One common feature of all the HPLC methods to a greater or lesser extent is the complexity of equipment and the stringent requirements, such as type of reagents, concentrations, pHs, ionic strength, stationary phases and other operating conditions, involved in the different steps which form the protocol and upon all of which the analytical results critically depend. Frequently, all these aspects make it very difficult to follow entirely the exact conditions as described in the original procedures, thus leading to difficulties in reproducing previously reported performances (Van Egmond *et al.*, 1994) or even causing discrepancies among laboratories. This is particularly notorious when the separation capability between some PSP toxins can be affected by very slight changes in operation conditions, as happens for example with saxitoxin and dcsaxitoxin in postcolumn oxidation-based systems. It must be kept in mind that even instrumental factors, such as the type of fluorescence detector, can affect critically the sensitivity of the detection system. In general, filter detectors are up to ten times less sensitive than dual monochromator detectors, or can give too great a rise in the base line under gradient conditions (Sullivan and Wekell, 1987). In addition, great differences in sensitivity have been reported between monochromator detectors from different manufacturers (Lawrence *et al.*, 1991).

Despite their drawbacks, the HPLC procedures have been widely used, although still offering a great potential for further improvement (Figures 6.2, 6.3, 6.4 and 6.5). The Lawrence, Thielert and (in part) Oshima methods were found adequate for the determination of saxitoxin in PSP-positive extracts of shellfish (Van Egmond *et al.*, 1994). Adoption of any of the cited HPLC methods must be made by considering not only the availability of trained personnel and equipment, but also the purpose of the study, differentiating between routine, occasional monitoring and research, and even taking into account the PSP toxins expected to be predominant in the samples to be studied.

Electrochemical oxidation. Electrochemical oxidation of PSP toxins has also been investigated as an alternative approach to replace chemical postcolumn oxidation for their determination by HPLC. Briefly, electrochemical oxidation is based on an electric current flowing between a working electrode and its counter electrode, through the eluent moving between them. Oxidation is due to loss of electrons from the solute molecules, thus causing the passage of current (Leppard, 1984). Generally speaking, the more the potential is made positive, the more the solute compounds tend to be oxidised. Theoretically, almost any molecule can be electrochemically oxidised by increasing the working electrode potential sufficiently, the main limitations

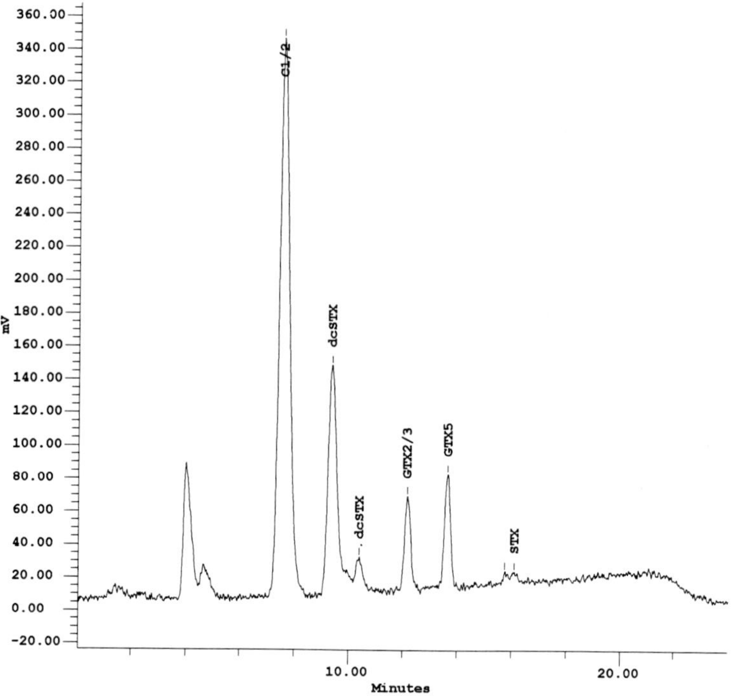

Figure 6.3 Chromatogram obtained in the analysis of contaminated mussel sample using the Lawrence method, with hydrogen peroxide as oxidant (Lawrence and Ménard, 1991b). This procedure served to quantify the levels of saxitoxin and decarbamoylsaxitoxin as 32 and 444 μg/100 g respectively, whereas the sample was found to contain 3035 μg saxitoxin equivalent/100 g by mouse bioassay and 4579 μg saxitoxin equivalent/100 g by fluorimetric assay (Burdaspal, 1994).

coming from decomposition of the solvent constituents or shortage in the lifetime of the electrode. Electrochemical cells can be designed so that the eluent flows by the electrode surface, this mode of operation being called amperometric, and the fraction of the electroactive compound that reacts is typically of the order of 5–15%. Greater effectiveness is achieved when the cell is designed so that the eluent flows through porous graphite electrodes. In this case, almost all the compound is oxidised and the cell is called a coulometric cell.

Early attempts using this technique demonstrated that saxitoxin and other PSP toxins were electrochemically active (Ragelis, 1985). This preliminary work indicated that N-1-hydroxy toxins and their sulphocarbamoyl analogues were much more easily oxidised in contrast to chemical oxidation, thus offering an interesting potential for identification of these compounds.

Application of this technique to the analysis of PSP toxins has been performed by replacing the entire postcolumn reaction system of the

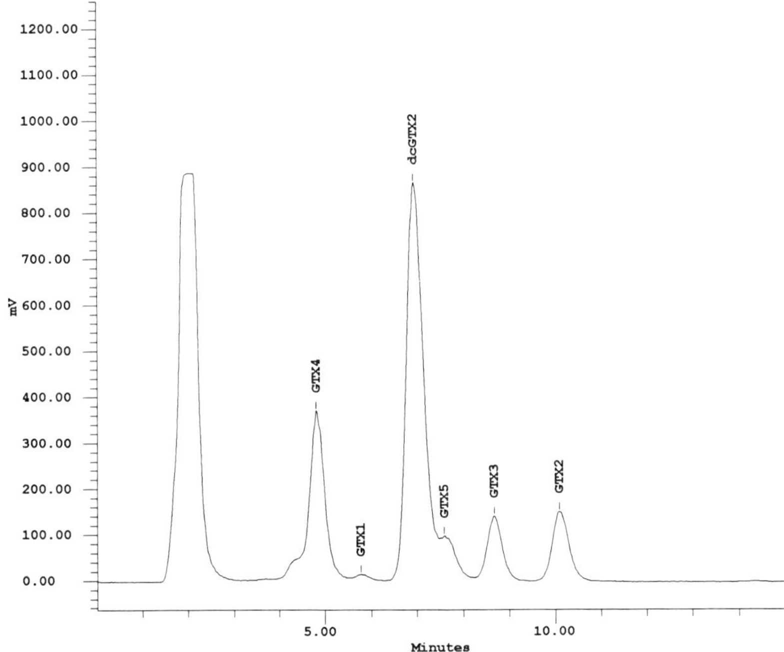

Figure 6.4 Chromatogram obtained in the analysis of contaminated mussel sample, using the Franco method, using the eluent for the separation of GTXs and dcGTXs (Franco and Fernandez-Vila, 1993). The sample was found to contain $3035\,\mu$g saxitoxin equivalent/100 g by mouse bioassay and $4579\,\mu$g saxitoxin equivalent/100 g by fluorimetric assay (Burdaspal, 1994).

standard Sullivan procedure (Sullivan and Wekell, 1987) with a coulometric electrochemical cell (Janiszewski and Boyer, 1993). The fluorescent oxidised toxins were detected by connecting the effluent directly to the fluorescence detector. Hydrodynamic voltammograms for saxitoxin showed significant oxidation at applied potentials greater than 0.8 V. A voltage of 0.8 V was chosen in order to preserve the lifetime of the porous graphite electrode. Under these conditions, the detection limits for purified saxitoxin and GTX-1, GTX-2, GTX-3, B-1 and C toxins, from crude dinoflagellate extracts were not very different from those reported in the literature using the traditional postcolumn chemical oxidation, neosaxitoxin being the only exception in the study mentioned.

Further studies have been conducted in order to confirm those findings and to try to optimise the operating conditions. Oxidation potential, pH, type of column, composition and flow of the mobile phase are factors which affect the performance of the electrochemical oxidation system. By using a similar gradient to that described in the Sullivan procedure (Sullivan and Wekell, 1987) with a modified mobile phase, it was found that at pH 7.0

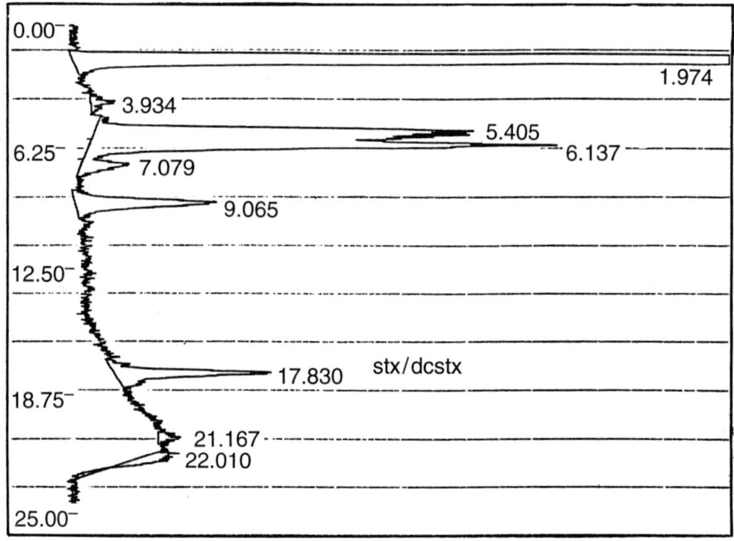

Figure 6.5 Chromatogram obtained for the analysis of contaminated mussel sample, by the Sullivan method, using a C_{18} HPLC column and a filter fluorescence detector (Sullivan and Wekell, 1987). The sample was found to contain 3035 µg saxitoxin equivalent/100 g by mouse bioassay and 4579 µg saxitoxin equivalent/100 g by fluorimetric assay (Burdaspal, 1994).

and 0.75 V the limit of detection for saxitoxin was 0.10 ng. Sensitivity for saxitoxin, GTX-2 and GTX-3 was an order of magnitude greater than that for neoStx and B-1, and two orders of magnitude greater than that for B-2 (Lawrence and Wong, 1995).

Electrochemical oxidation represents an important simplification to the conventional HPLC procedures based on postcolumn chemical oxidation by eliminating secondary pumps, reaction equipment and reagents. It is much more easily set up and maintained and it could offer additional advantages, such as the possibility of being directly coupled with a mass spectrometer, in the case of being incorporated into an HPLC system which can work with no salt eluents (Kirschbaum *et al.*, 1995).

Mass spectrometry (MS) can be used as an alternative detection system to fluorescence spectrometry when coupled to HPLC. Ion-spray mass spectrometry (see Chapter 7) coupled to precolumn periodate oxidation HPLC has shown about five times better sensitivity for saxitoxin than mouse bioassay (Quilliam *et al.*, 1989), its potentially generalised use being limited mainly by its high cost (Van Egmond *et al.*, 1993). Other LC–MS techniques, such as chemical ionisation with the moving belt interface, atmospheric-pressure chemical ionisation, thermospray and continuous-flow fast-atom bombardment, failed to provide spectra of the oxidised products (Quilliam *et al.*, 1993).

6.4.3 Electrophoretic methods

Electrophoresis is a separation technique which has been also applied in the PSP toxins field, although to a limited extent. Paper electrophoresis has been used in the study of naturally contaminated samples following purification of extracts by gel-permeation chromatography through Bio-Gel P-2 and further fractionation on Bio-Rex 70 (Ikawa *et al.*, 1985). Chromatography through Bio-Gel P-2 separates PSP toxins from other interfering compounds such as salts, proteins and carbohydrates. High-voltage paper electrophoresis of purified fractions from extracts of biological samples could separate PSP toxins into three groups according to their net charge, STX and neoSTX (2+), GTX toxins (1+) and C toxins (no net charge), the last-named migrating with the electro-endosmotic flow (Thibault *et al.*, 1991). PSP toxins were detected as blue fluorescent bands after spraying the paper with 1% hydrogen peroxide and further heating at 100°C for 5 min.

Capillary zone electrophoresis (CZE) is a relative new technique which offers a high-resolution separation capability. It was applied, in combination with either ultraviolet detection at 200 nm or mass spectrometric detection using an in-house constructed ion-spray CZE–MS interface, to the analysis of marine samples (Thibault *et al.*, 1991). CZE demonstrated a good separation efficiency for STX, neoSTX and other PSP toxins, its mass detection limit being much lower than that of the mouse bioassay, but due to the very much reduced amount of sample that can be loaded on to the capillary, the actual detection limits in terms of micrograms per 100 g of shellfish meat seem to be too high for practical monitoring purposes.

Improvement in detection limits was achieved by the application of on-column sample preconcentration with capillary isotachophoresis and a discontinuous buffer system prior to conventional capillary zone electrophoresis (Locke and Thibault, 1994). Nevertheless, the authors recognised that the analysis of complex biological extracts using ultraviolet detection had limited capability for practical identification of some PSP toxins, being more advantageous for the coupling with ion-spray mass spectrometry. The method provides detection limits comparable to those of HPLC but not requiring conversion to fluorescent derivatives, the only exception to its well-known separation efficiency being for the non-charged C toxins.

6.5 Conclusions

As has been discussed, there exists a wide collection of analytical methods which have been utilised more or less extensively for the determination of PSP toxins in phytoplankton, shellfish and other marine animals. A number of these methods are currently in use in many laboratories. The variation in the available procedures is so large that it covers the full range of techniques available in the field of food analysis. They include both biological and chemical approaches and range from the simple and cheap thin-layer chromatography to the sophisticated and expensive capillary electrophoresis or LC–MS. Some of these methods can be considered as unconventional, such as the house-fly test, and others as classical, such as the simple colorimetric assays.

The efforts exerted for years by scientific groups trying to develop such varied methods of analysis indicate that none can be considered as fully satisfactory (Table 6.1). The reasons behind this apparent failure are not only the intrinsic analytical challenge of the separation and detection of more than 20 compounds with molecular structures that are sometimes very closely related, but also their own molecular instability under certain environmental conditions such as temperature, pH and presence and levels of other influencing factors that can be found in such a complex and variable matrix as the different types of shellfish.

It seems that the pattern of PSP toxins in a single mollusc responds to a dynamic reality, being subjected to metabolic transformations that are so far difficult to predict. The number of possible transformations that can take place among PSP toxins owing to storage conditions prior to the analysis and through the analytical operations begs questions about the correspondence between analytical results and the true composition of PSP toxin content in the shellfish in their marine environment. From a public-health point of view, it might be important to consider also the possible degradation or transformation as a consequence of the conditions of the industrial processing and culinary treatments.

It is obvious that the selection of the appropriate method of analysis relies not only on the technical capability to perform it, but on the objectives of the study or programme in which it is to be employed. So, for research purposes a method will be required to be able to give the most precise and detailed information about the qualitative and quantitative composition in a number of samples which is normally limited. The complexity of the equipment, the high level of analyst's skill, and even the time-consuming nature of each analysis are generally secondary factors compared to achievement of the main goals of the study.

On the other hand, monitoring programmes usually deal with a large number of samples, the analytical results being needed in a short period of time. Methods of analysis or assays for routine control must ideally be robust, simple and not excessively expensive, and yet capable of detecting at the regulatory level all of the main toxins according to their abundance or toxic properties.

At present, it is clearly recognised that there is a pressing need to replace the mouse bioassay by other alternative procedures for monitoring. Nevertheless, the limitations underlying all the methods and the lack of international validation mean that most of the countries with specific PSP

Table 6.1 Summary of possibilities and limitations of current analytical techniques to determine paralytic shellfish poisons (Van Egmond *et al.*, 1993)

Method	Rate of analysis per day	Cost	Interferences and problems	Limit of detection	Applicability
Mouse bioassay	10–30	Low equipment, high labour	Salt. Strain dependence. Non-linearity. Death determination time-consuming.	$40\ \mu g/100\ g \pm 20\%$	Most shellfish algae
Other bioassays, e.g. oyster embryo	?	?	Method not fully developed.	?	Under test for mussels
Algal cell enumeration		Low equipment, high labour	Presence does not prove toxicity.		
ELISA	Many	High	Cross reactions or poor response to range of toxins present.	$5\text{–}10\ \mu g/100\ g \pm 7\%$	Bivalves
Sodium channel	Many using microplate reader	Low	Specificity unknown. Method not fully developed.	$1\text{–}10\ \mu g/100\ g$	Mussels
Spectrometry	Few, laborious clean-up	Low	Metals. Toxins have variable fluorescence.	$\sim\!4\ \mu g/100\ g$, but recovery can be low	Bivalves
TLC	Many	Low	Presently semi-quantitative.		
HPLC	20	Medium	Incomplete resolution of toxins. Co-extractants may give false positives.	$10\text{–}30\ \mu g/100\ g \pm 10\%$	Algae bivalves (crustacea have been tested but database is poor)
Electrophoresis	Many	Low	Semi-quantitative		Algae bivalves
Capillary electrophoresis	Probably as HPLC	Medium (UV) to high (laser fluorescence)	Method not fully developed.	$500\ \mu g/g$ (UV) $6.5\ \mu g/100\ g$ (fluorescence)	Algae bivalves
Mass spectrometry	As HPLC	High	Theoretically none.	$10\ \mu g/100\ g$	Standards

regulations indicate the use of the standard mouse bioassay as the method of analysis (Van Egmond *et al.*, 1992). It may be significant that the European Union stipulates that the biological method of analysis can be associated with a chemical method for saxitoxin detection or any other method accepted according to a given procedure; however, in case of discrepancy, the reference method will be the biological one (Council of the European Communities, 1991).

HPLC methods able to give practical information in a single run (Lawrence *et al.*, 1991; Thielert *et al.*, 1991) seem to be well placed as alternative methods, even though they do not fulfil all the desired requirements. In addition, the lack of standards and some difficulties attributed to the inadequacy of factors available for converting the individual chromatographic peaks into specific toxicity values (Jellet *et al.*, 1992) may delay their validation and worldwide adoption for routine analysis.

Other very promising methods which may have a real chance are the HPLC methods using electrochemical oxidation (Janiszewski and Boyer, 1993) and also the tissue-culture assay (Gallacher *et al.*, 1992; Jellet *et al.*, 1992; Manger *et al.*, 1993).

Acknowledgements

I thank J. Gilbert, H.P. Van Egmond and J.F. Lawrence for their criticisms, comments and valuable suggestions.

References

Adams, W.N. and Furfari, S.A. (1984) Evaluation of laboratory performance of the AOAC method for PSP toxin in shellfish. *Journal of the Association of Official Analytical Chemists*, **67**, 1147–1148.

Association of Official Analytical Chemists (1995) Paralytic shellfish poison, biological method. *Official Methods of Analysis*, 16th edition, Washington, DC: AOAC, method 959.08.

Barber, K.G., Kitts, D.D., Townsley, P.M. and Smith, D.S. (1988) Appearance and partial purification of a high molecular weight protein in crabs exposed to saxitoxin. *Toxicon*, **26**, 1027–1034.

Bates, H.A. and Rapoport, H. (1975) A chemical assay for saxitoxin, the paralytic shellfish poison. *Journal of Agricultural and Food Chemistry*, **23**, 237–239.

Bates, H.A., Kostriken, R. and Rapoport, H. (1978) A chemical assay for saxitoxin. Improvements and modifications. *Journal of Agricultural and Food Chemistry*, **26**, 252–254.

Berenguer, J.A., González, L., Jiménez, I., Legarda, T.M., Olmedo, J.B. and Burdaspal, P.A. (1993) The effect of commercial processing on the paralytic shellfish poison (PSP) content of naturally-contaminated *Acanthocardia tuberculatum* L. *Food Additives and Contaminants*, **10**, 217–230.

Bose, R.J. and Reid, J.E. (1979) Evidence for the heterogeneity of paralytic shellfish toxin in clam and mussel samples gathered near Prince Rupert, British Columbia. In *Toxic Dinoflagellate Blooms*, edited by D.L. Taylor and H.H. Siger, Elsevier/North Holland Inc., pp. 399–402.

Boyer, G.L., Fix Whicmann, C., Mosser, J., Schantz, E.J. and Schnoes, H.K. (1979) Toxins isolated from Bay of Fundy scallops. In *Toxic Dinoflagellate Blooms*, edited by D.L. Taylor and H.H. Siger, Elsevier/North Holland Inc., pp. 373–376.

Buckley, L.J., Ikawa, M. and Sasner, J.J. Jr. (1976) Isolation of *Gonyaulax tamarensis* toxins from soft shell clams (*Mya arenaria*) and a thin-layer chromatographic–fluorometric method for their detection. *Journal of Agricultural and Food Chemistry*, **24**, 107–111.

Buckley, L.J., Oshima, Y. and Shimizu, Y. (1978) Construction of a paralytic shellfish toxin analyser and its application. *Analytical Biochemistry*, **85**, 157–164.

Bundesgesundheitsamt (BGA) (1989) Determination of the algal toxin content in molluscs and mollusc products. Fluorometric method. *Amtliche Sammlung von Untersuchungsverfahren nach §35 LMBG*, 12.03/04-1, Berlin, Köln: Beuth Verlag GmbH.

Burdaspal, P.A. (1994) Unpublished data.

Burdaspal, P.A., Legarda, T.M. and Berenguer, J. (1991) Application of the fluorometric method to the routine analysis of PSP in shellfish. *Alimentaria*, March, 23–27.

Catterall, W.A. (1985) The voltage sensitive sodium channel: A receptor for multiple neurotoxins. In *Toxic Dinoflagellates*, edited by D.M. Anderson, A.W. White and D.G. Baden, Elsevier Science Publishing Co., pp. 329–342.

Catterall, W.A. and Nirenberg, M. (1973) Sodium uptake associated with activation of action potential ionophores of cultured neuroblastoma and muscle cells. *Proceedings of the National Academy of Sciences U.S.A.*, **70**, 3759–3763.

Cembella, A.D. and Lamoureux, G. (1993) A competitive inhibition enzyme-linked immunoassay for the detection of paralytic shellfish toxins in marine phytoplankton. In *Toxic Phytoplankton Blooms in the Sea*, edited by T.J. Smayda and Y. Shimizu, Elsevier Science Publishers B.V., pp. 857–862.

Childress, W.L. (1979) Chemical determination of PSP. IUPAC Collaborative Program, unpublished.

Chu, F.S. and Fan, T.S.N. (1985) Indirect enzyme-linked immunosorbent assay for saxitoxin in shellfish. *Journal of the Association of Official Analytical Chemists*, **68**, 13–16.

Council of the European Communities (1991) *Official Journal of the European Communities*, no. L268, pp. 1–14.

Doucette, G.J. (1995) Interactions between bacteria and harmful algae: A review. *Natural Toxins*, **3**, 65–74.

Franco, J.M. and Fernandez-Vila, P. (1993) Separation of paralytic shellfish toxins by reversed phase high performance liquid chromatography, with postcolumn reaction and fluorimetric detection. *Cromatographia*, **35**, 613–620.

Franco, J.M., Fernández, P. and Reguera, B. (1994) Toxin profiles of natural populations and cultures of *Alexandrium minutum* Halim from Galician (Spain) coastal waters. *Journal of Applied Phycology*, **6**, 275–279.

Franco, J.M., Fraga, S., Zapata, M., Bravo, I., Fernandez, P. and Ramilo, I. (1995) Comparison between different strains of genus *Alexandrium* of the minutum group. In *Harmful Marine Algal Blooms*, edited by P. Lassus, G. Arzul, E. Erard-Le Denn, P. Gentien and C. Marcaillou-Le Baut, Intercept Ltd., pp. 53–58.

Gallacher, S. and Birkbeck, T.H. (1992) A tissue culture assay for direct detection of sodium channel blocking toxins in bacterial culture supernates. *FEMS Microbiology Letters*, **92**, 101–107.

Gallacher, S., Evans, K., Blackburn, M., Waldock, M. and Birkbeck, T.H. (1993) Tissue culture assay for paralytic shellfish poisoning toxins. In *New Techniques in Food and Beverage Microbiology*, Society for Applied Bacteriology, pp. 203–210.

Gershey, R.M., Nevé, R.A., Musgrave, D.L. and Reichardt, P.B. (1977) A colorimetric method for determination of saxitoxin. *Journal of Fisheries Research Board of Canada*, **34**, 559–563.

Hall, S. and Reichardt, P.B. (1984) Cryptic paralytic shellfish toxins. In *Seafood Toxins*, ACS Symposium Series no. 262, edited by E.P. Ragelis, Washington, DC: American Chemical Society, pp. 113–123.

Hall, S., Nevé, R.A., Reichardt, P.B. and Swisher, G.A. (1979) Chemical analysis of paralytic shellfish poisoning in Alaska. In *Toxic Dinoflagellate Blooms*, edited by D.L. Taylor and H.H. Seliger, Elsevier/North Holland Inc., pp. 345–350.

Hellwig, E. and Petuely, F. (1980) Bestimung von Saxitoxin in Muschelkonserven. *Zeitschrift für Lebensmitteluntersuchung und Forschung*, **171**, 165–169.

Hurst, J.W., Selvin, R., Sullivan, J.J., Yentsch, C.M. and Guillard, R.L. (1985) Intercomparison of various assay methods for the detection of shellfish toxins. In *Toxic Dinoflagellates*, edited by D.M. Anderson, A.W. White and D.G. Baden, Elsevier Science Publishing Co., pp. 427–432.

Ikawa, M., Auger, K., Mosley, S.P., Sasner, J.J. Jr., Noguchi, T. and Hashimoto, K. (1985) Toxin profiles of the blue–green alga *Aphanizomenon flos-aquae*. In *Toxic Dinoflagellates*, edited by D.M. Anderson, A.W. White and D.G. Baden, Elsevier Science Publishing Co., pp. 299–304.

Janecek, M. and Quilliam, M.A. (1993) Analysis of paralytic shellfish poisoning toxins by automated pre-column oxidation and microcolumn liquid chromatography with fluorescence detection. *Journal of Chromatography*, **644**, 321–331.

Janiszewski, J. and Boyer, G.L. (1993) The electrochemical oxidation of saxitoxin and derivatives: its application to the HPLC analysis of PSP toxins. In *Toxic Phytoplankton Blooms in the Sea*, edited by T.J. Smayda and Y. Shimizu, Elsevier Science Publishers B.V., pp. 889–894.

Jellet, J.F., Marks, L.J., Stewart, J.E., Dorey, M.L., Watson-Wright, W. and Lawrence, J.F. (1992) Paralytic shellfish poison (saxitoxin family) bioassays: Automated endpoint determination and standardization of the *in vitro* tissue culture bioassay, and comparison with the standard mouse bioassay. *Toxicon*, **30**, 1143–1156.

Jonas-Davies, J., Sullivan, J.J., Kentala, L.L., Liston, J., Iwaoka, W.T. and Wu, L. (1984) Semiautomated method for the analysis of PSP toxins in shellfish. *Journal of Food Science*, **49**, 1506–1516.

Kirschbaum, J., Hummert, C. and Luckas, B. (1995) Determination of paralytic shellfish poisoning (PSP) toxins by application of ion-exchange HPLC, electrochemical oxidation, and mass detection. In *Harmful Marine Algal Blooms*, edited by P. Lassus, G. Arzul, E. Erard-Le Denn, P. Gentien and C. Marcaillou-Le Baut, Lavoisier, Intercept Ltd., pp. 309–314.

Kogure, K., Tamplin, M.L., Simidu, U. and Colwell, R.R. (1988) A tissue culture assay for tetrodotoxin, saxitoxin and related toxins. *Toxicon*, **26**, 191–197.

Lawrence, J.F. and Ménard, C. (1991) Liquid chromatographic determination of paralytic shellfish poisons in shellfish after prechromatographic oxidation. *Journal of the Association of Official Analytical Chemists*, **74**, 1006–1012.

Lawrence, J.F. and Wong, B. (1995) Evaluation of a postcolumn electrochemical reactor for oxidation of paralytic shellfish poison toxins. *Journal of the Association of Analytical Chemists International*, **78**, 698–704.

Lawrence, J.F., Ménard, C. and Charbonneau, C.F. (1991) A study of ten toxins associated with paralytic shellfish poison using prechromatographic oxidation and liquid chromatography with fluorescence detection. *Journal of the Association of Official Analytical Chemists*, **74**, 404–409.

Lawrence, J.F., Ménard, C. and Cleroux, C.H. (1995) An evaluation of prechromatographic oxidation for the liquid chromatographic determination of paralytic shellfish poisons in shellfish. *Journal of the Association of Official Analytical Chemists International*, **78**, 514–520.

Leppard, J.P. (1984) High pressure liquid chromatography. In *Food Analysis, Principles and Techniques. Volume 4. Separation Techniques*, edited by D.W. Gruenwedel and J.R. Whitaker, Marcel Dekker, Inc., pp. 365–407.

Locke, S.J. and Thibault, P. (1994) Improvement in detection limit for the determination of paralytic shellfish poisoning toxins in shellfish tissues using capillary electrophoresis/electrospray mass spectrometry and discontinuous buffer systems. *Analytical Chemistry*, **66**, 3436–3446.

Luckas, B. (1990) The occurrence of decarbamoyl-saxitoxin in canned mussels. In *Public Health Aspects of Seafood-Borne Zoonotic Diseases*, edited by E.M. Bernoth, Vet. Med. Hefte 1/91, pp. 89–97.

Luckas, B., Thielert, G. and Peters, K. (1990) Zur Problematik der selektiven Erfassung von PSP-Toxinen aus Muscheln. *Zeitschrift für Lebensmittel-Untersuchung und Forschung*, **190**, 491–495.

Luckas, B., Hummert, C., Thielert, G., Kirschbaum, J. and Boenke, A. (1994) Improvement of HPLC/fluorescence and HPLC/MS procedures for the determination of DSP and PSP toxins in contaminated shellfish. In *BCR Information, Chemical Analyses*, report EUR 15339 EN, European Commission.

Manger, R.L., Leja, L.S., Lee, S.Y., Hungerford, J.M. and Wekell, M.M. (1993) Tetrazolium-based cell bioassay for neurotoxins active on voltage-sensitive sodium channels: Semiautomated assay for saxitoxins, brevetoxins, and ciguatoxins. *Analytical Biochemistry*, **214**, 190–194.

McCulloch, A.W., Boyd, R.K., de Freitas, S.W., Foxall, R.A., Jamieson, M.V., Laycock, M.A., Quilliam, M.A., Wright, J.L.C., Boyko, V.J., McLaren, J.W., Miedema, M.R., Pocklington, R., Arsenault, E. and Richard, D.J.A. (1989) Zinc from oyster tissue as causative factor in mouse deaths in official bioassay for paralytic shellfish poison. *Journal of the Association of Official Analytical Chemists*, **72**, 384–386.

McFarren, E.F. (1959) Report on collaborative studies of the bioassay for paralytic shellfish poison. *Journal of the Association of Official Analytical Chemists*, **42**, 263–271.

McFarren, E.F., Schantz, E.J., Campbell, J.E. and Lewis, K. (1959) A modified Jaffé test for determination of paralytic shellfish poison. *Journal of the Association of Official Analytical Chemists*, **42**, 399–404.

Mold, J.D., Bowden, J.P., Stanger, D.W., Maurer, J.E., Lynch, J.M., Wyler, R.S., Shantz, E.J. and Rigel, B.J. (1957) Paralytic shellfish poison. VII. Evidence for the purity of the poison isolated from toxic clams and mussels. *Journal of the American Chemical Society*, **79**, 5235–5238.

Mosley, S., Ikawa, M. and Sasner, J.J. Jr. (1985) A combination fluorescence assay and Folin–Ciocalteau phenol reagent assay for the detection of paralytic shellfish poisons. *Toxicon*, **23**, 375–381.

Onoue, Y., Nobuchi, T., Nagashima, Y., Hashimoto, K., Kanoh, S., Ito, M. and Tsukada, K. (1983) Separation of tetrodotoxin and paralytic shellfish poisons by high-performance liquid chromatography with a fluorometric detection using *o*-phthalaldehyde. *Journal of Chromatography*, **257**, 373–379.

Oshima, Y. (1995a) Chemical and enzymatic transformation of paralytic shellfish toxins in marine organisms. In *Harmful Marine Algal Blooms*, edited by P. Lassus, G. Arzul, E. Erard-Le Denn, P. Gentien and C. Marcaillou-Le Baut, Lavoisier, Intercept Ltd., pp. 481–486.

Oshima, Y. (1995b) Postcolumn derivatization liquid chromatographic method for paralytic shellfish toxins. *Journal of the Association of Analytical Chemists International*, **78**, 528–532.

Oshima, Y., Machida, M., Sasaki, K., Tamaoki, Y. and Yasumoto, T. (1984) Liquid chromato-graphic–fluorometric analysis of paralytic shellfish toxins. *Agricultural Biological Chemistry*, **48**, 1707–1711.

Oshima, Y., Sugino, K. and Yasumoto, T. (1989) Latest advances in HPLC analysis of paralytic shellfish toxins. In *Mycotoxins and Phycotoxins '88*, edited by S. Natori, K. Hishimoto and Y. Ueno, Elsevier Science Publishers B.V., Amsterdam, pp. 319–326.

Oshima, Y., Sugino, K., Itakura, H., Hirota, M. and Yasumoto, T. (1990) Comparative studies on paralytic shellfish toxin profile of dinoflagellates and bivalves. In *Toxic Marine Phyto-plankton*, edited by E. Granéli, B. Sundstrom, L. Edler and D. Anderson, Elsevier Science Publishing Co., Inc., pp. 391–396.

Oshima, Y., Itakura, H., Lee, K., Yasumoto, T., Blackburn, S. and Hallegraeff, G. (1993) Toxin production by the dinoflagellate *Gymnodinium catenatum*. In *Toxic Phytoplankton Blooms in the Sea*, edited by T.J. Smayda and Y. Shimizu, Elsevier Science Publishers B.V., pp. 907–912.

Park, D.L., Adams, W.N., Graham, S.L. and Jackson, R.C. (1986) Variability of mouse bioassay for determination of paralytic shellfish poisoning toxins. *Journal of the Association of Official Analytical Chemists*, **69**, 547–550.

Pleasance, S., Quilliam, M.A. and Marr, J.C. (1992) Ionspray mass spectrometry of marine toxins. IV. Determination of diarrhetic shellfish poisoning toxins in mussels tissue by liquid chromatography/mass spectrometry. *Rapid Communications in Mass Spectrometry*, **6**, 121–127.

Price, R.J. and Lee, J.S. (1971) Interaction between paralytic shellfish poison and melanin obtained from butter clam (*Saxidomus giganteus*) and synthetic melanin. *Journal of the Fisheries Research Board of Canada*, **28**, 1789–1794.

Quilliam, M.A., Thomson, B.A., Scott, G.J. and Siu, K.W.M. (1989) Ion-spray mass spectrometry of marine neurotoxins. *Rapid Communications in Mass Spectrometry*, **3**, 145–150.

Quilliam, M.A., Janacek, M. and Lawrence, J.F. (1993) Characterization of the oxidation pro-ducts of paralytic shellfish poisoning toxins by liquid chromatography/mass spectrometry. *Rapid Communications in Mass Spectrometry*, **7**, 482–487.

Ragelis, E.P. (1982) Report on seafood toxins. *Journal of the Association of Official Analytical Chemists*, **65**, 327–331.

Ragelis, E.P. (1985) Report on seafood toxins. *Journal of the Association of Official Analytical Chemists*, **68**, 252–254.

Reichardt, P.B., Nevé, R.A., Gershey, R.M., Hall, S., Musgrave, D.L., Seaton, P.J. and Swisher, G.A. Jr. (1978) Chemical investigations of paralytic shellfish poisoning in Alaska. In *Sea Grant Report 78-3; IMS Report 78-4*, pp. 1–41.

Ross, M.R., Siger, A. and Abbott, B.C. (1985) The house fly: An acceptable subject for paralytic shellfish toxin bioassay. In *Toxic Dinoflagellates*, edited by D.M. Anderson, A.W. White and D.G. Baden, Elsevier Science Publishing Co., pp. 433–438.

Salter, J.E., Timperi, R.J., Hennigan, L.J., Sefton, L. and Reece, H. (1989) Comparison evaluation of liquid chromatographic and bioassay methods of analysis for determination of paralytic shellfish poisons in shellfish tissues. *Journal of the Association of Official Analytical Chemists*, **72**, 670–673.

Sato, S., Ogata, T., Kodama, M. and Kogure, K. (1988) Determination of paralytic shellfish toxins by neuroblast culture assay. In *Proceedings of the Japanese Association of Mycotoxicology*, supplement no. 1, edited by K. Aibara *et al.*, Tokyo: Japanese Association of Mycotoxicology, pp. 3–6.

Schantz, E.J., McFarren, E.F., Schafer, M.L. and Lewis, K.H. (1958) Purified shellfish poison for bioassay standardization. *Journal of the Association of Official Analytical Chemists*, **41**, 160–168.

Shimizu, Y., Buckley, L.J., Alam, M., Oshima, Y., Fallon, W.E., Kasai, H., Miura, I., Gullo, V.P. and Nakanishi, K. (1976) Structures of gonyautoxin II and III from the East Coast toxic dinoflagellate *Gonyaulax tamarensis*. *Journal of the American Chemical Society*, **98**, 5414–5416.

Shoptaugh, N.H., Carter, P.W., Foxall, T.L., Sasner, J.J. Jr. and Ikawa, M. (1981) Use of fluorometric for the determination of *Gonyaulax tamarensis* Var. Scabata toxins in New England shellfish. *Journal of Agricultural and Food Chemistry*, **29**, 198–200.

Siger, A., Abbott, B.C. and Ross, M. (1984) Response of the house fly to saxitoxins and contaminated shellfish. In *Seafood Toxins*, ACS Symposium no. 262, edited by E.P. Ragelis, Washington, DC: American Chemical Society, pp. 193–195.

Sommer, H. and Meyer, K.F. (1937) Paralytic shellfish poison. *Archives of Pathology*, **24**, 560–598.

Sullivan, J.J. and Iwaoka, W.T. (1983) High pressure liquid chromatographic determination of toxins associated with paralytic shellfish poisoning. *Journal of the Association of Official Analytical Chemists*, **66**, 297–303.

Sullivan, J.J. and Wekell, M.M. (1984) Determination of paralytic shellfish poisoning toxins by high pressure liquid chromatography. In *Seafood Toxins*, ACS Symposium no. 262, edited by E.P. Ragelis, Washington, DC: American Chemical Society, pp. 197–205.

Sullivan, J.J. and Wekell, M.M. (1987) The application of high performance liquid chromatography in a paralytic shellfish poisoning monitoring program. In *Seafood Quality Determination*, edited by D.E. Kramer and J. Liston, Elsevier/North Holland, New York, pp. 357–371.

Sullivan, J.J., Simon, M.G. and Iwaoka, W.T. (1983) Comparison of HPLC and mouse bioassay methods for determining PSP toxins in shellfish. *Journal of Food Science*, **48**, 1312–1314.

Sullivan, J.J., Wekell, M.M. and Kentala, L.L. (1985a) Application of HPLC for the determination of PSP toxins in shellfish. *Journal of Food Science*, **50**, 26–29.

Sullivan, J.J., Jonas-Davies, J. and Kentala, L.L. (1985b) The determination of PSP toxins by HPLC and autoanalyzer. In *Toxic Dinoflagellates*, edited by D.M. Anderson, A.W. White and D.G. Baden, Elsevier Science Publishing Co. Inc., pp. 275–280.

Tamplin, M.L. (1990) A bacterial source of tetrodotoxins and saxitoxins. In *Marine Toxins, Origin, Structure and Molecular Pharmacology*, ACS Symposium no. 418, edited by S. Hall and G. Strichartz, Washington, DC: American Chemical Society, pp. 78–86.

Thibault, P., Pleasance, S. and Laycock, M.V. (1991) Analysis of paralytic shellfish poison by capillary electrophoresis. *Journal of Chromatography*, **542**, 483–501.

Thielert, G., Kaiser, I. and Luckas, B. (1991) HPLC determination of PSP toxins. In *Proceedings of Symposium on Marine Biotoxins*, edited by J.M. Fremy, C.N.E.V.A., pp. 121–125.

Usleber, E., Schneider, E. and Terplan, G. (1991) Direct enzyme immunoassay in microtritation plate and test strip format for the detection of saxitoxin in shellfish. *Letters in Applied Microbiology*, **13**, 275–277.

U.S. Department of Health and Human Services (1991) Paralytic shellfish poisoning—Massachusetts and Alaska, 1990. *Morbidity and Mortality Weekly Report*, **40**, 157–161.

Van Egmond, H.P., Speyers, G.J.A. and Van den Top, H.J. (1992) Current situation on worldwide regulations for marine phycotoxins. *Journal of Natural Toxins*, **1**, 67–85.

Van Egmond, H.P., Aune, T., Lassus, P., Speijers, G.J.A. and Waldock, M. (1993) Paralytic and diarrhoeic shellfish poisons: Occurrence in Europe, toxicity, analysis and regulation. *Journal of Natural Toxins*, **2**, 41–83.

Van Egmond, H.P., Van den Top, H.J., Paulsch, W.E., Goenaga, X. and Vieytes, M.R. (1994) Paralytic shellfish poison reference materials: an intercomparison of methods for the determination of saxitoxin. *Food Additives and Contaminants*, **11**, 39–56.

Wong, J.L., Osterlin, R. and Rapoport, H. (1971) The structure of saxitoxin. *Journal of the American Chemical Society*, **93**, 7344–7345.

Yndestad, M. and Underdal, B. (1985) Survey of PSP in mussels (*Mytilus edulis* L.) in Norway. In *Toxic Dinoflagellates*, edited by D.M. Anderson, A.W. White and D.G. Baden, Elsevier Science Publishing Co. Inc., pp. 457–460.

7 Analysis of food contaminants by combined liquid chromatography–mass spectrometry (LC–MS)

JOHN GILBERT

Summary

Over the past ten years there have been significant advances in LC–MS both in terms of the development of new interfaces and in the availability of more reliable, robust and lower cost instrumentation. Thermospray, particle beam, electrospray and atmospheric pressure chemical ionisation interfaces are briefly described and compared, and advances in coupling of MS with supercritical fluid chromatography and capillary electrophoresis are outlined. The two major areas where LC–MS applications have increased in recent years have been for pesticide and veterinary drug residues, and these are reviewed, citing in particular examples where LC–MS now offers the capability of analysing polar and higher molecular weight residues which have previously proved less amenable to other techniques. More isolated examples of applications of LC–MS are found in the areas of mycotoxins, natural toxicants, food additives and packaging migration, which go some way to demonstrating the potential for further new applications.

Abbreviations

APCI, atmospheric pressure chemical ionisation
API, atmospheric pressure ionisation
CE, capillary electrophoresis
CE–MS, combined capillary electrophoresis mass spectrometry
CI, chemical ionisation
CID, collision-induced decomposition
CV, coefficient of variation
DON, 4-deoxynivalenol mycotoxin
DLI, direct liquid introduction LC–MS
DSP, diarrhetic shellfish poisoning
EI, electron ionisation
ETU, ethylene thiourea
FAB, fast atom bombardment
GC, gas chromatography
GC–MS, combined gas chromatography–mass spectrometry

HPLC, high performance liquid chromatography
ITP, isotachophoresis
LC–MS, combined liquid chromatography–mass spectrometry
MRL, maximum residue limit
MS, mass spectrometry
MS–MS, tandem mass spectrometry–mass spectrometry
MWt, molecular weight
NI, negative ionisation (or ion)
NICI, negative ion chemical ionisation
NIV, nivalenol mycotoxin
OCs, organochlorine pesticides
OPs, organophosphorus pesticides
PET, poly(ethylene terephthalate)
PI, positive ionisation (or ion)
PSP, paralytic shellfish poisoning
SFC, supercritical fluid chromatography
SIM, selected ion monitoring
SMC, sterimatocystin
TLC, thin-layer chromatography
UV, ultraviolet

7.1 Introduction

In the field of food contaminant residue monitoring, the two dominant areas of activity based on the numbers of samples analysed are undoubtably pesticide and veterinary drug residues. For example, in the UK in 1994, official monitoring was carried out on 60 000 samples for pesticide residues and 52 000 samples for veterinary drug residues, compared with only 1200 samples analysed for inorganic contaminants and 2600 samples for natural toxicants (Ministry of Agriculture, Fisheries and Food, 1995). Both pesticides and veterinary drugs are characterised by a wide range of analytes of differing characteristics and a need to monitor large numbers of samples in order to obtain statistically representative data. Confirmation of positive results by unequivocal means is also essential. In both areas, but particularly pesticides, benchtop systems have made GC–MS a routine operation instead of a specialist one, and selected ion monitoring has frequently replaced sequential GC analysis, which would previously have taken place on the same sample extracts on GC columns of differing polarity and using various selective detectors. A number of veterinary drugs have to date proved to be analytically fairly intractable, and although HPLC has been the method of choice, sensitive detection (through lack of an adequate chromophore) and confirmation of positive results have proved problematic. This is one of the areas where recent developments in LC–MS, such as electrospray and atmospheric pressure chemical ionisation (APCI), are beginning to have an impact, and the introduction of dedicated LC–MS systems will begin to revolutionise monitoring over the next few years as benchtop GC–MS systems have over the past 5–10 years.

Broadly the use of mass spectrometry for food contaminant analysis falls into three areas:

(a) Confirmation of the presence of known target analytes—this usually follows some form of preliminary screening.
(b) Monitoring or surveillance of contaminants directly without preliminary screening.
(c) Identification and characterisation of unknown contaminants in foods.

For (a) and (b), which probably comprises about 90% of the use of MS for food contaminant analysis, invariably selected ion monitoring (SIM) is required. Sensitivity demands range over about six orders of magnitude— from lowest levels of a few ng/kg (ppt) in the exceptional case of dioxins up to around 0.5 mg/kg (ppm) which is the maximum residue limit (MRL) for certain pesticides. In some instances these sensitivity demands have previously restricted widespread application of LC–MS to food contaminant problems. However, with some of the newer developments in interfacing and the use of negative ion chemical ionisation (NICI), sensitivity of LC–MS comparable to UV and fluorescence detection has been attainable, although dependent on the compound type. In both (a) and (b), quantitation is required particularly for regulatory work, and where there is a lack of signal stability (as is the case, for example, with particle beam LC–MS) there are good arguments for internal standardisation, with staple-isotope internal standards where possible. However, although there has been an increased use of stable isotopes as internal standards and these are indicated where appropriate in this chapter, this remains an area which still needs development. Probably the most interesting application of LC–MS is in the identification and characterisation of unknown contaminants. However, examples are really quite few, and in most instances monitoring is based on a prejudgement of what contaminants are to be expected.

7.2   Developments in LC–MS interfacing

In the early days of LC–MS, the main interfaces were the moving belt and the direct liquid introduction systems, both of which had intrinsic disadvantages in terms of sensitivity and compound-dependent limitations, as well as requiring skill and experience to operate. The moving belt system is restricted to a maximum flow rate of about 2 ml/min limited to organic mobile phase solvents and 0.5 ml/min with reverse-phase solvents but without buffers being present. In contrast, direct liquid introduction systems are restricted

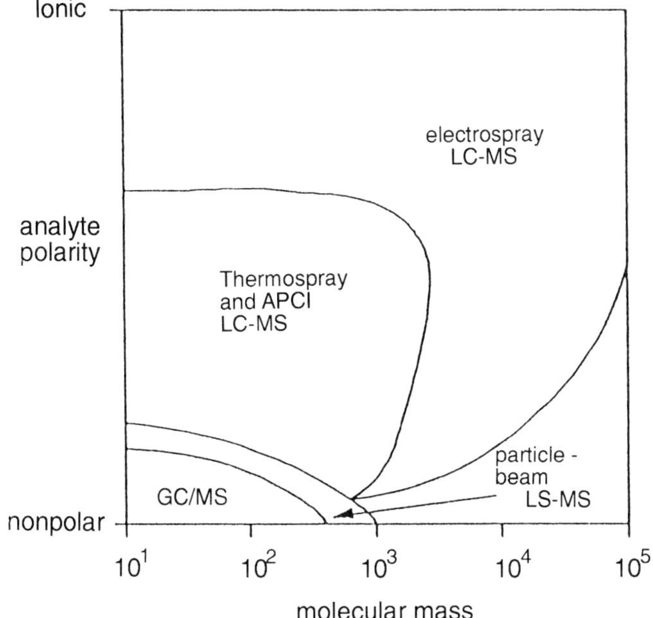

Figure 7.1 Schematic diagram showing the application areas of various LC–MS interfaces in terms of molecular weight and analyte polarity (reproduced from Niessen and Tinke, 1995, with permission from Elsevier Science Publishers).

to flow rates of about 0.05 ml/min and similarly are not suitable for use with buffers. The introduction of thermospray LC–MS was a significant advance, enabling analyses to be carried out with flow rates up to 2 ml/min and with reverse-phase solvents with organic buffers. The more recent developments of interfaces based on API instrumentation have extended application to a wider range of molecular mass than previously, as well as application to more polar analytes. A schematic diagram indicating the application areas of the various LC–MS interfaces in terms of analyte polarity and molecular mass is shown in Figure 7.1 (Niessen and Tinke, 1995). Recent reviews cover in depth the developments of instrumentation for LC–MS (Niessen and Tinke, 1995) and other reviews cover applications of LC–MS in the area of food chemistry (Games, 1987), pesticides (Slobodnik *et al.*, 1995) and veterinary drug residue monitoring (Van der Greef *et al.*, 1993).

7.2.1 Thermospray LC–MS

Until about 1990, of the various approaches to LC–MS, thermospray appeared to be the most robust and to offer the best possibility of routine operation. A typical LC–MS interface is shown in schematic form in Figure 7.2 and comprises a heated zone (typically a copper block) through

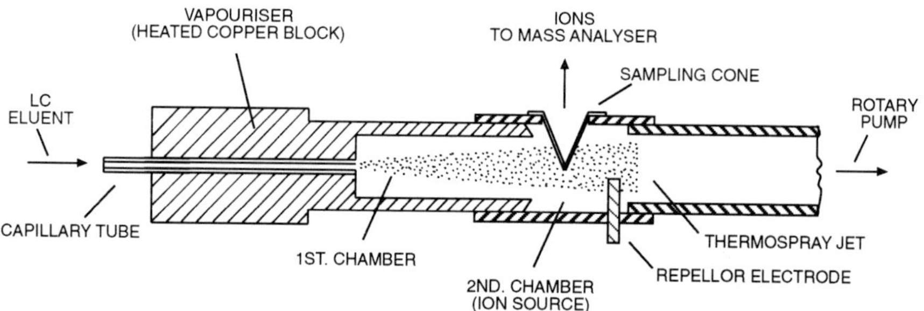

Figure 7.2 Schematic diagram showing a typical thermospray LC–MS interface.

which the LC eluent is introduced, a vaporisation zone where an aerosol is generated from the LC eluent, and a desolvation chamber which is maintained at fore-vacuum pressure by a high capacity pump. Essentially, thermospray involves introducing mobile phase containing electrolyte (e.g. 0.1 M ammonium acetate) through a heated stainless steel capillary into the MS source. At an appropriate temperature a supersonic jet of vapour is produced which contains fine droplets or particles. As the droplets continue to travel through the heated source (first chamber) they continue to vaporise, and ions are produced by a combination of ion evaporation and conventional chemical ionisation (CI). Ions are sampled (second chamber) through a conical exit aperture into the mass analyser, and can be introduced into either quadrupole or magnetic sector instruments. A number of publications are concerned with optimisation of thermospray, for example Voyksner and Haney (1985).

7.2.2 Particle beam LC–MS

A typical particle beam LC–MS interface is shown in Figure 7.3 in schematic form. The system comprises a nebuliser into which the eluent from the HPLC column and a stream of helium enter together, with the HPLC solvent

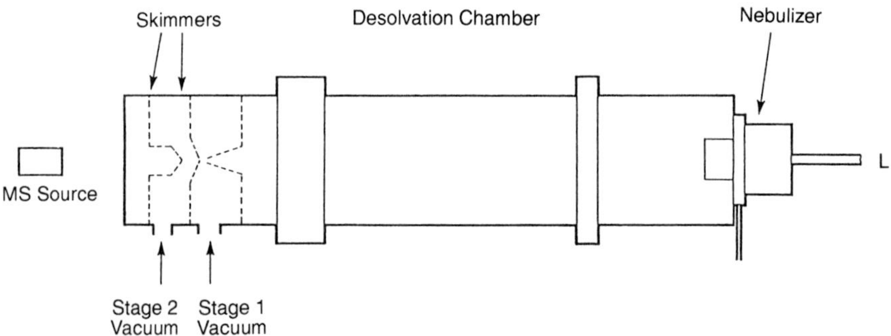

Figure 7.3 Schematic diagram showing a typical particle beam LC–MS interface.

(and analyte) thereby forming an aerosol. During transmission across the desolvation chamber the aerosol droplets begin to lose organic solvent by evaporation and essentially become small dry suspended particles. At the exit to the desolvation chamber the stream of particles enters a momentum separator where they are accelerated into a supersonic jet. The particles then form a well-collimated beam, and any remaining solvent molecules and nebulising gas are pumped away. Finally, the particle beam enters the source of the mass spectrometer where the analyte is evaporated and is ionised by electron impact in the conventional manner.

Spectra from particle beam LC–MS have the advantage of being similar to conventional electron ionisation spectra in that, as opposed to most other forms of LC–MS, there is significant fragmentation. The fragmentation is, however, variable, being highly dependent on operating conditions, which can mean that this technique is problematic for quantitative work, although these difficulties can be overcome with good internal standardisation. In general, particle beam LC–MS is also not particularly sensitive, which may be a limitation in some areas of trace analysis of foods. A number of publications concern optimisation of particle beam LC–MS, for example Voyksner *et al.* (1990).

7.2.3   *Electrospray LC–MS*

Various commercial electrospray LC–MS systems are available, often marketed and referred to in the literature by their branded name, such as the Perkin Elmer Sciex 'Ionspray interface'. Whilst in general the principles of operation are similar, the designs of the various commercial systems do differ significantly, and each will claim benefits in terms of sensitivity or mode of operation associated with the specific commercial system. Commercial systems also offer the possibility of rapid change between electrospray operation and APCI with the same interface and MS source, and again the design features here will differ insofar as different manufacturers have adopted different practices. A typical design of an electrospray interface is shown in schematic form in Figure 7.4.

The characteristics of electrospray spectra are determined by the mode of ionisation which occurs as a result of a strong electric field acting on the surface of the sample solution as it elutes from the nebuliser. An aerosol of charged droplets is produced, which diminish in size by solvent evaporation. Ions produced in electrospray are singly or multiply charged depending on the size and number of basic or acidic sites in the analyte molecule. Small molecules give singly charged $(M + H)^+$ or $(M - H)^-$ ions, whilst large biomolecules produce multiply charged ions. Whilst simple quasi-CI type spectra offer some advantages for selected ion monitoring for unequivocal characterisation, some fragmentation is an advantage, and this can be achieved in the electrospray mode of operation by inducing fragmentation.

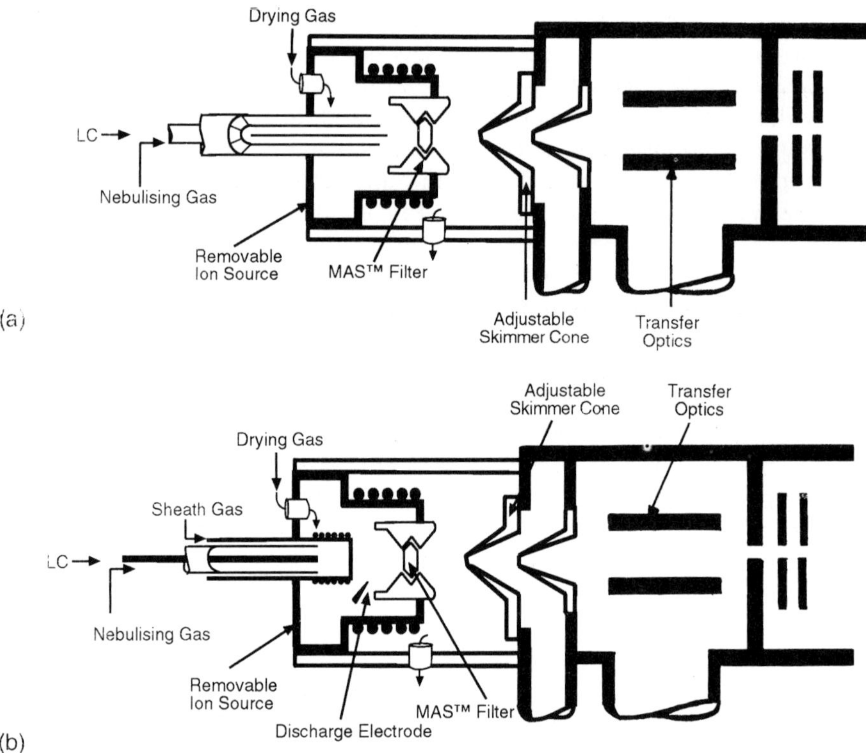

Figure 7.4 Schematic diagrams showing typical atmospheric pressure chemical ionisation interfaces. (a) Electrospray interface; (b) atmospheric pressure ionisation interface. Interface designs are those from a VG Fisons Platform instrument and are reproduced with permission from Fisons Instruments.

This collision induced decomposition (CID) mode of operation is made possible by applying an accelerating voltage to the sampling cone of the interface, and is frequently employed when the option of tandem mass spectrometry (MS–MS) is not otherwise available.

7.2.4 Atmospheric pressure chemical ionisation (APCI)

The main difference between APCI and electrospray is that in APCI a corona discharge in the source forms a CI reagent gas plasma composed of solvent derived ions, rather than an electric field being applied directly to the nebuliser (see Figure 7.4). This corona voltage has a threshold value generally around 3 kV, the exact value being dependent more on mobile-phase composition than on analyte structure. As the sample molecules are eluted from the HPLC they are swept into the gas plasma and become ionised by proton transfer or proton abstraction. APCI is thus a gas-phase ion–molecule reaction process which leads to ionisation of analyte molecules under atmospheric pressure conditions. The process is analogous to chemical ionisation, and because of the atmospheric pressure conditions there is a high frequency of analyte–reactant ion collisions, which ensures high sample ionisation efficiency. The ionisation is soft and results predominantly in protonated molecules $[M + H]^+$ in the positive-ion mode or $[M - H]^-$ in the negative-ion mode.

7.3 Coupling of other chromatographic methods

7.3.1 Supercritical fluid chromatography (SFC)

Supercritical fluids resemble gases in physical properties and are intermediate in terms of diffusivity between gases and liquids, making SFC complementary to both GC and HPLC. The principles of SFC are covered in a book by Smith (1988), applications have been recently reviewed by Chester et al. (1994), and coupling to MS is outlined in reviews by Wright et al. (1986), Games et al. (1988) and Arpino and Haas (1995). Capillary SFC eluent at typical flow rates of 0.5–5 μl/min can be introduced directly into the MS using restrictors and various heated frits for interfacing, from which both EI and CI spectra have been demonstrated as being obtainable. Applications using this approach have in general been restricted to either the analysis of low-molecular-mass substances or to such problems related to high-molecular-mass substances where low detection limits are not needed. For packed-column SFC with higher flows, interfacing has been achieved by eluent splitting, connection to a moving-belt LC system or introduction through a thermospray (Games et al., 1988). However, the most exciting developments in coupling of SFC–MS have been in the use of API sources, where it has proved possible to interface SFC to LC–MS by using commercial equipment without modification, requiring only restrictor introduction or improved heating of the restrictor and transfer lines. Both APCI (Sadoun et al., 1993) and electrospray have been successfully employed in combination with SFC (Arpino and Haas, 1995).

Applications of SFC–MS in the food contaminant field have to date been limited and in the main restricted to demonstrations of the attractiveness of the approach rather than for routine use. For example, for some eleven carbamates (including aldicarb and carbofuran) and six acid pesticides (including 2,4-D and Silvex), ammonia and methane CI spectra have been generated for sub-nanogram amounts of material (Kalinoski et al., 1986a). Chlorinated pesticides such as dieldrin, DDT and aldrin were analysed as standards by capillary SFC coupled to a double-focusing mass spectrometer (Jablonska et al., 1993) at nanogram levels in both positive- and negative-ion modes. A benchtop MS was coupled to capillary SFC and employed for the analysis of thermally labile pesticides such as aldicarb, diuron, methiocarb, alachlor, bendiocarb and carbaryl (Murugaverl *et al.*, 1993). A number of fat-soluble vitamins and polycyclic aromatic hydrocarbons have been determined by SFC with packed capillary columns coupled to LC–MS employing APCI (Matsumoto *et al.*, 1992). It is, however, in many instances difficult to see what advantage SFC–MS does offer over GC–MS for a number of the proposed applications, and the approach remains unproven for routine analysis of real sample extracts. Other applications of SFC–MS have included analysis of trichothecene mycotoxins such as deoxynivalenol and T-2 toxin (Kalinoski *et al.*, 1986b), where the advantages over GC–MS seem debatable, analysis of macrocyclic trichothecene mycotoxins, where there may be some attraction (Smith *et al.*, 1985), and veterinary drug residues including sulphonamides, where again the advantages seem debatable, as there are many adequate simpler alternative assay approaches (Perkins *et al.*, 1991a,b).

Although the analytes are not food contaminants and therefore outside the scope of this chapter, it is worth noting that supercritical fluid extraction (as opposed to chromatography) on-line to MS shows particular promise for analysis of higher-molecular-weight components, and for example has been successfully employed for the identification and quantitation of the presence of synthetic triacylglycerols in foods such as cookies, candies and ice cream (Huang *et al.*, 1994) using particle beam LC–MS.

7.3.2 Capillary electrophoresis (CE)

Separation by CE is based on differences in charge to mass ratios of analytes, as opposed to partition of solutes between a mobile and stationary phase, and thus CE is complementary to HPLC, providing a completely different selectivity. CE offers exceptional resolving power (theoretical plate values of several hundred thousand being frequently reported) and is particularly suited to the separation of polar substances readily ionisable in solution. The coupling of capillary CE with mass spectrometry has recently been reviewed (Cai and Henion, 1995), where it has been indicated that the major advances in CE–MS have been in the coupling to electrospray and continuous-flow fast atom bombardment (FAB). Interfaces between CE and MS are relatively simple and are described as 'sheath-flow', 'sheathless' or 'liquid junction' systems.

In the 'sheath-flow' type the cathode end of the CE terminates within a stainless steel capillary sheath where electrical contact is made, thus completing the CE circuit. A thin sheath of liquid flow passes down the stainless steel capillary over the CE tip into the electrospray source. This system does not lose separation efficiency, but the unavoidable addition of ionic and neutral species in the sheath flow does have the disadvantage of lowering maximum sensitivity achievable.

In the 'sheathless' interface, the CE outlet into the electrospray takes the form of a sharp capillary tip which enables formation of a spray of aqueous solution to occur. The outside of the needle may be coated with a conductive material such as silver or gold to maintain electrical contact. This interface has been found to offer better sensitivity, lower required flow rates and better long-term stability than the 'sheath-flow' types.

The 'liquid-junction' interface is used for coupling CE using a pneumatically assisted electrospray approach ('Ionspray') involving a stainless steel T-junction design. The cathode end of the CE capillary and the end of the electrospray needle are positioned in the centre of the T, separated by a gap of 10–25 μm, and the top end of the T is fitted with a make-up liquid reservoir. The advantage of this interface design is that it can compensate for the different flow rates required for the electrospray and the CE. Optimisation of this system appears to be more difficult than for the sheath-flow configuration, which overall is the more robust (Cai and Henion, 1995; Pleasance et al., 1992b).

Although there are many reports of the use of CE–MS, it has yet to be used routinely in the area of food contaminants and to date there are only isolated reports of applications in the literature. The analysis of some eight sulphonylurea herbicides (including chlorsulfuron), ranging in molecular weight from 364 to 410, using a liquid junction ion-spray CE–MS interface, was accomplished in 5 min (Garcia and Henion, 1992), although the detection limits achieved were not adequate for residue monitoring. Methanol extracts from herbal medicinal products have been analysed by CE–MS using a co-axial sheath-flow interface (Henion et al., 1994), where the viability of this approach both for identification of these alkaloids and for quantitation using tetrahydroberberine as an internal standard was demonstrated. Detectability enhancement has been investigated by using isotachophoresis for analyte focusing prior to CE–MS (Tinke et al., 1992). The combination of isotachophoresis and capillary zone electrophoresis (ITP–CZE) has also been used to improve sample loadability and concentration sensitivity prior to on-line MS (Lamoree et al., 1994), and the viability of this approach to 'real' samples was demonstrated for the analysis of β-agonist veterinary drugs (such as clenbuterol) in calf urine. Antimicrobial drugs such as trimethoprin, erythromycin and various sulphonamides of concern to the fish aquaculture industry have also been analysed by CE–MS (Pleasance et al., 1992b).

There have been a number of attempts to exploit the benefits of CE–MS in the analysis of phycotoxins such as PSP toxins, the pufferfish toxin tetrodotoxin (Pleasance et al., 1992a), and amnesic and diarrhetic shellfish toxins (Pleasance et al., 1992b). Improvements in the detection limits of at least two orders of magnitude over conventional capillary zone electrophoresis was demonstrated for PSP toxins by judicious choice of leading and terminating electrolytes for the preconcentration step (discontinuous buffer) prior to CE–electrospray MS (Locke and Thibault, 1994). Applications of LC–MS to the analysis of phycotoxins are discussed in more detail in section 7.4.3.

7.4   Applications of LC–MS to food contaminants

7.4.1   Pesticides

(i) General.   There have been a number of reviews concerning the analysis of pesticide residues in foods (for examples see Stan, 1990, and Shibamoto and Bjeldanes, 1993), and more specifically applications of mass spectrometry to pesticide analysis in the monograph by Rosen (1987), a treatise on LC–MS applications in environmental chemistry (Brown, 1990), and a compilation of pesticide and pollutant spectra by Safe and Hutzinger (1979). A recent review (Slobodnik *et al.*, 1995) covers the use of LC–MS for pesticides in the analysis of environmental samples. The area of pesticide analysis is complicated by the large numbers of possible residues, which are frequently classified in terms of functionality—insecticides, fungicides or herbicides—as well as being classified on the basis of chemical structure such as organochlorine, organo-phosphorus, carbamates, phenols, sulphonylureas, dinitroanilides, triazines, pyrethoids, etc. In general terms, of the major classes of pesticides the organo-chlorine pesticides (OCs) tend to be analysed by GC (electron capture detection) and the organophosphorus pesticides (OPs) by GC (flame photo-metric and alkali flame ionisation) with confirmation in both instances by GC–MS. Many of the OPs do not have chromophores or double bonds, making HPLC inappropriate due to difficulties in achieving sensitive and selective detection. LC–MS, however, offers an alternative mode of detection and use in a complementary manner while GC–MS offers a powerful integrated approach to the multi-residue determination of pesticides. For example, for fruit and vegetables Mattern *et al.* (1991) have demonstrated an integrated GC–MS and LC–MS approach for the analysis of some 20 pesticides in various crops (apples, peaches, potatoes, tomatoes, peppers, spinach, lettuce, snap beans and sweet corn), with limits of determination of 0.05 to 0.1 mg/kg achieved with average coefficients of variation of 13%.

There has been significant development of thermospray LC–MS for the analysis of a number of pesticides; for example, Volmer *et al.* (1993) have developed an LC–MS method for the extraction, separation and deter-mination of some 130 pesticides including anilides, carbamates, phenylureas, phenoxy acids, oximes, triazines, organophosphorus and quaternary ammo-nium compounds. However, these developments have been predominantly for the analysis of water and environmental samples (see various chapters, for example, in Brown, 1990), and application has been somewhat slower

Table 7.1 LC–MS analysis of pesticides in food and water

Pesticide	Foodstuff/matrix	LC–MS mode	Limit	Reference
ETU[1]	Fruits/lettuce	Particle beam	5 ng (20 μg/kg)	Doerge and Miles (1991)
Chlorophenoxy acids	Water	Particle beam	0.5 ng (1.1 μg/l)	Mattina (1991)
Benomyl	Tomato/peach	Thermospray	25 μg/kg	Liu *et al.* (1990)
Paraquat, diquat	Soil, water	Thermospray	400–600 ng (0.1–0.2 μg/kg)	Barcelo *et al.* (1993)
OPs/metabolites[2]	Beef	Thermospray	20–100 ng	Ioerger and Smith (1994)
OPs[3]	Fish	Thermospray	0.2–0.5 ng	Barcelo (1988)
Phenylureas, carbamates[4]	Fruits and vegetables	Thermospray	0.025–1.0 mg/kg	Liu *et al.* (1991)
OPs/phenylureas	Sediment	Thermospray	100–200 μg/kg	Barcelo and Albaiges (1989)
Herbicides[5]	Water	Thermospray	0.01 μg/l	Hammond *et al.* (1989)
Diflubenzuron	Mushrooms	Negative ion APCI	17 μg/kg	Barnes *et al.* (1995a)

[1] Ethylene thiourea.
[2] Organophosphorus pesticides (17) including parathion, fenthion, coumaphos and chlorpyrifos.
[3] Organophosphorus pesticides (18) including malathion, dimethoate, fenitrothion and temephos.
[4] Total of 19 thermally labile and non-volatile pesticides including aldicarb, diuron, methomyl, fenvalerate, folpet, iprodione and oryzalin.
[5] Chlophenoxy acids and uron herbicides.

in the analysis of foods. A comparison of the performance of some of the published LC–MS methodology for pesticide analysis in terms of limits of determination and precision is summarised in Table 7.1, which can be usefully contrasted with the performance of immunoassays for pesticides given in Chapter 5 (Tables 5.3–5.5).

The analysis of herbicides (urons and triazines), which can be problematic by GC due to their thermal instability, has been achieved by thermospray LC–MS in river water (Hammond *et al.*, 1989), although non-linear calibration due to ion-intensity variation with concentration limited usefulness for quantitation purposes, and this method has also been used for sulphonylurea herbicides in crops such as wheat (Shalaby and George, 1990). Where metabolites or degradation products are anticipated, LC–continuous-flow fast atom bombardment MS has proved to be helpful in providing unequivocal molar mass information and useful fragment ions to aid in characterisation (Reiser *et al.*, 1991). Open-tubular LC–MS and direct liquid introduction LC–MS have been shown to be viable for the analysis of metribuzin (a 1,2,4-triazine herbicide) and its metabolites in water and soyabean plant tissue (Escoffier *et al.*, 1989), and carbamates (39 compounds), triazines and phenylurea pesticides have been analysed in surface and drinking waters by thermospray LC–MS with limits of determination typically in the range 2–90 ng/l (Bagheri *et al.*, 1993). The characterisation of a variety of quaternary amine pesticides such as chlormequat, difenzoquat, mepiquat, diquat and paraquat, again in water and environmental samples such as soils, has been determined by positive ion thermospray LC–MS (Barcelo *et al.*, 1993). Chlorophenoxy acid herbicides (such as 2,4-D, 2,4,5-T and 'Silvex') require derivatisation if GC–MS is employed for analysis, whereas using LC–MS this can be avoided, and, using solid-phase extraction, rapid screening of water samples has been demonstrated (Mattina, 1991). In this report (Mattina, 1991), particle beam LC–MS was used operating under methane-enhanced electron-capture negative-ionisation conditions, which gave limits of determination in the low μg/l range and precision of 6–25%.

Benomyl is a widely used fungicide employed for the control of a variety of pest diseases, which rapidly degrades to carbendazim after application and can also give rise to thiophanate-methyl. All three compounds are included in the overall definition of the MRL. Using thermospray LC–MS employing methyl 2-benzimidazolecarbamate as internal standard, carbendazim has been determined in tomatoes, peaches and apples at a limit of determination of 0.025 mg/kg with a precision between 2.5 and 13.3% (Liu *et al.*, 1990). The internal standard has a structure similar to carbendazim and a close retention time, which was useful in improving the precision when operating in thermospray LC–MS.

A number of pesticides, such as the carbamates, produce as degradation products sulphoxides and sulphones which are moderately polar substances and have proved to be fairly intractable with no obvious and reliable approaches for their determination in foods. Although some of these residues can be analysed by GC, recent availability of relatively low-cost LC–MS systems offering electrospray and APCI has opened up new possibilities for multi-residue analysis, and publications are now beginning to appear. Again, examples initially have appeared predominantly in the environmental area, with electrospray being demonstrated as offering an attractive approach for determining pesticides such as propoxur, carbofuran and aldicarb sulphone in water (Lin and Voyksner, 1993) at levels of 1–10 μg/l, with confirmation by generation of CID spectra at increased repeller voltages. Using thermospray LC–MS a multi-residue approach has been demonstrated for some 19 thermally labile and non-volatile pesticides (Liu *et al.*, 1991), with limits of determination ranging from 0.025 to 1.0 mg/kg in commodities such as apples, beans, lettuce, peppers, potatoes and tomatoes.

(ii) Organophosphorus pesticides.   Organophosphorus pesticides (OPs) are widely used in agricultural production, and as with other insecticides need to be separated both as a class and individually. Although it has been demonstrated that GC–MS can be employed for the analysis of as many as 72 OPs in foods (Stan and Kellner, 1989), there can be attractions in some instances to pursuing the possibilities of LC–MS.

The mass-spectral behaviour of OPs has been studied by direct liquid introduction LC–MS (Barcelo *et al.*, 1987), comparing positive-ion (PI) and negative-ion (NI) conditions for ten selected compounds. Under thermospray conditions, OPs show gas-phase thermal degradation reactions that are strongly dependent on the gas-phase temperature. These reactions are mainly due to 'chemical dissociations' of neutral precursors followed by gas-phase protonation or ammonium addition. Decreasing gas-phase temperature and ion-source temperature to optimum values leads to an increase in quasi-molecular ion abundances (Volmer *et al.*, 1993). The characterisation of phosphorothionates (five compounds), phosphorothiolates (two compounds), phosphorodithioates (five compounds) and a phosphonate (trichlorfon) has been carried out by thermospray LC–MS (Barcelo, 1988; Farran *et al.*, 1988) utilising positive- and negative-ion 'filament-on' modes. In the PI mode, the base peak was $[M + NH_4]^+$ for all the compounds studied, while in the NI mode the OPs exhibited different fragmentation behaviour, such as electron capture, dissociative electron capture and anion attachment. The PI mode showed, for all the OPs studied, higher sensitivity than the NI mode. Sensitivity was claimed to be similar for PI thermospray LC–MS for the same compounds to GC–NCI–MS. This was confirmed in subsequent work (Barcelo and Albaiges, 1989), where for an environmental sample spiked with vamidothion, azinphos-methyl and paraoxon-methyl at 0.1 mg/kg the sensitivity in NI mode was about 1000 times poorer than in PI mode, which was subsequently employed for the analysis.

Betowski and Jones (1988) have compared GC–MS and thermospray LC–MS for the analysis for some ten OPs of interest as environmental contaminants, the pesticides examined being selected on the basis of being difficult compounds to analyse by conventional approaches. Whilst most of the OPs were more stable under LC–MS analysis conditions than GC–MS, problems still remained with trichlorfon, which showed 16% conversion to dichlorvos under thermospray conditions. For some 17 individual OPs and their metabolites in beef tissue, recoveries ranged from 65 to 126% at the 1 mg/kg level with coefficients of variation averaging 32% (Ioerger and Smith, 1994). Regression analysis showed that thermospray LC–MS response to OPs and degradation compounds was linear over the range of interest for all compounds detected in the NI mode. The apparent variability in recovery was attributed to fluctuating instrumental operating conditions (particularly interface temperatures) intrinsic to the thermospray LC–MS operating environment (Ioerger and Smith, 1994).

(iii) Carbamate pesticides.   Owing to thermal instability in some instances, a number of carbamate pesticides are not readily determined by GC or GC–MS. Most analyses are based on postcolumn derivatisation fluorescence detection after hydrolysis and formation of a suitable fluorescent derivative.

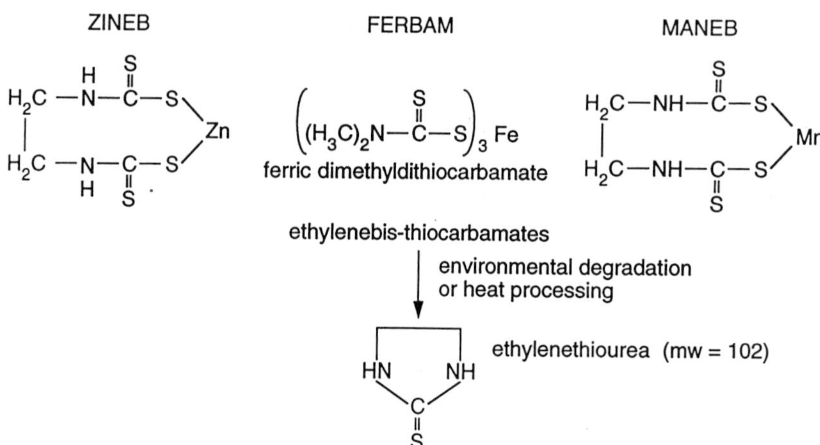

Figure 7.5 Structures of ethylenebisdithiocarbamate fungicides which may contain ETU as a formulation contaminant or give rise to ETU by degradation.

Employing carbofuran as a model analyte to exemplify the approach for determining carbamate pesticides, Rule *et al.* (1994) combined on-line immunoaffinity column clean-up with reversed-phase LC by column switching on-line to an APCI quadrupole MS (specially constructed), equipped with pneumatically assisted electrospray. Using carbofuran spiked into a crude potato extract, a limit of determination of $2.5\,\mu g/kg$ was obtained by monitoring ions at m/z 222, 165 and 123. Although in principle this approach was demonstrated as viable, further extensive work is still required to test the robustness of the method with routine sample throughput and to extend application to a wider range of sample matrices.

(iv) Ethylenethiourea (ETU).   Ethylenebisdithiocarbamate fungicides are widely used on fruits, vegetables, cereals and other food crops. These fungicides (shown in Figure 7.5) may contain ethylenethiourea (ETU) as a formulation contaminant, or it may be formed as an environmental degradation product or during food processing as a heat-induced degradation product. As ETU is a suspect human thyroid and liver carcinogen there is considerable interest in monitoring residues in food products, with a sensitivity requirement of the order of $10\,\mu g/kg$. Although ETU can be determined in foods by GC–MS, its molar mass of 102 does not lead to the production of characteristic MS ions. There can be problems with interferences in GC–MS, particularly with fatty processed food products such as potato crisps. Using particle beam LC–MS, determination limits of $5\,\mu g/kg$ were obtained for lettuce, apple sauce, banana pulp and papaya pulp (Doerge and Miles, 1991), better than limits obtainable by UV or electrochemical detection. An internal standard of ^{13}C-ETU was employed, being added at the stage of extraction from the commodity, to correct for recovery and improve

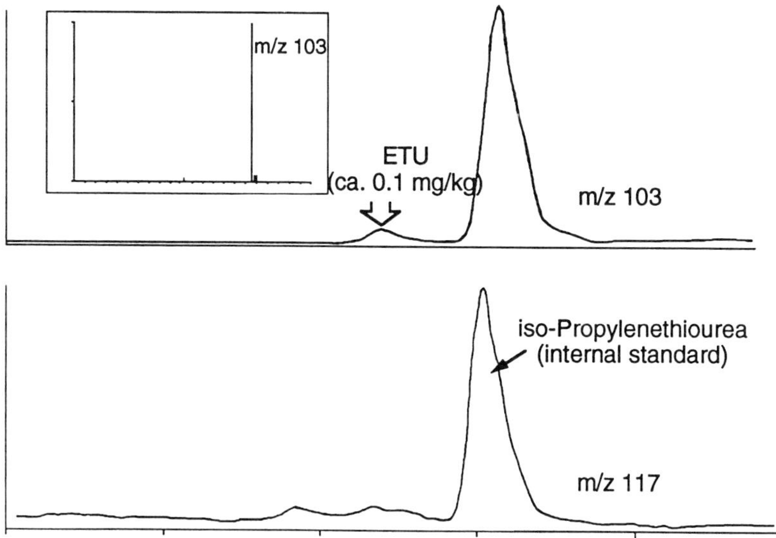

Figure 7.6 Thermospray LC–MS with selected ion monitoring of m/z 103 for ETU and m/z 117 for isopropylenethiourea internal standard. Sample extract from potato crisps containing 100 μg/kg ETU. Inset shows thermospray scanned spectrum for ETU with $[M + H]^+$ base peak. Taken from Wilkins (1993), unpublished results.

quantitation. No interferences were experienced in monitoring m/z 102 for ETU and m/z 103 for the internal standard.

In contrast, when thermospray LC–MS was used for monitoring ETU in oily foods such as potato crisps, although there was no evidence of interference of m/z 103 at the elution time of ETU, there was a large potential interfering component which eluted after ETU (Wilkins, 1993a). The SIM trace for m/z 103 is shown in Figure 7.6 for ETU at a concentration of about 100 μg/kg, and the later eluting internal standard isopropylenethiourea, detected by monitoring m/z 117. Due to the simplicity of the thermospray spectrum of ETU, comprising only the $[M + H]^+$ ion, there is an inevitable risk of interference and no means of achieving improved confidence through multiple ion monitoring. Thus, in this case operating with an MS–MS system by multiple reaction ion monitoring, the CID of m/z 103 $\rightarrow$ m/z 44 was monitored for ETU and the corresponding CID for the internal standard. The significant enhancement of the ETU peak (shown in Figure 7.7) compared to that of the later running interference peak gave improved confidence in the correct identification of ETU.

(v) Diflubenzuron. Diflubenzuron is an insecticide with a broad range of applications, including use on mushrooms, for which a number of analytical procedures have been reported. Diflubenzuron is not amenable to GC as it degrades thermally in the injector, and HPLC analysis using UV detection

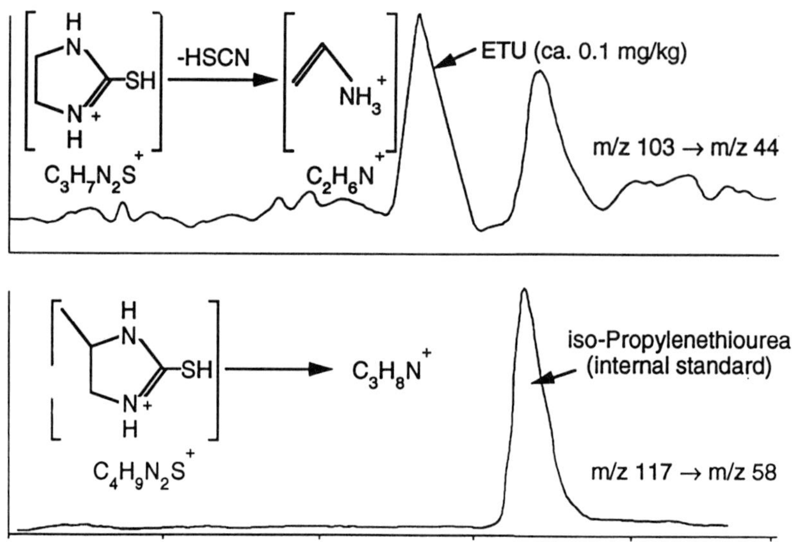

Figure 7.7 Thermospray LC–MS–MS with multiple reaction ion monitoring of m/z 103 → m/z 44 (loss of HSCN from ETU) and corresponding m/z 117 → m/z 58 (loss of HSCN from internal standard). Taken from Wilkins (1993), unpublished results.

has been found to be prone to interferences with some matrices such as chillies and plums. Wilkins (1993b) has reported a thermospray LC–MS method for diflubenzuron, monitoring $[M + H]^+$ ion at m/z 311, for which a limit of determination equivalent to 0.25 mg/kg was achieved in various foodstuffs. In recent work using APCI in NI mode in methanol–water mobile phase (Barnes *et al.*, 1995a), diflubenzuron was determined in mushrooms at a limit of determination of 0.017 mg/kg with an average recovery of 85% and a relative standard deviation of 14.5% ($n = 56$). This method employed m/z 309 for quantification (deprotonated diflubenzuron), which, as base peak, had a sensitivity increase of at least an order of magnitude over the thermospray method, which relied on less abundant fragment ions.

7.4.2 Veterinary drug residues

Veterinary drugs, like pesticides, encompass a number of different classes based on their end purpose, such as growth promotors, hormones, antibiotics, anthelmintics, tranquillisers and coccidiostats, and within each group are a large number of individual compounds, the type and structural diversity being great across the whole breadth of drugs that are approved for use in treatment of food-producing animals. An overview of the whole area is provided in a book by Crosby (1991), a review of analytical aspects by Shepherd (1991), and a perspective on the prospects of LC–MS in drug residue analysis by Van der Greef *et al.* (1993). A number of reviews have covered the chromatographic analysis of specific groups of veterinary drugs such as antibiotics (Bobbitt and Ng, 1992), tetracyclines (Barker and Walker, 1992), anabolics (Van Ginkel *et al.*, 1992), β-agonists (Rhijn *et al.*, 1993) and sulphonamides (Guggisberg *et al.*, 1992). A summary of LC–MS applications in veterinary drug residue analysis is given in Table 7.2.

(i) Antimicrobial agents—sulphonamides.   Sulphonamides are a class of antimicrobial drugs widely employed therapeutically and frequently used as feed additives. The target tissue for sulphonamide residue monitoring is the kidney or liver, and MRLs are frequently set at a level of 100 μg/kg or better for total sulphonamides. Methods therefore need to be multi-analyte to monitor both total and individual residue levels and to have a sensitivity for individual sulphonamides of better than 20 μg/kg (Shepherd, 1991). Sulphonamides provide an example of a veterinary drug residue where almost every possible mass-spectrometric approach has been employed at one time or another—GC–MS, LC–MS and MS–MS have each been applied by various workers, which enables the relative merits of these different approaches to be contrasted (Carignan and Carrier 1991; Guggisberg *et al.*, 1992). GC–MS is the longest established methodology and works well, although some care is required in derivatisation and some methods call for NH and NH$_2$ to be separately derivatised, which lengthens the analytical procedure. A typical GC–MS method using ^{13}C-labelled internal standards would have a CV of about 4–6% and a limit of 100 μg/kg total sulphonamides, and be capable of screening for a number of sulphonamides individually. Clearly, for an LC–MS method to compete against GC–MS, it has to offer comparable performance plus some further benefit, such as being simpler or faster through not necessitating derivatisation.

A thermospray LC–MS method has been published for determining up to ten sulphonamide antibiotic residues in milk at levels as low as 10 μg/l (Abian *et al.*, 1993). In all cases the $[M + H]^+$ ion was the base peak, except for sulphanilamide which showed an $[M + NH_4]^+$ ion. In a 14-min chromatographic run, complete separation was obtained for all the sulphonamides except sulphamethazine/sulphamethizole and sulphapyridine/sulphaquinoxaline, although these were easily distinguished by their different $[M + H]^+$ base peaks. Confirmation of positive results was achieved by collision induced decomposition of the $[M + H]^+$ ion using a Finnigan TSQ70 mass spectrometer, which gave rise to m/z 92, 108 and 156 as common ions for all the sulphonamides. Balizs *et al.* (1993) employed [^{13}C]-sulphadimidine as internal standard for a thermospray approach to the analysis of some four sulphonamides and their N-4-acetyl metabolites in pig muscle, achieving a CV of 6.4–9.8% at limits of quantitation below the MRL for sulphonamides.

There is considerable interest in veterinary drug residue monitoring in reducing time spent on sample preparation and clean-up and in automating wherever possible. Capitalising on the specificity of MS as a detection system

Table 7.2 LC–MS analysis of veterinary drug residues

Veterinary drug	Animal product	LC–MS mode	Limit (on-column or μg/kg)	Reference
Sulphonamides[1]	Milk	Thermospray	0.4 ng	Abian *et al.* (1993)
Sulphonamides[2]	Meat/blood	Thermospray MS–MS	2–10 μg/kg	Kristiansen *et al.* (1994)
Sulphonamides[3]	Meat	Thermospray	<75 μg/kg	Balizs *et al.* (1993)
Sulphonamides[4]	Salmon	Electrospray	25 μg/kg	Pleasance *et al.* (1991)
Clenbuterol	Urine	Thermospray	1 μg/l	Blanchflower and Kennedy (1989b)
Ivermectin	Milk/liver	Particle beam	2 μg/l or 15 μg/kg	Heller and Schenck (1993)
Nitroxynil	Meat/liver/kidney	Thermospray	2 μg/kg	Blanchflower and Kennedy (1989a)
Nicarbazin	Chicken tissue	Thermospray	20 μg/kg	Lewis *et al.* (1989); Leadbetter and Matusik (1993)
β-lactams[5]	Milk	Electrospray	0.1 μg (100 μg/l)	Tyczkowska *et al.* (1994)
β-lactams[6]	Milk	Electrospray	10 μg/l	Straub and Voyksner (1993)
Cephapirin	Milk	Thermospray	100 μg/kg	Tyczkowska *et al.* (1991)
Aminoglycosides	Kidney	Electrospray MS–MS	20 μg/kg	McLaughlin *et al.* (1994)
Polyether ionophores	Liver	Electrospray	30 μg/kg	Schneider *et al.* (1991)
Spiramycin	Meat	Particle beam	20 μg/kg	Sanders and Delepine (1994)
Furazolidone	Meat	Thermospray	0.6 μg/kg	Blanchflower *et al.* (1993a)
Tetracyclines	Honey	LC–FAB–MS	100 μg/kg	Oka *et al.* (1994a)
Tetracyclines	Milk	Particle beam	100 μg/l	Kijak *et al.* (1991)
Chloramphenicol	Milk/tissue	Thermospray	1–2 μg/kg	Blanchflower *et al.* (1993b)
Trenbolone	Bile/faeces	Thermospray	0.5 μg/kg	Hewitt *et al.* (1993)
Oxolinic acid	Fish tissue	Thermospray	0.1 ng	Horie *et al.* (1993)
Rafoxanide	Tissue	Thermospray	20 μg/kg	Blanchflower *et al.* (1990)
Moxidectin	Cattle fat	Thermospray	25 g/kg	Khunachak *et al.* (1993)
Ipronidazole, dimetridazole	Turkey tissue	Thermospray MS–MS	2 μg/kg	Matusik *et al.* (1992)

[1] Total of 10 sulphonamides.
[2] Total of 6 sulphonamides.
[3] Total of 4 sulphonamides plus their N-4-acetyl metabolites.
[4] Total of 21 sulphonamides.
[5] Total of 6 β-lactam antibiotics.
[6] Total of 6 β-lactam antibiotics.

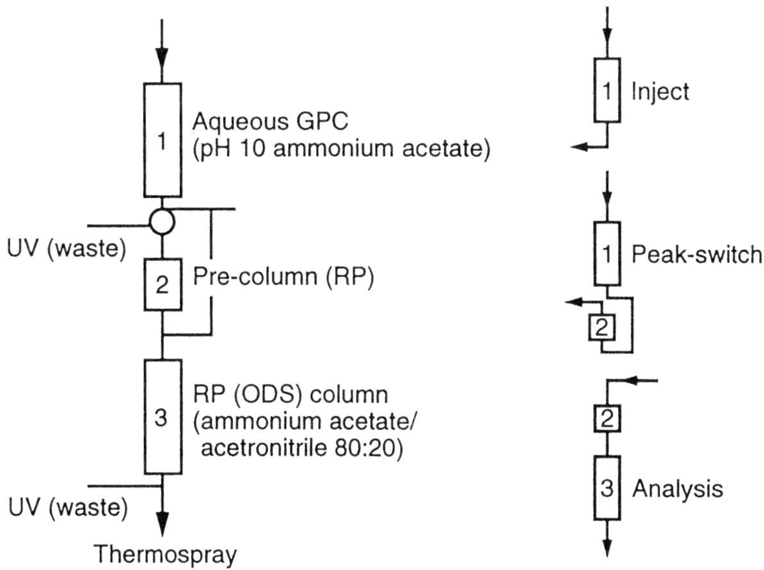

Figure 7.8 On-line combined LC–LC–MS system used for the analysis of sulphamethazine in pig kidney (see also Figure 7.9). Analytical sequence: (i) column flow to waste from GPC loaded with sample extract; (ii) at point of elution of sulphamethazine peak, switched from column 1 to precolumn 2 and run to waste; (iii) reverse-flow elution of sulphamethazine as sharp peak on to analytical ODS column for LC–MS thermospray analysis.

in this laboratory we developed a thermospray LC–MS procedure for sulphonamide analysis using LC–LC with automated peak switching to minimise the normal sample preparation. The system, which is illustrated in Figure 7.8, comprises an aqueous GPC (size exclusion) column operating with ammonium acetate as mobile phase, coupled to a C_{18}-ODS column running isocratically with ammonium acetate–acetonitrile. The sample is injected with the eluent from column 1 running to waste and solvent flowing on to the ODS column coupled to the HPLC using the thermospray interface. Immediately prior to the elution of the peak of interest, the column eluent is switched to the precolumn, trapping just the narrow region of interest. The precolumn is used to obtain a sharp injection by loading on one end of the precolumn and sweeping off with reversed flow. The analyte then passes directly into the thermospray MS. The advantage of this approach is that the sample clean-up is minimal—the homogenised tissue sample is equilibrated initially with deuterated internal standard, and then only solvent extraction with ethyl acetate is required prior to LC–LC separation and isotope dilution LC–MS.

Figure 7.9 illustrates a typical analysis for sulphamethazine in pig kidney using this LC–LC–MS approach. The sample is clearly seen to be positive (containing $80\,\mu g/kg$) by monitoring M + 1 for sulphamethazine at m/z 279 and M + 1 at m/z 283 for the d_4-sulphamethazine internal standard.

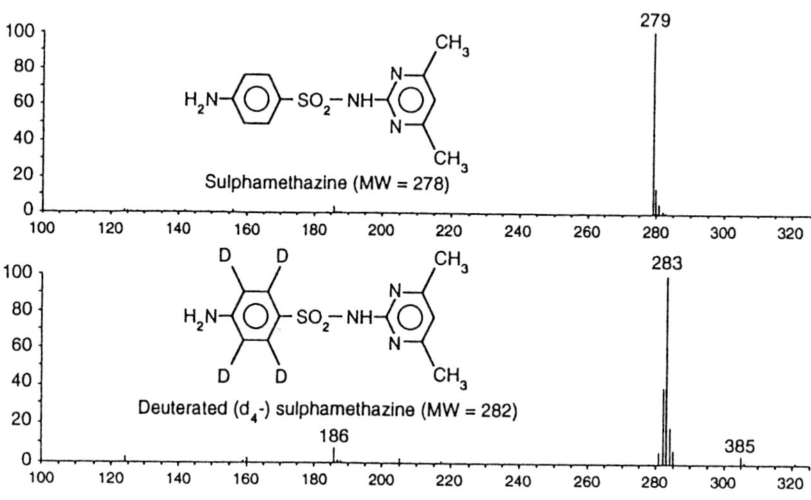

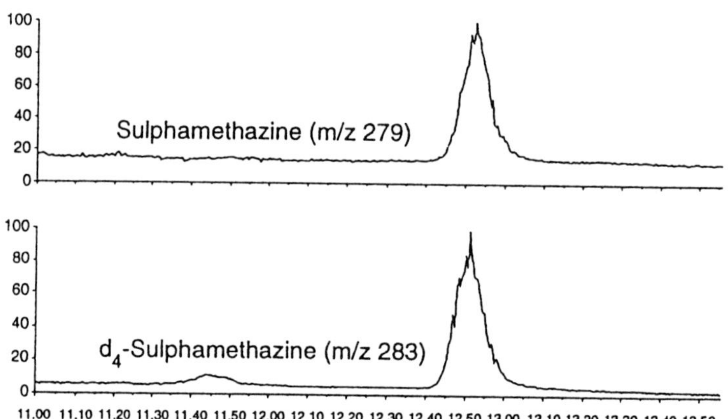

Figure 7.9 Thermospray spectrum of sulphamethazine and d4-sulphamethazine internal standard. Selected ion monitoring of *m/z* 279 for sulphamethazine and *m/z* 283 for d4-sulphamethazine from sample of pig kidney containing 80 μg/kg.

The limit of detection of the method was 50 μg/kg and the procedure gave a recovery of 80% overall. This approach, whilst attractive from the standpoint of routine screening for a single target analyte, is difficult to modify to become a multi-sulphonamide procedure unless all the sulphonamides are sufficiently close in GPC elution volume to be collected in a single one-peak switching operation. Also, lack of fragmentation in the thermospray spectrum might argue for MS–MS, with multiple reaction monitoring to provide the basis of unequivocal confirmation.

The application of MS–MS to sulphonamide analysis has been seen as offering the possibility of very rapid sample throughput by utilising the specificity of the detection mode. Kristiansen *et al.* (1994) compared a flow injection MS–MS approach with an LC–thermospray MS–MS approach for the analysis of sulphonamides in both meat and blood samples. Extraction of the matrix was achieved simply by shaking with ethyl acetate and evaporating off the solvent, followed by direct injection of the re-dissolved residue with no further clean-up. Mass analysis of the flow-injected samples was by parent ion detection, and it was shown that it was possible to analyse some 40–60 extracts per hour. It was claimed that no contamination problems or loss of sensitivity was evident even after as many as 1000 analyses of crude extracts. In contrast, the LC–thermospray MS–MS approach could only cope with 6–7 samples per hour, although the limit of detection of the latter approach was lower than for flow-injection analysis. Despite the use of sulphapyridine as internal standard, the CVs achieved were as high as 35% in some instances, significantly poorer than most GC–MS methods, which could limit application to a screening mode (Kristiansen *et al.*, 1994).

An exhaustive study has been carried out on the analysis of some 21 sulphonamides using a Sciex triple quadrupole instrument equipped with an API source and an Ionspray interface (Pleasance *et al.*, 1991). Some 16 of the sulphonamides were separated as a test mixture by HPLC, whilst further characterisation of isomeric and isobaric sulphonamides was achieved by CID using the MS–MS capability. Using a sample of farmed salmon suspected of containing sulphadimethoxine, it was demonstrated that it was possible to confirm this sulphonamide at a level of 10 μg/kg. Packed-column SFC–MS using moving belt and modified thermospray interfaces has been employed for sulphonamide analysis (Perkins *et al.*, 1991b), although in view of the range of alternative approaches that work well for these analytes, this particular example is best regarded as a demonstration using model compounds rather than a sensible viable option for sulphonamides.

(ii) Antibiotics—β-lactams. Methods of analysis for β-lactam antibiotics have suffered from being generally insensitive and slow, and have mostly focused on penicillin residues in milk. MRLs for penicillins range from 4–30 μg/l in milk to 50–300 μg/kg in meat, with methodology being centred on HPLC approaches (Boison, 1992). Initial applications of LC–MS have involved thermospray, which has been widely employed for determining a range of penicillins (cloxacillin, ampicillin, hetacillin and amoxicillin) in milk with limits of 100–200 μg/l, spectra exhibiting $[M+H]^+$ and $[M+Na]^+$ ions along with several fragment ions (Voyksner *et al.*, 1991). Positive-ion detection was employed, as in all cases it was several times more sensitive than the negative-ion mode. The precision of the LC–MS was not reported, although good correlation was obtained between LC–MS and LC–UV. Ampicillin and its metabolites have been identified in

urine by thermospray LC–MS (Suwanrumpha and Freas, 1989) using MS–MS for structural elucidation, and cephapirin and its metabolite have been determined in milk at 100 and 500 μg/l respectively by thermospray LC–MS (Tyczkowska *et al.*, 1991).

More recent work (Straub and Voyksner, 1993; Tyczkowska *et al.*, 1994) has employed electrospray LC–MS for the simultaneous determination of some six β-lactams in milk at a limit of 100 μg/l, with monitoring of $[M + H]^+$ and the two most intense fragment ions above m/z 150 to provide best specificity for each antibiotic. The use of perfusive particle capillary columns (320 μm ID) and on-column concentration of β-lactam antibiotics in milk ultrafiltrate have been shown to be a way of lowering detection limits to 10 μg/l in milk (Straub *et al.*, 1994). This methodology offers not only good detection limits but also rapid analysis (13 min/sample), with no loss in column performance or increase in back pressure occurring after the analysis of some 100 milk samples spiked at low ppb levels with β-lactam antibiotics.

(iii) Antibiotics—aminoglycosides. Aminoglycoside antibiotics comprise a group of compounds produced biosynthetically which includes streptomycin, dihydrostreptomycin, neomycin B, spectinomycin and the gentamicin C complex and which range in molecular weight from 332 to 614 and are all highly polar substances containing multiple free hydroxyl and amino groups. The US FDA level of concern for these antibiotics in animal products ranges from 100 μg/kg for spectinomycin to 500 μg/kg for streptomycin. Gentamycin, when analysed by thermospray LC–MS (Getek *et al.*, 1991), could be separated into its three major components as well as a minor compound, and under full scan conditions a limit of detection of 400 ng on-column was achieved. More recent work (McLaughlin *et al.*, 1994), using a Sciex TAGA triple quadrupole with an API source operating in a PI mode, has demonstrated that multi-residue analysis of aminoglycosides is possible in kidney samples at limits from 30 to 500 μg/kg depending on the compound. With the exception of spectinomycin, the other five aminoglycosides produced electrospray spectra with the ion current concentrated into one principal multiply charged ion, which because of its consequent lower mass was not sufficiently selective for SIM at the limits required for monitoring against tolerance levels. However, by using MS–MS, the multiply charged ions became parents of ions of higher mass, which were adequate for confirmation purposes, e.g. for dihydrostreptomycin (MWt = 583) the electrospray spectrum had a base peak at m/z 293, ascribed to $[M + 2H]^{2+}$, which under MS–MS conditions gave rise to m/z 409, 263 and 176, which were employed for subsequent SIM (McLaughlin *et al.*, 1994).

(iv) Antibiotics—polyether ionophores. Polyether ionophores are fermentation-derived biologically active compounds characterised by the presence

of a carboxy acid group and several cyclic ether units, and include for example monensin, narasin, lasalocid, salinomycin, and semduramicin. Using pneumatically assisted electrospray LC–MS, Schneider *et al.* (1991) have demonstrated that, to avoid multiple low-intensity molecular ion formation from direct analysis of these ionophores, it was better to shift the molecular ion to specific metal adducts by selective addition of a metal salt to the mobile phase. Thus, for the analysis of semduramicin in chicken liver with the addition of sodium acetate to the mobile phase to form exclusively the sodium adduct molecular ion (m/z 896), using MS–MS to monitor $[M - CO_2]$ and $[M - CO_2 - H_2O]$, it was possible to confirm the presence of this antibiotic at a level of 30 μg/kg.

(v) Antibiotics—tetracyclines. Tetracyclines are important antibiotics for which MRLs range from 10 to 600 μg/kg depending on the sample matrix. Analytical methodology has been reviewed by Barker and Walker (1992), who indicate that they see improved immunological procedures, multi-residue methods and more rapid HPLC procedures as the way forward for the future. Although LC–MS does therefore seem an attractive prospect, most LC conditions for tetracyclines require mobile phases containing non-volatile compounds such as oxalic acid and citric acid to aid improvement of chromatographic resolution, and these are not generally suitable for LC–MS. Rather than modify LC conditions for LC–MS, Oka *et al.* (1994b) have preferred to carry out primary analysis by LC, and confirmation by TLC–fast atom bombardment (FAB), involving concentration of the TLC spot, application of thioglycerol matrix and analysis using a TLC–FAB ion source. Confirmation of tetracyclines at 50 μg/l was achieved, based on the presence of $[M + H]^+$ ions for oxytetracycline and chlortetracycline and additionally the presence of $[M + H - NH_3]^+$ ions for tetracycline and doxycycline. FAB has similarly been employed for the analysis of the same four tetracyclines in honey (Oka *et al.*, 1994a), but this time with frit FAB–LC–MS, employing a laboratory-made flow-splitter interface. The detection limit, based on the presence of two or more characteristic ions including $[M + H]^+$, was claimed to be 200 μg/kg.

Using particle beam LC–MS, a method has been developed for the confirmation of oxytetracycline, tetracycline and chlortetracycline in milk (Kijak *et al.*, 1991). Sample preparation involved centrifugation of the milk, C_{18} cartridge clean-up, elution with 0.1 M oxalic acid in methanol, and separation using methanol–oxalic acid–acetonitrile mobile phase. Some problems were encountered with clogging of the particle beam skimmer with oxalic acid, which were minimised by switching the LC column eluent to waste for a portion of the LC run. This method was shown to be capable of achieving a limit of detection of 100 μg/l (20 ng injected), with monitoring of four ions for each tetracycline. The inherent response variability of particle beam was overcome by analysing standards bracketing each sample, and confirmation was taken as having been achieved if relative abundances for the sample fell within 15% of values obtained for the standards (Kijak *et al.*, 1991). Tetracyclines have also been analysed by thermospray LC–MS (Porter, 1993), although no indications were given of method performance with real samples, and Riond *et al.* (1989) have employed thermospray to study the pharmacokinetics of doxycycline in calves.

(vi) Antibiotics—chloramphenicol.   Chloramphenicol is a broad-spectrum antibiotic with a provisional MRL in meat of $10\,\mu g/kg$. There are both statutory and surveillance requirements for monitoring residues not only in animal tissues but also in milk and eggs. Particle beam LC–MS has proved useful for determining chloramphenicol metabolites in the plasma of chickens to which the antibiotic was administered (Delepine *et al.*, 1993), with four of the seven metabolites being identified on the basis of their molecular ions and fragmentation patterns. Using thermospray LC–MS, chloramphenicol has been determined in milk at a limit of $2\,\mu g/l$ and in tissue at $1\,\mu g/kg$ (Blanchflower *et al.*, 1993b). This method employed d_5-chloramphenicol as an internal standard, which enabled CVs of between 2.9 and 6.5% to be obtained for spiking levels of 4.1 and $8.6\,\mu g/kg$ in liver.

(vii) Anthelmintics.   Ivermectin, which falls within the class of anthelmintics, is a macrocyclic lactone and is the common name given to a mixture of two structural homologues. This drug is approved for the treatment of parasitic infections in beef cattle and has an MRL of $15\,\mu g/kg$ in liver (Shepherd, 1991). The particle beam LC–MS spectrum of the macrocyclic lactone ivermectin obtained in negative-ion chemical ionisation mode is shown in Figure 7.10 (Heller and Schenck, 1993). The spectrum shows a molecular ion at m/z 874 and ions at m/z 856 and 838 arising from the loss of one and two water molecules respectively. The ion at m/z 764 was due to a retro-Diels–Alder reaction combined with macrolide ring cleavage, and significant variability in the relative abundance of this ion was detected (from 1% to 200% relative to m/z 874) depending on the extent of any co-eluting materials. The ion at m/z 585 is due to the loss of sugar moieties from ivermectin.

There are few methods available for determining ivermectin residues, and these are somewhat time-consuming and involve dehydration of ivermectin to form a fluorescent derivative for HPLC analysis. For the particle beam LC–MS analysis of milk for ivermectin (Heller and Schenck, 1993), a simplified sample preparation involved liquid–liquid extraction (hexane–acetone), centrifugation, and evaporation to give an oil, which was dissolved in hexane and then extracted further with acetonitrile. A further solid-phase extraction step was included if thought necessary. HPLC analysis was carried out using a C_{18} column operated with a mobile phase of acetonitrile–ammonium acetate (80:20) at $0.4\,ml/min$. In NICI mode using a particle beam interface,

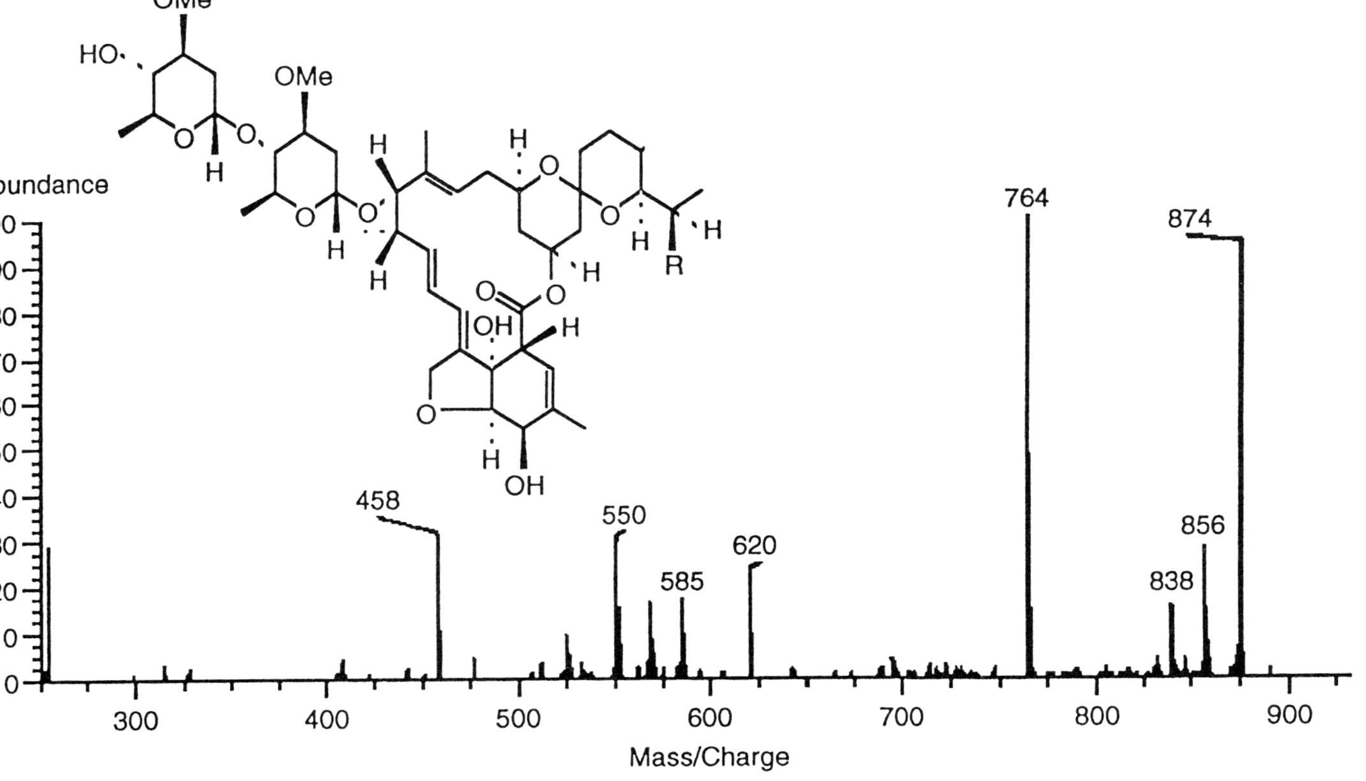

Figure 7.10 Particle beam LC–MS: negative-ion chemical ionisation spectrum of H_2B_{1a} homologue of ivermectin extracted from milk fortified at $100\,\mu g/l$ (background subtracted spectrum). Spectrum obtained on a Hewlett-Packard 5988A MS equipped with a 59980A particle beam interface. Chromatographic separation using acetonitrile–50 mM ammonium acetate adjusted to pH 4 (80:20) at 0.4 ml/min (figure redrawn from Heller and Schenck, 1993, with permission from John Wiley and Sons Ltd).

SIM was carried out of five high-mass ions. Figure 7.11 shows the traces for m/z 856 and m/z 874 for the analysis of a milk sample spiked at $2\,\mu\text{g}/\text{l}$ (equivalent to 4 ng of ivermectin on-column) superimposed over the trace of the control milk (unspiked). There was clearly no detectable response in the blank milk for either of the two ions being monitored. To the right of Figure 7.11 is a similar trace for a milk sample containing naturally incurred ivermectin at a level of contamination of $2.6\,\mu\text{g}/\text{l}$. Matrix effects were evident with this analysis of milk, which necessitated all standards being analysed after spiking into milk, as co-eluting compounds caused considerable changes in the relative abundances of the fragment ions. Spiking of standards was carried out at a similar concentration to that of the unknown, and standards and unknowns were run as consecutive samples. However, even with this considerable attention to detail and averaging of responses, it was still suggested that agreement to $\pm 15\%$ is the best precision likely to be attainable by particle beam LC–MS (Heller and Schenck, 1993).

Levamisole is another anthelmintic (de-worming) drug for which there is an MRL of $10\,\mu\text{g}/\text{kg}$ for muscle, kidney and fat and $100\,\mu\text{g}/\text{kg}$ for liver, and for which liver is the target organ for monitoring. Screening for the presence of levamisole can routinely be carried out by HPLC with UV detection, and in most instances the chromatogram is free of interferences and it can be assumed that levamisole is not present as a residue. However, in the few cases of positive samples exceeding the MRL there is a need for MS confirmation, which is also needed in some few cases where there are interferences in the HPLC chromatogram occurring at the elution volume for levamisole.

A thermospray LC–MS method was developed in 1992, and this has proved to be very robust and has been in routine operation for the analysis of a substantial number of samples (Barnes, 1995). The thermospray spectrum for levamisole is shown in Figure 7.12, which has a base peak at m/z 205, the $[\text{M} + \text{H}]^+$ ion, and little else except a small $[\text{M} + \text{Na}]^+$ ion. Samples were run under HPLC conditions with ammonium acetate in the mobile phase and with no discharge voltage applied in thermospray. In Figure 7.13 a typical 'problem' extract from a sheep liver sample is shown, where by HPLC with UV detection there were interferences obscuring any possible peak at the levamisole elution volume. When the same sample extract was run by thermospray LC–MS, with monitoring of m/z 205 and 206, there was clear confirmation of the presence of levamisole at a level of $30\,\mu\text{g}/\text{kg}$. Unfortunately, for this assay there was no suitable stable isotope-labelled internal standard available for levamisole. Ion intensities under thermospray conditions are not always stable and quantitation via calibration graphs can be unreliable if the instrument response changes. Therefore, in this assay quantitation was carried out by analysing the sample extract in question bracketed by two standards and ratioing for quantification purposes on the average response of the two standards. In all cases, the unknown has to be diluted to approximately the same concentration as the standard so

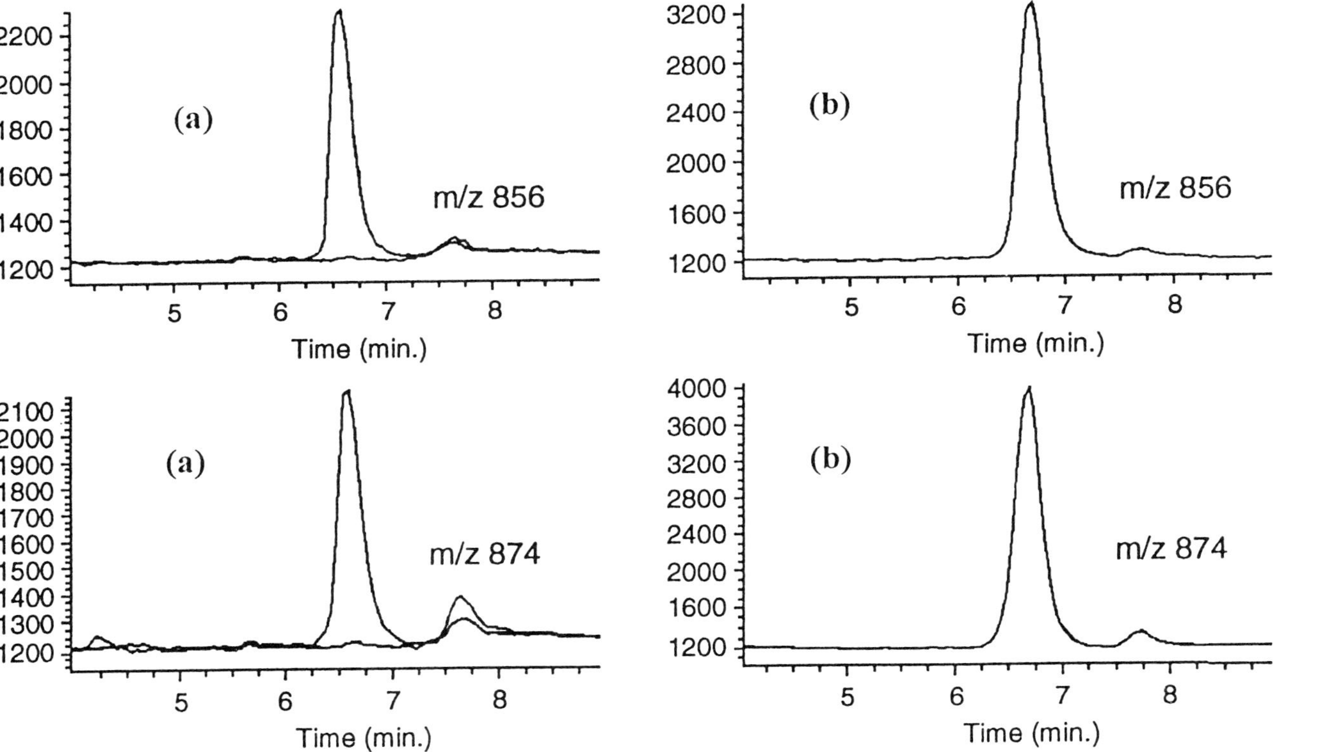

Figure 7.11 Selected ion monitoring LC–MS of ivermectin in milk. Analysis using a Hewlett-Packard 5988A MS equipped with a 59980A particle beam interface. Peak width of 0.5 units for SIM of masses m/z 856.5 and m/z 874.51 for a dwell time of 500 ms per ion. Chromatographic separation using acetonitrile–50 mM ammonium acetate adjusted to pH 4 (80:20) at 0.4 ml/min. (a) Milk spiked with ivermectin at 2.0 μg/l; overlying trace of blank control milk. (b) Milk containing naturally incurred ivermectin at 2.6 μg/l (figure redrawn from Heller and Schenck, 1993, with permission from John Wiley and Sons Ltd).

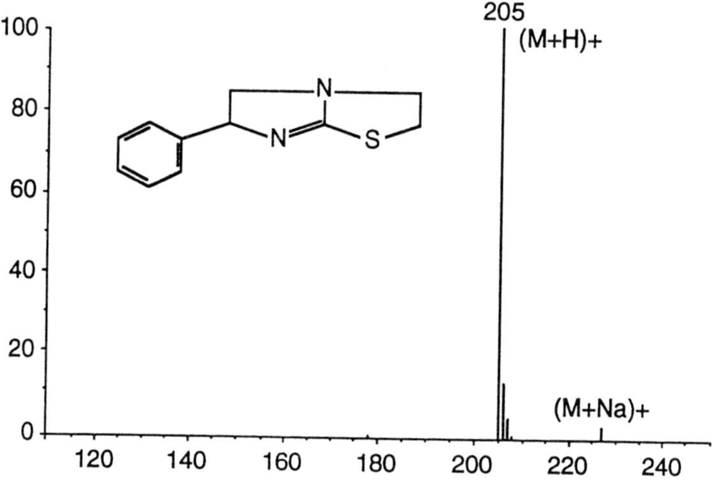

Figure 7.12 Thermospray LC–MS spectrum of anthelmintic drug levamisole. HPLC conditions: 300 mm × 3.9 mm ID C$_{18}$ column operated with 50:50 0.1 M ammonium acetate–acetonitrile at a flow rate of 1 ml/min. Analysis using a VG Fisons 12-250 quadrupole MS with the thermospray capillary at 250°C and no discharge voltage applied.

that peak areas of similar size are being compared by this procedure. With this procedure, the method had a precision of about 10% and the limit of detection was shown to be 1 μg/kg (Barnes, 1995).

There are other examples of applications of thermospray LC–MS to the analysis of anthelmintics. Blanchflower and Kennedy (1989a) have determined nitroxynil residues in muscle, liver and kidney at a detection limit of 2 μg/kg and with a sample throughput of 15 samples/operator/day. Similarly, two other anthelmintics, in this case members of the group of benzimidazoles

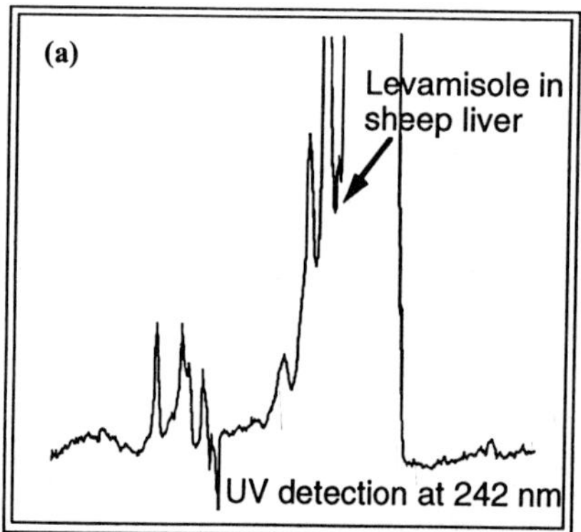

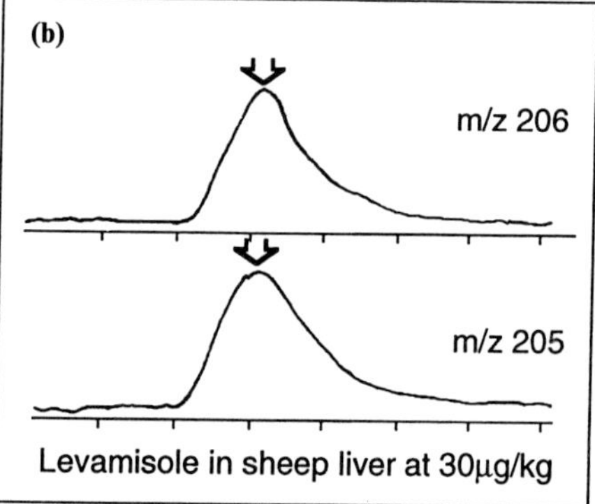

Figure 7.13 Analysis of levamisole in sheep liver: (a) by HPLC with UV detection at 242 nm, indicating problems of interference; and (b) by selected ion monitoring of m/z 205 and 206 using thermospray LC–MS (Barnes, 1995).

(fenbendazole and oxfendazole), have been analysed in sheep's liver at a detection limit of 0.05–0.1 μg/kg and with a CV of 7–10% (Blanchflower *et al.*, 1993c).

7.4.3   *Mycotoxins, phycotoxins and natural toxicants*

(i) Mycotoxins.   Although the primary method of analysis for most mycotoxins in foods and animal feeds is HPLC, good clean-up procedures and the specificity of fluorescence detection have been generally found to be adequate, and there has surprisingly been little pressure to pursue MS confirmation of positive results. Also, with the exception of the trichothecenes, where levels of interest range from 0.2 mg/kg upwards, in general for mycotoxins in foods high sensitivity has been required for confirmation, which has been outside the attainable range of LC–MS. However, whilst sensitivity is still an important consideration, as can be seen from Table 7.3, with developments in thermospray, electrospray and APCI, routine LC–MS analysis of mycotoxins in foods has become a reality.

In some of the early LC–MS work, a comparison was made of direct introduction LC–MS for the trichothecenes nivalenol and deoxynivalenol (Tiebach *et al.*, 1985) with derivatisation and GC–MS. Through use of NICI, detection limits of 5–10 μg/kg were achievable by LC–MS, although it was recognised that this approach at that time was not robust enough for routine operation, with limiting factors being the effects of impurities in the highly concentrated sample extracts and fluctuations in the eluent transfer from the HPLC into the ion source, leading to significant decreases in instrument sensitivity. The application of thermospray LC–MS has proved to be

Table 7.3 LC–MS analysis of mycotoxins in food and feeds

Mycotoxin	Foodstuff/matrix	LC–MS mode	Limit (ng)	Reference
Aflatoxins	Peanuts	Thermospray	0.06–0.1	Hurst *et al.* (1991)
Sterigmatocystin	Bread/cheese	APCI (positive)	2–5 μg/kg	Scudamore *et al.* (1995)
T-2 toxin	Plasma, urine	Thermospray	0.5	Voyksner *et al.* (1985)
Zearalenone	Plasma, urine	Thermospray	10.0	Voyksner *et al.* (1985)
DON/NIV	Wheat	DLI (NICI)	5–10 μg/kg	Tiebach *et al.* (1985)
Macrocyclic	Standards	SFC–MS	0.001	Smith *et al.* (1985)
trichothecenes	Plants	Thermospray	2.0–5.0	Krishnamurthy *et al.* (1989)
Ochratoxin A	Barley	DLI (NICI)	3 μg/kg	Abramson (1987)
Ochratoxin A	Wheat	Thermospray	30.0–60.0	Rajakyla *et al.* (1987)
Patulin	Wheat	Thermospray	30.0–60.0	Rajakyla *et al.* (1987)
Fumonisin B_1	Standards	Electrospray	0.005	Korfmacher *et al.* (1991)
Fumonisin B_1	Maize	Electrospray	20.0 mg/kg	Calda *et al.* (1995)

DON = 4-deoxynivalenol and NIV = nivalenol trichothecene mycotoxins.

more successful again for *Fusarium* mycotoxins such as T-2 toxin, HT-2 toxin, diacetoxyscirpenol, T-2 tetraol, deoxynivalenol and zearalenone (Voyksner *et al.*, 1985). The advantage of LC–MS analysis over GC–MS was claimed to be the avoidance of the need for derivatisation. The thermospray spectra of these mycotoxins obtained with ammonium acetate buffer were in all cases simple, consisting of an $[M + NH_4]^+$ adduct ion indicating molecular weight, and only deoxynivalenol and zearalenone producing $[M + H]^+$ ions. Even though spiked urine samples were analysed to demonstrate the specificity of selected ion monitoring LC–MS, it is doubtful that the sensitivity of 1–5 ng on-column for each mycotoxin would be adequate for diagnostic purposes for real samples where mycotoxicoses were implicated.

For the determination of ochratoxin A, HPLC is the only viable approach, and LC–MS has been evaluated for the analysis of samples of grain for this mycotoxin by direct liquid introduction (Abramson, 1987) and by thermospray (Rajakyla *et al.*, 1987). By direct liquid introduction using aqueous formic acid–acetonitrile mobile phase as reagent gas, LC–MS spectra were obtained with an $[M + H]^+$ at m/z 404 in positive-ion mode and an M^- at m/z 403 in negative-ion mode, in both instances showing characteristic isotope cluster patterns for the chlorine atom present in ochratoxin A. When analysing a barley extract using SIM for m/z 403.1, it was shown that a limit of detection of 3 µg/kg was achievable, although the eluting peak showed significant tailing and eluted on the tail of the background interference (Abramson, 1987). Recent work (unpublished) has demonstrated the feasibility of analysis of ochratoxin A by APCI, where, using methanol–water (50:50), the spectrum shown in Figure 7.14 was obtained for 200 ng of standard. The spectrum shows an $[M + H]^+$ ion at m/z 404 with associated isotope peak at m/z 406, and substantial fragmentation, which could be reduced by lower focus and extraction voltages or perhaps a softer ionisation approach such as electrospray.

Thermospray LC–MS perhaps offers the biggest advantage over HPLC or GC–MS, where the universal nature of the detection mode offers the possibility of multi-mycotoxin analysis, particularly where the mycotoxins of interest would normally be analysed by GC–MS (say T-2 toxin) or HPLC (say patulin or zearalenone). However, whilst the feasibility of this multi-mycotoxin approach can be readily demonstrated by spiking a range of mycotoxins into sample extracts such as wheat (Rajakyla *et al.*, 1987), in reality commodity type is usually a determinant of which mycotoxin should be looked for, and universal extraction and clean-up procedures are difficult to envisage in view of the range of structure and polarity of mycotoxins. Nevertheless, the potential sensitivity of thermospray LC–MS has been demonstrated for T-2 toxin, where, for a signal to noise ratio of 20:1, it was possible to detect 300 pg of this mycotoxin on-column by SIM thermospray LC–MS, equivalent to between 3 and 40 µg/kg. The feasibility of

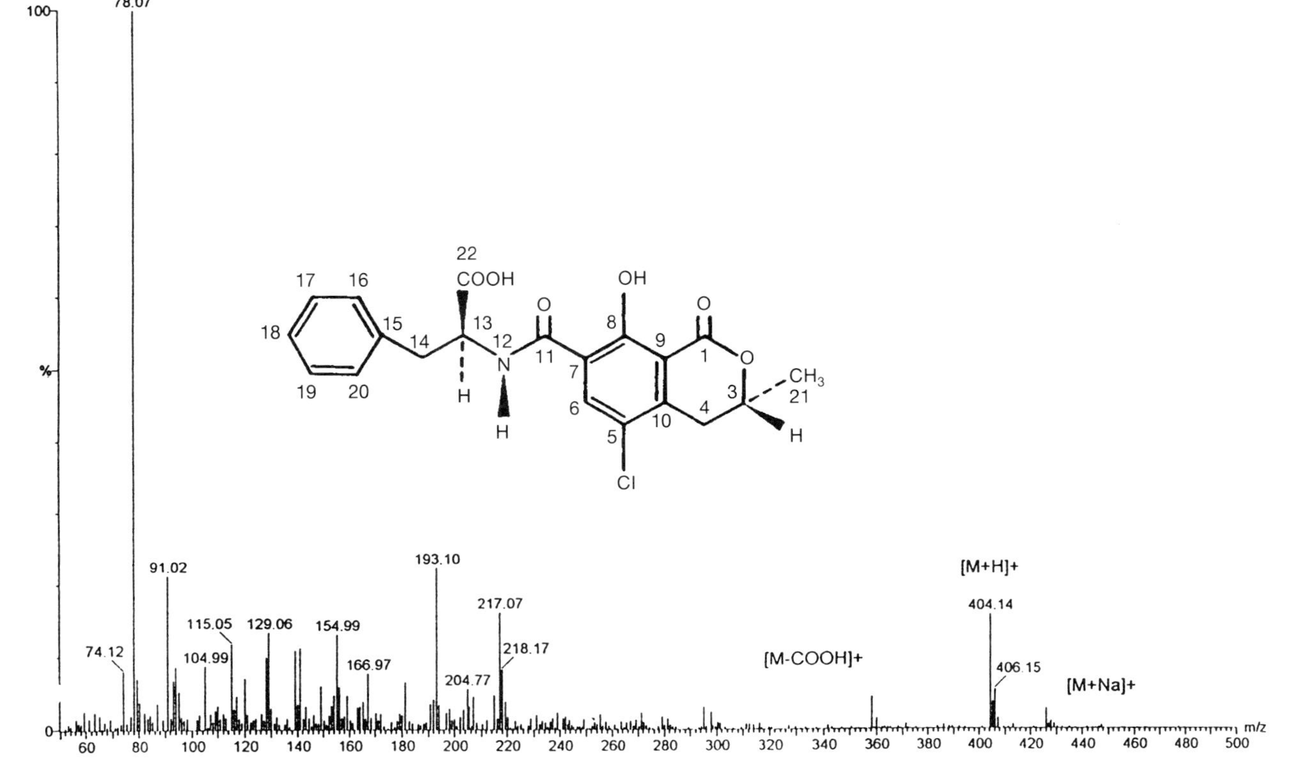

Figure 7.14 APCI spectrum of ochratoxin A (200 ng on column). Analysis performed on VG Fisons Platform LC–MS using a mobile phase of 50 : 50 methanol–water at a flow rate of 1 ml/min. APCI probe temperature of 400°C, corona discharge = 3.24 kV and high voltage lens 0.21 kV.

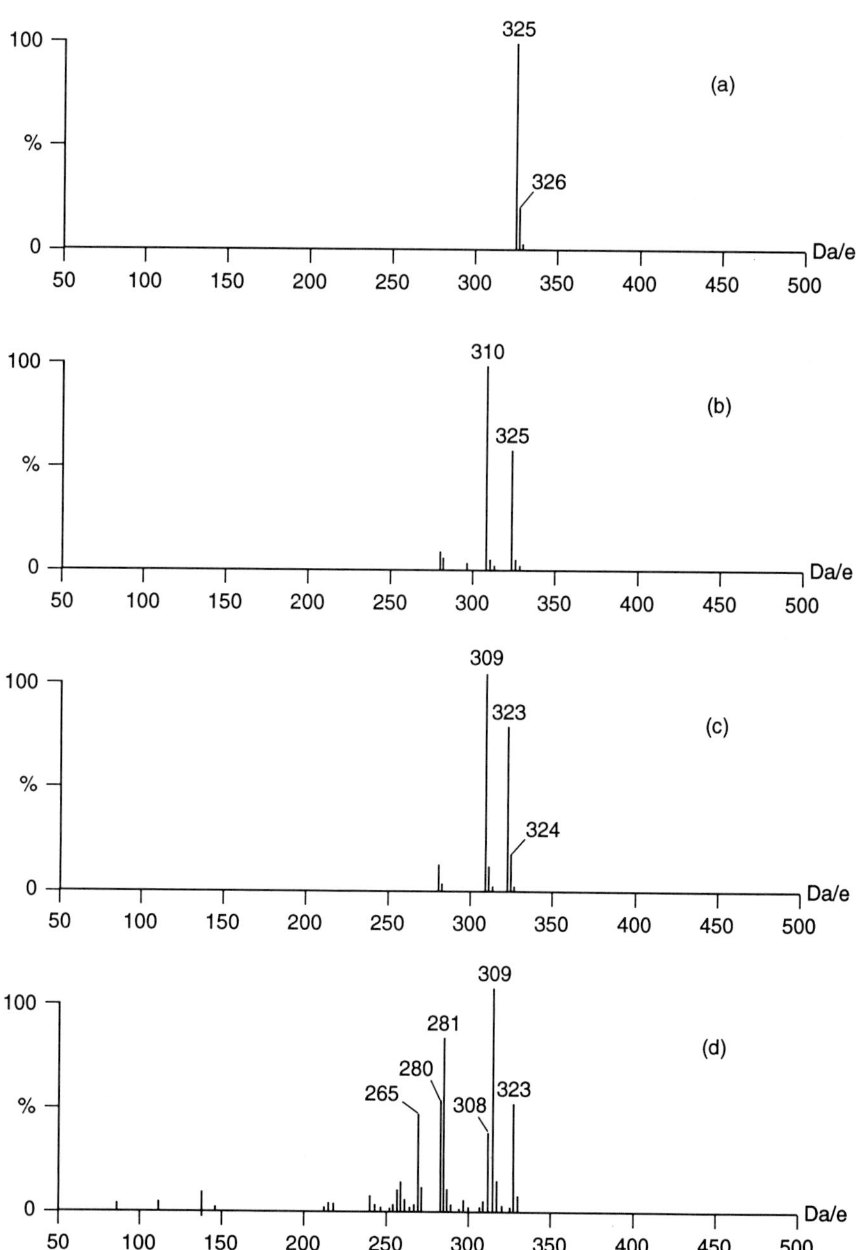

Figure 7.15 Effect of varying cone voltage on APCI LC–MS spectrum of sterigmatocystin mycotoxin (MWt = 324). (a) Positive APCI, cone voltage of 20 V, solution of sterigmatocystin concentration 10 ng/μl. (b) Positive APCI, cone voltage of 45 V, solution of sterigmatocystin concentration 10 ng/μl. (c) Negative APCI, cone voltage 10 V, solution of sterigmatocystin concentration 100 ng/μl. (d) Negative APCI, cone voltage 45 V, solution of sterigmatocystin concentration 100 ng/μl (taken from Scudamore *et al.*, 1995).

analysing for aflatoxins by thermospray LC–MS has been demonstrated (Hurst *et al.*, 1991), although the lowest limit of 60 pg on-column for aflatoxin B_1 is still higher than the 20 pg that can be achieved by fluorescence HPLC (Shepherd and Gilbert, 1984), and the levels for an extract appeared to be substantially higher. With many papers, such as Hurst *et al.* (1991), whilst potential can be readily demonstrated for one or two samples, substantially more work is needed to show the robustness of a technique required for routine operation. However, thermospray LC–MS has proved useful in characterising the reaction products of the postcolumn iodination of aflatoxins B_1 and G_1 for fluorescence detection (Holcomb *et al.*, 1991), where it was shown that the derivatives were formed by the addition of an iodine atom and a methoxy group across the double bond located on the furan rings of these two molecules.

Analysis for the mycotoxin sterigmatocystin has in the past proved to be difficult because, although structurally similar to the aflatoxins, this compound lacks a chromophore for HPLC detection or an easy mode of derivatisation, whilst monitoring at high sensitivity is nevertheless required. Using APCI, the optimum conditions for LC–MS analysis of sterigmatocystin were established (Scudamore *et al.*, 1995). Figure 7.15 shows the effect of varying the cone voltage on the spectrum of sterigmatocystin in both positive and negative APCI, from which it was concluded that positive APCI gave the better limit of detection (0.01 ng/μl from an injection of 10 μl of standard solution). In positive-ion APCI, the $[M + H]^+$ ion is evident at m/z 325 as base peak at 20 V, whilst $[M - 14 + H]^+$ is base peak at the higher cone voltage. More fragmentation was evident in the negative-ion mode, and solutions of higher concentration were required to obtain the same sensitivity. LC–MS has been applied to the analysis of foods, e.g. maize, bread and cheese, which are commodities for which sterigmatocystin contamination has been previously implicated. Figure 7.16 shows SIM traces for a sterigmatocystin standard, for bread spiked at 20 μg/kg and for blank material. Using positive-ion APCI and monitoring m/z 325, limits of detection in the range 1–5 μg/kg were achievable, with a precision of 15% at a concentration of 0.02 ng/μl. Calibration curves were linear over the range 0.001–0.100 ng/l, with correlation coefficients in the region of 0.999 (Scudamore *et al.*, 1995).

The fumonisin mycotoxins are of considerable interest as potential contaminants of human food and animal feeds, and a significant amount of work has been carried on method development (see Chapter 3). Apart from possible occurrence in processed foods such as corn flakes and snack food products, the level of interest is generally higher than is the case, for example, for aflatoxins, which makes these mycotoxins particularly suitable for LC–MS analysis. Thermospray LC–MS, fast atom bombardment (FAB) and electrospray LC–MS have been compared for the analysis of fumonisin B_1 (Korfmacher *et al.*, 1991). Under thermospray flow-injection analysis conditions, about 1 μg of fumonisin B_1 was required for a spectrum, and

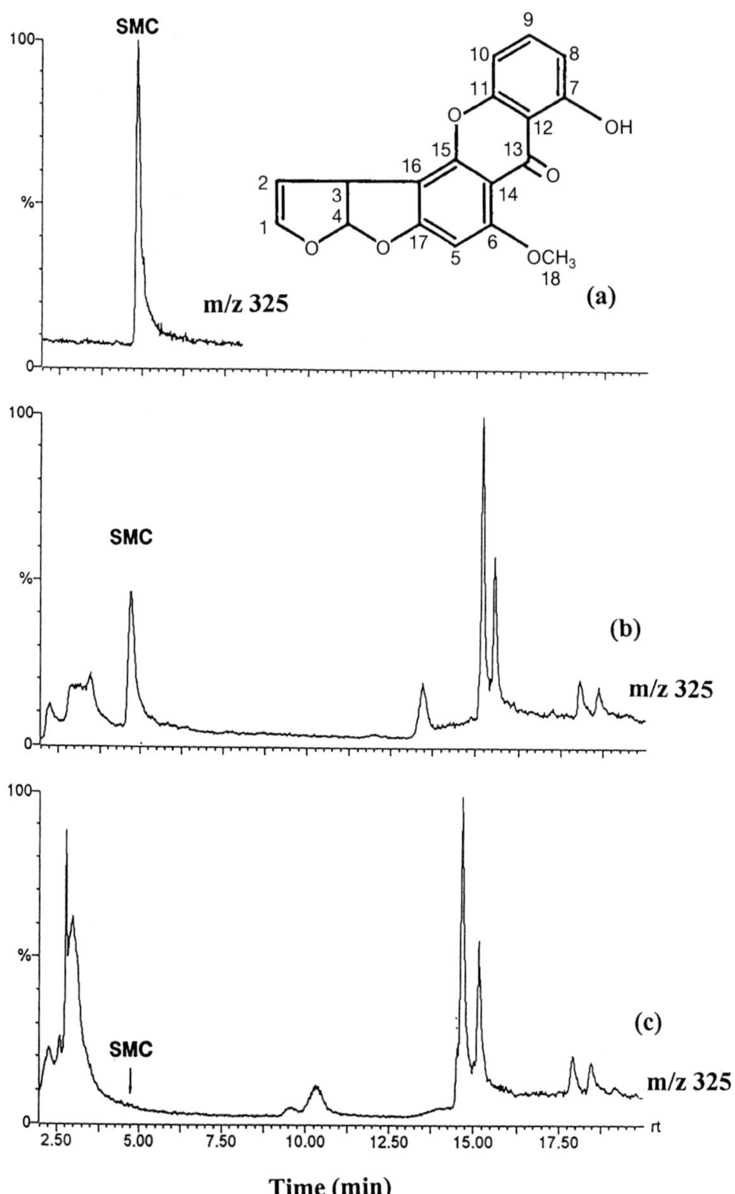

Figure 7.16 Positive-ion APCI LC–MS determination of the mycotoxin sterigmatocystin (SMC) in samples of bread. Analysis using VG Fisons Platform LC–MS system operating with mobile phases of solvent A: 10 mM ammonium acetate and solvent B: 50:50 acetonitrile–methanol, at a flow rate of 0.5 ml/min with a gradient held at 80% B then raised to 100% B. APCI probe temperature 550°C, corona discharge 3.10 V and high voltage lens 0 kV. Selected ion monitoring of *m/z* 325. (a) Sterigmatocystin standard at 0.02 ng/μl; (b) sterigmatocystin spiked at 20 μg/kg into extract from bread; (c) blank extract from bread (taken from Scudamore *et al.*, 1995).

even then the intensity of the $[M + H]^+$ ion was less than 10% of the base peak. Under electrospray conditions the sensitivity was better, and spectra with essentially no fragmentation were obtained for low nanogram amounts of material. Although no work was performed in electrospray for sample extracts, it was anticipated that similar performance to that of work with FAB should be achievable with SIM limits of a few picograms of fumonisin B_1 (Korfmacher *et al.*, 1991). Electrospray LC–MS was employed to characterise fumonisin B_1 in culture material from a novel fungal source, using MS–MS to confirm the proposed identification (Chen *et al.*, 1992). More recent work has been carried out using electrospray LC–MS with sphinganine analogue mycotoxins (Caldas *et al.*, 1995), which has included fumonisins B_1 and B_2. The relative merits of using positive and negative electrospray were assessed, and it was concluded that the negative mode offered advantages, in that spectra were less likely to be complicated by alkali metal cationisation and the formation of $[M - H]^-$ ions was more reliable than the formation of protonated molecules. With corn extracts, the limit of detection was about 20 mg/kg, which was rather too high to be of practical use for foodstuff monitoring, and as matrix effects were evident it was concluded that more attention would be required to sample clean-up before electrospray could be employed for routine foodstuff surveillance (Caldas *et al.*, 1995).

The macrocyclic trichothecenes are thermally labile, polar, biologically active mycotoxins for which there have been no robust methods available for routine analysis. Whilst on the one hand there is no evidence of these compounds being human food contaminants, the lack of adequate methodology makes this judgement somewhat presumptious, and there is anyway a need for methodology for monitoring animal feeds, where there have been incidents of mycotoxicoses attributable to macrocyclic trichothecenes. Although these mycotoxins are amenable to HPLC analysis, lack of a suitable chromophore has been a limiting factor for some of the macrocyclics, and methods have tended to rely on concurrent use of both GC and HPLC. However, thermospray LC–MS has proved to be a particularly attractive approach for analysing some ten macrocyclics (including some isomeric compounds) of the toxic roridin and biologically active baccharinoid types (Krishnamurthy *et al.*, 1989). The procedure, which was applied to crude plant extracts, involved using a synthetically modified macrocyclic trichothecene (8-ketoverrucarin) as internal standard for quantitation purposes, and was shown to be applicable at levels of 2–5 ng (signal to noise ratio of at least 10:1), with monitoring of between two and six ions depending on the macrocyclic trichothecene in question. Smith *et al.* (1985) tackled the problem of the analysis of macrocyclic trichothecenes, employing SFC–MS with capillary SFC and columns ranging from 1.5 m to 15 m and coated with cross-linked polymethylsiloxane phase. This work employed supercritical carbon dioxide mobile phase, and ammonia as reagent gas for CI

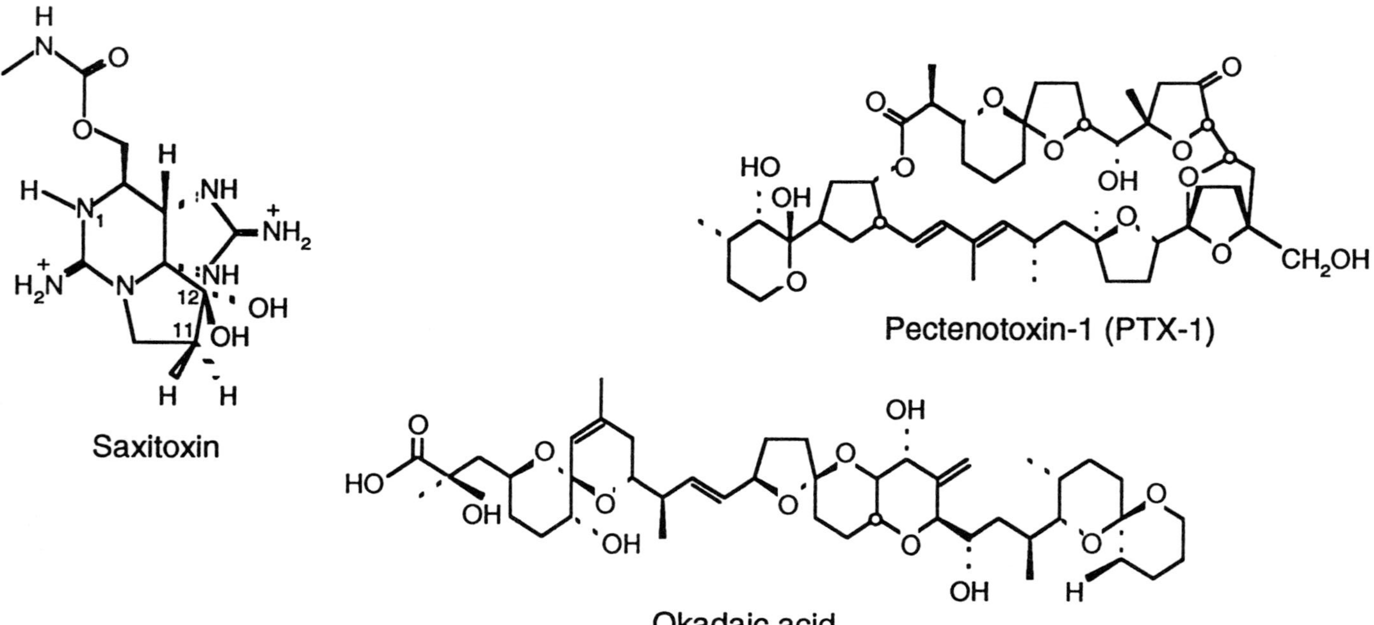

Figure 7.17 Structures of some selected phycotoxins chosen to illustrate the diversity of molecular weight and functionality of these compounds.

mass spectrometry, whence simple spectra were obtained with essentially no fragmentation. Using rapid capillary SFC it was demonstrated that with SIM it was possible in a 4-min chromatographic run to monitor diacetoxy-scirpenol, verrucarin J, verrucarin A, roridin E and roridin A when three of the last four were essentially unresolved chromatographically.

(ii) Phycotoxins.   Two types of algae-related seafood poisoning toxins have attracted considerable attention for human health and economic reasons—paralytic shellfish poisoning (PSP) and diarrhetic shellfish poisoning (DSP) (see Chapter 6). Reliance on bioassays, such as the mouse bioassay, for monitoring of these toxins, despite being the official method in many countries, has not been satisfactory, and there have been significant moves towards development of appropriate HPLC methodology, which has involved use of postcolumn derivatisation procedures (Hungerford and Wekell, 1992; Luckas, 1992). In view of the need for rigorous confirmation of PSP/DSP screening results, and because of evidence from bioassays of the presence of previously unidentified toxins, there has been considerable incentive to develop MS-based procedures. The presence of readily ionisable functionalities makes PSP and DSP toxins ideal candidates not only for separation techniques based on ionic mobilities in aqueous buffers (such as capillary electrophoresis) but also for electrospray LC–MS.

The chemical structures of a number of phycotoxins are shown in Figure 7.17, and these can be seen to be of a moderate molecular weight, for example okadaic acid has a molecular weight of 804. These molecules are polar, containing a large number of free hydroxyl groups, and in the case of saxitoxin also contain free amine groups. The major difficulty with HPLC analysis is the lack of a suitable chromophore, which means that derivatisation (for example postcolumn) is necessary.

Electrospray has been found to be particularly attractive for phycotoxins, the spectrum of okadaic acid, for example, shown in Figure 7.18, displaying an $[M + H]^+$ ion at m/z 805 being base peak with the only other ions evident at m/z 787 attributed to $[M + H - H_2O]^+$ (10%) and m/z 827 attributed to $[M + Na]^+$. Samples were run by injection from a solution of acetonitrile–water (40:60) containing 0.1% trifluoroacetic acid. Figure 7.18 also demonstrates quantitation using electrospray by duplicate injection of a series of dilutions of a 100 μg/ml stock solution of okadaic acid with monitoring of m/z 805 (Pleasance *et al.*, 1990). Ion-spray was also investigated for the analysis of three further marine toxins: domoic acid, saxitoxin and tetrodotoxin (Quilliam *et al.*, 1989). Detection limits by flow injection analysis were estimated to be 30 pg for saxitoxin, 100 pg for domoic acid and 200 pg for tetrodotoxin. In addition to electrospray LC–MS, there have been a number of publications evaluating capillary electrophoresis, which seems to have some potential in this area (Pleasance *et al.*, 1992a; Locke and Thibault, 1994) (see also section 7.3.2).

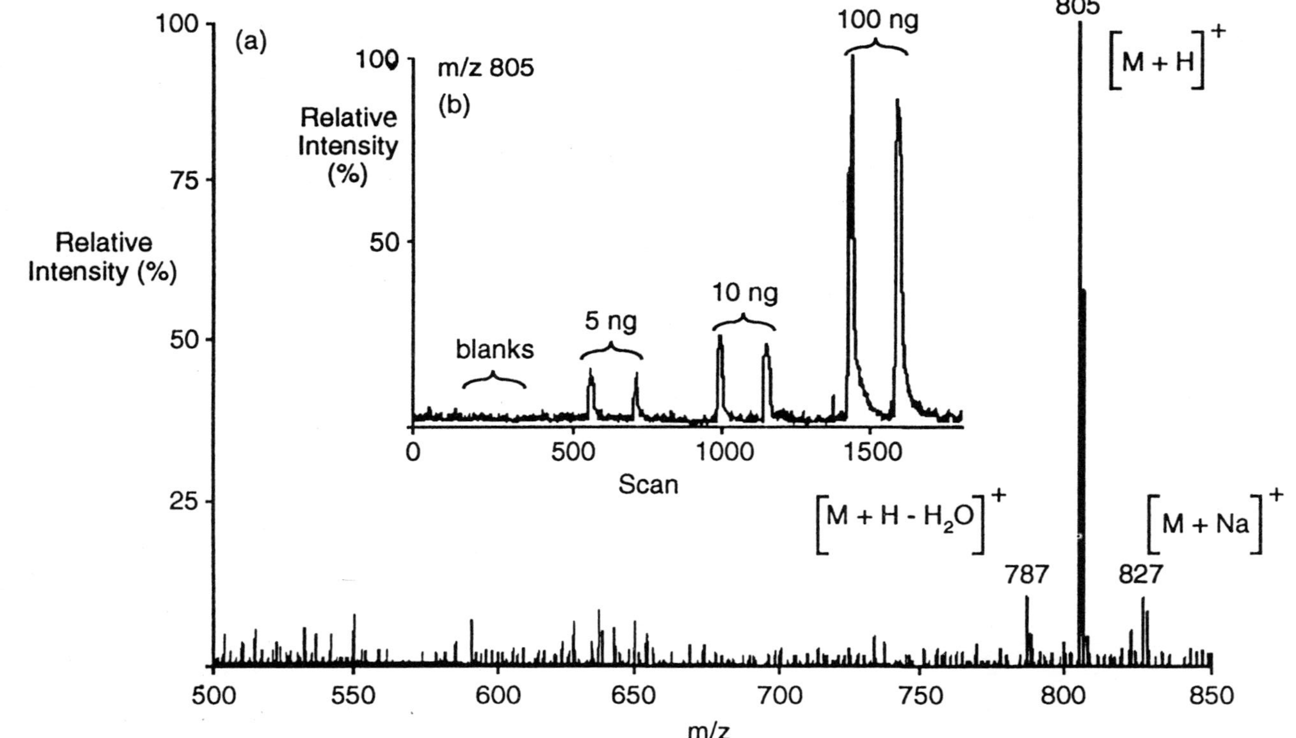

Figure 7.18 Electrospray positive ion spectrum of okadaic acid. Insert shows flow injection analysis with SIM of molecular ion (m/z 805) for duplicate injections of a standard (100 μg/ml stock solution). Conditions: mobile phase consisting of (1:1) water–acetonitrile with 0.15 M trifluoroacetic acid at a flow rate of 100 μl/min (reproduced from Pleasance *et al.*, 1990, with permission from John Wiley and Sons Ltd).

(iii) Other natural toxicants. There is considerable interest in the presence of biologically active constituents in plant materials, particularly those sold as 'health' foods which are commonly available not only as leaves, roots or flowers but also as tablets, tinctures or drinks with claimed medicinal effects. Although strictly not 'food contaminants', these components are frequently classed as natural toxicants and thus monitored and controlled alongside, for example, mycotoxins. Thermospray LC–MS has been employed for the identification of various flavonol glycosides (Pietta *et al.*, 1994) from *Ginkgo biloba, Calendula officinalis* and *Tilia cordata*. Thermospray provided information on molecular weight (ranging from 594 to 741), sugar sequence and glycosylation sites of these glycosides which would only otherwise have been obtainable by the far more time-consuming preparative isolation followed by hydrolysis and HPLC or GC of the resulting aglycones and sugars.

Pyrrolizidine alkaloids are important natural toxicants in terms of being potential animal feed components which can enter the food chain, and as constituents of herbal products such as comfrey used in teas and herbal remedies. LC–MS in the APCI mode has been used to determine a number of pyrrolizidine alkaloids in honey obtained from bees with access to ragwort pollen potentially containing jacoline, jacobine, seneciphylline and senecionine alkaloids (Crews *et al.*, 1995). Spectra of a number of pyrrolizidine alkaloids are shown in Figure 7.19, in each case giving base peak $[M + H]^+$ ions, which were employed for selected ion monitoring of honey extracts. A non-naturally occurring alkaloid, monocrotaline, was used as an internal standard, and the method enabled a limit of detection of 2 μg/kg to be obtained when spiked into honey.

7.4.4 Food packaging materials

There is a need to be able to analyse both plastics and paperboard food packaging materials for the presence of components including additives that might give rise to migration into foods. Extracts from packaging materials will in general contain relatively high concentrations of these components and sensitivity will not therefore necessarily be a critical parameter. Allen *et al.* (1994) have used particle beam LC–MS to characterise the electron beam irradiation transformation products of an antioxidant (Irganox 1010) used in polyolefin food packaging. The plastic was extracted with chloroform and the concentrated extract analysed directly by gradient elution HPLC on-line to VG Masslab Trio 1 quadrupole MS with a particle beam LC–MS interface. Of some 56 transformation products that were detected, a number were identified on the basis of their mass spectra, typical examples of these being illustrated in Figure 7.20 (Allen *et al.*, 1993). Some 18 different plastics additives of different structural types of differing polarity, volatility and functionality, encompassing UV absorbers, antioxidants and

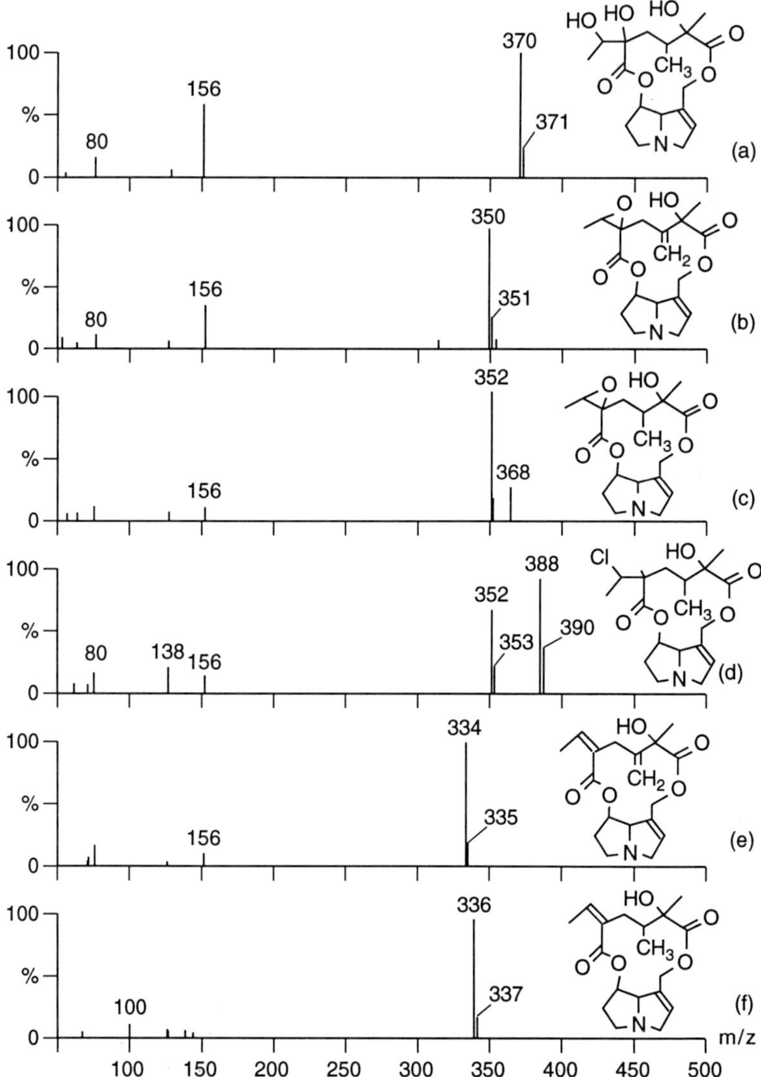

Figure 7.19 APCI spectra of a pyrrolizidine alkaloids: (a) jacoline; (b) jacozine; (c) jacobine; (d) jaconine; (e) seneciphylline; (f) senecionine. Analysis using a VG Fisons Platform LC–MS operated in APCI mode with scanning from m/z 60 to 600. LC mobile phase of 0.1 M ammonium hydroxide and acetonitrile with a gradient of 10 to 90% acetonitrile over 30 min at a flow rate of 1 ml/min (taken from Crews *et al.*, 1995, unpublished results).

stabilisers, with molecular weights ranging from 214 to 1176, have been analysed by particle beam and APCI (Clarke *et al.*, 1995). With particle beam LC–MS, the source temperature exerted a critical effect on spectra and sensitivity, and only in the case of *N*,*N*-dimethyloctylamine was it not possible to obtain a spectrum, whereas using APCI under PI conditions all

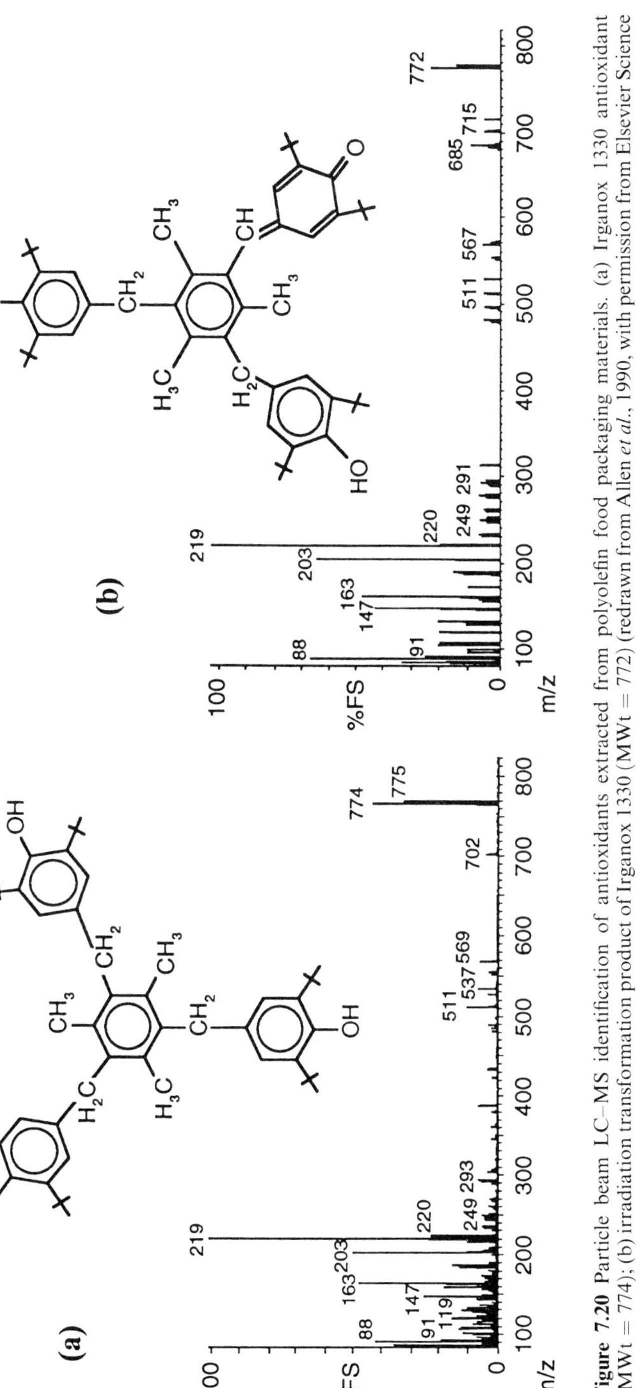

Figure 7.20 Particle beam LC–MS identification of antioxidants extracted from polyolefin food packaging materials. (a) Irganox 1330 antioxidant (MWt = 774); (b) irradiation transformation product of Irganox 1330 (MWt = 772) (redrawn from Allen *et al.*, 1990, with permission from Elsevier Science Publishers).

18 compounds produced acceptable spectra with $[M + H]^+$ or $[M]^+$ plus identifiable fragments.

An area that is of particular interest with regard to the use of poly(ethylene terephthalate) (PET) for food contact applications is the presence of oligomers in the plastic, which may be important components as potential contaminants for carbonated beverages, for high temperature uses and for plastics recycling. By operating the mass spectrometer at unusually high temperatures (370°C for the thermospray tip) it was possible to obtain spectra of oligomers from PET (Milon, 1991) identified as cyclic trimers and tetramers (MWts of 576 and 768, respectively). Further work using thermospray (Guarini *et al.*, 1993) has been carried out on the products of depolymerisation of PET with ethylene glycol. However, improved performance of the LC–MS in this instance was achieved by derivatisation of the mixture with perfluoroanhydrides, allowing the identification of linear and cyclic PET oligomers, diethylene glycol-containing oligomers and monoacetyl derivatives of linear oligomers. The use of positive-ion APCI for the analysis of PET oligomers from food-grade PET resin and beverage bottles (Barnes *et al.*, 1995b) has demonstrated this approach to be more sensitive than thermospray, giving good spectra with sub-microgram quantities of oligomers. Using this approach, characteristic spectra of the cyclic oligomers from trimer (MWt = 576) to the heptamer (MWt = 1344) were obtained, together with spectra from a homologous series 44 mass units higher, identified as oligomers with one monoethylene glycol unit replaced by a diethylene glycol unit.

7.4.5 *Miscellaneous contaminants*

Outside of the major areas indicated above, the full potential of LC–MS in the area of food contaminants has not yet been fully exploited. A number of isolated examples do indicate where there is scope for future development. Polyacylamide is used as a flocculant in the sugar industry, and as it may contain low levels of residual acrylamide monomer there is interest in determining the possible presence of acrylamide at ppt levels in sugar products. Cutie and Kallos (1986) have showed that by derivatisation of acrylamide with bromine and multi-dimensional LC clean-up prior to thermospray LC–MS, most interferences were eliminated, and with postcolumn addition of trifluoroacetic acid it was possible to detect the dibromopropionamide derivative at 0.2 μg/kg in sugar.

There is interest in the specifications of food colours in terms of identification of impurities, particularly in the case of natural food colours, and also in the characterisation of decomposition and degradation products of food colours. Using pseudo-electrochromatography, which involves a pressure-driven liquid passing through a packed bed with an axial potential difference, coupled with negative-ion electrospray MS, water-soluble food colours were chosen as model compounds to demonstrate the potential of the technique (Hugener *et al.*, 1993). For these sulphonated azo dyes the base peaks were the triply charged ions, and the singly charged $[M - H]^-$ ion could hardly be distinguished from the background.

As examples of artificial sweeteners, the potential of APCI and electrospray was evaluated for the analysis of saccharin and acesulfame-K (Clarke *et al.*, 1995). Poor spectra were obtained for acesulfame-K under both PI and NI conditions in APCI at all probe temperatures. No spectra were obtained for either compound in PI electrospray, whereas both compounds produced good NI spectra in electrospray, which was selected for subsequent LC–MS analysis.

A series of potent heterocyclic amines that are both mutagenic and carcinogenic have been recognised for some time as contaminants that may be formed most notably in some protein-rich heated foods, such as broiled meat and fish. HPLC and GC (with derivatisation) are frequently used for monitoring these compounds (Knize *et al.*, 1992) although derivatisation can be problematic at low concentrations, and high-temperature operation of GC columns can give rise to instability problems. Although fluorescence HPLC is good for screening, LC–MS has been employed for confirmation and quantitation of these mutagens in salmon, sardine and beef (Yamaizumi *et al.*, 1986) and in beef and beef extracts (Turesky *et al.*, 1988). Both groups of workers used stable-isotope (deuterated) internal standard analogues of three of the imidazoquinoxaline mutagens, which not only corrected for recovery losses during sample work-up but also improved quantitation during LC–MS. The LC–MS procedure enabled low $\mu g/kg$ amounts of the mutagens to be detected in cooked beef, with as little as 10 g of the foodstuff taken for analysis (Turesky *et al.*, 1988).

7.5 Conclusions

A major development in the last few years has been the availability of API for LC–MS and the introduction by manufacturers of modestly priced instruments dedicated to LC–MS which offer electrospray and APCI facilities. LC–MS has thus moved from being a specialised and somewhat problematic technique to one akin to GC–MS which can be applied routinely. Although there have been extensive applications of thermospray LC–MS in the area of food contaminants, it has had limitations dependent on compound type which have restricted usage. Electrospray and APCI have, however, opened up LC–MS for a number of contaminants which were previously difficult to analyse by other methods, due to fairly high molecular weight, lack of a chromophore and polarity. Sensitivity, which has also previously been a limitation of LC–MS, is now less of a problem, and the technique can offer limits of detection well within the demanding requirements of surveillance work. For the future a continuation of the trend towards greater use of these techniques is to be expected, with perhaps more concentration on improvement of quantitative aspects.

References

Abian, J., Churchwell, M.I. and Korfmacher, W.A. (1993) High-performance liquid chromatography–thermospray mass spectrometry of ten sulphonamide antibiotics. Analysis in milk at the ppb level. *J. Chromatogr.*, **629**, 267–276.

Abramson, D. (1987) Measurement of ochratoxin A in barley extracts by liquid chromatography–mass spectrometry. *J. Chromatogr.*, **391**, 315–320.

Allen, D.W., Clench, M.R., Crowson, A. and Leathard, D.A. (1993) Identification by particle-beam liquid chromatography–mass spectrometry of transformation products of the antioxidant Irganox 1330 in food-contact polymers subjected to electron beam irradiation. *J. Chromatogr.*, **629**, 283–290.

Allen, D.W., Clench, M.R., Crowson, A., Leathard, D.A. and Saklatvala, R. (1994) Characterisation of electron beam generated transformation products of Irganox 1010 by particle beam liquid chromatography–mass spectrometry with on-line diode array detection. *J. Chromatogr. A*, **679**, 285–297.

Arpino, P.J. and Haas, P. (1995) Review. Recent developments in supercritical fluid chromatography–mass spectrometry coupling. *J. Chromatogr.*, **703**, 479–488.

Bagheri, H., Brouwer, E.R., Ghijsen, R.T. and Brinkman, U.A.Th. (1993) On-line low-level screening of polar pesticides in drinking and surface waters by liquid chromatography–thermospray mass spectrometry. *J. Chromatogr.*, **647**, 121–129.

Balizs, G., Benesch-Girke, L. and Hewitt, S.A. (1993). Determination of four sulphonamides and their N-4-acetyl metabolites in swine muscle using liquid chromatography/thermospray mass spectrometry. In *Proceedings of the EuroResidue II Conference*, vol. 1. Veldhoven, The Netherlands, May 1993 (edited by N. Haagama, A. Ruiter and P.B. Czedik-Eysenberg), pp. 155–159.

Barcelo, D. (1988) Application of thermospray liquid chromatography/mass spectrometry for determination of organophosphorus pesticides and trialkyl and triaryl phosphates. *Biomed. and Environ. Mass Spectrom.*, **17**, 363–369.

Barcelo, D. and Albaiges, J. (1989) Characterization of organophosphorus compounds and phenylurea herbicides by positive and negative ion thermospray liquid chromatography–mass spectrometry. *J. Chromatogr.*, **474**, 163–173.

Barcelo, D., Maris, F.A., Geerdink, R.B., Frei, R.W., De Jong, G.I. and Brinkmann, U.A.Th. (1987) Comparison between positive, negative and chlorine-enhanced negative chemical ionization of organophosphorus pesticides in on-line liquid chromatography–mass spectrometry. *J. Chromatogr.*, **394**, 65–76.

Barcelo, D., Durand, G. and Vreeken, R.J. (1993) Determination of quaternary amine pesticides by thermospray mass spectrometry. *J. Chromatogr.*, **647**, 271–277.

Barker, S.A. and Walker, C.C. (1992) Chromatographic methods for tetracycline analysis in foods. *J. Chromatogr.*, **624**, 195–209.

Barnes, K.A. (1995) Unpublished results.

Barnes, K.A., Startin, J.R., Thorpe, S.A., Reynolds, S.L. and Fussell, R.J. (1995a) Determination of the pesticide diflubenzuron in mushrooms by high performance liquid chromatography/mass spectrometry with atmospheric pressure chemical ionisation. *J. Chromatogr.*, **712**, 85–93.

Barnes, K.A., Damant, A.P., Startin, J.R. and Castle, L. (1995b) Qualitative liquid chromatography–atmospheric pressure ionisation–mass spectrometry analysis of polyethylene terephthalate oligomers. *J. Chromatogr. A*, **712**, 191–199.

Betowski, L.D. and Jones, T.L. (1988) Analysis of organophosphorus pesticide samples by high-performance liquid chromatography/mass spectrometry and high-performance liquid chromatography/mass spectrometry/mass spectrometry. *Environ. Sci. Technol.*, **22**, 1430–1434.

Blanchflower, W.J. and Kennedy, D.G. (1989a) Determination of nitroxynil residues in tissues using high-performance liquid chromatography–thermospray mass spectrometry. *Analyst*, **114**, 1013–1015.

Blanchflower, W.J. and Kennedy, D.G. (1989b) A rapid screening procedure for the detection and quantification of clenbuterol in bovine urine using thermospray liquid chromatography/mass spectrometry. *Biomed Environ. Mass Spectrom.*, **18**, 935–936.

Blanchflower, W.J., Kennedy, D.G. and Taylor, S.M. (1990) Determination of rafoxanide in plasma using high performance liquid chromatography (HPLC) and in tissue using HPLC–thermospray mass spectrometry. *J. Liq. Chromatogr.*, **13**, 1595–1609.

Blanchflower, W.J., McCracken, R.J. and Kennedy, D.G. (1993a) Detection of furazolidone in muscle using liquid chromatography/thermospray mass spectrometry. In *Proceedings of the EuroResidue II Conference*, vol. 1, Veldhoven, The Netherlands, May 1993 (edited by N. Haagama, A. Ruiter and P.B. Czedik-Eysenberg), pp. 201–205.

Blanchflower, W.J., Cannavan, A., McCracken, R.J., Hewitt, S.A. and Kennedy, D.G. (1993b) Determination of chloramphenicol in plasma, milk and tissue using liquid chromatography/ thermospray mass spectrometry. In *Proceedings of the EuroResidue II Conference*, vol. 1, Veldhoven, The Netherlands, May 1993 (edited by N. Haagama, A. Ruiter and P.B. Czedik-Eysenberg), pp. 196–200.

Blanchflower, W.J., Cannavan, A. and Kennedy, D.G. (1993c) Determination of fenbendazole and oxfendazole in tissues using liquid chromatography/thermospray mass spectrometry. In *Proceedings of the EuroResidue II Conference*, vol. 1, Veldhoven, The Netherlands, May 1993 (edited by N. Haagama, A. Ruiter and P.B. Czedik-Eysenberg), pp. 191–195.

Bobbitt, D.R. and Ng, K.W. (1992) Chromatographic analysis of antibiotic materials in food. *J. Chromatogr.*, **624**, 153–170.

Boison, J.O. (1992) Chromatographic methods of analysis for penicillins in food-animal tissues and their significance in regulatory programs for residue reduction and avoidance. *J. Chromatogr.*, **624**, 171–194.

Brown, M.A. (ed.) (1990) *Liquid Chromatography/Mass Spectrometry—Applications in Agricultural, Pharmaceutical and Environmental Chemistry.* ACS Symposium Series, no. 420. American Chemical Society: Washington DC.

Cai, J. and Henion, J. (1995) Review. Capillary electrophoresis–mass spectrometry. *J. Chromatogr.*, **703**, 667–692.

Caldas, E.D., Jones, A.D., Winter, C.K., Ward, B. and Gilchrist, D.G. (1995) Electrospray ionization mass spectrometry of sphinganine analog mycotoxins. *Anal. Chem.*, **67**, 196–207.

Carignan, G. and Carrier, K. (1991) Quantitation and confirmation of sulfamethazine residues in swine muscle and liver by LC and GC/MS. *J. Assoc. Off. Anal. Chem.*, **74**, 479–482.

Chen, J., Mirocha, C.J., Xie, W., Hoggie, L. and Olson, D. (1992) Production of the mycotoxin fumonisin B_1 by *Alternaria alternata* f. sp. *lycopersici*. *Appl. Environ. Microbiol.*, **58**, 3928–3931.

Chester, T.L., Pinkston, J.D. and Raynie, D.E. (1994) Supercritical fluid chromatography and extraction. *Anal. Chem.*, **66**, 106R–130R.

Clarke, P.A., Barnes, K., Crews, C. and Startin, J.R. (1995) Unpublished results.

Crews, C., Clarke, P.A. and Startin, J.R. (1995) Unpublished results.

Crosby, N.T. (1991) *Determination of Veterinary Residues in Food.* Ellis Horwood: New York.

Cutie, S.S. and Kallos, G.J. (1986) Determination of acrylamide in sugar by thermospray liquid chromatography/mass spectrometry. *Anal. Chem.*, **58**, 2425–2428.

Delepine, B., Sanders, P., Dagorn, M. and Laurentie, M. (1993) Study of chloramphenicol metabolites in chicken plasma using a particle beam interface for combining liquid chromatography with mass spectrometry. In *Proceedings of the EuroResidue II Conference*, vol. 1, Veldhoven, The Netherlands, May 1993 (edited by N. Haagama, A. Ruiter and P.B. Czedik-Eysenberg), pp. 267–271.

Doerge, D.R. and Miles, C.J. (1991) Determination of ethylenethiourea in crops using particle beam liquid chromatography/mass spectrometry. *Anal. Chem.*, **63**, 1999–2001.

Escoffier, B.H., Parker, C.E., Mester, T.C., Dewit, J.S.M., Corbin, F.T., Jorgensen, J.W. and Tomer, K.B. (1989). Comparison of open-tubular liquid chromatography and direct liquid introduction liquid chromatography–mass spectrometry for the analysis of metribuzin and its metabolites in plant tissue and water samples. *J. Chromatogr.*, **474**, 301–316.

Farran, A., De Pablo, J. and Barcelo, D. (1988) Identification of organophosphorus insecticides and their hydrolysis products by liquid chromatography in combination with UV and thermospray–mass spectrometric detection. *J. Chromatogr.*, **455**, 163–172.

Games, D.E. (1987) Applications of high performance liquid chromatography/mass spectrometry (LC/MS) in food chemistry. In *Applications of Mass Spectrometry in Food Science* (edited by J. Gilbert), Elsevier Applied Science: London, pp. 193–237.

Games, D.E., Berry, A.J., Mylchreest, I.C., Perkins, J.R. and Pleasance, S. (1988) Supercritical fluid chromatography–mass spectrometry. In *Supercritical Fluid Chromatography* (edited by R.M. Smith), Royal Society of Chemistry Chromatography Monograph, RSC: London, pp. 159–174.

Garcia, F. and Henion, J. (1992) Fast capillary electrophoresis–ion spray mass spectrometric determination of sulfonylureas. *J. Chromatogr.*, **606**, 237–247.

Getek, T.A., Vestal, M.L. and Alexander, T.G. (1991) Analysis of gentamicin by high-performance liquid chromatography combined with thermospray mass spectrometry. *J. Chromatogr.*, **554**, 191–203.

Guarini, A., Guglielmetti, G. and Po, R. (1993) Application of liquid chromatography–thermospray mass spectrometry to the analysis of polyester oligomers. *J. Chromatogr.*, **647**, 311–318.

Guggisberg, D., Mooser, A.E. and Koch, H. (1992) Review. Methods for the determination of sulphonamides in meat. *J. Chromatogr.*, **624**, 425–437.

Hammond, I., Moore, K., James, H. and Watts, C. (1989) Thermospray liquid chromatography–mass spectrometry of pesticides in river water using reversed-phase chromatography. *J. Chromatogr.*, **474**, 175–180.

Heller, D.N. and Schenek, F.J. (1993) Particle beam liquid chromatography/mass spectrometry with negative ion chemical ionization for the confirmation of ivermectin residue in bovine milk and liver. *Biomed. Mass Spectrom.*, **22**, 184–193.

Henion, J.D., Mordehai, A.V. and Cai, J. (1994) Quantitative capillary electrophoresis–ion spray mass spectrometry on a benchtop ion trap for the determination of isoquinoline alkaloids. *Anal. Chem.*, **66**, 2103–2109.

Hewitt, S.A., Blanchflower, W.J., McCaughey, W.J., Elliott, C.T. and Kennedy, D.G. (1993) Liquid chromatography–thermospray mass spectrometric assay for trenbolone in bovine bile and faeces. *J. Chromatogr.*, **639**, 185–191.

Holcomb, M., Korfmacher, W.A. and Thompson, H.C. (1991) Characterization of iodine derivatives of aflatoxin B_1 and G_1 by thermospray mass spectrometry. *J. Anal. Toxicol.*, **15**, 289–292.

Horie, M., Saito, K., Nose, T., Tera, M. and Nakazawa, H. (1993) Confirmation of residual oxolinic acid, nalidixic acid and piromidic acid in fish by thermospray liquid chromatography–mass spectrometry. *J. Liq. Chromatogr.*, **16**, 1463–1472.

Huang, A.S., Robinson, L.R., Gursky, L.G., Profita, R. and Sabidong, C.G. (1994) Identification and quantification of SALATRIM 23CA in foods by the combination of supercritical fluid extraction, particle beam-LC–mass spectrometry, and HPLC with light scattering detector. *J. Agric. Food Chem.*, **42**, 468–473.

Hugener, M., Tinke, A.P., Niessen, W.M.A., Tjaden, U.R. and van der Greef, J. (1993) Pseudo-electrochromatography–negative-ion electrospray mass spectrometry of aromatic glucuronides and food colours. *J. Chromatogr.*, **647**, 375–385.

Hungerford, J.M. and Wekell, M.M. (1992) Analytical methods for marine toxins. In *Handbook of Natural Toxins*, vol. 7, *Food Poisoning* (edited by A.T. Tu), Marcel Dekker Inc.: New York, pp. 415–473.

Hurst, W.J., Martin, R.A. and Vestal, C.H. (1991) The use of HPLC/thermospray MS for the confirmation of aflatoxins in peanuts. *J. Liq. Chromatogr.*, **14**, 2541–2550.

Ioerger, B.P. and Smith, J.S. (1994) Determination and confirmation of organophosphate pesticides and their metabolites in beef tissue using thermospray/LC–MS. *J. Agric. Food Chem.*, **42**, 2619–2624.

Jablonska, A., Hansen, M., Ekeberg, D. and Lundanes, E. (1993) Determination of chlorinated pesticides by capillary supercritical fluid chromatography–mass spectrometry with positive- and negative-ion detection. *J. Chromatogr.*, **647**, 341–350.

Kalinoski, H.T., Wright, B.W. and Smith, R.D. (1986a) Ammonia and methane chemical ionization mass spectra of acid and carbamate pesticides using direct supercritical fluid injection. *Biomed. Environ. Mass Spectrom.*, **13**, 33–45.

Kalinoski, H.T., Udseth, H.R., Wright, B.W. and Smith, R.D. (1986b) Supercritical fluid extraction and direct fluid injection mass spectrometry for the determination of trichothecene mycotoxins in wheat samples. *Anal. Chem.*, **58**, 2421–2425.

Khunachak, A., Dacunha, A.R. and Stout, S.J. (1993) Liquid chromatographic determination of moxidectin residues in cattle tissues and confirmation in cattle fat by liquid chromatography/mass spectrometry. *J. Assoc. Offic. Anal. Chem. Internat.*, **76**, 1230–1235.

Kijak, P.J., Leadbetter, M.G., Thomas, M.H. and Thompson, E.A. (1991) Confirmation of oxytetracycline, tetracycline and chlortetracycline residues in milk by particle beam liquid chromatography/mass spectrometry. *Biomed. Mass Spectrom.*, **20**, 789–795.

Knize, M.G., Felton, J.S. and Gross, G.A. (1992) Chromatographic methods for the analysis of heterocyclic amine food mutagens/carcinogens. *J. Chromatogr.*, **624**, 253–265.

Korfmacher, W.A., Chiarelli, M.P., Lay, J.O., Bloom, J., Holcomb, M. and McManus, K.T. (1991) Characterization of the mycotoxin fumonisin B_1: Comparison of thermospray, fast-atom bombardment and electrospray mass spectrometry. *Rapid Commun. Mass Spectrom.*, **5**, 463–468.

Krishnamurthy, T., Beck, D.J., Isensee, R.K. and Jarvis, B.B. (1989) Mass spectral investigations on trichothecene mycotoxins. VII. Liquid chromatographic–thermospray mass spectrometric analysis of macrocyclic trichothecenes. *J. Chromatogr.*, **469**, 209–222.

Kristiansen, G.K., Brock, R. and Bojesen, G. (1994) Comparison of flow injection/thermospray MS/MS and LC/thermospray MS/MS methods for determination of sulphonamides in meat and blood. *Anal. Chem.*, **66**, 3253–3258.

Lamoree, M.H., Reinhoud, N.J., Tjaden, U.R., Niessen, W.M.A. and van der Greef, J. (1994) On-capillary isochophoresis for loadability enhancement in capillary zone electrophoresis/mass spectrometry of β-agonists. *Biomed. Mass Spectrom.*, **23**, 339–345.

Leadbetter, M.G. and Matusik, J.E. (1993) Liquid chromatographic determination and liquid chromatographic–thermospray mass spectrometric confirmation of nicarbazin in chicken tissues: Interlaboratory study. *J. Assoc. Offic. Anal. Chem. Internat.*, **76**, 420–423.

Lewis, J.L., Macy, T.D. and Garteiz, D.A. (1989) Determination of nicarbazin in chicken tissues by liquid chromatography and confirmation of identity by thermospray liquid chromatography/mass spectrometry. *J. Assoc. Offic. Anal. Chem.*, **72**, 577–581.

Lin, H.-Y. and Voyksner, R.D. (1993) Determination of environmental contaminants using an electrospray interface combined with an ion trap mass spectrometer. *Anal. Chem.*, **65**, 451–456.

Liu, C.-H., Mattern, G.C., Xiaobing, Y. and Rosen, J.D. (1990) Determination of benomyl by high performance liquid chromatography/mass spectrometry selected ion monitoring. *J. Agric. Food Chem.*, **38**, 167–171.

Liu, C.-H., Mattern, G.C., Xiaobing, Y., Rosen, R.T. and Rosen, J.D. (1991) Multiresidue determination of nonvolatile and thermally labile pesticides in fruits and vegetables by thermospray liquid chromatography/mass spectrometry. *J. Agric. Food Chem.*, **39**, 718–723.

Locke, S.J. and Thibault, P. (1994) Improvement in detection limits for the determination of paralytic shellfish poisoning toxins in shellfish tissues using capillary electrophoresis/electrospray mass spectrometry and discontinuous buffers. *Anal. Chem.*, **66**, 3436–3446.

Luckas, B. (1992) Phycotoxins in seafood—toxicological and chromatographic aspects. *J. Chromatogr.*, **624**, 439–456.

Matsumoto, K., Nagata, S., Hattori, H. and Tsuge, S. (1992) Development of directly coupled supercritical fluid chromatography with packed capillary column–mass spectrometry with atmospheric pressure chemical ionization. *J. Chromatogr.*, **605**, 87–94.

Mattern, G.C., Liu, C.-H., Louis, J.B. and Rosen, J.D. (1991) GC/MS and LC/MS determination of 20 pesticides for which dietary oncogenic risk has been estimated. *J. Agric. Food Chem.*, **39**, 700–704.

Mattina, M.J.I. (1991) Determination of chlorophenoxy acids using high-performance liquid chromatography–particle beam mass spectrometry. *J. Chromatogr.*, **542**, 385–395.

Matusik, J.E., Leadbetter, M.G., Barnes, C.J. and Sphon, J.A. (1992) Identification of dimetridazole, ipronidazole, and their alcohol metabolites in turkey tissues by thermospray tandem mass spectrometry. *J. Agric. Food Chem.*, **40**, 439–443.

McLaughlin, L.G., Henion, J.D. and Kijak, P.J. (1994) Multi-residue confirmation of aminoglycoside antibiotics and bovine kidney by ion spray high-performance liquid chromatography/tandem mass spectrometry. *Biolog. Mass Spectrom.*, **23**, 417–429.

Milon, H. (1991) Identification of poly(ethylene terephthalate) cyclic oligomers by liquid chromatography–mass spectrometry. *J. Chromatogr.*, **554**, 305–309.

Ministry of Agriculture, Fisheries and Food (1995) *Steering Group on Chemical Aspects of Food Surveillance—Annual Report 1994*. Food Surveillance Paper no. 45, HMSO: London.

Murugaverl, B., Voorhees, K.J. and DeLuca, S.J. (1993) Utilization of a benchtop mass spectrometer with capillary supercritical fluid chromatography. *J. Chromatogr.*, **633**, 195–205.

Niessen, W.M.A. and Tinke, A.P. (1995) Review. Liquid chromatography–mass spectrometry. General principles and instrumentation. *J. Chromatogr.*, **703**, 37–57.

Oka, H., Ikai, Y., Hayakawa, J., Harada, K., Asukabe, H., Suzuki, M., Himei, R., Horie, M., Nakazawa, H. and MacNeil, J.D. (1994a) Improvement of chemical analysis of antibiotics. 22. Identification of residual tetracyclines in honey by frit FAB/LC/MS using a volatile mobile phase. *J. Agric. Food Chem.*, **42**, 2215–2219.

Oka, H., Ikai, Y., Hayakawa, J., Masuda, K., Harada, K.-I. and Suzuki, M. (1994b) Improvement of chemical analysis of antibiotics. Part XIX. Determination of tetracycline antibiotics in milk by liquid chromatography and thin-layer chromatography/fast atom bombardment mass spectrometry. *J. Assoc. Offic. Anal. Chem. Internat.*, **77**, 891–895.

Perkins, J.R., Games, D.E., Startin, J.R. and Gilbert, J. (1991a) Analysis of veterinary drugs using supercritical fluid chromatography and supercritical fluid chromatography–mass spectrometry. *J. Chromatogr.*, **540**, 257–270.

Perkins, J.R., Games, D.E., Startin, J.R. and Gilbert, J. (1991b) Analysis of sulphonamides using supercritical fluid chromatography and supercritical fluid chromatography–mass spectrometry. *J. Chromatogr.*, **540**, 239–256.

Pietta, P., Facino, R.M., Carini, M. and Mauri, P. (1994) Thermospray liquid chromatography–mass spectrometry of flavonol glycosides from medicinal plants. *J. Chromatogr.*, **661**, 121–126.

Pleasance, S., Quilliam, M.A., Freitas, A.S.W., Marr, J.C. and Cembella, A.D. (1990) Ion-spray mass spectrometry of marine toxins. II. Analysis of diarrhetic shellfish toxins in plankton by liquid chromatography/mass spectrometry. *Rapid Commun. Mass Spectrom.*, **4**, 206–213.

Pleasance, S., Blay, P., Quilliam, M.A. and O'Hara, G. (1991) Determination of sulphonamides by liquid chromatography, ultraviolet diode array detection and ion-spray tandem mass spectrometry with application to cultured salmon flesh. *J. Chromatogr.*, **558**, 155–173.

Pleasance, S., Ayer, S.W., Laycock, M.V. and Thibault, P. (1992a) Ionspray mass spectrometry of marine toxins. III. Analysis of paralytic shellfish poisoning toxins by flow-injection analysis, liquid chromatography/mass spectrometry and capillary electrophoresis/mass spectrometry. *Rapid Commun. Mass Spectrom.*, **6**, 14–24.

Pleasance, S., Thibault, P. and Kelly, J. (1992b) Comparison of liquid-junction and coaxial interfaces for capillary electrophoresis–mass spectrometry with application to compounds of concern to the aquaculture industry. *J. Chromatogr.*, **591**, 325–339.

Porter, S. (1993) The confirmation of tetracycline residues in kidney by thermospray liquid chromatography–mass spectrometry. In *Proceedings of the EuroResidue II Conference*, vol. 2, Veldhoven, The Netherlands, May 1993 (edited by N. Haagama, A. Ruiter and P.B. Czedik-Eysenberg), pp. 533–537.

Quilliam, M.A., Thomson, B.A., Scott, G.J. and Siu, K.W.M. (1989) Ion-spray mass spectrometry of marine neurotoxins. *Rapid Commun. Mass Spectrom.*, **3**, 145–150.

Rajakyla, E., Laasasenaho, K. and Sakkers, P.J.D. (1987) Determination of mycotoxins in grain by high performance liquid chromatography and thermospray liquid chromatography–mass spectrometry. *J. Chromatogr.*, **348**, 391–402.

Reiser, R.W., Barefoot, A.C., Dietrich, R.F., Fogiel, A.J., Johnson, W.R. and Scott, M.T. (1991) Application of microcolumn liquid chromatography–continuous-flow fast atom bombardment mass spectrometry in environmental studies of sulfonylurea herbicides. *J. Chromatogr.*, **554**, 91–101.

Rhijn, van, J.A., Traag, W.A., Heskamp, H.H. and Zuidema, T. (1993) LC–MS–MS analysis of β-agonists, a multi-component approach. In *Proceedings of the EuroResidue II Conference*, vol. 2, Veldhoven, The Netherlands, May 1993 (edited by N. Haagama, A. Ruiter and P.B. Czedik-Eysenberg), pp. 572–575.

Riond, J.-L., Tyczkowska, K. and Riviere, J.E. (1989) Pharmokinetics and metabolic inertness of doxycline in calves with mature or immature rumen function. *Am. J. Vet. Res.*, **50**, 1329–1333.

Rosen, J.D. (1987) *Applications of New Mass Spectrometry Techniques in Pesticide Chemistry.* Wiley Interscience: New York.

Rule, G.S., Mordehai, A.V. and Henion, J. (1994) Determination of carbofuran by on-line immunoaffinity chromatography with coupled-column liquid chromatography/mass spectrometry. *Anal. Chem.*, **66**, 230–235.

Sadoun, F., Virelizier, H. and Arpino, P.J. (1993) Packed-column supercritical fluid chromatography coupled with electrospray ionization mass spectrometry. *J. Chromatogr.*, **647**, 351–359.

Safe, S. and Hutzinger, O. (1979) *Mass Spectrometry of Pesticides and Pollutants*. CRC Press Inc., Boca Raton, Florida.

Sanders, P. and Delepine, B. (1994) Confirmatory analysis for spiramycin residue in bovine muscle by liquid chromatography/particle beam mass spectrometry. *Biolog. Mass Spectrom.*, **23**, 369–375.

Schneider, R.P., Lynch, M.J., Ericson, J.F. and Fouda, H.G. (1991) Electrospray ionization mass spectrometry of semduramicin and other polyether ionophores. *Anal. Chem.*, **63**, 1789–1794.

Scudamore, K.A., Hetmanski, M.T. and Clarke, P.A. (1996) Development of analytical methods for the determination of sterigmatocystin in cheese, bread and corn products. *Food Addit. Contam.*, **13**, 343–358.

Shalaby, L.M. and George, S.W. (1990) Multiresidue analysis of thermally labile sulfonylurea herbicides in crops by liquid chromatography/mass spectrometry. In *Liquid Chromatography/ Mass Spectrometry—Applications in Agricultural, Pharmaceutical and Environmental Chemistry*, ACS Symposium Series no. 20 (edited by M.A. Brown), American Chemical Society: Washington, DC, pp. 75–91.

Shepherd, M.J. (1991) Analysis of veterinary drug residues in edible animal products. In *Food Contaminants—Sources and Surveillance* (edited by C.S. Creaser and R. Purchase), Royal Society of Chemistry: Cambridge, pp. 109–176.

Shepherd, M.J. and Gilbert, J. (1984) An investigation of HPLC post-column iodination conditions for the enhancement of aflatoxin B_1 fluorescence. *Food Addit. Contam.*, **1**, 325–335.

Shibamoto, T. and Bjeldanes, L.F. (1993) Pesticide residues in food. In *Introduction to Food Toxicology*, Academic Press: San Diego, pp. 141–156.

Slobodnik, J., van Baar, B.L.M. and Brinkman, U.H.Th. (1995) Column liquid chromatography–mass spectrometry: selected techniques in environmental applications for polar pesticides and related compounds. *J. Chromatogr. A*, **703**, 81–121.

Smith, R.D., Udseth, H.R. and Wright, B.W. (1985) Rapid and high resolution capillary supercritical fluid chromatography (SFC) and SFC/MS of trichothecene mycotoxins. *J. Chromatogr. Sci.*, **23**, 192–199.

Smith, R.M. (ed.) (1988) *Supercritical Fluid Chromatography*. Royal Society of Chemistry Chromatography Monographs, RSC: London.

Stan, H.-J. (1990) Pesticides. In *Principles and Applications of Gas Chromatography in Food Analysis* (edited by M.H. Gordon), Ellis Horwood: New York, pp. 261–325.

Stan, H.-J. and Kellner, G. (1989) Confirmation of organophosphorus pesticide residues in food applying gas chromatography/mass spectrometry with chemical ionization and pulsed positive negative detection. *Biomed. Environ. Mass Spectrom.*, **18**, 645–651.

Straub, R.F. and Voyksner, R.D. (1993) Determination of penicillin G, ampicillin, amoxicillin, cloxacillin, and cephapirin by high-performance liquid chromatography–electrospray mass spectrometry. *J. Chromatogr.*, **647**, 167–181.

Straub, R., Linder, M. and Voyksner, R.D. (1994) Determination of β-lactam residues in milk using perfusive-particle liquid chromatography combined with ultrasonic nebulization electrospray mass spectrometry. *Anal. Chem.*, **66**, 3651–3658.

Suwanrumpha, S. and Freas, R.B. (1989) Identification of metabolites of ampicillin using liquid chromatography/thermospray mass spectrometry and fast atom bombardment tandem mass spectrometry. *Biomed. Mass Spectrom.*, **18**, 983–994.

Tiebach, R., Blass, W., Steinmeyer, S. and Weber, R. (1985) Confirmation of nivalenol and deoxynivalenol by on-line liquid chromatography–mass spectrometry and gas chromatography–mass spectrometry. Comparison of methods. *J. Chromatogr.*, **318**, 103–111.

Tinke, A.P., Reinhoud, N.J., Niessen, W.M.A., Tjaden, U.R. and van der Greef, J. (1992) On-line isotachophoretic analyte focusing for improvement of detection limits in capillary electrophoresis/electrospray mass spectrometry. *Rapid Commun. Mass Spectrom.*, **6**, 560–563.

Turesky, R.J., Huynh-Ba, T., Aeschbacher, H.U. and Milon, H. (1988) Analysis of mutagenic heterocyclic amines in cooked beef products by high-performance liquid chromatography in combination with mass spectrometry. *Food Chem. Toxic.*, **26**, 501–509.

Tyczkowska, K.L., Voyksner, R.D. and Aronson, A.L. (1991) Development of an analytical method for cephapirin and its metabolite in bovine milk and serum by liquid chromatography with UV-VIS detection and confirmation by thermospray mass spectrometry. *J. Vet. Pharmacol. Therap.*, **14**, 51–60.

Tyczkowska, K.L., Voyksner, R.D., Straub, R.F. and Aronson, A.L. (1994) Simultaneous multi-residue analysis of β-lactam antibiotics in bovine milk by liquid chromatography with ultra-violet detection and confirmation by electrospray mass spectrometry. *J. Assoc. Offic. Anal. Chem. Internat.*, **77**, 1122–1131.

Van der Greef, J., Arts, C.J.M., Van Baak, M. and Verheij, E.R. (1993) Possibilities of liquid chromatography-mass spectrometry in drug residue analysis. In *Proceedings of the Euro-Residue II Conference*, vol. 1, Veldhoven, The Netherlands, May 1993 (edited by N. Haagama, A. Ruiter, and P.B. Czedik-Eysenberg), pp. 35–40.

Van Ginkel, L.A., Jansen, E.H.J.M., Stephany, R.W., Zoontjes, P.W., Schwillens, P.L.W.J., van Rossum, H.J. and Visser, T. (1992) Liquid chromatographic purification and detection of anabolic compounds. *J. Chromatogr.*, **624**, 389–401.

Volmer, D., Preiss, A., Levsen, K. and Wunsch, G. (1993) Thermospray mass spectral studies of pesticides. Temperature and salt concentration effects on the ion abundancies in thermospray mass spectra. *J. Chromatogr.*, **647**, 235–259.

Voyksner, R.D., Hagler, W.M., Tyczkowska, K. and Haney, C.A. (1985) Thermospray high performance liquid chromatographic/mass spectrometric analysis of some *Fusarium* mycotoxins. *J. High Res. Chromatogr. Chromatogr. Commun.*, **8**, 119–125.

Voyksner, R.D. and Haney, C.A. (1985) Optimization and application of thermospray high-performance liquid chromatography/mass spectrometry. *Anal. Chem.*, **57**, 991–996.

Voyksner, R.D., Smith, C.S. and Knox, P.C. (1990) Optimization and application of particle beam high-performance liquid chromatography/mass spectrometry to compounds of pharmaceutical interest. *Biomed. Environ. Mass Spectrom.*, **19**, 523–534.

Voyksner, R.D., Tyczkowska, K. and Aronson, A.L. (1991) Development of analytical methods for some penicillins in bovine milk by ion-paired chromatography and confirmation by thermospray mass spectrometry. *J. Chromatogr.*, **567**, 389–404.

Wilkins, J.P.G. (1993a) Unpublished results.

Wilkins, J.P.G. (1993b) Determination of residues of the insecticide diflubenzuron in foodstuffs using liquid chromatography–thermospray mass spectrometry and identification of some natural components found in chillies. *Anal. Proc.*, **30**, 396–397.

Wright, B.W., Kalinoski, H.T. and Smith, R.D. (1986) Critical review: Capillary supercritical fluid chromatography–mass spectrometry. *J. High Res. Chromatogr. Chromatogr. Commun.*, **9**, 145–153.

Yamaizumi, Z., Kasai, H., Nishimura, S., Edmonds, C.G. and McCloskey, J.A. (1986) Stable isotope dilution quantification of mutagens in cooked foods by combined liquid chromatography–thermospray mass spectrometry. *Mutation Res.*, **173**, 1–7.

8 Analysis of foods and biological samples for dioxins and PCBs by high resolution gas chromatography–mass spectrometry

DONALD J. HANNAH, LAWRENCE J. PORTER and SIMON J. BUCKLAND

Summary

Analysis of food and biological matrices for polychlorinated dibenzo-*p*-dioxins (PCDDs), dibenzofurans (PCDFs) and biphenyls (PCBs) by high-resolution gas chromatography and high-resolution mass spectrometry has developed rapidly in recent years. The understanding of relative toxicity of the various congeners has led to the development of selective and sensitive methods that are able to provide relevant data cost-effectively, and with a high level of quality assurance. The methods are based on stable-isotope dilution mass spectrometry with the mass spectrometer operating in the selected ion recording (SIR) mode, and have been developed into regulatory methods such as the USEPA methods 1613 and 8290. Key to the methods are the clean-up steps, where the matrix is concentrated while retaining the key analytes. Concentration factors of up to 1000 are obtained with analyte recovery generally exceeding 80%. Few new techniques have been developed for the clean-up of sample matrices; however, existing techniques have been automated successfully. Capillary gas chromatography columns have been improved, in both the design of the stationary phases and the robustness of the columns. However, full isomer-specific separation of all the 2,3,7,8-substituted PCDDs and PCDFs, and the co-planar PCBs, still requires the use of two columns. Effective compromise can be achieved with those columns designed specifically for dioxin analysis, especially for the separation of the tetrachlorinated dioxins and furans. High-resolution SIR mass spectrometry generally operates with 10 000 resolving power, with some applications requiring up to 22 000 resolving power. Modern data systems enable result calculations to be readily made, with significant increases in quality assurance data being achieved at the sub-picogram per gram (parts per trillion, ppt) level.

8.1 Introduction

This contribution updates and covers in more detail the chapter *Applications of Quantitative Mass Spectrometry in Food Science* by Gilbert (1987). At that time high resolution (HR) selected ion recording (SIR) gas chromatography–mass spectrometry (GC–MS) of analytes was a rapidly developing technique, and Gilbert (1987) described in some detail the rationale and limitations of the method. HRGC–HRMS SIR is now well established and is the method of choice for dioxin laboratories, with over 100 high-resolution mass spectrometers worldwide dedicated to dioxin and other similar target environmental contaminant analyses.

The following topics will be discussed in more detail in this chapter:

- Nomenclature of dioxins, dibenzofurans and PCBs, with reference to allied components present in samples as contaminants or substitutes.
- Extraction and clean-up methodologies for isomer-specific dioxin, dibenzofuran and PCB analysis in foods and other biological samples.
- Analysis of these analytes by GC–MS, including method validation and quality assurance.
- Some examples of the determination of these analytes in foods and other biological samples.

To place a reasonable limit on the scope of this chapter, biological samples are taken to exclude those from human origin (adipose tissue, breast milk, etc.) and materials with possible biological activity (sediments, soil, water, air, etc.).

8.2 Nomenclature

This chapter covers the analysis of polychlorinated isomers of three-ring systems: biphenyl **(1)**, dibenzo-*p*-dioxin **(2)** and dibenzofuran **(3)** usually called 'dioxins' in the general literature. Some of these chlorinated isomers possess demonstrable toxicity, the most toxic of them being the well-known 2,3,7,8-tetrachlorodibenzo-*p*-dioxin **(4)**. Generally, toxicities of members of these classes of compounds are all expressed relative to compound **4**, and isomers from all classes demonstrating such toxicities are known collectively as 'dioxin-like' compounds (USEPA, 1994). When present in

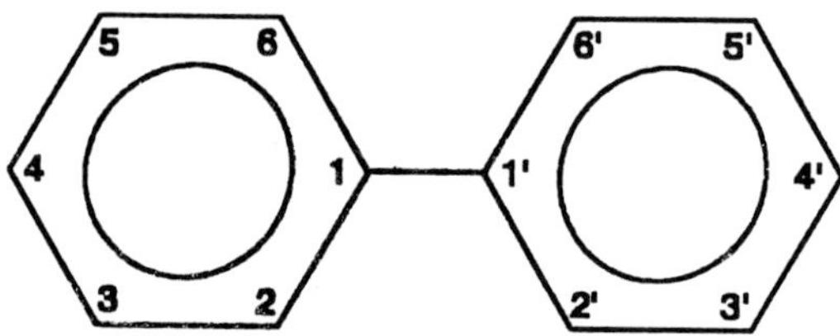

1

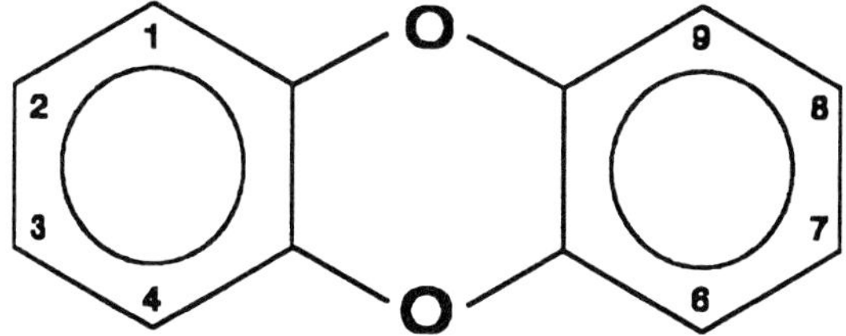

2

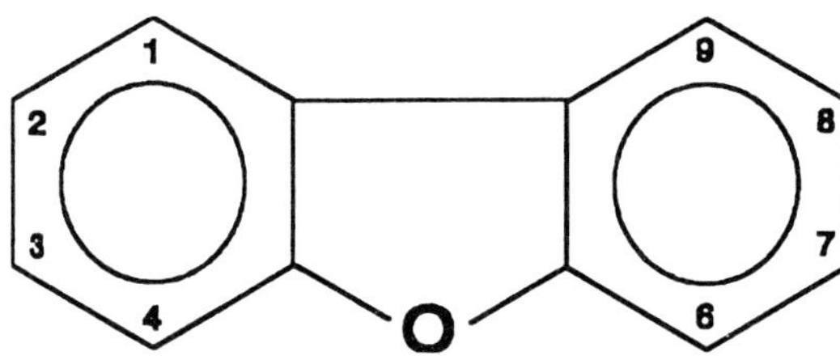

3

4

an extract or matrix their concentrations are commonly expressed in terms of total 2,3,7,8-TCDD toxic equivalents (see later).

Substituted PCBs are named from the numbering system in structure (1). They are generally identified in the environmental chemistry literature by utilising the IUPAC numbering system, which assigns all PCBs a unique sequential number according to increasing chlorine substitution from PCB#0 (biphenyl) to the fully substituted decachlorobiphenyl, PCB#209. This system, originally developed by Ballschmiter and Zell (1980), is utilised in Table 8.1, which refers to selected PCBs by both their systematic name and their IUPAC PCB number.

Polychlorinated dibenzo-*p*-dioxins (PCDDs) and dibenzofurans (PCDFs) are simply identified by the number and placement of chlorine substituents, using the numbering in structures (2) and (3). Table 8.2 gives selected examples.

Table 8.1 Structures, IUPAC numbers and toxicity equivalent factors for PCBs

Type	Congener		TEF
	IUPAC no.	Structure	
Non-*ortho*	77	3,3′,4,4′-TCB	0.0005
	126	3,3′,4,4′,5-PeCB	0.1
	169	3,3′,4,4′,5,5′-HxCB	0.01
Mono-*ortho*	105	2,3,3′,4,4′-PeCB	0.0001
	114	2,3,4,4′,5-PeCB	0.0005
	118	2,3′,4,4′,5-PeCB	0.0001
	123	2′,3,4,4′,5-PeCB	0.0001
	156	2,3,3′,4,4′,5-HxCB	0.0005
	157	2,3,3′,4,4′,5′-HxCB	0.0005
	167	2,3′,4,4′,5,5′-HxCB	0.00001
	189	2,3,3′,4,4′,5,5′-HpCB	0.0001
Di-*ortho*	170	2,2′,3,3′,4,4′,5-HpCB	0.0001
	180	2,2′,3,4,4′,5,5′-HpCB	0.00001

The prime purpose for the analysis of dioxin-like compounds is to measure their concentration in matrices so as to assess the degree of health or environmental risk presented by such samples. The fact that dioxin-like compounds normally occur in environmental or biological samples as complex mixtures of congeners has led to the concept of toxicity equivalency factors (TEF) so as to simplify risk assessment and expedite regulatory control. TEF (used to calculate the total toxic equivalents, TEQ, in a sample) are estimated by measuring the toxicities of dioxin-like compounds

Table 8.2 Structures and toxicity equivalent factors for 2,3,7,8-substituted PCDDs and PCDFs

Congener	TEF	
	NATO/CCMS	Nordic
2,3,7,8-TCDD	1	1
1,2,3,7,8-PeCDD	0.5	0.5
1,2,3,4,7,8-HxCDD	0.1	0.1
1,2,3,6,7,8-HxCDD	0.1	0.1
1,2,3,7,8,9-HxCDD	0.1	0.1
1,2,3,4,6,7,8-HpCDD	0.01	0.01
OCDD	0.001	0.001
2,3,7,8-TCDF	0.1	0.1
1,2,3,7,8-PeCDF	0.5	0.1
2,3,4,7,8-PeCDF	0.5	0.5
1,2,3,4,7,8-HxCDF	0.1	0.1
1,2,3,6,7,8-HxCDF	0.1	0.1
2,3,4,6,7,8-HxCDF	0.1	0.1
1,2,3,7,8,9-HxCDF	0.1	0.1
1,2,3,4,6,7,8-HpCDF	0.01	0.01
1,2,3,4,7,8,9-HpCDF	0.01	0.01
OCDF	0.001	0.001

relative to the toxicity of 2,3,7,8-tetrachlorodibenzo-*p*-dioxin (2,3,7,8-TCDD), structure **4**, as determined in the same *in vitro* and *in vivo* studies.

There has been considerable debate over the TEF values for individual PCDDs and PCDFs other than 2,3,7,8-TCDD. This has led to a number of systems being used for risk assessment. Two of these, NATO/CCMS (1988), and Nordic (Ahlborg, 1988) are shown in Table 8.2. The NATO/CCMS set of TEF values, referred to as the International (I-TEF) values, are now the most widely used. It is of note that all the PCDD and PCDF isomers assigned TEF values have four or more chlorine atoms and contain, *inter alia*, the 2,3,7,8-substitution pattern. For this reason PCDD and PCDF analyses are generally confined to tetra- through to octa-chlorinated isomers.

The most authoritative set of TEF values for PCBs appears to be that of the WHO-ECEH and IPCS consultative committee (Ahlborg *et al.*, 1994). The PCB isomers considered currently to have TEF values are listed in Table 8.1. It is of note that these toxic isomers also contain four or more chlorine substituents, contain two pairs of vicinal chlorine atoms (as do the 2,3,7,8-substituted PCDD and PCDF isomers) and contain no more than one chlorine atom per ring adjacent to the biphenyl ring junction (1–1′) bond. It is considered that the rationale for these effects is that toxic PCBs must be able to adopt a coplanar conformation and so mimic PCDD and PCDF binding to the Ah receptor protein (Ahlborg *et al.*, 1994).

There are a number of other halogenated environmental contaminants that may mimic PCDDs and PCDFs and so contribute to the dioxin-like toxicity of a matrix. These include the brominated, or mixed brominated/chlorinated, analogues of PCDDs and PCDFs. Others are chlorinated naphthalenes **(5)**, diphenylethers **(6)** and terphenyls **(7)**. TEF values have been suggested for most of these compounds (Safe, 1992) based on structure–activity relationships, but such values have yet to be widely accepted and used.

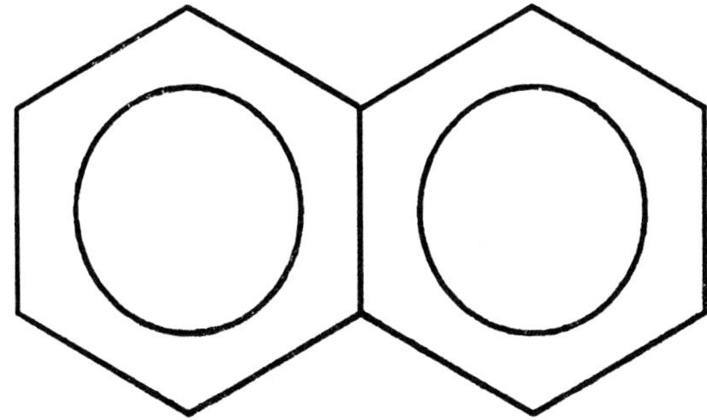

5

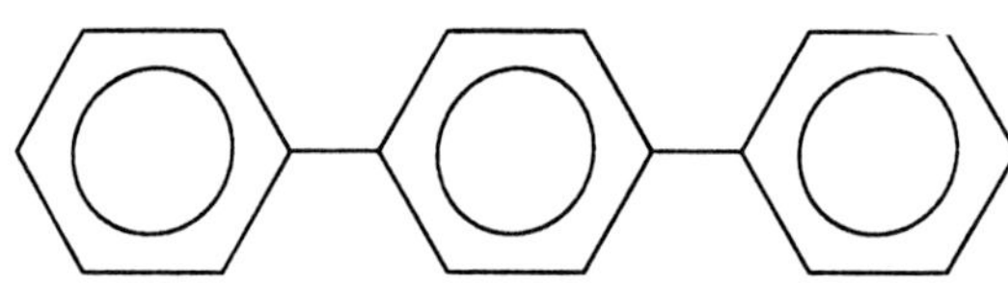

6

7

8.3 Sample preparation and clean-up

The methods applied to the preparation, clean-up and enrichment of samples for PCB, PCDD and PCDF analysis have undergone few modifications in the last eight years. PCBs, PCDDs and PCDFs concentrate in the lipid fraction of biological samples, and the analytical methods are still primarily based on extraction of the lipid components of the sample by solvents, often with some form of acid or alkali treatment. The analytes are removed from the lipid components by a variety of techniques such as acid treatment, gel permeation chromatography, and dialysis using semipermeable membranes. The resultant extracts undergo final enrichment of the analytes using a range of solid-phase chromatographic techniques prior to analysis by high-resolution capillary gas chromatography and high-resolution mass spectrometry.

The traditional methods of lipid removal, such as partition against concentrated sulphuric acid or the use of gel-permeation chromatography with supports such as BioBeads SX-3, are widely reported in the literature. Recently, a more innovative method has been described (Meadows *et al.*, 1993) using semipermeable dialysis membranes. When the dialytic procedure is optimised, very satisfactory and highly reproducible recovery of chlorinated contaminants can be obtained in about three days with over 90% of the lipid material removed in a passive, single operation. The technique is suited to situations where lipid removal to effect the first stages of analyte enrichment requires as little treatment of the sample as possible.

Enrichment techniques are critical to PCB, PCDD and PCDF analysis, to enable the often very low levels of these compounds to be detected. These use low-pressure chromatography on various combinations of alumina, neutral and acidified silica, Florisil and carbon supports. There are some methods that employ normal- and reverse-phase HPLC, but these are now used less as they do not allow for the full range of PCBs, PCDDs and PCDFs to be as easily separated as on other support materials.

The key material for enabling enrichment of 5–50 g of sample to a final volume of 10 μl is carbon. Several applications of this are noted widely in the literature, such as charcoal, Carbopak® or Carbosphere® on diatomaeous earth, APX-21® on glass fibres, or Carbosieve® material. In each case the technique is adapted to enable the planar, chlorinated hydrocarbon classes to be selectively, specifically and sequentially eluted from the support material with the minimum of solvent. Since this step is usually the most difficult to perfect, laboratories which have developed their own applications tend to be reluctant to change to other methods without good cause. A flow chart of a typical sequence of clean-up and enrichment steps, as used in our laboratory, is outlined in Figure 8.1.

In recent years the clean-up and sample enrichment processes have been further developed by incorporation into automated systems (Turner *et al.*, 1992). A commercial version of this system (FMS DioxinPrep System™) enables the defatted extracts from up to six samples to be treated, in unattended operation, by alumina, neutral and acidified silica, and carbon, to prepare final extracts for GC–MS analysis. Modifications to the basic PCDD and PCDF system allow for the coplanar PCBs to be separated as well.

8.4　GC–MS methodology

8.4.1　*Chromatographic requirements*

The GC–MS analysis of dioxin-like compounds places considerable demands on both chromatographic and MS performance. Firstly, a considerable number of congeners for each class must be separated completely: PCDD (49, tetra to octa), PCDF (87, tetra to octa) and PCB (172, tetra to deca)— number of congeners and degree of chlorine substitution in parentheses. In particular, those isomers with known toxicity must be unequivocally separated and quantified. Secondly, the chromatographic delivery system must transfer analytes to the mass spectrometer with minimum losses, as all target analytes must be detected at the sub-picogram (sub-femtomole) level.

In common with most other environmentally significant organic compounds, it is now universally accepted that the best chromatographic performance is delivered by high-performance (i.e. high resolution) capillary columns. In a detailed study involving chromatography of all 136 tetra- to octa-chlorinated PCDD and PCDF isomers, Ryan *et al.* (1991) assessed

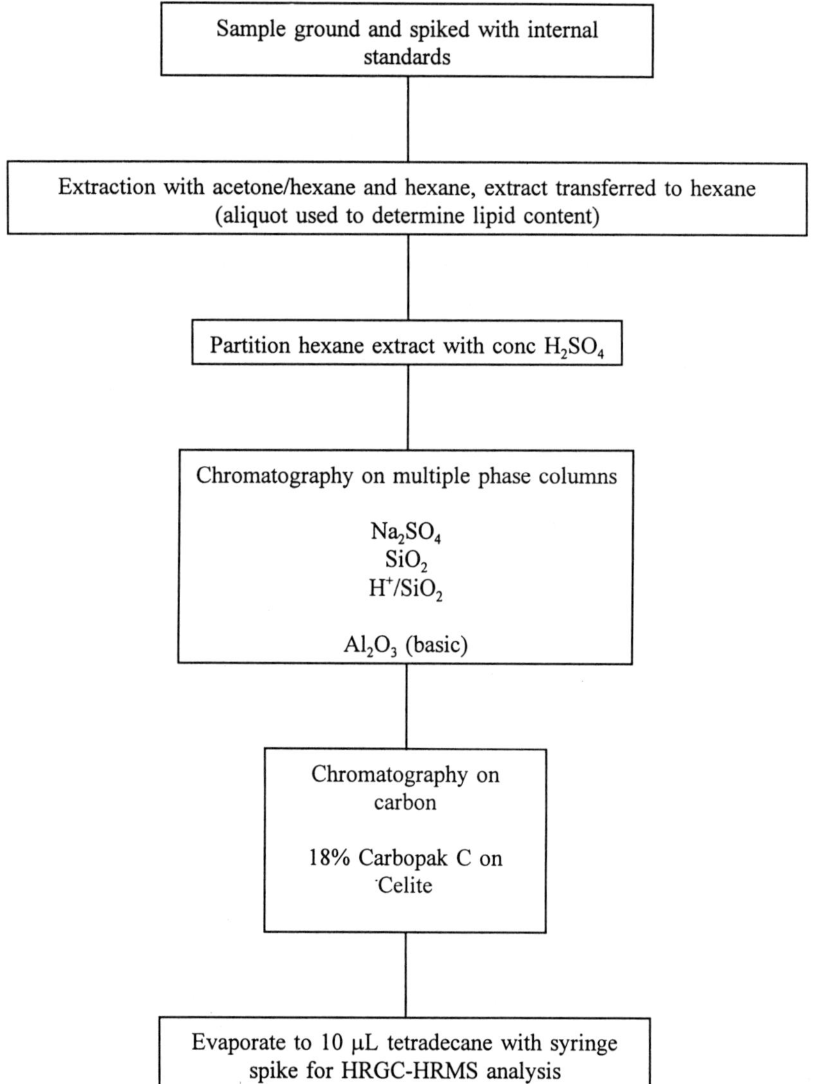

Figure 8.1 Diagram of typical clean-up procedure.

the comparative performance of ten commercially available bonded-phase capillary columns. They concluded that no single column would separate all isomers, but most could be separated by various combinations of two different columns. In particular, such an approach will provide unequivocal separation of all toxic 2,3,7,8-chlorinated PCDD and PCDF congeners. Currently most dioxin laboratories use a combination of a non-polar 5% phenyl methyl silicone column (i.e. DB-5, HP-5, Ultra 2, Rtx-5, etc.) for their primary analytical separation, supplemented with a polar cyanopropyl silicone column (i.e. SP-2331, SP-2330, or Rtx-2330) to effect the separation of all the tetra- to hexa-2,3,7,8-substituted PCDD and PCDF isomers completely.

A 5% phenyl methyl phase column is very useful in the analysis of PCDDs and PCDFs, as it separates each homologue group (tetra, penta, hexa, hepta, octa) cleanly and sequentially from each other, thus expediting ready estimation of the total concentration of each class. The supplementary separation on a polar cyanopropyl (SP-2331) phase column must be used to confirm the concentration of 2,3,7,8-TCDD, 2,3,7,8-TCDF, 2,3,4,7,8-PeCDF, and 1,2,3,7,8,9-HxCDD in complex mixtures. Finally the 1,2,3,4,7,8-HxCDF isomer is not completely separated from other isomers on either column type, co-eluting with 1,2,3,4,6,7-HxCDF on the 5% phenyl methyl silicone column and with 1,2,3,4,7,9-HxCDF on the cyanopropyl silicone column. A suitable correction calculation must therefore be performed to determine the concentration of 1,2,3,4,7,8-HxCDF in the sample.

Various other columns have been suggested for PCDD and PCDF analyses, including DB-DIOXIN and DB5MS columns. While these have their place in PCDD and PCDF analyses, they have not as yet supplanted the above combination of columns in achieving full isomer-specific separation. This is not to say that the alternatives do not have their advantages; for instance, the DB5MS column gives superior separation of tetrachlorinated isomers.

Our own approach is to use a 25 m HP Ultra 2 column for the primary dioxin analysis supplemented with a 60 m SP2331 column separation. The Ultra 2 column appears to give the best transmission and chromatography of the more highly chlorinated (hexa to octa) PCDD and PCDF isomers and a very low column bleed background ionisation, which is important for achieving low detection limits.

Various columns have been suggested for PCB analysis, but the general consensus is that a 5% phenyl methyl silicone column is the most generally useful (Duinker and Hillebrand, 1983; Mullin *et al.*, 1984; Erickson, 1986; Schultz *et al.*, 1989; Draper *et al.*, 1991; Storr-Hansen, 1991). We generally use a 25 m Ultra 2 column for PCB analyses, as for PCDDs and PCDFs.

The requirement for maximum sensitivity to yield the lowest possible detection limits for dioxin-like compounds can only be achieved by splitless or on-column injection, as these methods deliver, theoretically, all the analyte solution injected on the GC column. The former technique is most commonly used. Grob (1994) has recently reviewed the critical parameters involved to achieve efficient injection on to capillary columns. Typically injection liners have 2–4 mm i.d. and corresponding volumes of 0.25–1 ml. Injection of $2\,\mu l$ of a solvent typically used for dioxin analyses leads to expanded volumes of vapour in the liner of 0.2–0.5 ml, which obviously may exceed the volume of a narrow-bore liner. An obvious solution is to

Table 8.3 Typical chromatographic conditions for PCDDs and PCDFs on a 25 m Ultra 2 capillary column

Parameter	Value
Injection mode	Splitless
Injector temperature	250°C
MS transfer line	300°C
Injector purge time	0.9 min
Temperature #1	170°C
Time #1	4.0 min
Rate #1	10.0°C/min
Temperature #2	210°C
Time #2	0.1 min
Rate #2	3.0°C/min
Temperature #3	275°C
Time #3	12.0 min
Rate #3	0.0°C/min

confine injections to wide-bore liners, but this must be balanced against possible loss of chromatographic resolution and even sensitivity.

Grob (1994) points out that there are considerable advantages in placing a small amount (approximately 1 mg) of silanised glass wool in the liner, which results in improved chromatographic peak shapes and even sensitivity. This view is supported by our own experience, where considerably improved peak shapes for hexa- to octa-PCDDs and -PCDFs are obtained with glass wool present in the liner. As a general rule of thumb, injections should be limited to 1 μl for narrow-bore liners, which should contain a small amount of evenly-packed silanised glass wool. Wider-bore liners should be used for 2 μl injections.

We generally use tetradecane to prepare PCDD and PCDF analytical sample solutions because of its low volatility, enabling working solutions with 10–20 μl volumes to be prepared with little danger of significant solvent loss subsequent to puncturing the sample vial septum. A solvent with greater volatility (hexane or toluene) is normally used for PCB analyses because of the much wider range of analyte volatilities involved, i.e. lower chlorinated PCB congeners would not be separated from tetradecane.

Typical chromatographic conditions for PCDD and PCDF analyses on an Ultra 2 column are given in Table 8.3. The retention times listed for target analytes in these classes in Table 8.4 were obtained under these conditions. The temperature program used to separate PCBs is similar but starts at 60°C. Typical retention times are given in Table 8.5.

8.4.2 *Mass spectral requirements*

In spite of many studies exploring the efficacy of quadrupole MS (in both the positive- and the negative-ion mode) and various MS–MS configurations for PCDD and PCDF analyses, high resolution GC–high resolution MS (HRGC–HRMS) operated in the SIR (selected ion recording, also known as SIM, selected ion monitoring, or SICP, selected ion current profile, in the USEPA, 1990, Method 1613 document) mode, remains the method of choice for dioxin laboratories. This analytical configuration is still the only one that can produce reliable analytical data at the required detection level (sub-picogram amounts of individual PCDD and PCDF isomers injected on column) against a background of other persistent chlorinated analytes which often co-occur with PCDDs and PCDFs (PCBs, diphenyl ethers, etc.). Table 8.6 gives a list of compounds which may co-occur with PCDDs and PCDFs and the resolution required to achieve separation between the target ion(s) of the analyte and contaminant. The internationally accepted resolving power (RP) used for PCDD and PCDF analyses is 10 000 (10% valley definition) and it may be observed that this is sufficient to resolve most contaminants. There are documented examples where a higher RP (22 000) may be necessary (Hart and Patterson, 1993).

Methods of PCDD and PCDF analysis used by most dioxin laboratories follow USEPA Methods 1613 or 8290, or are at least very close to these protocols. These require that quantitation is achieved by stable-isotope dilution HRGC–HRMS using $^{13}C_{12}$ analogues of the 2,3,7,8-substituted isomers. Such isomers are available for all the 17 targeted 2,3,7,8-substituted isomers, but in practice, 15 are used to quantify the PCDDs and PCDFs. Those excluded are $^{13}C_{12}$-OCDF which has ions which interfere with the identification and quantification of the native OCDD, and $^{13}C_{12}$-1,2,3,7,8,9-HxCDD, which is used as an internal standard, i.e. is added after sample extraction, together with $^{13}C_{12}$-1,2,3,4-TCDD. The MS experiment proceeds by monitoring two ions per analyte selected from the molecular-ion isotope cluster (generally the most abundant pair). Analytes are identified by comparison of their GC retention times and the experimentally observed ion abundance ratios for the characteristic pairs of m/z values for each congener, with the retention time ranges of authentic standards and the theoretical ion abundance ratios of the exact m/z values.

By using a GC column or columns capable of resolving all the 2,3,7,8-substituted isomers from all other isomers (see previous section), these isomers are identified when the retention time and m/z abundance ratios agree within predefined limits of the retention times and exact m/z ratios of authentic standards. Table 8.4 illustrates these points for the analysis of PCDDs and PCDFs on an Ultra 2 column. This gives the retention time and relative abundance tolerances for the PCDD and PCDF target analytes and characteristic m/z values for each analyte. The 'function' value (Fn) 1–5 defines the retention time window for each homologue group of analytes (tetra- to octa-chlorinated). Table 8.7 illustrates the equivalent information for an SP2331 column acquisition.

Table 8.4 Parameter table for the processing of PCDD and PCDF data acquired on a Hewlett-Packard Ultra 2 capillary column and VG Analytical VG70 using the VG Analytical OPUSquan^{TM} data processing system.

Ent	Type	Name	Method	Std	Multiply by factor			RT	Win.	Std	Fn	m1	m2	m1/m2	Tol.
					Amo	1	2								
1	Unk	2,3,7,8-TCDF	rel_int	18	1.0	y	n	16:15	10.0	18	1	305.899	303.902	1.30	0.20
2	Unk	1,2,3,7,8-PeCDF	rel_int	19	1.0	y	n	19:53	10.0	18	2	339.860	337.863	1.52	0.25
3	Unk	2,3,4,7,8-PeCDF	rel_int	20	1.0	y	n	20:43	10.0	18	2	339.860	337.863	1.52	0.25
4	Unk	1,2,3,4,7,8-HxCDF	rel_int	21	1.0	y	n	24:11	10.0	23	3	373.821	375.818	1.24	0.15
5	Unk	1,2,3,6,7,8-HxCDF	rel_int	22	1.0	y	n	24:20	5.0	21	3	373.821	375.818	1.24	0.15
6	Unk	2,3,4,6,7,8-HxCDF	rel_int	21	1.4	y	n	25:05	10.0	33	3	373.821	375.818	1.24	0.15
7	Unk	1,2,3,7,8,9-HxCDF	rel_int	23	1.0	y	n	26:01	10.0	33	3	373.821	375.818	1.24	0.15
8	Unk	1,2,3,4,6,7,8-HpCDF	rel_int	24	1.0	y	n	28:19	10.0	33	4	407.782	409.779	1.04	0.15
9	Unk	1,2,3,4,7,8,9-HpCDF	rel_int	25	1.0	y	n	30:15	10.0	33	4	407.782	409.779	1.04	0.15
10	Unk	OCDF	rel_int	33	0.8	y	n	34:50	10.0	33	5	443.740	441.743	1.12	0.15
11	Unk	2,3,7,8-TCDD	rel_int	26	1.0	y	n	16:50	10.0	18	1	321.894	319.897	1.30	0.20
12	Unk	1,2,3,7,8-PeCDD	rel_int	28	1.0	y	n	21:08	10.0	18	2	355.855	353.858	1.52	0.25
13	Unk	1,2,3,4,7,8-HxCDD	rel_int	29	1.0	y	n	25:16	10.0	31	3	389.816	391.813	1.24	0.15
14	Unk	1,2,3,6,7,8-HxCDD	rel_int	30	1.0	y	n	25:24	5.0	29	3	389.816	391.813	1.24	0.15
15	Unk	1,2,3,7,8,9-HxCDD	rel_int	31	1.0	y	n	25:47	10.0	33	3	389.816	391.813	1.24	0.15
16	Unk	1,2,3,4,6,7,8-HpCDD	rel_int	32	1.0	y	n	29:46	10.0	33	4	423.777	425.774	1.04	0.15
17	Unk	OCDD	rel_int	33	1.0	y	n	34:41	10.0	33	5	459.735	457.738	1.12	0.15
18	IS/RT	^{13}C-2,3,7,8-TCDF	abs_int		11.0	y	n	16:14	10.0	33	1	317.939	315.942	1.30	0.20
19	IS	^{13}C-1,2,3,7,8-PeCDF	abs_int		11.3	y	n	19:52	10.0	18	2	351.900	349.903	1.52	0.25
20	IS	^{13}C-2,3,4,7,8-PeCDF	abs_int		11.0	y	n	20:42	10.0	18	2	351.900	349.903	1.52	0.25
21	IS	^{13}C-1,2,3,4,7,8-HxCDF	abs_int		9.6	y	n	24:10	10.0	23	3	385.861	387.858	1.23	0.15
22	IS	^{13}C-1,2,3,6,7,8-HxCDF	abs_int		9.7	y	n	24:19	5.0	21	3	385.861	387.858	1.23	0.15
23	IS	^{13}C-1,2,3,7,8,9-HxCDF	abs_int		12.2	y	n	26:00	10.0	33	3	385.861	387.858	1.23	0.15
24	IS	^{13}C-1,2,3,4,6,7,8-HpCDF	abs_int		7.1	y	n	28:18	10.0	33	4	419.822	421.819	1.04	0.15
25	IS	^{13}C-1,2,3,4,7,8,9-HpCDF	abs_int		7.1	y	n	30:15	10.0	33	4	419.822	421.819	1.04	0.15

26	IS	^{13}C-2,3,7,8-TCDD	abs_int		11.9	y	n	16:50	10.0	18	1	333.934	331.937	1.30	0.20
27	SS	^{13}C-1,2,3,4-TCDD	abs_int		1.0	y	n	16:40	10.0	18	1	333.934	331.937	1.30	0.20
28	IS	^{13}C-1,2,3,7,8-PeCDD	abs_int		9.0	y	n	21:07	10.0	18	2	367.895	365.898	1.52	0.25
29	IS	^{13}C-1,2,3,4,7,8-HxCDD	abs_int		9.7	y	n	25:15	10.0	31	3	401.856	403.853	1.24	0.15
30	IS	^{13}C-1,2,3,6,7,8-HxCDD	abs_int		12.2	y	n	25:23	5.0	29	3	401.856	403.853	1.24	0.15
31	IS	^{13}C-1,2,3,7,8,9-HxCDD	abs_int		11.1	y	n	25:46	10.0	33	3	401.856	403.853	1.24	0.15
32	IS	^{13}C-1,2,3,4,6,7,8-HpCDD	abs_int		7.8	y	n	29:45	10.0	33	4	435.817	437.814	1.04	0.15
33	IS/RS	^{13}C-OCDD	abs_int		31.5	y	n	34:40	60.0		5	471.775	469.778	1.12	0.15
34	Tot	Total non-tetra-furans	rel_int	18	0.0	n	n	—	—		1	305.899	303.902	1.30	0.20
35	Tot	Total non-tetra-dioxins	rel_int	26	0.0	n	n	—	—		1	321.894	319.897	1.30	0.20
36	Tot	Total non-penta-furans	rel_int	19	0.0	n	n	—	—		2	339.860	337.863	1.52	0.25
37	Tot	Total non-penta-dioxins	rel_int	28	0.0	n	n	—	—		2	355.855	353.858	1.52	0.25
38	Tot	Total non-hexa-furans	rel_int	21	0.0	n	n	—	—		3	373.821	375.818	1.24	0.15
39	Tot	Total non-hexa-dioxins	rel_int	29	0.0	n	n	—	—		3	389.816	391.813	1.24	0.15
40	Tot	Total non-hepta-furans	rel_int	25	0.0	n	n	—	—		4	407.782	409.779	1.04	0.15
41	Tot	Total non-hepta-dioxins	rel_int	32	0.0	n	n	—	—		4	423.777	425.774	1.04	0.15

Table 8.5 Parameter table for the processing of PCB data acquired on a Hewlett-Packard Ultra 2 capillary column and VG Analytical VG70 using the VG Analytical OPUSquan^{TM} data processing system.

Ent	Type	Name	Method	Std	Multiply by factor			RT	Win.	Std	Fn	m1	m2	m1/m2	Tol.
					Amo	1	2								
1	Unk	DI_PCB#8	rel_int	23	1.0	n	n	8:05	5.0	23	1	222.000	223.997	1.54	0.20
2	Unk	DI_PCB#15	rel_int	23	1.0	n	n	9:17	5.0	23	1	222.000	223.997	1.54	0.20
3	Unk	TRI_PCB#30	rel_int	24	1.0	n	n	9:12	5.0	24	1	255.961	257.958	1.04	0.20
4	Unk	TRI_PCB#28	rel_int	24	1.0	n	n	10:34	5.0	24	2	255.961	257.958	1.04	0.20
5	Unk	TET_PCB#52	rel_int	25	1.0	n	n	11:36	5.0	25	2	289.922	291.919	0.77	0.20
6	Unk	TET_PCB#80	rel_int	26	1.0	n	n	13:38	5.0	26	2	289.922	291.919	0.77	0.20
7	Unk	TET_PCB#77	rel_int	27	1.0	n	n	15:46	5.0	27	2	289.922	291.919	0.77	0.20
8	Unk	PENT_PCB#101	rel_int	28	1.0	n	n	14:29	5.0	28	2	325.880	327.877	1.56	0.20
9	Unk	PENT_PCB#99	rel_int	28	1.0	n	n	14:40	5.0	28	2	325.880	327.877	1.56	0.20
10	Unk	PENT_PCB#118	rel_int	30	1.0	n	n	16:50	5.0	29	2	325.880	327.877	1.56	0.20
11	Unk	PENT_PCB#105	rel_int	30	1.0	n	n	17:36	5.0	29	3	325.880	327.877	1.56	0.20
12	Unk	PENT_PCB#126	rel_int	29	1.0	n	n	18:45	5.0	29	3	325.880	327.877	1.56	0.20
13	Unk	HEX_PCB#153	rel_int	30	1.0	n	n	17:27	5.0	30	3	359.841	361.838	1.25	0.20
14	Unk	HEX_PCB#141	rel_int	31	1.0	n	n	17:53	5.0	31	3	359.841	361.838	1.25	0.20
15	Unk	HEX_PCB#138	rel_int	30	1.0	n	n	18:26	5.0	30	3	359.841	361.838	1.25	0.20
16	Unk	HEX_PCB#169	rel_int	32	1.0	n	n	21:42	5.0	32	3	359.841	361.838	1.25	0.20
17	Unk	HEP_PCB#187	rel_int	33	1.0	n	n	19:01	5.0	33	3	393.802	395.799	1.04	0.20
18	Unk	HEP_PCB#183	rel_int	33	1.0	n	n	19:11	5.0	33	3	393.802	395.799	1.04	0.20
19	Unk	HEP_PCB#180	rel_int	33	1.0	n	n	20:51	5.0	33	3	393.802	395.799	1.04	0.20
20	Unk	HEP_PCB#170	rel_int	33	1.0	n	n	22:04	5.0	33	3	393.802	395.799	1.04	0.20
21	Unk	OCT_PCB#202	rel_int	34	1.0	n	n	20:08	5.0	34	3	427.763	429.760	0.89	0.20
22	Unk	OCT_PCB#194	rel_int	34	1.0	n	n	25:23	30.0	34	3	427.763	429.760	0.89	0.20

23	IS	^{13}C-DI_PCB#15	abs_int	2.6	y	n	9 : 16	10.0	30	1	234.040	236.037	1.54	0.20
24	IS	^{13}C-TRI_PCB#28	abs_int	5.3	y	n	10 : 33	10.0	30	2	268.001	271.995	1.04	0.20
25	IS	^{13}C-TET_PCB#52	abs_int	6.3	y	n	11 : 35	10.0	30	2	301.962	303.959	0.77	0.20
26	IS	^{13}C-TET_PCB#80	abs_int	5.0	y	n	13 : 38	10.0	30	2	301.962	303.959	0.77	0.20
27	IS	^{13}C-TET_PCB#77	abs_int	8.0	y	n	15 : 45	10.0	30	2	301.962	303.959	0.77	0.20
28	IS	^{13}C-PENT_PCB#101	abs_int	4.8	y	n	14 : 28	10.0	30	2	337.902	339.917	1.56	0.20
29	IS	^{13}C-PENT_PCB#126	abs_int	4.8	y	n	18 : 44	10.0	30	3	337.902	339.917	1.56	0.20
30	IS/RS	^{13}C-HEX_PCB#153	abs_int	34.4	y	n	17 : 26	20.0		3	371.881	373.878	1.25	0.20
31	IS	^{13}C-HEX_PCB#l41	abs_int	5.0	y	n	17 : 53	10.0	30	3	371.881	373.878	1.25	0.20
32	IS	^{13}C-HEX_PCB#169	abs_int	4.7	y	n	21 : 41	10.0	30	3	371.881	373.878	1.25	0.20
33	IS	^{13}C-HEP_PCB#180	abs_int	22.1	y	n	20 : 50	10.0	30	3	405.843	407.839	1.04	0.20
34	IS	^{13}C-OCT_PCB#202	abs_int	8.4	y	n	20 : 08	10.0	30	3	439.803	441.800	0.89	0.20

Table 8.6 Common chemical interferences in the GC-MS determination of 2,3,7,8-TCDD (Campana *et al.*, 1989)

Compound	Molecular ion	Interfering ion	Exact mass	Resolution required
2,3,7,8-TCDD	$C_{12}H_4{}^{35}Cl_3{}^{37}ClO_2$	—	321.8936	—
Cl_9PCB	$C_{12}H{}^{35}Cl_8{}^{37}Cl$	$M^{+\cdot} - 4{}^{35}Cl$	321.8491	7223
Cl_7PCB	$C_{12}H_3{}^{35}Cl_7$	$M^{+\cdot} - 2{}^{35}Cl$	321.8678	12 476
Cl_4Xanthene	$C_{13}H_6{}^{35}Cl_2{}^{37}Cl_2O$	$M^{+\cdot}$	321.9114	18 084
Cl_5PCBPE	$C_{13}H_7{}^{35}Cl_3{}^{37}Cl_2O$	$M^{+\cdot} - H{}^{35}Cl$	321.9114	18 084
DDT	$C_{14}H_9{}^{35}Cl_2{}^{37}Cl_3$	$M^{+\cdot} - H{}^{35}Cl$	321.9292	9042
DDE	$C_{14}H_8{}^{35}Cl{}^{37}Cl_3$	$M^{+\cdot}$	321.9292	9042
Cl_4PCBPE	$C_{13}H_8{}^{35}Cl_3{}^{37}ClO$	$M^{+\cdot}$	321.9299	8868
Cl_4MeOPCB	$C_{13}H_8{}^{35}Cl_8{}^{37}ClO$	$M^{+\cdot}$	321.9299	8868

Table 8.8 gives an example of the mass spectral parameters for data acquisition for the TCDD and TCDF isomers. It is of note that each ion has a dwell time of 40 ms in which the mass is sampled on the peak top and an inter-channel jump time of 10 ms in which the electrostatic analyser (ESA) and accelerating (ACV) voltages change to sample the next mass. Two of the masses are non-analyte and correspond to the peak at m/z 316.9824 in the reference compound perfluorokerosene (PFK) which is used both as a lock mass (to correct continuously the magnet current) and a check mass (which should always appear to be the same height as the lock mass). In this example the total cycle time to sample all ions is 585 ms. This is a sufficiently high sampling rate to give adequate chromatographic peak shapes (typically 10–20 sample points across a peak) and allow accurate chromatographic quantitation to be effected.

Another important feature of the experimental conditions is that the source is operated at low electron energy (typically 25–35 eV). This minimises the contribution of the ionisation of the helium carrier gas to the background ionisation and aids achievement of low detection limits through maximising the signal to noise ratio of the chromatographic peaks obtained from the selected ion traces. These experimental conditions are equally valid for PCB analyses.

As stated before, USEPA Methods 1613 and 8290 require that an MS operates in the SIR peak top (PT) monitoring mode at a resolving power of 10 000. This is a compromise between a resolution sufficient to eliminate interferences (see Table 8.6) and giving sufficient ion transmission to achieve sub-picogram detection limits. This is easily achieved by the current generation of instruments (VG Autospec Ultima and Finnigan MAT-95, and Kratos Concept) which are 5–10 times more sensitive than the previous generation of instruments, which provided the bulk of PCDD and PCDF analyses in the 1980s to early 1990s (VG Analytical VG70 series, Finnigan MAT-90 and Kratos MS-80). Indeed, an example has been provided where the VG Ultima achieved sub-picogram detection limits in the PT SIR mode at an RP of 22 000 (Hart and Patterson, 1993).

Table 8.7 Parameter table for the processing of PCDD and PCDF data acquired on a Supelco SP2331 capillary column and VG Analytical VG70 using the VG Analytical OPUSquan[TM] data processing system.

Ent	Type	Name	Method	Std	Multiply by factor			RT	Win.	Std	Fn	m1	m2	m1/m2	Tol.
					Amo	1	2								
1	Unk	2,3,7,8-TCDF	rel_int	7	1.0	n	n	28:27	10.0	7	1	305.899	303.902	1.30	0.20
2	Unk	2,3,4,7,8-PeCDF	rel_int	8	1.0	n	n	36:02	10.0	7	2	339.860	337.863	1.52	0.25
3	Unk	1,2,3,4,7,8-HxCDF	rel_int	9	1.0	n	n	34:58	10.0	7	2	373.821	375.818	1.24	0.15
4	Unk	1,2,3,6,7,8-HxCDF	rel_int	10	1.0	n	n	35:17	10.0	7	2	373.821	375.818	1.24	0.15
5	Unk	2,3,7,8-TCDD	rel_int	11	1.0	n	n	24:33	10.0	7	1	321.894	319.897	1.30	0.20
6	Unk	1,2,3,7,8,9-HxCDD	rel_int	13	1.0	n	n	39:57	10.0	7	3	389.816	391.813	1.24	0.15
7	IS/RT	^{13}C-2,3,7,8-TCDF	abs_int		11.0	y	n	28:26	20.0		1	317.939	315.942	1.30	0.20
8	IS	^{13}C-2,3,4,7,8-PeCDF	abs_int		11.0	y	n	36:01	10.0	7	2	351.900	349.903	1.52	0.25
9	IS	^{13}C-1,2,3,4,7,8-HxCDF	abs_int		9.6	y	n	34:56	10.0	7	2	385.861	387.858	1.24	0.15
10	IS	^{13}C-1,2,3,6,7,8-HxCDF	abs_int		9.7	y	n	35:15	10.0	7	2	385.861	387.858	1.24	0.15
11	IS	^{13}C-2,3,7,8-TCDD	abs_int		11.9	y	n	24:32	10.0	7	1	333.934	331.937	1.30	0.20
12	RS	^{13}C-1,2,3,4-TCDD	abs_int		1.0	n	n	24:51	10.0	7	1	333.934	331.937	1.30	0.20
13	IS	^{13}C-1,2,3,7,8,9-HxCDD	abs_int		11.1	y	n	39:55	10.0	7	3	401.856	403.853	1.24	0.15

Table 8.8 Mass spectrometer acquisition parameters for TCDD and TCDF.

Instrument	VG 70S with OPUS/SIOS
Type	SIR Voltage
Calibration file	CAL14FEB
High mass	375.8
Low mass	303.9
Resolution	10 000
Ionisation mode	EI+
Accelerating voltage	8000 V
Scan law	Voltage—down
Magnet control	Current
Number of channels	12
Cycle time (ms)	585

Channel	Mass	Channel dwell time (ms)	Interchannel time (ms)
1	303.9016	40	10
2	305.8987	40	10
3	315.9419	40	10
4 (Lock mass)	316.9824	25	10
5 (Check mass)	316.9824	40	10
6	317.9389	40	10
7	319.8965	40	10
8	321.8936	40	10
9	327.8847	40	10
10	331.9368	40	10
11	333.9339	40	10
12	375.8364	40	10

Modern instruments offer an alternative scanning mode called mass profile scanning where 'mini' ESA scans, performed across a small mass window (typically 300 ppm at 10 000 RP), are able to detect small deviations in mass from that anticipated for the target analyte. The resulting mass profiles may be used to resolve a chromatographic peak into its constituents, which would have appeared as a single peak in the PT mode (Tong *et al.*, 1992). While there is some loss in sensitivity when compared with the PT mode it is still sufficient to achieve the required detection limits on modern instruments.

Method 1613 (USEPA, 1990) also offers the possibility that PCDDs and PCDFs can be determined by HRGC–LRMS, i.e. on a quadrupole instrument. The nominal detection limits are about 10–100 times higher than for the HRGC–HRMS method, reflecting the relative performance of quadrupole and double focusing mass spectrometers. While this method can be used for preliminary screening of samples or for analyses where there is confidence that there are extremely low levels of background contamination, we have found it to be most useful for quantifying hepta- and octa-chlorinated PCDDs and PCDFs in association with pentachlorophenol contamination. This approach is valid because of the relatively high m/z values of the target ions, making interference from other compounds less likely, and the relatively high concentrations involved.

HRGC–LRMS is usually the method of choice for PCB analyses. While the earlier trend was to report PCBs as Arochlor (or equivalent) mixtures, this approach is not applicable to environmental or food samples, where the sample PCB profiles are significantly different from the original Arochlor mixture. With the increasing recognition of the dioxin-like toxicity of the PCB isomers, the tendency now is to focus on the analysis of the dioxin-like PCBs and to report them as target analytes with accompanying information on their total toxicity. The best approach in this case is to analyse them, as for PCDDs and PCDFs, by stable-isotope dilution using $^{13}C_{12}$ standards. Native standards are available for all the toxic PCB isomers in Table 8.1 and $^{13}C_{12}$ standards for most of these compounds as well. Usually quite satisfactory analyses with acceptable detection limits are obtained, as PCBs are usually present at 10–100 times higher concentration than co-occurring PCDDs and PCDFs in food or environmental samples.

Application of HRGC–HRMS is often an advantage for PCB analysis, especially for samples with significant hydrocarbon contamination, as PCBs are often difficult to separate cleanly enough to give good data for the lower (mono- to tetra-) chlorinated congeners. Table 8.5 gives a typical set of data for processing PCB HRGC–HRMS data—in this case for samples separated on a 25 m Ultra 2 column. It may be noted that PCBs do not separate as conveniently on this column as PCDDs and PCDFs, owing to there being overlap between PCB isomers from different homologue groups.

HRGC–HRMS is required for the analysis of the three coplanar PCBs, 77, 120 and 169, which are often present in biological and food samples at low to sub-picogram levels. The sensitivity of HRMS is required to measure the levels of these compounds adequately, and to resolve them from potentially interfering co-eluting compounds.

8.5 Examples of analysis

Figure 8.2 shows the TCDF chromatogram from a sample of milk packaged in a carton manufactured from chlorine-bleached kraft pulp, analysed on a Hewlett-Packard Ultra 2 column. The pattern of 1,2,7,8-TCDF, 2,3,7,8-TCDF and 1,2,8,9-TCDF is typical of PCDFs derived from the chlorine-based bleaching of pulp. Improvements in bleaching technology in recent years have resulted in the levels of TCDDs and TCDFs in pulp products being minimised, and modern cartons do not leach these compounds into milk packaged in them.

Figure 8.3 shows the TCDD pattern and Figure 8.4 the HxCDD pattern determined in the fillets from carp sampled downstream from a pulp mill, analysed on a Hewlett-Packard Ultra 2 column. The carp has only bio-accumulated the PCDDs and PCDFs that are 2,3,7,8-chlorinated, with the three HxCDD congeners clearly separated.

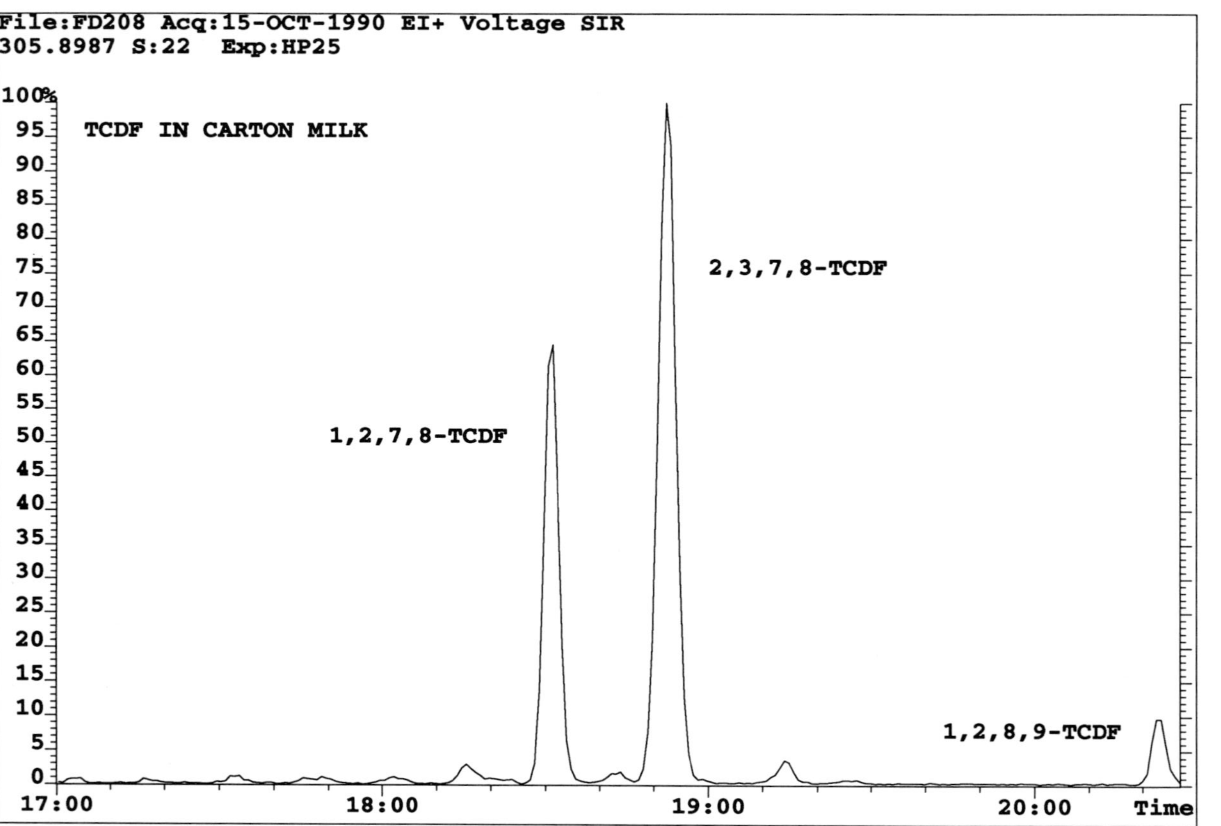

Figure 8.2 TCDF chromatogram on HP Ultra 2 column of milk packaged in chlorine-bleached carton.

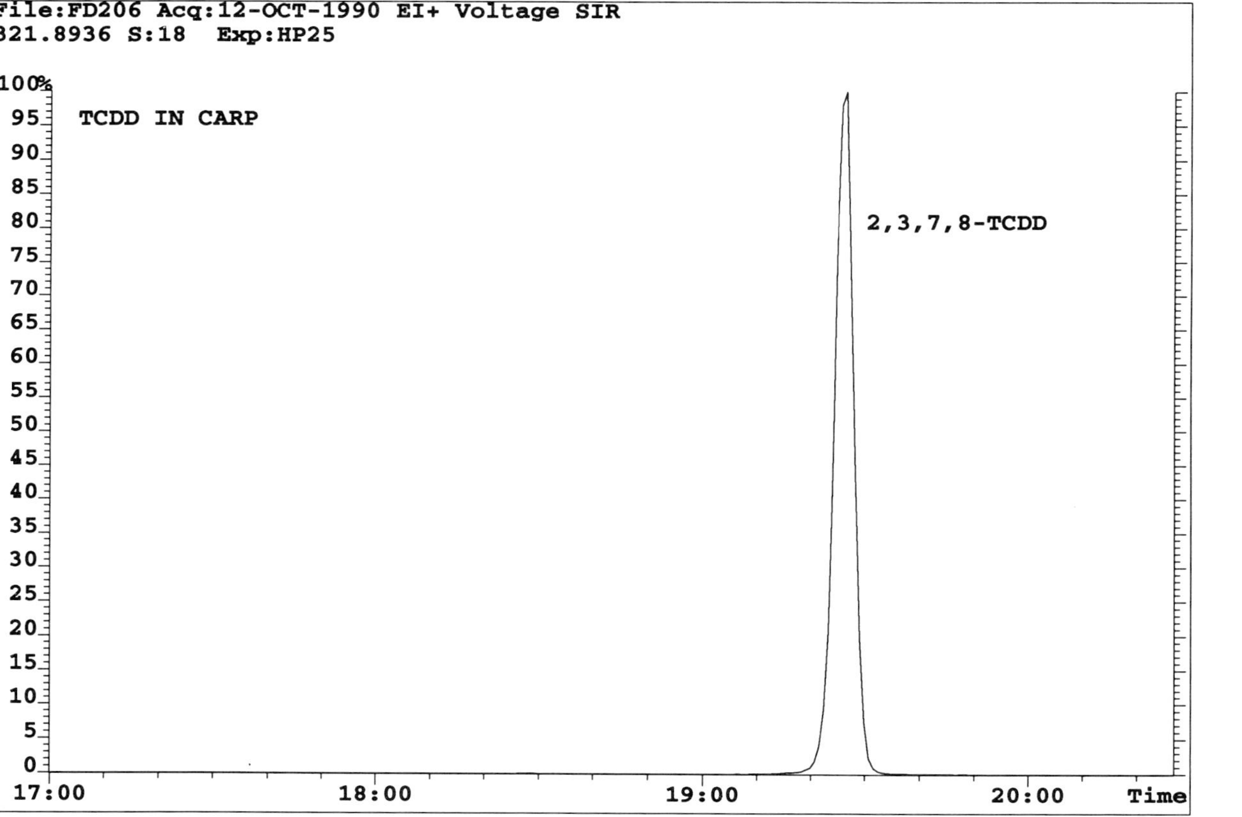

Figure 8.3 TCDD chromatogram on HP Ultra 2 column of carp fillet sampled downstream of a pulp mill.

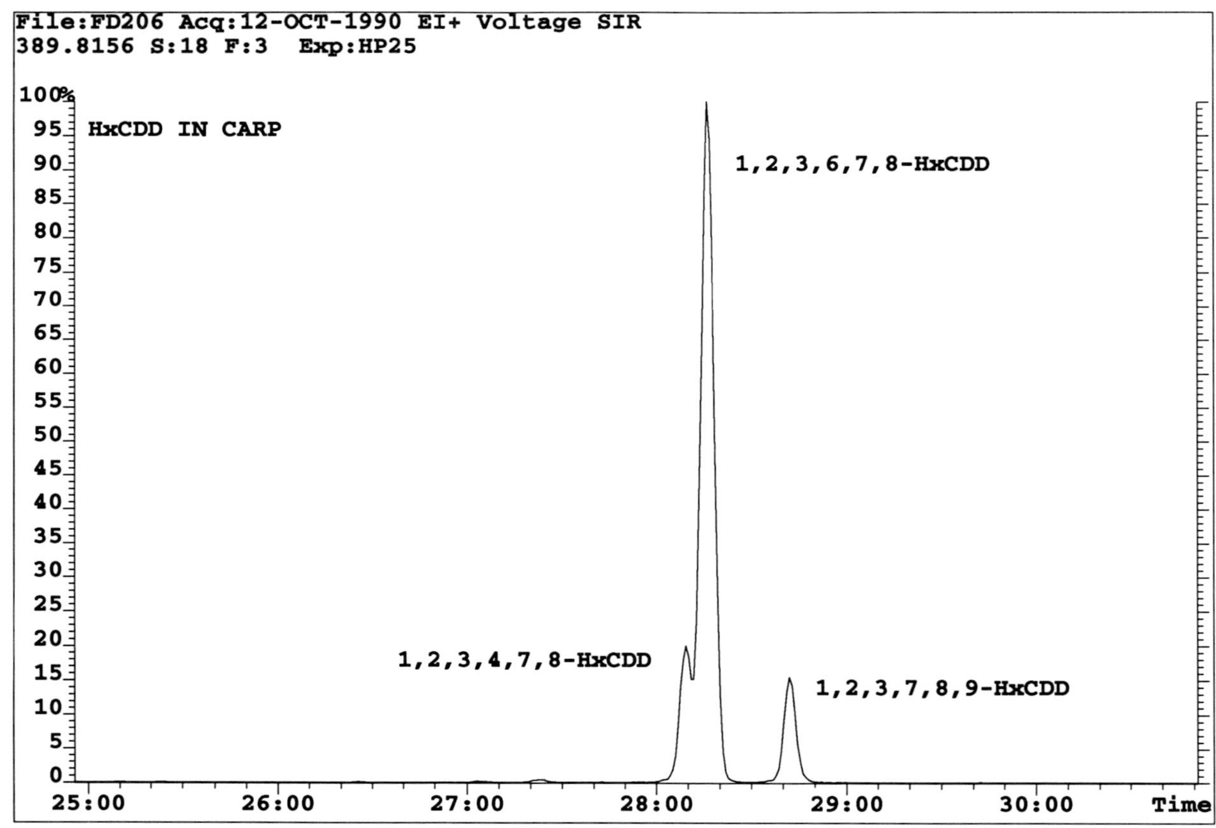

Figure 8.4 HxCDD chromatogram on HP Ultra 2 column of carp fillet sampled downstream of a pulp mill.

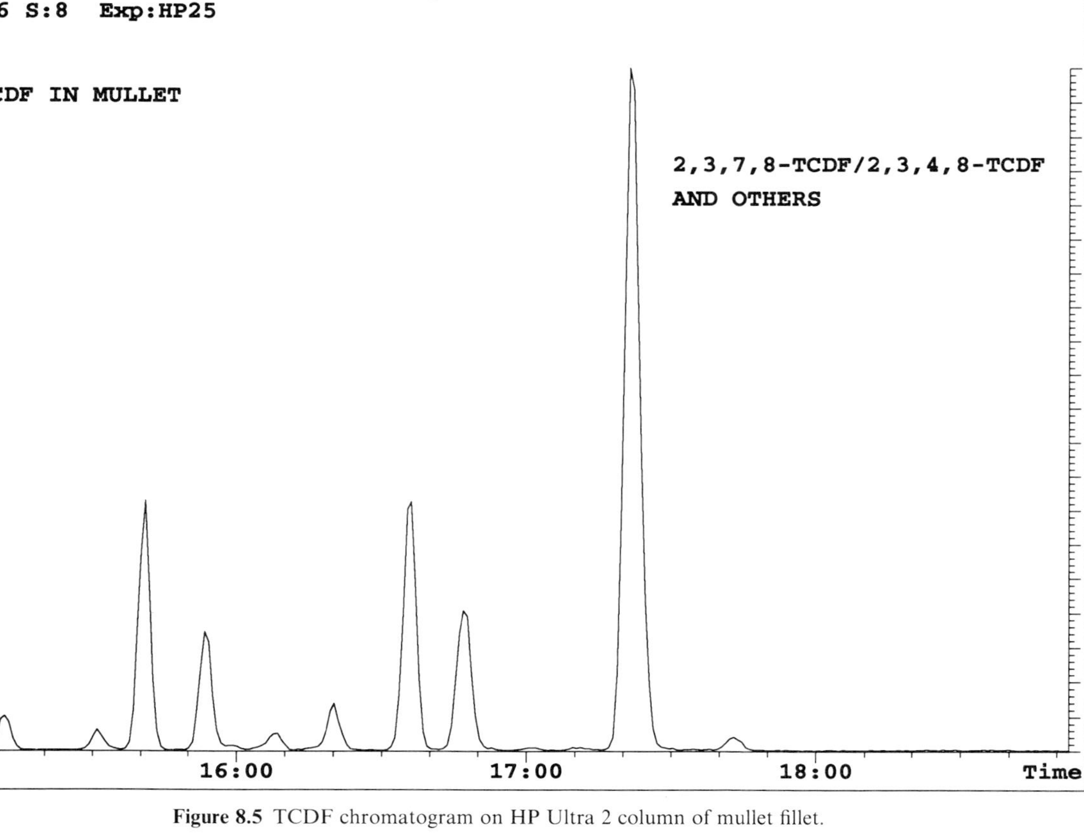

Figure 8.5 TCDF chromatogram on HP Ultra 2 column of mullet fillet.

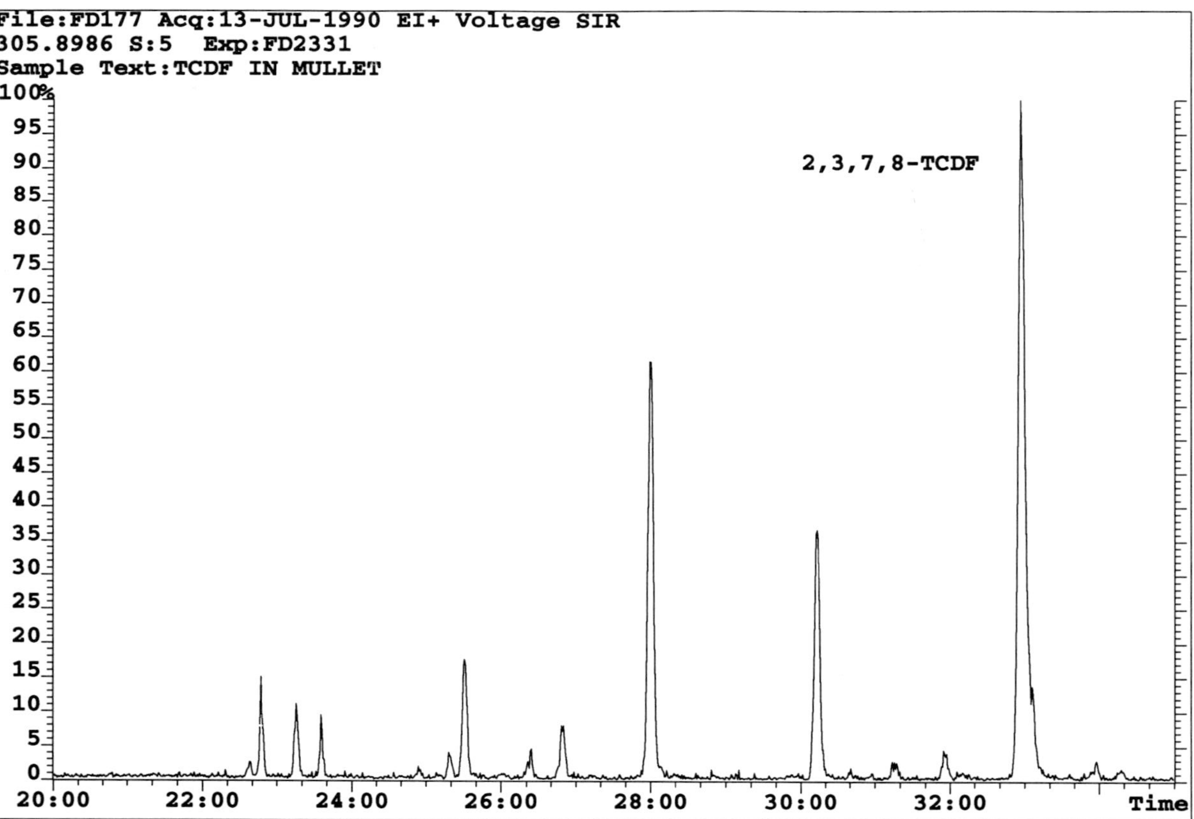

Figure 8.6 TCDF chromatogram on SP2331 column of mullet fillet.

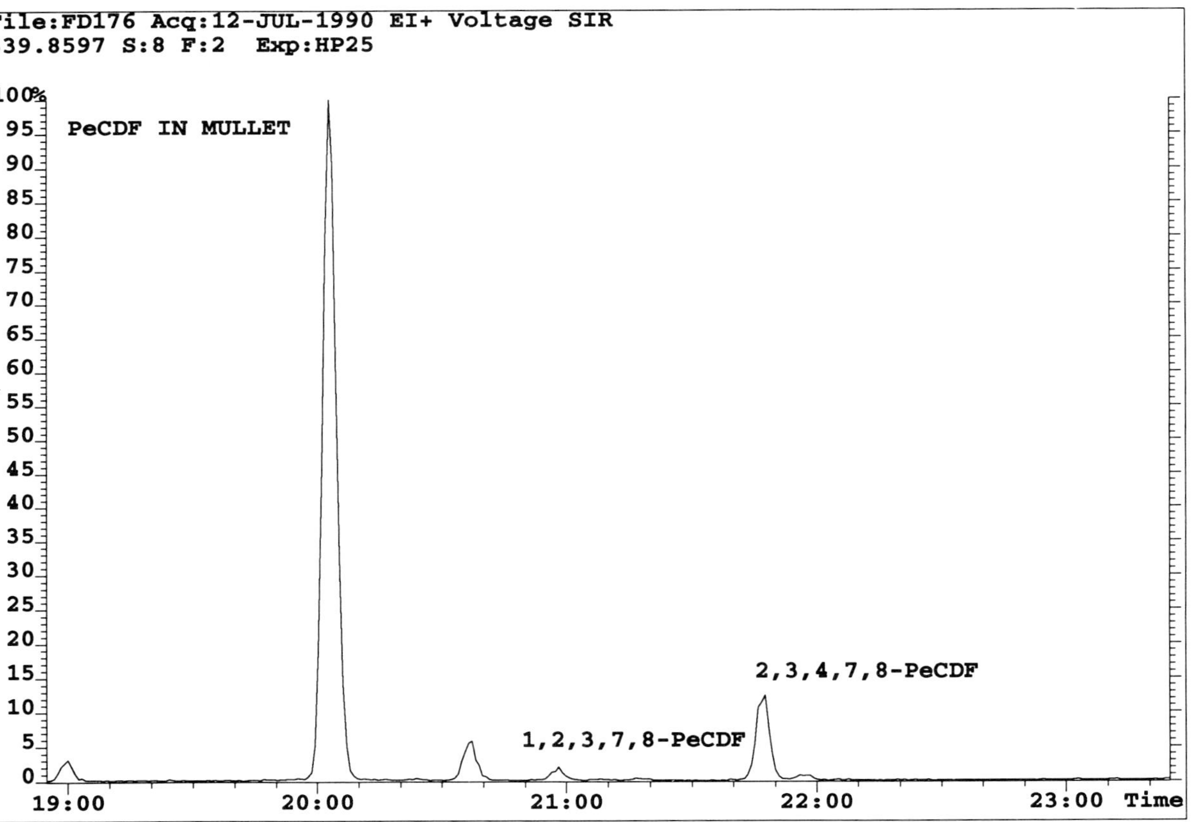

Figure 8.7 PeCDF chromatogram on HP Ultra 2 column of mullet fillet.

Figure 8.5 shows the TCDF pattern in mullet analysed on a Hewlett-Packard Ultra 2 column. Unlike carp and other fin fish, mullet appears to bioaccumulate a wider profile of PCDDs and PCDFs. Of concern are the 2,3,7,8-chlorinated congeners. The Ultra 2 column cannot unequivocally separate 2,3,7,8-TCDF from 2,3,4,8-TCDF and other TCDF isomers. This separation can be effected on an SP2331 column as shown in Figure 8.6, where the 2,3,7,8-TCDF is clearly separated from other interfering TCDF isomers.

Figure 8.7 shows the presence of 1,2,3,7,8-PeCDF and 2,3,4,7,8-PeCDF in the same mullet sample analysed on an Ultra 2 column. This again shows that mullet has bioaccumulated not only the 2,3,7,8-chlorinated isomers, but also other PeCDFs, and that these non-2,3,7,8-chlorinated isomers have dominated the congener group.

References

Ahlborg, U.G. (1988) *Nordisk Dioxinriskbedömning* (Nordic risk assessment of dioxins). Stockholm: National Institute of Environmental Medicine (with English summary).

Ahlborg, U.G., Becking, G.C., Birnbaum, L.S., Brouwer, A., Derks, H.J.G.M., Feeley, M., Golor, G., Hanberg, A., Larsen, J.C., Liem, A.K.D., Safe, S.H., Schlatter, C., Wærn, F., Younes, M. and Yrjänheikki, E. (1994) Toxic equivalency factors for dioxin-like PCBs— Report on a WHO-ECEH and IPCS consultation, December 1993. *Chemosphere*, **28**, 1049–1067.

Ballschmiter, K. and Zell, M. (1980) Analysis of polychlorinated biphenyls (PCB) by glass capillary chromatography. *Fresenius Z. Anal. Chem.*, **302**, 20–31.

Campana, J.E., Schoengold, D.M. and Butler, L.C. (1989) An environmental reference materials program: I. Dioxin performance evaluation materials. *Chemosphere*, **18**, 169–176.

Draper, W.M., Wijekoon, D. and Stephens, R.D. (1991) Speciation and quantitation of Arochlors in hazardous wastes based on PCB congener data. *Chemosphere*, **22**, 147–163.

Duinker, J.C. and Hillebrand, M.T.J. (1983) Characterisation of PCB components in Clophen formulations by capillary GC–MS and GC–ECD techniques. *Environ. Sci. Technol.*, **17**, 449–456.

Erickson, M.D. (1986) Chapter 7 in *Analytical Chemistry of PCBs*. Boston: Butterworths, pp. 171–266.

Gilbert, J. (1987) Applications of quantitative mass spectrometry in food analysis. Chapter 2 in *Applications of Mass Spectrometry in Food Science* (ed. J. Gilbert) London: Elsevier Applied Science, pp. 73–140.

Grob, K. (1994) Injection Techniques in Capillary GC. *Anal. Chem.*, **66**, 1009A–1018A.

Hart, J.R. and Patterson, D.G. Jr. (1993) The analysis of TCDD in human serum using very high resolution SIR. In *Abstracts: 41st ASMS Conference on Mass Spectrometry*, pp. 129–130.

Meadows, J., Tillit, D., Huckins, J. and Schroeder, D. (1993) Large-scale dialysis of sample liquids using a semipermeable membrane device. *Chemosphere*, **26**, 1993–2006.

Mullin, M.D., Puchini, C.M., McCrindle, S., Rumkes, M., Safe, S.H. and Safe, L.M. (1984) High resolution PCB analysis: Synthesis and chromatographic properties of all 209 PCB congeners. *Environ. Sci. Technol.*, **18**, 468–476.

NATO/CCMS (1988) North Atlantic Treaty Organisation Committee on Challenges in the Modern Society. *International Toxicity Equivalency Factor (I-TEF) Method of Risk Assessment for Complex Mixtures of Dioxins and Related Compounds*. Report no. 176. Brussels: North Atlantic Treaty Organisation.

Ryan, J.J., Conacher, H.B.S., Panopio, L.G., Lau, B.P., Hardy, J.A. and Masuda, Y. (1991) Gas chromatographic separations of all 136 tetra- to octa-polychlorinated dibenzo-*p*-dioxins and polychlorinated dibenzofurans on nine stationary phases. *J. Chromatogr.*, **541**, 131–183.

Safe, S.H. (1992) Development, validation and limitations of toxicity equivalency factors. *Chemosphere*, **25**, 61–64.

Schultz, D.E., Petrick, G. and Duinker, J.C. (1989) Complete characterisation of polychlorinated biphenyl congeners in commercial Arochlor and Clophen mixtures by multidimensional gas chromatography–electron capture detection. *Environ. Sci. Technol.*, **23**, 852–859.

Storr-Hansen, E. (1991) Comparative analysis of thirty polychlorinated biphenyl congeners on two capillary columns of different polarity with non-linear multi-level calibration. *J. Chromatogr.*, **558**, 375–391.

Tong, H., Gross, M., Giblin, D., Monson, S., Huang, L., Moore, C., Moncur, J. and Ryan, P. (1992) HRGC/HRMS with mass profile monitoring for the analysis of dioxin and related compounds. *Chemosphere*, **25**, 21–24.

Turner, W.E., Isaacs, S.G. and Patterson, D.G. Jr. (1992) An automated sample clean-up apparatus used in the procedure for measuring polychlorinated dibenzo-*p*-dioxins, dibenzofurans and *ortho*-unsubstituted (planar) biphenyls in human serum and adipose tissue. *Chemosphere*, **25**, 805–810.

USEPA (1990) *Method 1613: Tetra- through Octachlorinated Dioxins and Furans by Isotope Dilution HRGC/HRMS. Revision A* (1990), United States Environmental Protection Agency, Office of Water Regulations and Standards, Industrial Division, Washington, DC, 1–44.

USEPA (1994) *Estimating Exposure to Dioxin-Like Compounds. Review Draft.* United States Environmental Protection Agency EPA/600/6-88/005Cc, Washington, DC, vol. II.

9 Approaches to evaluating high-temperature food packaging materials as sources of food contamination

HENRY C. HOLLIFIELD and TIMOTHY H. BEGLEY

Summary

We have developed several new analytical protocols to provide data for evaluating the safety of high-temperature food-contact articles like microwave susceptors and dual ovenable trays used in cooking and reheating. The improved protocols simulate high-temperature use conditions encountered in both microwave and conventional ovens, and are conducted using food substitutes and test cells appropriate for these environments. They are suitable for the identification and quantitation of multiple volatile chemicals as well as non-volatile adjuvants, oligomers and contaminants. These procedures lend themselves to analysis of various food-contact articles containing poly(ethylene terephthalate), polypropylene, polycarbonate and nylon. Gas chromatography–mass spectrometry, high-performance liquid chromatography and supercritical fluid extraction and chromatography are the techniques used in these tests for analysis of food simulants and extracts.

These tests are complemented by the use of theoretical migration models, which predict polymer barrier properties over a wide range of temperatures using diffusion data obtained under convenient experimental conditions. These techniques are particularly helpful for comparative studies or when high-temperature experimental measurements are impractical or extremely difficult to obtain. Application of these protocols and models provides a much improved capability to assess packaging-derived food contamination in high-temperature applications.

9.1 Introduction

Prior to 1987, most regulatory and food safety investigations of food–package interactions dealt with temperatures typically below 121°C, including those packages required for hot filling, pasteurization or retort sterilization as well as most low-temperature food applications. Exceptions to this generalization are some paper products used in conventional ovens for baking and dual ovenable trays used in microwave ovens for reheating foods. Conventional wisdom at the time assumed that these products would not experience temperatures above 100°C. Food–package interaction studies typically involved developing analytical techniques that were capable of detecting food contaminants in packaging materials at parts per billion (ppb) levels and in foods and simulants at parts per trillion levels, using enhanced gas-chromatographic and liquid-chromatographic techniques, new test cell designs and multi-residue protocols. Practical applications of mathematical modeling of migration theory to describe food–package interactions were typically limited to a few real-world applications where Fickian diffusion could be distinguished from partition effects on substances moving from polymers to food stimulants (Hollifield *et al.*, 1988).

By 1987 the food industry offered a variety of food packaging intended for food preparation in both conventional and microwave ovens, including microwave susceptors for cooking popcorn, pizza, waffles, pot pies and french fries. With the introduction of microwave-active susceptors, it was readily apparent that a new era of food packaging had arrived. Susceptor packaging incorporates a microwave-interactive structure into a container or some other cooking aid, such as a crisping sleeve or heating pad. These devices, typically made of metalized polymer film, poly(ethylene terephthalate) (PET), laminated with adhesive to paper or paperboard, can rapidly heat to temperatures exceeding 200°C and actually cook foods contained within. DeLassus *et al.* (1988) demonstrated how strongly temperature affects the permeabilities and diffusivities of flavor and aroma compounds in barrier films. The permeability and diffusivity are more than 1000 times greater at retort temperatures (121–135°C) than at room temperature. Obviously, given that susceptor materials are constructed of ordinary poly-mers, paper and adhesives and are subject to more severe temperatures than those studied by DeLassus *et al.* (1988), legitimate questions arise about the integrity of such a package, and the possible formation of by-products from degradation of the package components and their consequent migration to food.

A number of investigators have since studied the performance of high-temperature food packaging and cookware. Castle *et al.* (1990) extensively studied the migration testing of non-volatiles from various cookware, packaging and microwave-active materials for high-temperature food-use applications, and proposed standardized test conditions for use in the European Community. Jickells *et al.* (1990) similarly developed headspace techniques for benzene in polymer food packaging and monitored its migration to foods. Sackett *et al.* (1991) studied the migration of volatile and non-volatile deuterated chemicals from fortified susceptors to foods and simulants, showing that actual migration may be much lower to foods than to simulants. A similar conclusion was arrived at by Risch *et al.* (1991), who used a specially designed cell to collect volatiles formed during susceptor popping of popcorn.

Our laboratory has addressed many of these same subjects and in doing so has developed a number of analytical protocols to determine both volatile and non-volatile migrants, including decomposition products and additives from the paper, polymer and adhesives in selected susceptors, foods and food simulants. Earlier work at the US Food and Drug Administration (FDA) in this area has been previously reported (Hollifield, 1991). This paper updates those earlier studies and provides a brief description of a highly sensitive supercritical fluid chromatographic method that was developed initially for aliphatic adhesive migrants from high-temperature packaging, but was found to be more generally useful for all types of polymer additives.

In addition, because high-temperature experimental test methods quickly exceed the practical limits of available laboratory apparatus, we have found it increasingly convenient to complement experimental studies with theoretical modeling. Since modeling of migration depends so heavily on the use of diffusion coefficient data, we precede the discussion on migration modeling with a brief reference to use of lagtime measurements for determining diffusion coefficients in various polymers at elevated temperatures. This is followed by a description of an extended model to predict diffusion-limited migration over a broad range of temperatures in polyolefins and PET. The section concludes with a modeling application predicting initial contamination levels in recycled PET polymer which would ultimately restrict contaminant migration to <0.5 ppb dietary exposure, the level below which most known carcinogens pose less than a one in a million lifetime risk if present in the daily diet, as defined by FDA's threshold of regulation policy (FDA, 1993). (The definition of the threshold limit has been used here for convenience. FDA has not officially adopted this migration limit in defining functional barriers which prevent migration of contaminants to foods.)

Although the work described here focuses on characterizing and analyzing contaminants from microwave susceptors, many of these procedures are also useful in evaluating microwave, conventional or dual ovenable packaging for which cooking temperatures are generally less severe. Also, one should consider the models proposed here in the strict sense of the word; they have all the limitations of models, and we believe provide reasonably conservative estimates of migration, but best serve as guides to probable migration behavior against which experimental observations may be compared.

9.2 Microwave susceptors—sizzling hot food packaging materials

Several forms of microwave susceptors are in use today, but most can be classified as either bilaminate microwave susceptor boards, like those used for pizzas, or trilaminate microwave susceptor bags, like those for popcorn. The active component of the susceptor is typically a lightly metalized polymer film. PET is most often used for this application because it melts at a fairly high temperature, 265°C, has good electrical resistance and low moisture absorption, resists combustion and is self-extinguishing. Other plastic films such as polyetherimides or modified polyesters have occasionally been substituted for PET. In bilaminated susceptors, a metalized plastic film acts as the food contact surface. This film is laminated to paper or paperboard by means of an adhesive layer. The bags contain the susceptor cemented between two layers of paper, one of which is coated with a grease-resistant coating that serves as the food contact surface. The adhesives used in these applications are generally acrylate- or vinyl acetate-based. The paper components vary with the manufacturing source. Other susceptor items such as sleeves and trays may vary in structure but also typically contain paper, plastic and adhesive components. When subjected to microwave radiation, the metalized films get very hot, achieving temperatures required to pop popcorn, brown or crisp food surfaces and cook pot pie crusts. Some measurements have indicated spot temperatures above 260°C. Photomicrographs of early commercial products displayed evidence of occasional melting, cracking and breaking of the polymer. Because of the fragile nature of the materials used to construct the susceptor, the wide variation in the quality of the metalized films and the consumer's inexperience with the new technology, the initial encounter with these products was more likely than not to lead to product abuse as a result of overheating. Newer commercial susceptor products now available are much improved, and performance in the hands of an educated consumer is less likely to result in abuse or overcooking of food. Still, these products consist of essentially fragile paper, adhesive and polymer constructions, and rapid migration of packaging residues occurs.

The hot metalized film, coated on one or both sides with adhesive and in intimate contact with paper, frees significant amounts of volatile and non-volatile substances that have a wide range of chemical functional groups (FDA, 1989). Therefore, several analytical techniques have been required for separating, identifying and quantifying this mixture of chemical contaminants. Headspace gas chromatography (GC) with flame ionization and mass spectrometric (MS) detection has been the method of choice to determine the more volatile components. High-performance liquid chromatography (HPLC) with UV detection is used extensively to determine UV-absorbing non-volatiles. Supercritical fluid chromatography has proven partially successful in the analysis of aliphatic non-volatile susceptor residues.

9.3 Volatile chemicals produced during susceptor heating

9.3.1 Capillary headspace GC and GC–MS for determination of volatiles

A headspace capillary GC procedure has proven most effective for determining the volatile substances emitted from high-temperature food packaging

(McNeal and Hollifield, 1993a). Although this method preceded and served as a basis for the American Society for Testing Materials Standard F-1308-90 (ASTM, 1990), it differs significantly in sample preparation details. The primary point of departure is its use of constant microwave heating conditions, so that all susceptor products can be readily compared with each other. The ASTM method allows the analyst to choose different microwaving conditions for each test, to mimic a time–temperature profile that would be obtained by monitoring food-susceptor interface temperatures when cooking specific food types. While the FDA protocol has been used primarily to characterize and compare various materials used in susceptor constructions, different sampling protocols have been applied to test for migrants retained in food and fatty food simulants (McNeal, 1988; Castle *et al.*, 1990; Jickells *et al.*, 1991; Risch *et al.*, 1991; Rose, 1991; Sackett *et al.*, 1991).

Experimental susceptor sample preparation.   Samples of susceptor materials were prepared for analysis by placing $6.45\,cm^2$ strips of each susceptor in a 22-ml headspace vial and sealing it with a Teflon-lined silicone rubber septum and an aluminum crimp cap. The sample and vial were microwaved for 5 min along with 250 ml of water in the oven. The water acted as an energy sink, preventing the susceptor from burning by absorbing excess microwave energy. These microwave exposure conditions were experimentally chosen so that after heating, the test bilaminate susceptor strips were similar in appearance to a susceptor that was actually used in food cooking applications. It was arbitrarily decided to test trilaminate susceptor under the same conditions. For quantitation by the method of standard additions, fortified samples were prepared in a manner similar to that described above, with 10 μl of mixed fortification standards added to each vial.

Susceptor materials to be used in migration testing were handled differently. In this case the susceptor was cut to fit a polytetrafluorethylene (PTFE) migration cell. A food simulant, Miglyol^{TM}, was added to the cell to achieve a weight-to-surface area ratio comparable to the intended food application. The cell containing the susceptor disk and oil was then microwaved according to instructions provided by the food processor. Following microwave heating, a 2 g amount of oil was transferred to a purge and trap sample tube. The oil was purged of retained volatile chemicals with nitrogen by using a Tekmar LSC 2000 purge and trap apparatus equipped with Vocarb 4000 mixed carbon solid absorbent bed. Trapped chemicals were desorbed from this trap into the gas chromatograph–mass spectrometer (McNeal and Hollifield, 1993b).

The microwave oven used for these experiments was a typical 700 W model available to consumers at most appliance stores. The oven was calibrated weekly, using ASTM procedure F1317-90. The oven was considered acceptable for the tests as long as the variability of the output wattage did not exceed 10% relative standard deviation. The magnetron was preconditioned by heating 2 l of water on high power for 10 min and allowing the oven to cool for 10 min before commencing tests.

After microwave heating of the susceptor, the resulting volatiles retained in the sealed vials were analyzed by headspace GC or GC–MS. Two instruments were used in this study. Qualitative analysis was performed using a Hewlett-Packard 5890A gas chromatograph equipped with an HP 5970B mass selective detector and a 5% phenylmethyl silicone capillary column. This instrument was also equipped for sub-ambient oven operation and had a purge–trap interface for measurement of very volatile compounds. This system was sufficiently sensitive, so that even low concentrations of minor volatile chemicals could be identified with a high degree of confidence. The second instrument, which was used primarily for quantitation, was a Perkin-Elmer Sigma 2000 gas chromatograph equipped with an HS-100 automated headspace sampler equipped with a flame ionization detector. A 30 m, 0.25 mm i.d., 5% phenylmethyl silicone capillary column provided adequate resolution of chromatographic peaks.

Separation and identification of the gas-chromatographic peaks were relatively straightforward. Mass spectra of the unknown peaks were matched with those in the National Institute of Standards and Technology (NIST) mass spectra library. The numbers of unknown chemicals recovered from individual susceptors varied significantly from product to product (Figure 9.1). The number of volatile residues in the susceptors tested ranged from 34 to 105. Both their number and their gross amounts were intimately related to the stability of the paper and adhesive components at elevated temperatures. The quality and quantity of paper also varied from product to product, as indicated by variations in gross amounts of alcohols, aldehydes, ketones and furans produced on heating. The use of acrylic-based adhesives tends to yield more volatile residue chemicals than does use of vinyl acetate-based adhesives. Table 9.1 lists the identities of a number of volatile compounds recovered from a limited survey of susceptor products performed when the above method was first developed. Table 9.2 lists those chemicals found in analyses of various products thought to result primarily from the heating of adhesives.

The capillary gas chromatograms of volatiles from conventionally heated trays were quite similar to those obtained from microwave susceptors. The chromatograms were dominated by peaks of paper decomposition products like those in Table 9.1. The only adhesive-derived chemicals seen in significant amounts were acetic acid (typical of vinyl acetate-based adhesives), glycerol triacetate (used as a plasticizer in the popcorn bag susceptor), and *n*-butanol (typical of butyl acrylate-based adhesives).

Quantifying the volatile chemicals is a bit more of a challenge. For example, early eluting chemicals can be separated only with difficulty. Using the sub-ambient GC oven is helpful in performing this task. Even

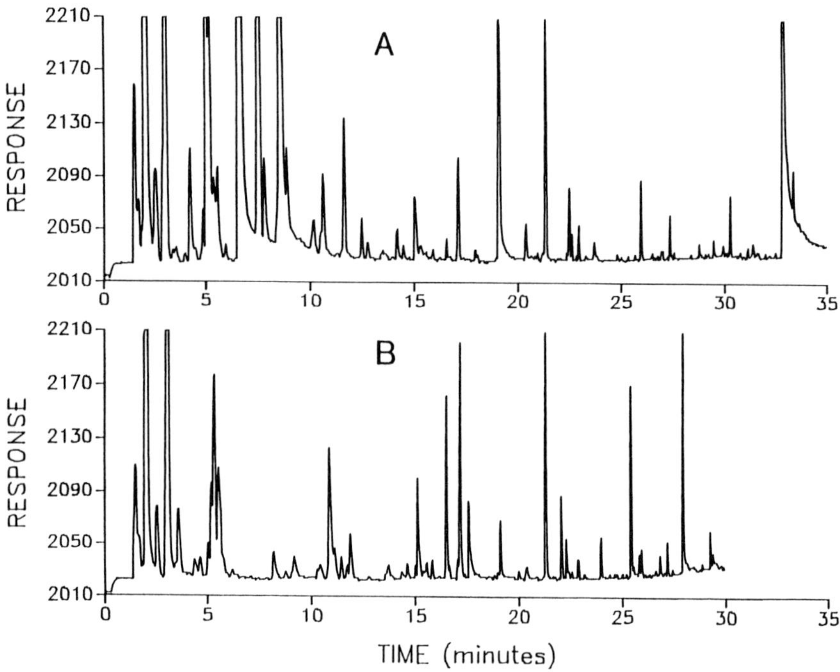

Figure 9.1 AOAC P1274 chromatogram A: FID chromatogram of bilaminate-type susceptor containing an acrylic-based adhesive; chromatogram B: FID chromatogram of bilaminate-type susceptor containing a vinyl acetate-based adhesive.

so, the abundance of free carboxylic acids, such as acetic acid, a breakdown product of vinyl adhesives, interferes with the chromatographic resolution of some GC peaks and subsequent estimations of area or peak height necessary for quantitation. Because of the strong adsorption of polar chemicals by the susceptor cellulose, the procedure is not very sensitive for trace levels of these substances. Also, water, which may be present in the susceptor paper at levels up to 12%, can have two opposing influences on the determination. First, it can lessen the adsorption of polar substances on the susceptor paper; second, if present in the condensed phase in the headspace vial, it may solubilize the polar substances and reduce the sensitivity of the headspace procedure for these chemicals. The only quantitative technique found to compensate for these problems is standard additions. This solution assumes that the GC peaks of interest have been identified and that authentic standards are available. Obviously, the problem is unresolved for unidentified peaks.

9.3.2 *Migration of adhesive and paper volatiles*

The exact quantitation of volatiles formed when a susceptor is heated has little significance outside of comparing one product with another, unless

Table 9.1 Volatile chemicals released from 11 different susceptors during microwave cooking[a]

Compound	MSD response[b]	No. of occurrences	Concn range, $\mu g/in^2$
		Standard additions method[c]	
Acetone–furan (co-eluters)	×	11	0.06–36
Isopropanol		1	
Vinyl acetate	×	6	0.01–0.88
Methyl ethyl ketone		1	
Formic acid		2	
2-Methyl-1-propanol	×	7	16.4–105
1,1,1-Trichloroethane	×	4	1.05–75
Acetic acid	×	8	
Crotonaldehyde	×	7	0.10–4.5
Benzene	×	8	<0.01–0.22
Butanol	×	8	2.04–72
Methyl methacrylate	×	7	0.02–4.2
Toluene	×	5	0.06–1.1
Hexanal	×	11	0.22–8.5
Furfural	×	10	0.52–32
Furan methanol		7	0.07–18
Butyl ether		1	0.35–0.39
Styrene	×	7	0.04–2.0
Butyl acrylate		4	0.05–0.68
2-(Butoxy)ethanol		6	0.02–5.8
2-Ethyl-1-hexanal	×	1	
Benzaldehyde	×	9	0.02–7.04
2-Ethyl-1-hexanol	×	3	0.26–6.8
o-Dichlorobenzene		1	
2-(2-Butoxyethoxy) ethanol	×	4	21.44–120
		Identified by using MS library[d]	
Butanal		2	
Methyl vinyl ketone	×	8	
Pentanal	×	9	0.35–3.1
Methyl furan		2	
Ethyl furan		1	
Butanoic acid		1	
3-Methyleneheptane	×	2	0.24–1.8
2-(Propoxy)ethanol	×	1	0.81–1.3
Heptanal	×	5	
2-Heptanone		2	
3-Heptanone		1	
Octanal	×	5	
1-Phenylpropanedione	×	2	
Nonanal	×	4	
Octyl acetate		1	
Benzoic acid		1	
Decanal		1	
5-Hydroxymethylfurfural		1	

[a]Reproduced from McNeal and Hollifield (1993a), p. 1273.
[b]Compound occurs at least once having MSD response >20% of that of the internal standard.
[c]Identification is based on coincident retention times of authentic standards plus agreement within 5% intensity of the ten most abundant ions.
[d]Identification of a susceptor volatile is based on a >90% match with the NBS (formerly National Bureau of Standards; now National Institute of Standards and Technology, NIST) library reference spectrum.

Table 9.2 Adhesive based degradation by-products released from susceptors during microwave heating

Chemical	Concentration $(\mu g/in^2)^{a,b}$
Isopropanol	0–6.2
Acetic acid	>50
t-Butanol	0–3.5
Vinyl acetate	0–4.6
Isobutanol	0–3.0
Methyl chloroform	>50
Benzene	0–0.3
n-Butanol	0–54.0
Ethyl acrylate	0–0.6
Toluene	0–0.5
Ethyl benzene	0–0.3
Xylenes	0–2.0
n-Butyl ether	0–1.4
Butyl acrylate	0–1.4
Styrene	0–1.8
Glycerol triacetate	>50

[a]Not all chemicals reported were found in every material tested.
[b]Limits of detection varied widely. Levels reported for acetic acid, glycerol triacetate and methyl chloroform are estimates.

the contaminants transfer to food and are eaten by a consumer. For this reason it is more meaningful to ascertain the extent to which volatile chemical residues from the susceptors are retained by foods or food simulants. A number of migration tests were run to investigate this issue. Results of these experiments are shown in Table 9.3. Volatiles migrated into the Miglyol™ food simulant from both heat susceptors and conventionally heated trays. In the case of a PET–paperboard tray containing no adhesive, 33 ng/ml of

Table 9.3 Volatile chemicals from susceptor packaging retained in Miglyol^{a,b}

Chemical	Concentration (ng/ml)
t-Butanol	0–8
Acetic acid	>100
Vinyl acetate	0–10
n-Butanol	0–20
Toluene	0–2
2-Furfuraldehyde	0–318
Xylenes	0–2
Benzaldehyde	0–9
Aliphatic aldehydes[c]	>100

[a]Quantitation based on external calibration.
[b]Data were obtained by headspace analysis of corn oil after cooking for 3 min by microwave susceptor at mass-to-area ratio of 7.1 g/in^2.
[c]Aliphatic aldehydes such as 2-methyl propanal, butyraldehyde, hexanal, and octanal were present and were most likely derived from the paper.

2-furfuraldehyde was the only detectable volatile migrant. However, with an adhesive-bound TV-dinner tray, 2-furfuraldehyde was not detected, but acetic acid and 9 ng/ml of vinyl acetate were found in the Miglyol®. Although it was estimated that $>50\,\mu g/in^2$ ($7.5\,\mu g/cm^2$) of glycerol triacetate is produced on heating the popcorn bag susceptor, the chemical was not detected in hot Miglyol® by purge and trap headspace GC. Migration from susceptors was similar to that of conventionally heated trays. Only those volatile chemicals formed in significant amounts migrated and were retained to a slight degree in hot Miglyol®.

These results confirm that only those volatile chemicals formed in the highest concentrations during high temperature use are likely to migrate and be retained in foods or food simulants. The amounts retained, in all cases, were in the low ppb range. The greater portion of volatiles produced appeared to escape to the oven cavity and were exhausted. No extremely potent toxins were identified among the volatiles found. For these reasons, it was concluded that the probability of any significant amount of volatile substances being retained by food as a result of being cooked on a microwave susceptor or in other high-temperature packaging tested is quite low. These findings are in agreement with other studies conducted by Risch *et al.* (1991) and Sackett *et al.* (1991).

9.4 Non-volatile chemical residues

Several investigators have made important contributions to the determination of PET oligomers and residues that migrate to foods and simulants (Kashtock and Breder, 1980; Ashby, 1988; Tice, 1988; Castle *et al.*, 1989). However, most of these methods measure terephthalic acid esters and are not specific for PET oligomers or other package migrants. FDA developed the first general HPLC method for the determination of specific PET oligomers and UV-absorbing compounds such as adhesive components from microwave susceptors. Subsequently, this methodology has been successfully applied to both the package materials and the foods prepared in them (Begley and Hollifield, 1989, 1990a, b; Begley *et al.*, 1990; Kashtock *et al.*, 1990).

Potential migrants of higher molecular weight, such as plasticizers and oligomers, have been arbitrarily classified as 'non-volatiles' and are generally determined by HPLC. However, some are certainly volatile enough to be determined by GC, and are referred to by some analysts as 'semi-volatiles'. In fact, a GC–MS total ion chromatogram of an acetonitrile extract of a pot pie-type susceptor construction revealed about 35 individual peaks of components ranging from C_{16} hydrocarbons to the PET cyclic trimer (Figure 9.2). Most of these substances were not extracted before the susceptor was heated in the microwave, suggesting that many are probably degradation products. So far, HPLC methods have been developed to

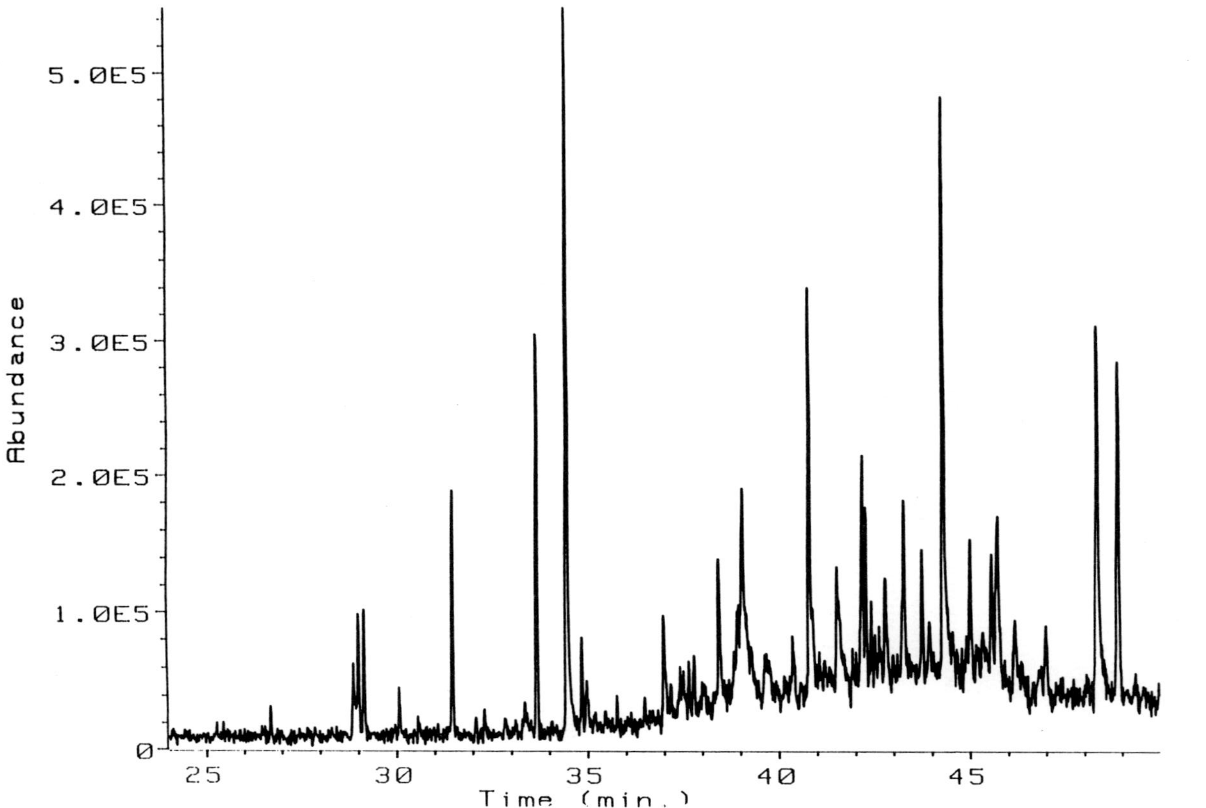

Figure 9.2 GC–MS total ion chromatogram of an acetonitrile extract of a pot pie-type susceptor construction reveals about 35 individual peaks of components ranging from C_{16} hydrocarbons to the PET cyclic trimer.

monitor the migration of only the UV-absorbing substances into foods and food simulants. Attempts to monitor migration of the non-UV-absorbing migrants have not met with much success because of the poor sensitivity and non-specificity of typical HPLC detectors. Some researchers have tried to determine total non-volatiles and the amounts of non-UV-absorbers by extraction and gravimetric procedures, but these approaches have yet to be validated.

9.4.1 High-temperature food-simulant considerations

The evaluation of contaminant migration from food packaging is often made easier by the use of food-simulating liquids (FSLs). In a migration experiment these liquids may simulate various food type characteristics and most important, generally have simpler matrices with far fewer analytical interferences than actual foods. Typical FSLs used in regulatory test specifications in the United States and Europe include water and 3% acetic acid for use as aqueous and acid food simulants; and 10, 15, 50 and 95% ethanol, and food oil (corn and olive oils) to simulate low alcohol and fatty foods. In many tests, food oil fatty food simulants are the most difficult FSLs in which to evaluate migration.

One of the difficulties of using corn and olive oils for high-temperature migration experiments is their tendency to degrade during the course of a migration experiment. This creates numerous analytical interferences that are difficult to avoid except when using isotope or radiolabeled analysis techniques. For example, when a UV detector is used to measure the migration of a non-volatile UV-absorbing additive into corn oil, the number and magnitude of the background components increase with the heating time and temperature of the experiment. These background components can inhibit detection of low migrant levels and bias analytical results. These and other difficulties have led researchers to seek alternatives to unsaturated food oils as viable food simulants. Figge (1972) and Figge and Koch (1973) found that a synthetic triglyceride mixture, similar to coconut oil and known as HB307, produced migration results consistent with those of other edible oils. HB307 is not without disadvantages, however. Chief among them are its limited distribution and high cost, especially in the USA. Moreover, HB307 is a solid at room temperature, and this introduces mass-transfer limitations in low-temperature experiments.

Others have sought perfection in single-component simulants. Lauryl alcohol and 1-octanol have been compared with corn oil and HB307 and resulted in three times more butylated hydroxytoluene (BHT) migration from high-density polyethylene (HDPE) than to the oils (Till *et al.*, 1982). In the case of styrene migration from polystyrene, hexadecane and decanol produced similar migration values to corn oil and HB307 at 40°C, but produced ten times more migration at 70°C (Snyder and Breder, 1985).

Iso-octane was found to simulate the migration from numerous polymers to edible oils at 40°C for 10 days when the temperature and the time of the experiment were reduced to 20°C for 2 days (De Kruijf and Rijk, 1988). Four- to eight-carbon alcohols (Lickly *et al.*, 1990) exaggerated the migration of Irganox 1010 from polyolefins at 100°C and above. The overall migration from several polymers at high temperatures (95–175°C) was simulated by using iso-octane at 60°C (De Kruijf and Rijk, 1994) and gave results similar to those with olive oil.

Nonetheless, for high-temperature testing, we found no other vegetable oil product to be as inexpensive and easy to use with the least analytical interference as a coconut oil distillation fraction described as Miglyol 812®. This product is liquid at room temperature and gave migration results consistent with those of HB307 and corn oil when used to evaluate extreme high-temperature migration (>200°C) from microwave susceptor packaging (Begley and Hollifield, 1990b). Its high temperature stability makes it particularly suited for such applications, as it produces relatively low UV interferences upon heating.

This discussion would not be complete without mentioning attempts to use oil–water emulsions and oil-coated supports as high-temperature food simulants. At temperatures approaching 100°C, water rapidly evaporates, disturbing emulsion equilibrium. As a consequence, analytical results are difficult to reproduce. Use of oil-coated supports may hold some promise when carefully prepared. At least one laboratory appears to have successfully used an oil-coated support in high-temperature migration tests (Castle *et al.*, 1990).

9.4.2 *Migration of non-volatiles from susceptors*

The FDA procedure for monitoring migration of UV-absorbing non-volatile residues from microwave susceptors uses Miglyol as the food simulant of choice. Refined naturally occurring corn oil was originally used for these studies because it was readily available and did not interfere with PET oligomer determination. Initially, it was mistakenly thought that the oligomers from the PET food contact layer were the only major migrants of interest. It was later learned that PET barrier properties diminished when the film was heated above its glass transition temperature, allowing additives and decomposition products from the adhesive and paper susceptor components to migrate readily through the PET to foods and food simulants. Corn oil decomposition products interfered with the determination of the adhesive and paper-based migrants. Other oils were examined and several possible alternative food simulants were identified. These included HB307, paraffin oil and Miglyol. All developed relatively few chromatographic interferences under the test conditions used (Begley and Hollifield, 1990b).

The migration experiments were conducted on unused susceptor stock or on actual susceptor products purchased in local supermarkets. Migration experiments to oil were performed using a special Teflon single-sided extraction cell designed at Waldorf Corp. (St. Paul, MN). All susceptor test materials were cut from flat areas of susceptors supplied with commercial food products. The cutouts were sized and shaped to fit the extraction cell. (For further details, see Begley and Hollifield, 1990b, 1991.)

9.5   Test cell considerations for microwave environments

Ideally, migration cells for use in microwave environments should be constructed of materials transparent to microwave radiation; they should be inert and able to withstand temperatures of at least 220°C. PTFE is a common material that readily meets these rigorous requirements. Figure 9.3 reveals the initial design of the early PTFE microwavable migration cell, originally developed by the Waldorf Corporation (St. Paul, MN). This cell played an important role in a collaborative study by the ASTM to measure non-volatile, UV-absorbing extractables from microwave susceptors (ASTM, 1991). Other cells were also used for testing migration in the

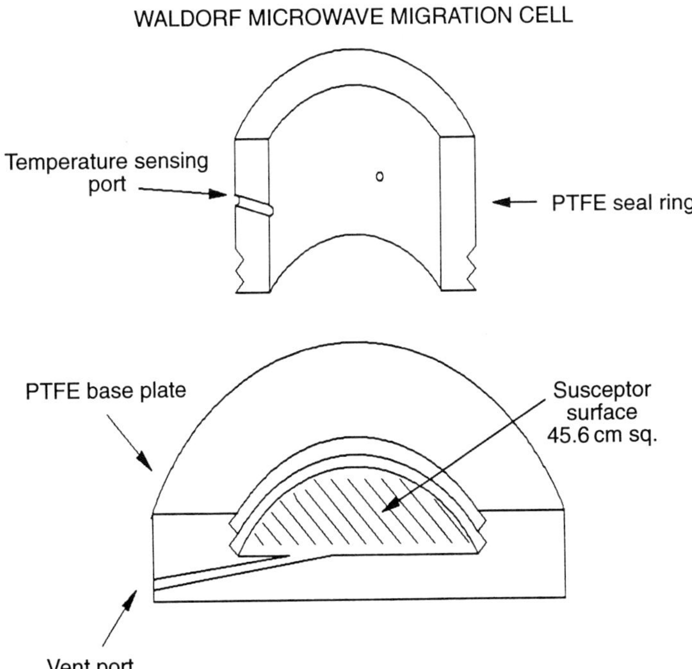

Figure 9.3 Diagram showing the initial design of the early PTFE microwavable migration cell.

microwave oven. A completely sealed PTFE cell was used by De Kruijf and Rijk (1994) to measure the migration into water, 3% acetic acid and 15% ethanol from microwave packaging.

Our test cell was originally designed with an exposed susceptor area of $44\,cm^2$ ($7.07\,in^2$). The ratio of oil weight to surface area was arbitrarily selected as $1.1\,g/cm^2$ ($7.1\,g/in^2$). The mixed weight and area units are attributable to early FDA regulation writers and appear in the Federal Code of Regulations (CFR, 1995, Section 21, Part 174–177, Indirect Food Additives). Small holes drilled through the bottom and sides of the cell permit introduction of fluoroptic temperature probes to monitor food-simulant temperatures during tests using a Luxtron Model 755 fluoroptic temperature-sensing instrument. Food oil simulants were typically cooked for 3 min. Actual foods were cooked according to package directions, usually for the maximum recommended time. Controls were prepared by cooking the oil or french fries in a Pyrex petri dish heated by a microwave susceptor board placed beneath the dish. The PTFE cell was not used for the control experiments. After microwaving, 3 g of oil or all of the cooked food was extracted and analyzed.

9.6   Recovery and analysis of migrating chemicals

Following heating by contact with the microwave susceptor, the migrating chemicals were recovered from the oil simulant or cooked foods by solvent extraction, and analyzed by HPLC. Extraction of the oil was proceeded by transferring a 3 g aliquot to a separatory funnel, diluting it with 50 ml of hexane, and extracting it with two 25 ml portions of acetonitrile (Begley and Hollifield, 1989, 1990b). Foods were extracted with hexane in an explosion-proof Waring blender, and the extracts were filtered and further extracted with acetonitrile in a similar manner (Begley *et al.*, 1990). The acetonitrile layer was concentrated in a Kuderna-Danish evaporative concentrator and analyzed by HPLC, using a Microsorb-C, $5\,\mu m$ particle size, 250 mm × 4.6 mm column, an acetonitrile–water mobile phase, and a Waters Model 480 Lambda Max variable wavelength detector.

A number of susceptor products have been analyzed by this technique. Table 9.4 shows clearly that several chemicals migrate from the susceptor to both oil and food (french fries). The levels found in the oil are generally higher than those found in the food. The concentrations of these non-volatile susceptor migrants in foods are also significantly higher than those of the volatile chemicals. The UV-absorbing, non-volatile chemicals found in the highest concentrations are typically PET oligomers and adhesive-based plasticizers, and other adhesive components. All of these substances have also been found in foods cooked using susceptors. The principal residues found in several of the foods include diethylene and dipropylene glycol

Table 9.4 Corresponding migration levels of non-volatile components in food and food-simulating vegetable oil[a]

Migrant	Food (mg/kg)	Oil (mg/kg)
PET oligomers	7.6	18.4
Diethyleneglycol dibenzoate	11.0	18.4
Dipropyleneglycol dibenzoate	7.8	12.7
Diglycidylether of bisphenol A	1.5	—

[a]Food cooked by microwave susceptor for 4 min, as recommended by food company; oil cooked for 3 min; all tests conducted at a mass-to-surface area ratio of $7.1\,g/in^2$ (Begley and Hollifield, 1990b).

dibenzoates. These compounds occur in food at levels comparable to those of the PET oligomers (about 15 mg/kg each). The diglycidyl ether of bisphenol A, an epoxy compound used in adhesives, has also been found in some packages. The highest level of oligomers and plasticizers found in a single food such as french fries has been about 45 mg/kg (PET oligomers plus the two plasticizers). However, this value does not include estimates for either volatile or non-UV-absorbing non-volatile migrants. We have found PET oligomers and various plasticizers in foods such as pizzas, waffles, sandwiches, pot pies and french fries at parts per million levels.

In addition to the typical bilaminate microwave construction used with the above products, studies have been conducted on four major brands of microwave popcorn bags, which are typical trilaminate constructions. Two plasticizers, diethylene glycol dibenzoate and dipropylene glycol dibenzoate, along with PET oligomers, were found to migrate into Miglyol fatty food simulant from at least one test bag. An unidentified alcohol migrated from two other bags. It was estimated that none of the popcorn bag migrants reached parts per million levels in food.

9.7 Possible food alteration products

The non-volatile migrants from microwave susceptors are thought to be made up of all types of chemical moieties. Several alcohol moieties are known to be either raw materials in susceptor components or formed as decomposition products during susceptor heating. An alcohol-specific procedure was developed to obtain more information on these chemicals. This procedure is based on the reaction of dinitrobenzoyl chloride with alcohol groups, and has been applied to alcohols in acetonitrile extracts of fatty food simulants heated in contact with microwave susceptors. Preliminary tests with alcohol-fortified acetonitrile solutions showed this approach to be feasible. In practice, however, recoveries from fortified Miglyol test extracts were quite low. However, when paraffin oil fatty food simulant was substituted for the vegetable oil in this experiment, recoveries were very good. This suggests that alcohols added to Miglyol and subjected to susceptor heating conditions very likely transesterify, resulting in the low recovery. It is probable that similar alcohols used in raw materials of the susceptor components when freed from this package matrix could also form esters with fats and vegetable oils.

A shortage of resources has delayed further exploration of this hypothesis. Given the large numbers of chemicals that migrate from susceptor food packaging and the very high temperatures at the food–package interface, the potential exists for the formation of many food alteration products. Although we believe the initial results of this investigation strongly support this theory, much additional work needs to be done to isolate and identify specific alteration products to show satisfactorily that this is more than theory.

9.8   Temperature measurements

One of the more controversial tasks associated with the study of high-temperature packaging is selecting an appropriate test temperature. The controversy arises in part because of the difficulty of measuring the maximum temperature attained by susceptor materials. These data are important because excessive temperatures lead to material failure and the formation of degradation products. Also, at elevated temperatures, chemical additives and other residues in the package will migrate more rapidly to the surface, where they can transfer to the food. Because internal package temperatures are not easily determined and may vary widely from side to side or top to bottom, scientists have tended to use food–package interface temperature measurements to describe the heat histories of packaging materials (Kashtock *et al.*, 1990). But these interface temperatures, moderated by contact with food, are typically lower than those attained by the susceptor substrate and not truly indicative of the diffusivity of migrating chemicals through the polymer surface to food. These temperatures can also be particularly variable in microwave fields within the oven cavity.

These conclusions are supported by infrared photographic observations of susceptor surfaces during microwave heating and by differential scanning calorimetric (DSC) measurements on previously heat-treated PET films. Infrared photography of the surface of a susceptor in a microwave oven shows a constantly changing heat pattern when the susceptor is subjected to microwave fields. The temperature pattern ranges from moderate to extremely hot. The key phrase is 'constantly changing heat pattern'. Such patterns are not accurately portrayed by averaged temperatures obtained by point source measurements. Moreover, when food is cooked and the temperature is monitored by infrared camera, generally higher temperatures are obtained in some areas of the container than are indicated by fluoroptic probe measurements (Lentz and Crossett, 1988). Obviously, any temperature probe at the food–package interface will produce an average temperature between that of the hot cooking surface and the cooler food mass that is being heated. A probe located between a very hot microwave surface and a frozen piece of breaded fish will indicate a temperature intermediate between the two. Assuming that the microwave energy source is reasonably constant, the susceptor temperature will always be hotter than the fluoroptic probe temperature at the food–package interface. If the food mass is largely made up of water, it will appear to be at a steady-state temperature as it loses water, but the interface temperature can be significantly lower than the internal susceptor temperature.

A variation of a DSC technique used by Moore *et al.* (1989) to study the heat history of amorphous PET has been used to provide data on the maximum temperature attained by the PET film in cooking experiments. By using DSC, endotherms were obtained on both heated and unheated test samples. Unheated amorphous film has a very characteristic DSC trace. On the other hand, when it is placed in contact with food and heated in a microwave or conventional oven, its post-test DSC curve shows an endotherm indicative of increased crystallization that has occurred as a result of the heat treatment. The appearance of the film after heating is typically a patchwork of clear and opaque areas, with the opaqueness representing increased crystallinity. Some areas become hotter than neighboring areas, possibly because of variations in the microwave field and in food composition and contact surface. DSC curves are selectively run on those portions of the film that display the greatest degree of crystallinity and, therefore, experience the highest temperature. Control samples were heated in a hot air oven set at 200°C. DSC curves were taken before and after heating. The indicated maximum temperature was $200 \pm 3°C$.

Table 9.5 shows temperatures obtained by both fluoroptic probes and DSC curves for a series of PET films heated in cooking experiments. Results are given for controls heated in the hot air oven and for susceptor cooking experiments on pizza and pot pie. In almost every case, the maximum temperature attained by the package (as determined by DSC) exceeded that recorded by the fluoroptic probe. These results suggest that the indicated fluoroptic probe temperatures may be significantly lower than actual temperatures experienced by the film and very different from measured temperature maxima.

These tests raise significant questions about the reliability of fluoroptic point source temperature measurements in microwave environments. It appears that point source food–package interface temperatures are misleading and not indicative of the internal package temperatures. Such temperature results would be poor indicators of package performance and additive migration. For this reason, it is felt that calculated migration data or simulated migration tests based on the use of the interface temperature

Table 9.5 Food–package interface temperature estimates obtained by differential scanning calorimetry (DSC) and fluoroptic probes

Product	DSC av. temp. (°C)	n[a]	Fluoroptic temp. (°C)[b]
Pizza	207	4	130, 185, 140
Pizza	184	1	137, 134, 143
Pizza	228	1	203, 212, 207
Pizza	220	6	112, 133, 156
Meat pot pie	196	2	104, 176
Pizza	216	3	204, 221, 219
Pizza	222	3	197, 201, 210
Pizza	220	5	154, 173, 174
Film (150°C)[c]	154	4	
Film (175°C)[c]	180	3	
Film (201°C)[c]	206	2	

[a] DSC determinations were conducted *n* times on the same specimen.
[b] Measured at three different locations at the food–susceptor interface.
[c] PET films heated in a temperature-controlled hot-air oven were used for controls.

measurements, such as time–temperature product profiles, are not reliable substitutes for actually measuring additive migration into foods under cooking conditions.

9.9 Supercritical fluid extraction and chromatography for assessing sources of food contamination

A dilemma associated with analysis of contaminants in polymeric food packaging concerns the generation of large volumes of hazardous solvent waste, the disposal of which is expensive. This is especially true when Soxhlet extraction and HPLC techniques are used for the analysis. An alternative approach to the use of conventional solvent extraction and HPLC is to use supercritical fluids for extraction and chromatography.

A supercritical fluid (SF) is a fluid at or above its critical temperature and pressure. These fluids can provide a unique and efficient way to extract and chromatograph components of food packaging, because they possess gas-like mass-transfer characteristics as well as liquid-like solvating characteristics. The most commonly used SF in analytical chemistry procedures is carbon dioxide (CO_2), which has the advantages of a relatively low temperature and pressure critical point, 31°C and 7.38 MPa (73 atm), and can be readily obtained in high purity.

The fundamental principles and advantages of using SFs for analytical extraction, SF extraction (SFE), were outlined by Rizvi *et al.* (1986) and more recently by Hawthorne (1990). Fast extraction times are possible because SFs have larger solute diffusivities and lower viscosities than liquid solvents. The extraction times using SFs tend to be in the minutes range, whereas liquid extractions may take hours. The solvent strength of the SF can be easily varied by using temperature and pressure or possible modifiers like methanol or N_2O_2 to extract components of different polarities selectively. Because many SFs are gases at ambient conditions, concentrating analytes is much simpler than concentrating analytes in liquid solvents. Extraction of thermally unstable compounds can be more successful using SFs, because SFs like CO_2 have low critical temperatures ($31\,^\circ C$), permitting extractions to be carried out at low temperatures. Finally, the non-toxic nature of CO_2 is environmentally friendly and reduces solvent waste.

Another attractive feature of CO_2 for the SF chromatographic (SFC) analysis of contaminants from plastics is that it allows use of a non-specific detector, such as a flame ionization detector or MS detector, for those compounds which may be difficult to analyze by GC and compounds which do not absorb UV light. These types of compounds, which are usually non-volatile and non-UV absorbing, can sometimes be determined by HPLC, using a refractive index (RI) detector or an evaporative light-scattering detector. These detectors generally have higher detection limits and other restrictions. Alternatively, non-UV absorbing, non-volatile compounds can be measured by HPLC–MS, but instruments for these techniques are not widely available.

A significant limitation of SFC for trace analytical chemistry work, however, is the small sample injection size imposed on the SF system by the columns used in this technique. These small sample volumes greatly limit the detection limits of the technique. Attempts to use larger injection volumes without solvent venting or splitting may result in poor chromatography. These shortcomings of SFC have been compensated for by the introduction of splitless injection (Buskhe *et al.*, 1988) and a large-volume injection valve and sample concentrating apparatus which enhances the detectability to $10\,\mu g/kg$ for a polymer antioxidant (Campbell *et al.*, 1992). Afforded the sensitivity available with the large-volume injection technique, SFC has been used successfully to monitor the migration of low concentrations of polymer additives to food-simulating liquids (Berg *et al.*, 1993).

The use of SFE–SFC for the determination of oligomers and additives in plastics began around 1988. Anton *et al.* (1988) demonstrated that SFC could be used for oligomer and polymer additive analysis in ethylene–propylene O-rings. An SFE coupled to GC–MS was used to determine bromophenols and diphenylcarbonate in polybutylene terephthalate (Schmidt *et al.*, 1989). Additionally, SFE and SFC have been used for slip agents, phosphite stabilizers, UV stabilizers and antioxidants in polypropylene and polyethylene (Kithinji *et al.*, 1990; Ryan *et al.*, 1990). Polymer additives and oligomers have also been determined by Cotton *et al.* (1991) in nylon, PET, and poly(ether ketone), using on-line SFE–SFC.

The unique capabilities of SFC seem particularly advantageous for analyzing food contaminants from multi-layered food packaging laminates like the paper–adhesive–polymer constructions encountered in microwave susceptors. Each of these layers contains chemicals that can potentially migrate to food when the container is heated to over 200°C. Paperboard, which decomposes at these temperatures, may also contain coatings or additives. The adhesive contains monomers, oligomers and other additives which may be semi-volatile or non-volatile but nonetheless become mobile under these conditions. The PET layer also contains monomers, oligomers and possibly additives. Determination of these chemicals by a single technique like GC or HPLC alone is generally not possible. On the other hand, SFC has been demonstrated to be useful for this kind of problem. Calvey *et al.* (1995) used SFC to show the presence of C_{31}–C_{35} alkyl ketones (paper additive), 2-butoxyethoxyethanol (adhesive additive), diglycidyl ether of bisphenol A (adhesive monomer), diethylene glycol dibenzoate (plasticizer), and PET oligomers in extracts from susceptor packages. In the future, when SFC equipment becomes more readily available, it will no doubt become more widely used to solve problems of this type.

9.10 Modeling additive migration from polymers to foods

Monitoring diffusion of additives through polymers and determining their migration to foods is experimentally challenging. Mass transfer at the polymer–simulant interface, partitioning of migrant between the polymer and food-simulating solvent, assuring homogeneity of the migrant distribution in the polymer, and edge and seam effects on migration, as well as many other factors, make accurate experimental measurements difficult. Nonetheless, with an increasing interest in subjecting food packaging materials to higher and higher temperatures (greater than 135°C), there is a growing need for reliable migration data to demonstrate the safety of such usage. But in the higher temperature range, availability of adequate food simulants and experimental apparatus is limited. Certainly, alternative means for obtaining migration data that describe the performance of materials over a range of temperatures are highly desirable. For example, an effective mathematical model might offer advantages such as:

- Shortening the time required to evaluate molecular diffusion in numerous polymer systems over broad temperature ranges.
- Saving time by focusing laboratory resources on the most promising adjuvant–polymer combinations.
- In some cases, possibly reducing or eliminating extensive migration testing required by regulatory agencies and shortening the regulatory approval process for new substances, materials and chemical additives.

Our interest in modeling additive diffusion in polymers and using these models to predict migration to foods and food simulants dates back to 1976, when contracts were awarded to Arthur D. Little, Inc. (Cambridge,

MA) to assess the suitability of traditional food-simulating solvents (Arthur D. Little, 1983) and to the National Bureau of Standards (now the National Institute for Standards and Technology) to investigate factors controlling migration (NBS, 1982). From these studies came a number of important results. For example, a new, more efficient migration cell for two-sided extractions was developed, a set of equations was derived to describe some of the migration results, and alternative food-simulating solvents were investigated. Snyder and Breder (1985) followed up that work, using computer-based programs to fit observed migration data to the proposed migration model. They used the new migration cell and a styrene in polystyrene system to evaluate possible improved fatty food simulating solvents.

Many others have either put forward or suggested diffusion models: Pace and Datyner (1979), Mauritz and Storey (1990), Lickly *et al.* (1993), Baner *et al.* (1994) and Földes (1995), among others. Each of these models has a different approach and priority to model the complex phenomenon of additive diffusion and, therefore, serves to meet a unique need(s). For example, Baner *et al.* (1994) used over 600 measured diffusion coefficients in polyolefins to construct a mathematical relationship for the diffusion in polyolefins as a function of temperature and molecular weight. This resulted in the following expression

$$D \leq 10\,000 \exp\left[A_p - aM_r - b\left(\frac{1}{T}\right)\right] \qquad (9.1)$$

where A_p is a polymer-specific constant and a and b are constants accounting for the dependence of diffusion on the penetrant's molecular weight (M_r). The values for a and b are 0.0101 and 10450, respectively. The values for A_p are 11, 8 and 6 for LDPE, HDPE and PP, respectively. Using these coefficients and the expression above, estimates of diffusion coefficients can be obtained for chemicals in polyolefins.

In polymer systems other than polyolefins, the amount of diffusion data is limited; therefore, constructing a model by statistical fitting of data is not possible. For this reason other approaches to characterizing migration/diffusion from food packaging polymers are needed. In an earlier paper (Limm and Hollifield, 1995a), we reviewed the basic concepts and mathematical equations of additive migration and used them to describe the effect of mixing on additive migration to foods. With renewed interest, we have more recently developed a semi-empirical model of additive diffusion and applied it to both polyolefin polymers (Limm and Hollifield, 1995b) and PET (Limm *et al.*, 1995).

9.10.1   Diffusion model

Our diffusion model (Limm and Hollifield, 1995b) is based on the simplification of existing diffusion theories by Pace and Datyner (1979) and the generalized trend found by Berens and Hopfenberg (1982) for the diffusion of small molecules in polymers. From these it is possible to arrive at the following relationships:

$$\ln D(MW,\ T) = \ln A + \alpha(MW)^{\frac{1}{2}} - \frac{K(MW)^{\frac{1}{3}}}{T} \qquad (9.2)$$

where D is the diffusion coefficient of a chemical/additive, MW is the molecular weight of a chemical/additive, T is the temperature in degrees Kelvin and A, α and K are constants determined from experimental data.

Some of the assumptions intrinsic in the model are:

1. That diffusion of chemicals is completely defined by their molecular weights; that is, the effect of molecular shape on the diffusion coefficient is ignored.
2. That the molecular weights of chemicals are proportional to the third power of their van der Waals diameters.
3. That, for large molecules, such as plasticizers or antioxidants, the relationships between molecular diameters and diffusion coefficients or the activation energy of diffusion are the same as those established for relatively small penetrants with molecular weights of less than 100 daltons (Berens and Hopfenberg, 1982).
4. That the activation energy of diffusion is constant in the temperature range of interest.

From a regulatory point of view, the failure of the model to take partition factors into consideration is not necessarily a major shortcoming. Typically, in the absence of migration data, 100% migration is assumed as a conservative estimate. In most cases, because diffusion in the polymer is a slow process, estimates of migration using Fickian diffusion coefficients will allow the calculation of more realistic migration levels than assuming that 100% of the adjuvant migrates in a specified time. Because partition factors very often influence the migration of adjuvants into non-fatty foods, the model would provide a conservative overestimate of actual migration in these cases.

Even so, availability of model diffusion data may be useful as a reference frame for experimental data which may not have been collected under ideal conditions, thus serving as a useful complement to very limited laboratory data. Alternatively, one has the option of using Piringer's (1994) model (Baner et al., 1994), which appears to have better low-temperature correlation with available data.

9.10.2   Determination of model parameters

Model parameters, A, α and K, are determined by using diffusion data for two compounds of different molecular weight (Limm and Hollifield, 1995b).

Table 9.6 Empirical constants for FDA's migration model[a]

Polymer	K	α	$\ln A$
PP	1335	0.597	−2.10
HDPE	1761	0.819	0.90
LDPE	1140	0.555	−4.16

[a]Reported by Limm and Hollifield (1995b).

One data set needs to cover diffusion in the polymer over a large temperature range. This data set is used to construct a $\ln D$ versus $1/T$ plot, where the slope of the linear regression line is K in the model expression. The second datum is used to solve the model equation for α and $\ln A$ by solving two simultaneous equations containing two different molecular weights and K as a constant slope (constant activation energy) for all compounds diffusing in the polymer.

Generally, migration data at high temperatures are scarce because of experimental difficulties in conducting migration measurements at these temperatures. This results in most migration data corresponding to temperatures less than 100°C. For polyolefins (the most common food packaging polymer) there are some experimental diffusion data above 100°C. To determine model parameters for low-density polyethylene (LDPE), high-density polyethylene (HDPE) and polypropylene (PP), migration data for two antioxidants (Goydan *et al.*, 1990; Limm and Hollifield, 1995b), Irganox 1010® (I-1010) and Irganox 1076® (I-1076) with molecular weights of 1178 and 531, respectively, were used. The results of modeling to determine the coefficients are tabulated in Table 9.6. The constants in Table 9.6 were used to calculate diffusion coefficients in HDPE for some chemicals of varying molecular weights and compared to the measured values in Table 9.7. The calculated values listed in Table 9.7 show good agreement with the measured values, except for BHT. The calculated BHT values are about 10–20 times faster than the measured diffusion values. A partial explanation for this difference may be that the experimental data for BHT were collected at 4, 20 and 40°C, temperatures at which partitioning factors are likely to slow migration. Since the model does not account for partition factors, it will overestimate migration in such cases. Work is in progress to validate the model further, and to explain these observed differences between experimental data and model predictions more satisfactorily.

The diffusion model described here represents a first step in systematically estimating additive migration over an extended temperature range based on limited experimental data, and is still undergoing validation. Although some of our simplifying assumptions perhaps may ignore potentially significant factors, they are designed to build in a margin of safety by overestimating migration, especially at the lower temperatures, where partitioning between the polymer surface and food simulants may be a factor. The model proposed here is necessarily semi-empirical because of the difficulty in accurately

Table 9.7 Comparison of measured and calculated diffusion coefficients for chemicals in HDPE

Chemical	Measured	Calculated
BHT[a]		
60°C	5.5×10^{-10}	6.8×10^{-9}
30°C	1.2×10^{-11}	2.9×10^{-10}
n-Dotriacontane[a]		
60°C	3.0×10^{-10}	2.4×10^{-10}
30°C	1.2×10^{-12}	4.4×10^{-12}
Octadecane[a]		
60°C	1.0×10^{-9}	3.5×10^{-9}
30°C	1.3×10^{-10}	1.3×10^{-10}
Butanal[b]		
25°C	3.6×10^{-8}	5.5×10^{-8}
Decanal[b]		
25°C	1.8×10^{-9}	1.1×10^{-9}
1-Pentanol[b]		
25°C	2.9×10^{-9}	2.1×10^{-8}
2-Heptanol[b]		
25°C	7.7×10^{-10}	5.2×10^{-9}
Irganox 1076[c]		
121°C	7.0×10^{-8}	8.1×10^{-8}

[a] NBS (1982).
[b] Johansson and Leufven (1994).
[c] Arthur D. Little, Inc. (1988).

quantifying polymer morphology and the use of limited data on penetrants that have rather large molecular dimensions. Early attempts to validate the model based on existing literature data have been mixed. This, we expect, is due to the difficulties others have also experienced in making reliable migration measurements. Consequently, validation experiments are planned in the near future. Meanwhile, initial results suggest that the model has the potential to be useful in estimating additive migration over a wide temperature range, especially when experimental data are scarce.

9.11   Experimental determination of diffusion coefficients

As indicated in the previous section, in order to use the modeling approach, one must know the diffusion coefficients for the migrating species in the polymer at the appropriate temperatures. For many chemicals in some food packaging polymers, such as polyolefins, many diffusion data have been published. However, there is little information on diffusion characteristics of large molecules in other polymers, such as PET. PET is a high-volume recycled polymer in the USA and is increasingly being used at elevated temperatures. Therefore, it is most helpful to be able to determine such data experimentally for large molecules, both to further validate theoretical model applications and to evaluate actual food packaging polymer applications.

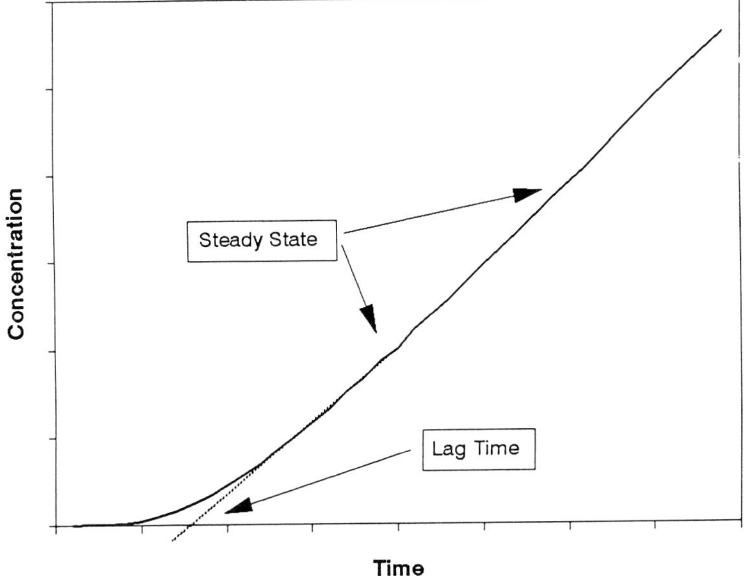

Figure 9.4 Ideal concentration versus time plot for diffusion through a film. Lagtime measurements are made by extrapolating the steady-state portion of the curve to zero concentration.

Three approaches are commonly used to measure diffusion coefficients in polymers. The first and probably most important is the lagtime method from permeation experiments. In a typical lagtime experiment, the amount of chemical transiting a polymer film of thickness l (cm) is measured and plotted versus time. For a polymer–penetrant system that obeys Fick's law, this type of plot is illustrated in Figure 9.4. The lagtime is determined by extrapolating the steady-state portion of this curve to a time when the concentration is zero. From the lagtime the diffusion coefficient can be calculated from the following relationship (Barrer, 1939):

$$D = \frac{l^2}{6 * \text{Lagtime}} \qquad (9.3)$$

The second approach is a static sorption experiment where the mass sorbed into a polymer is measured with an ultrasensitive balance. The amount of penetrant sorbed, M_t, is measured as a function of time. At long sorption times the amount sorbed will approach an equilibrium value M_∞. From the half-time of the sorption experiment, where $M_t/M_\infty = 0.5$, it has been shown that the diffusion coefficient is given by equation (9.4) (Ayres $et\ al.$, 1983):

$$D = 0.49 \frac{l^2}{t_{\frac{1}{2}}} \qquad (9.4)$$

Finally, the reverse of the sorption experiment, desorption, can be performed to determine diffusion coefficients. In desorption experiments, it is assumed that the migrant is uniformly distributed in the polymer. In these measurements the initial amount of migrant in the polymer does not have to be the equilibrium amount M_∞. The diffusion coefficient in this desorption experiment can be calculated by using equation (9.4) and the time corresponding to 50% of migrant loss from the polymer. Alternatively, at short times, the diffusion coefficient can be calculated from the slope of the curve of M_t versus the square root of the time for the desorption measurements by rearranging equation (9.5) (Crank, 1975). The experimental advantage of using sorption and desorption techniques for determining diffusion coefficients is that the results are not dependent on the seal integrity, as with permeation measurements. C_0 is the initial chemical concentration in the gas phase for sorption studies or in the polymer surface for desorption experiments.

$$M_t = 2C_0\sqrt{\frac{Dt}{\pi}} \qquad (9.5)$$

Of these three techniques, we have favored the use of lagtime experiments and have developed approaches for specifically measuring diffusion of non-volatile chemicals through biaxially oriented PET and nylon at elevated temperatures (Begley and Hollifield, 1995). Figure 9.5 illustrates a typical lagtime

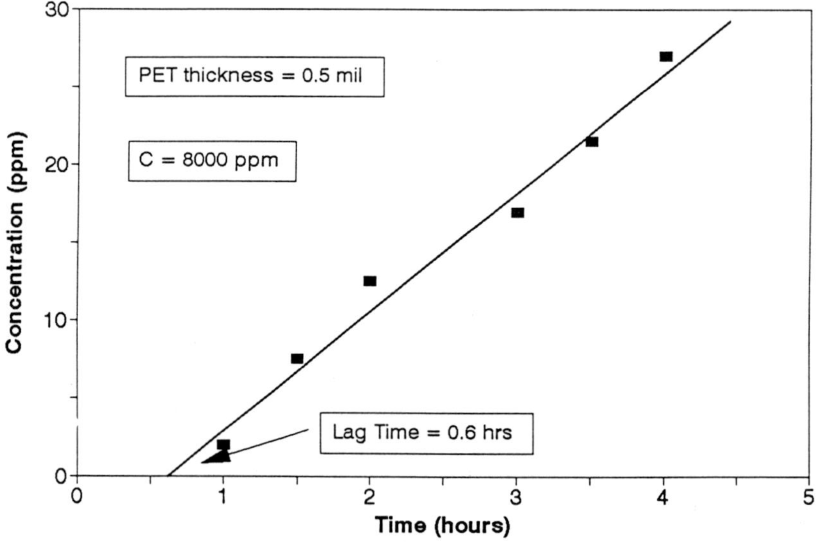

Figure 9.5 Illustration of a lagtime experiment for DEGDB permeating PET into Miglyol at 110°C.

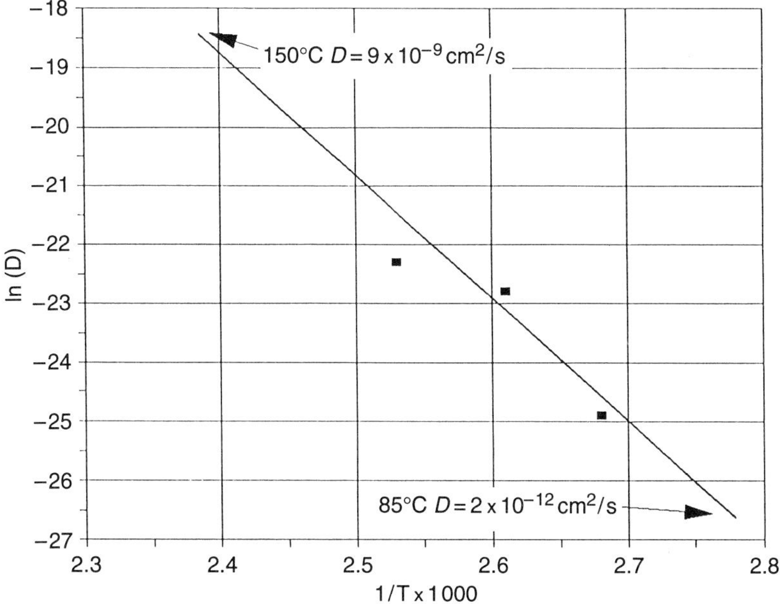

Figure 9.6 The Arrhenius behavior for the DEGDB/PET system.

experiment for diethylene glycol dibenzoate (DEGDB) permeating PET into Miglyol at 110°C. This figure suggests the permeation of DEGDB through PET exhibits Fickian type behavior. From the lagtime of 0.6 h, the effective diffusion coefficient for DEGDB in PET at 110°C is calculated with equation (9.1) to be 1.2×10^{-10} cm^2/s. The Arrhenius behavior for the DEGDB–PET system is illustrated in Figure 9.6. Extrapolating this data from 85°C to 150°C shows effective diffusion coefficients for DEGDB in PET to range from 2×10^{-12} cm^2/s to 9×10^{-9} cm^2/s. Having the means of generating these data over a range of temperatures not only permits the study of specific migration of individual additives and contaminants but also helps in obtaining valuable data with which to evaluate migration modeling predictions.

9.12 Functional barrier considerations in recycled polymer applications

Theoretical modeling also appears to be helpful in evaluating migration of contaminants from postconsumer recycled polymers (Begley and Hollifield, 1995). Over the past few years public pressure and local government mandates have put into place curbside collection of postconsumer recyclable glass, metal, paper and plastic wastes, and thus created an abundant supply of raw materials for recycling into new products or packaging materials. In 1993, the latest figures available, PET bottles were recycled at a USA national rate of 30%, up 23% over 1992, according to the National Association for Plastic Container Recovery, Charlotte, NC (Beck, 1994). In 1994, multi-layered PET bottles incorporating an inner layer of post-consumer recycled PET have been marketed in the USA to meet hot-fill and barrier requirements for juices, lemonades and teas. Recycled materials used in this manner for food packaging require some special considerations to ensure that non-regulated chemicals or contaminants either are not present in the food packaging material or do not migrate into the food (FDA, 1992). The potential food contamination may be greatly reduced by using alternative package designs, such as laminated structures. This section will discuss diffusion considerations for recycled polymers and the functional barrier characteristics of polymers (i.e. PET) as they apply to food packages made from recycled resins.

When contaminant levels in recycled polymers are difficult to predict because of variations in source raw materials, an alternate approach is the use of diffusion theory to estimate likely contaminant migration to food based on reasonable assumptions of source contamination. Evaluation of polymer diffusion properties as an alternative way to evaluate migration has been suggested elsewhere (Begley and Hollifield, 1993; Baner *et al.*, 1994). In this approach, the migration-limited case, the diffusion properties within each polymer must be characterized as a function of temperature. These diffusion properties can be evaluated by studying diffusion of surrogate chemicals likely to survive remanufacturing procedures. Additionally, the expected concentration of the test chemicals after reprocessing must be known. Once the diffusion properties of the polymer are known, then one can estimate the likelihood of a contaminant entering the food in amounts above a predetermined threshold level.

Food contamination can be estimated by using variations of previously developed mathematical relationships showing that diffusion of chemicals in polymers obeys Fick's law. For example, Till *et al.* (1982) demonstrated that the solution to Fick's law expressed in equation (9.5) gives reasonable estimates of chemical migration to food from the recycled polymer itself. From equation (9.5), migration (M_t) is directly related to the concentration of contaminant in the polymer (C_0, mass/cm^3) and to the square root of the diffusion coefficient (D, cm^2/s) and the time (t, s). Using equation (9.5) and an assumption that approximately 1.5 g of food contacts each cm^2 of packaging to convert food package surface area to food mass, the amount of migration over the shelf life of the package can be calculated. This calculation assumes that the concentration and the diffusion coefficient of the migrating species are known. By knowing the residual contamination level in the reprocessed polymer and its diffusion characteristics for the temperature conditions of use, an estimate of migration can be calculated.

If the recycled polymer is separated from the food by a functional barrier of virgin polymer, then, to a first approximation, the advantage in retarding migration of a specific migrant can be estimated by using the solution to Fick's law for the case of diffusion through a film where the initial concentration in the film is zero and the concentration on the outside of one side of the film is constant. This case is described mathematically in equation (9.6) (Crank, 1975), where l is the film thickness, C_0 is the surface concentration, and D, t are the same as described for equation (9.5).

$$M_t = \left[\frac{Dt}{l^2} - \frac{1}{6} - \frac{2}{\pi^2} \sum_{n=1}^{\infty} \frac{-1^n}{n^2} \exp\left(\frac{-Dn^2\pi^2 t}{l^2} \right) \right] C_0 l \qquad (9.6)$$

For comparative purposes let us assume the following scenario: the contaminant level in the recycled polymer is 100 mg/kg, the food was stored for 30 days in the package, and the diffusion coefficient for the contaminant in the polymer was 4×10^{-13} cm^2/s. Using these values, equation (9.5) calculates that 101 μg/kg of the contaminant would be in the food. For the case with a virgin layer 1 mil (0.0254 mm) thick, solution of equation (9.6) yields only 8 μg/kg of the contaminant in the food after the same 30 days. Obviously, the presence of the virgin layer has a significant advantage in reducing the possible exposure to specific contaminants. This effect was demonstrated experimentally by Franz *et al.* (1994) for recycled polypropylene (PP) covered with a virgin layer. They concluded that a PP virgin layer acted as a barrier to the migration of components from recycled PP under the intended conditions of use.

One of the shortcomings of using the diffusion through a film equation (equation 9.6) to estimate the amount of migration that might enter the food from a laminated structure is overestimation, which generally occurs when the value for D/l^2 exceeds 1.5. This situation is depicted in Figure 9.7, where the amounts of migration predicted using equations (9.5) and (9.6) are plotted versus time. This figure shows that after about 1 h, equation (9.6) predicts more migration from a layered package than would be expected from a monolayer package (equation 9.5), which is physically unreasonable. There cannot be a greater amount migrating from a package with a virgin layer than from a package without the layer. This overestimation results because equation (9.6) assumes that the concentration on the side containing the contaminant is always constant and that it is free to move directly into the film (i.e. a pure liquid always in contact with one side of the film). But in actuality, in a layered package, the total concentration that enters the film is governed by equation (9.5), which can result in a concentration gradient across the virgin layer. An analysis of this concentration gradient across the virgin layer was performed by Laoubi *et al.* (1995). Therefore, a more realistic approach to evaluating the overall migration through a layered package is to consider the layered package as an infinitely thick single layer, where the contamination is initially confined within a limited space. Migration can be estimated by evaluating the concentration–distance function given

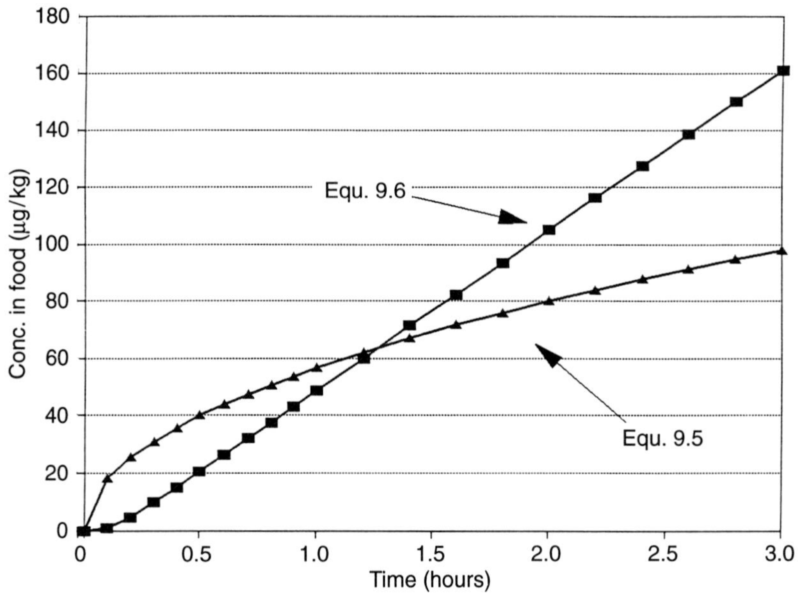

Figure 9.7 Comparison of the predicted migration using equations 9.5 and 9.6.

below in equation (9.7) (Crank, 1975) at a time equivalent to the package life-time.

$$C = \frac{1}{2} C_0 \left[\mathrm{erf}\left(\frac{h - x}{2\sqrt{(Dt)}} \right) + \mathrm{erf}\left(\frac{h + x}{2\sqrt{(Dt)}} \right) \right] \tag{9.7}$$

In equation (9.7), h is the thickness of the package material (i.e. 0.051 cm or 20 mil) and x is the distance from the center of the package material. An illustration of equation (9.7) with the perspective of a layered package is given in Figure 9.8. This figure illustrates the concentration profiles for two migrants, **A** and **B**, having diffusion coefficients equal to 5×10^{-8} cm^2/s and 5×10^{-9} cm^2/s respectively, and using a time of 1 h and a base package thickness of 20 mil (0.051 cm). The bar in this figure represents a 1 mil (0.025 mm) virgin layer and is positioned on the graph to indicate how the concentration–distance profiles cross a package with a virgin layer. This figure clearly illustrates that contaminants with these diffusion coefficients would diffuse past the 1 mil virgin layer in 1 h and would be expected to enter the food. The actual migration to food is also dependent on the mass-transfer characteristics at the food–package interface. For safety evaluations, it is assumed that all migrants at the food–package interface enter the food.

Migration from the layered package is evaluated by integrating the area under the curve at distances beyond the 1 mil (0.025 mm) virgin layer. For

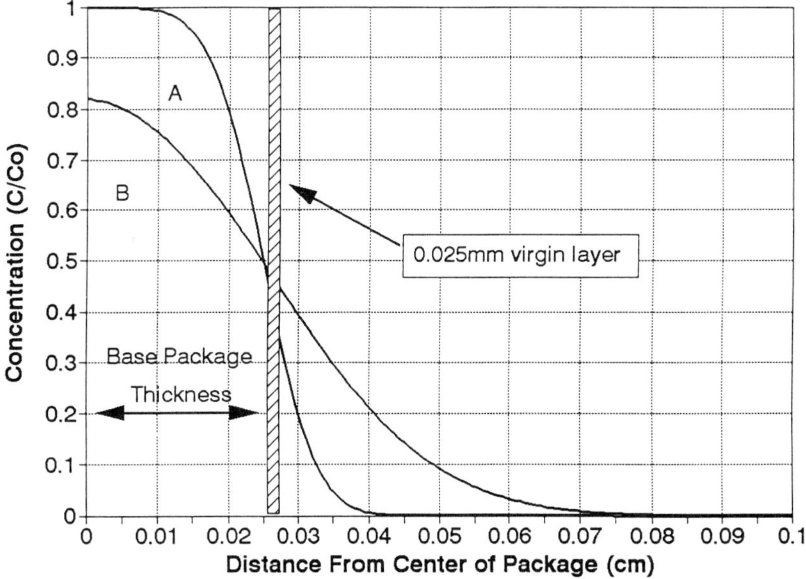

Figure 9.8 Concentration profile for migration through a layered package.

example, in Figure 9.8, the quantitative migration beyond the 1 mil layer would be 5% and 25% of the total amount in the package for contaminants with diffusion coefficients of $5 \times 10^{-9}\,\mathrm{cm^2/s}$ and $5 \times 10^{-8}\,\mathrm{cm^2/s}$, respectively.

Using data obtained from lagtime experiments on the permeation of PET by diethylene glycol dibenzoate (DEGDB) (Begley and Hollifield, 1995), one can evaluate the barrier effectiveness of various thicknesses of virgin PET film in combination with recycled polymer containing specified concentrations of contaminants (using equation 9.7). Figure 9.9 illustrates that migration from a 20-mil (0.051 cm) recycled layer containing 10 mg/kg contaminant and heated to 150°C for 1 h would result in less than 10 μg/kg migration through a 3-mil (0.076 mm) virgin PET food contact layer.

An excellent study on the effect of a virgin layer on the barrier performance of PET soft-drink bottles has been presented by Franz *et al.* (1995). In this work, a quantity of PET polymer was fortified with a mixture of trichloroethane, toluene, chlorobenzene, phenylcyclohexane, benzophenone, and phenyldecane. The contaminated polymer was used to prepare both single-ply and triple-ply bottles. In the triple-ply bottles, the doped polymer was sandwiched between two virgin layers of PET polymer, each 186 μm thick. Migration tests were performed on each bottle type using aqueous ethanol and 3% acetic acid as food simulants while the bottle contents were heated for 10 days at 40°C. The single-layered bottles made entirely with contaminated polymer exhibited considerably higher levels of migration; for

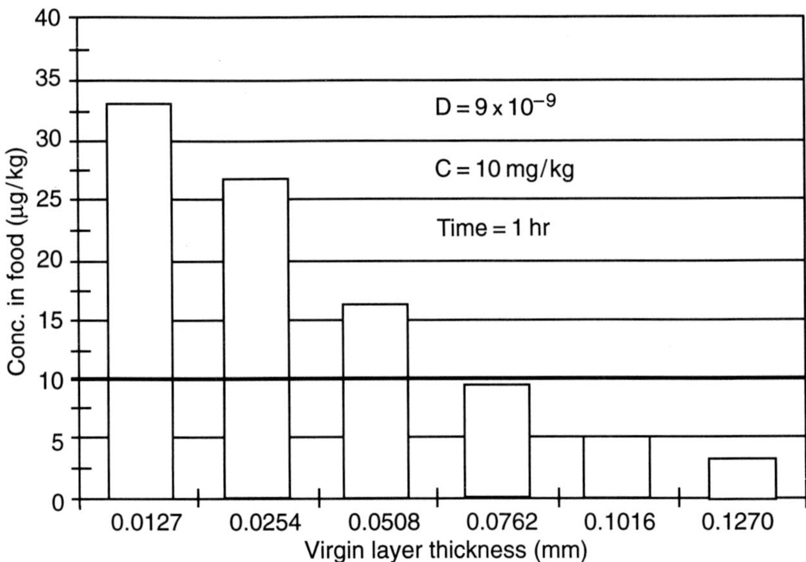

Figure 9.9 Migration from a 20-mil (0.051 cm) recycled layer containing 10 ppm contaminant and heated to 150°C for 1 h results in <10 ppb migration through a 3-mil (0.076 mm) virgin PET food contact layer.

example, toluene reached 56 μg/kg in some food-simulating liquids. On the other hand, migration from the triple-ply bottles was limited by the virgin polymer functional barrier layer to a maximum of 0.4 μg/kg.

These findings by Franz *et al.* (1995) are in accord with the theoretical functional barrier predictions described above. These studies greatly enhance our confidence in the use of theoretical diffusion modeling of large organic molecules. We anticipate that as scientists develop better techniques for measuring diffusion coefficients for various polymer systems, we will experience increased usage of theoretical models as replacement for more time-consuming experimental test procedures.

References

Anton, K., Menes, R. and Widmer, H.M. (1988) Direct coupling of CO_2 fluid extraction with capillary supercritical fluid chromatography. *Chromatographia*, **26**, 221–223.

Arthur D. Little, Inc. (1983) Final summary report: A study of indirect food additive migration. FDA Contract 223-77-2630.

Arthur D. Little, Inc. (1988) Final summary report: High temperature migration testing of indirect food additives. FDA contract 223-87-2162.

Ashby, R. (1988) Migration from polyethylene terephthalate under all conditions of use. *Food Additives and Contaminants*, **5**, 485–492.

ASTM Standard F 1349-91 (1991) Standard test method for nonvolatile UV absorbing extractables from microwave susceptor packaging. American Society for Testing and Materials, Philadelphia, PA.

ASTM Standard F 1308-90 (1990) Standard test method for quantitating volatile extractives in microwave susceptors used for food products. American Society for Testing and Materials, Philadelphia, PA.

Ayres, J.L., Osborne, J.L., Hopfenberg, H.B. and Koros, W.J. (1983) Effect of variable storage times on the calculation of diffusion coefficients characterizing small molecule migration in polymers. *Industrial and Engineering Chemistry Product Research and Development*, **22**, 86–89.

Baner, A.L., Franz, R. and Piringer, O.G. (1994) Alternative methods for the determination and evaluation of migration potential from polymeric food contact materials. *Deutsche Lebensmittel-Rundschau*, **90**(5), 137–143.

Barrer, R.M. (1939) Permeation, diffusion and solution of gases in organic polymers. *Transactions of the Faraday Society*, **35**, 628–643.

Beck, R.W. (1994) *1993 National Post-Consumer Plastics Recycling Rate Study*, American Plastics Council, Washington, DC.

Begley, T.H. and Hollifield, H.C. (1989) High-performance liquid chromatographic determination of migrating poly(ethylene terephthalate) oligomers in corn oil. *Journal of Agricultural and Food Chemistry*, **38**, 145–148.

Begley, T.H. and Hollifield, H.C. (1990a) Evaluation of polyethylene terephthalate cyclic trimer migration from microwave food packaging using temperature–time profiles. *Food Additives and Contaminants*, **7**, 339–346.

Begley, T.H. and Hollifield, H.C. (1990b) Migration of dibenzoate plasticizer and polyethylene terephthalate cyclic oligomers from microwave susceptor packaging into food-simulating liquids and food. *Journal of Food Protection*, **53**, 1062–1066.

Begley, T.H. and Hollifield, H.C. (1991) Application of a poly(tetrafluoroethylene) single-sided migration cell for measuring migration through microwave susceptor films. In *Food Packaging Interactions II*, ACS Symposium Series no. 473, American Chemical Society, Washington, DC, pp. 53–66.

Begley, T.H. and Hollifield, H.C. (1993) Recycled polymers in food packaging: migration considerations. *Food Technology*, **47**, 109–112.

Begley, T.H. and Hollifield, H.C. (1995) Food packaging made from recycled polymers: functional barrier considerations. ACS Symposium Series no. 609, American Chemical Society, Washington, DC pp. 445–457.

Begley, T.H., Dennison, J.L. and Hollifield, H.C. (1990) Migration into food of polyethylene terephthalate (PET) cyclic oligomers from PET microwave susceptor packaging. *Food Additives and Contaminants*, **7**, 797–803.

Berens, A.R. and Hopfenberg, H.B. (1982) Diffusion of organic vapors at low concentrations in glassy PVC, polystyrene and PMMA. *Journal of Membrane Science*, **10**, 283–303.

Berg, B.E., Hegna, D.R., Orlien, N. and Greibrokk, T. (1993) Determination of low levels of polymer additives migrating from polypropylene to food simulating liquids by capillary SFC and solvent venting injection. *Chromatographia*, **37**, 271–276.

Buskhe, A.F., Berg, B.E., Gyllenhaal, O. and Greibrokk, T. (1988) Splitless injection in capillary supercritical fluid chromatography. *Journal of High Resolution Chromatography and Chromatography Communications*, **11**, 16–20.

Calvey, E.M., Begley, T.H. and Roach, J.A.G. (1995) SFC–FTIR spectroscopy and SFC–MS of nonvolatile extrants of microwave susceptor packaging. *Journal of Chromatographic Science*, **33**, 61–65.

Campbell, R.M., Cortes, H.J. and Green, L.S. (1992) Large-volume injection in capillary supercritical fluid chromatography. *Analytical Chemistry*, **64**, 2852–2857.

Castle, L., Mayo, A., Crews, C. and Gilbert, J. (1989) Migration of poly(ethylene terephthalate) (PET) oligomers from PET plastics into foods during microwave and conventional cooking and into bottled beverages. *Journal of Food Protection*, **52**, 337–342.

Castle, L., Jickells, S.M., Gilbert, J. and Harrison, N. (1990) Migration testing of plastics and microwave-active materials for high-temperature food-use applications. *Food Additives and Contaminants*, **7**, 779–796.

Code of Federal Regulations (1989) Section 21, Part 174-177, Indirect Food Additives.

Code of Federal Regulations (1995) Section 21, Part 174.5, Indirect Food Additives.

Cotton, N.J., Bartle, K.D., Clifford, A.A., Ashraf, S., Moulder, R. and Dowle, C.J. (1991) Analysis of low molecular weight constituents of polypropylene and other polymeric materials using on-line SFE–SFC. *Journal of High Resolution Chromatography*, **14**, 164–168.

Crank, J. (1975) *The Mathematics of Diffusion*, 2nd edn, Oxford University Press, Oxford, UK.

De Kruijf, N. and Rijk, R. (1988) Iso-octane as fatty food simulant: possibilities and limitations. *Food Additives and Contaminants*, **5**, 467–483.

De Kruijf, N. and Rijk, R. (1994) Test methods to simulate high-temperature exposure. *Food Additives and Contaminants*, **11**, 197–220.

DeLassus, P.T., Strandburg, G. and Howell, B.H. (1988) Flavor and aroma permeation in barrier film: the effects of high temperature and high humidity. *Tappi Journal*, **71**, 177–181.

FDA (1989) Packaging materials for use under high temperature conditions in microwave ovens. *Federal Register*, **54**, 37340–37343.

FDA (1992) *Points to Consider for the Use of Recycled Plastics in Food Packaging: Chemistry Considerations*. FDA, Center for Food Safety and Applied Nutrition (HFS-245), Washington, DC 20204.

FDA (1993) Food additives: threshold of regulation for substances used in food-contact articles. *Federal Register*, **58**, 52719–52729.

FDA (1995) *Recommendations for Chemistry Data for Indirect Food Additive Petitions*. FDA, Center for Food Safety and Applied Nutrition (HFS-245), Washington, DC 20204.

Figge, K. (1972) Migration of additives from plastics films into edible oils and fat simulants. *Food and Cosmetics Toxicology*, **10**, 815–828.

Figge, K. and Koch, J. (1973) Effect of some variables on the migration of additives from plastics into edible fats. *Food and Cosmetics Toxicology*, **11**, 975–988.

Földes, E. (1995) Mobility of additives in ethylene polymers. *Polymer Bulletin*, **34**, 93–99.

Franz, R., Huber, M. and Piringer, O.G. (1994) Testing and evaluation of recycled plastics for food packaging use—possible migration through a functional barrier. *Food Additives and Contaminants*, **11**, 479–496.

Franz, R., Huber, M., Piringer, O., Damant, A.P., Jickells, S.M. and Castle, L. (1996) Study of functional barrier properties of multi-layer recycled poly(ethyleneterephthalate) (PET) bottles for soft drinks. *Journal of Agricultural and Food Chemistry*, **44**, 892–897.

Goydan, R., Schwope, A.D., Reid, R.C. and Cramer, G. (1990) High temperature migration of antioxidants from polyolefins. *Food Additives and Contaminants*, **7**, 323–337.

Hawthorne, S.B. (1990) Analytical-scale supercritical fluid extraction. *Analytical Chemistry*, **62**, 633A–642A.

Hollifield, H.C. (1991) FDA studies of high-temperature food packaging. In *Food Packaging Interactions II*, ACS Symposium Series no. 473, American Chemical Society, Washington, DC, pp. 22–36.

Hollifield, H.C., Snyder, R.C., McNeal, T.P. and Fazio, T. (1988) Advances in analytical methods for migrants. In ACS Symposium Series no. 365, American Chemical Society, Washington, DC, pp. 136–145.

Jickells, S.M., Crews, C., Castle, L. and Gilbert, J. (1990) Headspace analysis of benzene in food contact materials and its migration into foods from plastics cookware. *Food Additives and Contaminants*, **7**, 197–205.

Johansson, F. and Leufven, A. (1994) Food packaging polymer films as aroma vapor barriers at different relative humidities. *Journal of Food Science*, **59**, 1328–1331.

Kashtock, M.E. and Breder, C.V. (1980) Migration of ethylene glycol from polyethylene terephthalate bottles into 3% acetic acid. *Journal of the AOAC International*, **63**, 168–172.

Kashtock, M.E., Wurts, C.B. and Hamlin, R.N. (1990) A multilab study of food/susceptor interface temperatures measured during microwave preparation of commercial food products. *Journal of Packaging Technology*, **4**, 14–19.

Kithinji, J.P., Bartle, K.D., Raynor, M.W. and Clifford, A.A. (1990) Rapid analysis of polyolefin antioxidants and light stabilizers by supercritical fluid chromatography. *Analyst*, **115**, 125–128.

Laoubi, S., Feigenbaum, A. and Vergnaud, J.M. (1995) Safety of recycled plastics for food contact materials: testing to define a functional barrier. *Packaging Technology and Science*, **8**, 17–27.

Lentz, R.R. and Crossett, T.M. (1988) Food/susceptor interface temperature during microwave heating. *Microwave World*, **9**, 11–16.

Lickly, T.D., Bell, C.D. and Lehr, K.M. (1990) The migration of Irganox 1010 antioxidant from high-density polyethylene and polypropylene into a series of potential fatty-food simulants. *Food Additives and Contaminants*, **7**, 805–814.

Lickly, T.D., Markham, D.A. and McDonald, M.E. (1993) Migration of acrylonitrile from styrene/acrylonitrile copolymer into food-simulating liquids. *Journal of Agricultural and Food Chemistry*, **41**, 119–124.

Limm, W. and Hollifield, H.C. (1995a) Effect of temperature and mixing on polymer adjuvant migration into corn oil and water. *Food Additives and Contaminants* (in press).

Limm, W. and Hollifield, H.C. (1995b). Modeling of additive diffusion in polyolefins. *Food Additives and Contaminants* (in preparation).

Limm, W., Begley, T.H. and Hollifield, H.C. (1995) Modeling of additive diffusion in polyethylene terephthalate. *209th ACS National Meeting*, Anaheim, CA.

Mauritz, K.A. and Storey, R.F. (1990) A general free volume based theory for the diffusion of large molecules in amorphous polymers above T_g. 1. Application to di-n-alkyl phthalates in PVC. *Macromolecules*, **23**, 441–450.

McNeal, T.P. (1988) *FDA Protocol for Determination of Microwave Susceptor-Derived Volatile Chemicals*. FDA, Division of Food and Color Additives, HFF-330, Washington, DC 20204.

McNeal, T.P. and Hollifield, H.C. (1993a) Determination of volatile chemicals released from microwave-heat-susceptor food packaging. *Journal of the AOAC International*, **76**, 1268–1275.

McNeal, T.P. and Hollifield, H.C. (1993b) Determination of components and thermal decomposition products of adhesive formulations used in high temperature food packaging. *Pittsburgh Conference on Analytical Chemistry and Applied Spectroscopy*, Atlanta GA, poster #435.

Moore, J.T., Reeves, B.J. and Rothwell, L.A. (1989) Monograph: *A Technique for the Determination of the Maximum Temperature Reached by Foods and Food-Tray Interfaces During Cooking*. Eastman Kodak Co., Kingsport, TN.

National Bureau of Standards (1982) Final summary report: Migration of low molecular weight additives in polyolefins and copolymers. NBSIR82-2472.

Pace, R.J. and Datyner, A.J. (1979) Statistical mechanical model of diffusion of simple penetrants in polymers. I. Theory. *Journal of Polymer Science: Polymer Physics Edition*, **17**, 437–452.

Piringer, O.G. (1994) Evaluation of plastics for food packaging. *Food Additives and Contaminants*, **11**, 221–230.

Risch, S.J., Heikkila, K. and Williams, R. (1991) Analysis of volatiles produced in foods and packages during microwave cooking. ACS Symposium Series No. 473, Food Packaging Interactions II, 1–10.

Rizvi, S.S.H., Benado, A.L., Zollweg, J.A. and Daniels, J.A. (1986) Supercritical fluid extraction: fundamental principles and modeling methods. *Food Technology*, **40**, 55–65.

Rose, W.P. (1991) Determining volatile extractives from microwave susceptor food packaging. In *Food Packaging Interactions II*, ACS Symposium Series no. 473, American Chemical Society, Washington, DC, pp. 67–78.

Ryan, T.W., Yocklovich, S.G., Watkins, J.C. and Levy, E.J. (1990) Quantitative analysis of additives in polymers using coupled supercritical fluid extraction–supercritical fluid chromatography. *Journal of Chromatography*, **505**, 273–282.

Sackett, P.A., Wendt, D.J., Graff, J.T. and Pesheck, P.S. (1991) Examination of migration of selected contaminants from microwave susceptors to food. *TAPPI, Proceedings of Polymers, Laminations, and Coatings Conference*, vol. 2, pp. 789–799.

Schmidt, S., Blomberg, L. and Wannnman, T. (1989) Analysis of volatiles in polymers, Part II. Supercritical fluid extraction–open tubular GC–MS. *Chromatographia*, **28**, 400–404.

Snyder, R.C. and Breder, C.V. (1985) New FDA migration cell used to study migration of styrene from polystyrene into various solvents. *Journal of the AOAC International*, **68**, 770–775.

Tice, P.A. (1988) PIRA project on migration of monomers and overall migration. *Food Additives and Contaminants*, **5**, 373–380.

Till, D.E., Ehntholt, D.J., Reid, R.C., Schwartz, P.S., Sidman, K.R., Schwope, A.D. and Whelan, R.H. (1982) Migration of BHT antioxidant from high density polyethylene to foods and food simulants. *Industrial and Engineering Chemistry Product Research and Development*, **21**, 106–113.

10 Progress in developing European statutory methods of analysis

ROGER WOOD

Summary

The procedures that are required for a laboratory to produce consistently reliable data are described; such procedures include the need to implement an appropriate programme of quality assurance measures, i.e. to demonstrate that it is in statistical control, to participate in proficiency testing schemes which provide an objective means of assessing and documenting the reliability of the data it is producing, and to use methods of analysis which are 'fit for purpose'.

The requirements of official bodies which aid the production of reliable data in this area are described, as is the progress that has been made internationally in the preparation of protocols on internal quality control, proficiency testing and procedures for the collaborative testing of methods of analysis.

Methods of analysis which are being developed by the European Committee for Standardization (CEN) for food additives and contaminants are outlined; such procedures are likely to be of future importance in view of their possible direct incorporation in, or their use to enforce, future EU legislation.

10.1 Introduction

It is now internationally recognised that for a laboratory to produce consistently reliable data it must implement an appropriate programme of quality assurance measures. Amongst such measures, the laboratory must demonstrate that it is in statistical control, must participate in proficiency testing schemes which provide an objective means of assessing and documenting the reliability of the data it is producing, and must use methods of analysis which are 'fit for purpose'. These requirements may be summarised as follows.

10.1.1 Internal quality control (IQC)

IQC is one of a number of concerted measures that analytical chemists can take to ensure that the data produced in the laboratory are of known quality and uncertainty. In practice this is determined by comparing the results achieved in the laboratory at a given time with a standard. IQC therefore comprises the routine practical procedures that enable the analyst to accept a result or group of results or reject the results and repeat the analysis. IQC is undertaken by the inclusion of particular reference materials, 'control materials', into the analytical sequence and by duplicate analysis.

10.1.2   Proficiency testing

Participation in proficiency testing schemes provides laboratories with an objective means of assessing and documenting the reliability of the data they are producing. Although there are several types of proficiency testing schemes they all share a common feature: test results obtained by one laboratory are compared with those obtained by one or more testing laboratories.

The proficiency testing schemes must provide a transparent interpretation and assessment of results. Laboratories wishing to demonstrate their proficiency should seek and participate in proficiency testing schemes relevant to their area of work.

10.1.3   Analytical methods

Analytical methods should be validated as fit for purpose before use by a laboratory. Laboratories should ensure that, as a minimum, the methods they use are fully documented, laboratory staff are trained in their use, and control mechanisms are established to ensure that the procedures are under statistical control.

10.2   Legislation—the EU Food Control Directive

Although it is widely recognised that it is necessary for the above measures to be implemented, in the food sector control laboratories are also being required to do so by legislative requirements, as described below.

Methods of analysis have been prescribed by legislation for a number of foodstuffs since the UK acceded to the European Community in 1972. However, the Community new recognises that the quality of results from a laboratory is equally as important as the method used to obtain the results. This is best illustrated by consideration of the Council Directive on the Official Control of Foodstuffs (OCF) which was adopted by the Community in June, 1989 (EEC, 1989).

In Article 13 it is stated:

'In order to ensure that the application of this Directive is uniform throughout the Member States, the Commission shall, within one year of its adoption, make a report to the European Parliament and to the Council on the possibility of establishing Community quality standards for all laboratories involved in inspection and sampling under this Directive.'

Following that the Commission, in September 1990, produced a Report which recommended establishing Community quality standards for all laboratories involved in inspections and sampling under the OCF Directive. Proposals on this have now been adopted by the Community in the Directive on Additional Measures Concerning the Food Control of Foodstuffs (AMFC) (EEC, 1993).

The relevant Articles are:

Article 3, which states:

'1. Member States shall take all measures necessary to ensure that the laboratories referred to in Article 7 of Directive 89/397/EEC comply with the general criteria for the operation of testing laboratories laid down in European Standard EN 45001 supplemented by Standard Operating Procedures and the random audit of their compliance by quality assurance personnel, in accordance with the OECD principles Nos. 2 and 7 of good laboratory practice as set out in Section II of Annex 2 of the Decision of the Council of the OECD of 12 Mar 1981 concerning the mutual acceptance of data in the assessment of chemicals.

2. In assessing the laboratories referred to in Article 7 of Directive 89/397/EEC Member States shall:

(a) apply the criteria laid down in European Standard EN 45002; and
(b) require the use of proficiency testing schemes as far as appropriate.

Laboratories meeting the assessment criteria shall be presumed to fulfil the criteria referred to in paragraph 1.

Laboratories which do not meet the assessment criteria shall not be considered as laboratories referred to in Article 7 of the said Directive.

3. Member States shall designate bodies responsible for the assessment of laboratories as referred to in Article 7 of Directive 89/397/EEC. These bodies shall comply with the general criteria for laboratory accreditation bodies laid down in European Standard EN 45003.

4. The accreditation and assessment of testing laboratories referred to in this article may relate to individual tests or groups of tests. Any appropriate deviation in the way in which the standards referred to in paragraphs 1, 2 and 3 are applied shall be adopted in accordance with the procedure laid down in Article 8.'

and Article 4, which states:

'Member States shall ensure that the validation of methods of analysis used within the context of official control of foodstuffs by the laboratories referred to in Article 7 of Directive 89/397/EEC comply whenever possible with the provisions of paragraphs 1 and 2 of the Annex to Council Directive 85/591/EEC of 23 December 1985 concerning the introduction of Community methods of sampling and analysis for the monitoring of foodstuffs intended for human consumption.' (EEC, 1985).

As a result of the adoption of the above Directives, legislation is now in place to ensure that there is confidence not only in national laboratories but also those of the other Member States—thus facilitating the so-called 'mutual recognition' aspects.

The effect of the AMFC Directive is that organisations must consider the following aspects within the laboratory:

- the organisation of the laboratory,
- how well the laboratory actually carries out analyses, and
- the methods of analysis used in the laboratory.

All these aspects are interrelated, but in simple terms may be thought of as:

- becoming accredited to an internationally recognised standard; such accreditation is aided by the use of internal quality control procedures,
- participating in proficiency schemes, and
- using validated methods.

Although the legislative requirements apply only to food control laboratories, the effect of their adoption is that other food laboratories will have to achieve the same standard in order for their results to be recognised as equivalent.

10.3   Accreditation

The AMFC Directive requires that food control laboratories should be accredited to the EN 45000 series of Standards as supplemented by some of the OECD GLP principles. In the UK, government departments will nominate the United Kingdom Accreditation Service (UKAS) to carry out the accreditation of official food control laboratories for all the aspects prescribed in the Directive. However, as the accreditation agency will also be required to comply to EN 45003 Standard and to carry out assessments in accordance with the EN 45002 Standard, any other accreditation agencies that are members of the European Co-operation for Accreditation of Laboratories (EAL) may also be nominated to carry out the accreditation. Similar procedures will be followed in the other Member States, all having or developing equivalent organisations to UKAS.

It has been the normal practice for UKAS to accredit the scope of laboratories on a method-by-method basis. In the case of official food control laboratories undertaking non-routine or investigative chemical analysis, it is accepted that it is not practical to use an accredited fully documented method in the conventional sense, which specifies each sample type and analyte. In these cases a laboratory must have a protocol defining the approach to be adopted, which includes the requirements for validation and quality control. Full details of procedures used, including instrumental parameters, must be recorded at the time of each analysis in order to enable the procedure to be repeated in the same manner at a later date. It is therefore recommended that, for official food control laboratories undertaking analysis, appropriate methods are accredited on a generic basis with such generic accreditation being underpinned where necessary by specific method accreditation.

Food control laboratories seeking to be accredited for the purposes of the Directive should include, as a minimum, the following techniques in generic protocols: HPLC, GC, atomic absorption and/or ICP (and microscopy). A further protocol on sample preparation procedures (including digestion and solvent dissolution procedures) should also be developed. Other protocols for generic methods which are acceptable to UKAS may also be developed. Proximate analyses should be addressed as a series of specific methods including moisture, fat, protein and ash determinations.

Where specific Regulations are in force then the methods associated with the Regulations shall be accredited if the control laboratory wishes to offer enforcement of the Regulations to customers. Examples of these are methods of analysis for aflatoxins and methods of analysis for specific and overall migration for food contact materials.

By using the combination of specific method accreditation and generic accreditation it will be possible for laboratories to be accredited for all the analyses which they are capable and competent to undertake. Method performance validation data demonstrating that the method was fit for purpose shall be demonstrated before the test result is released, and method performance shall be monitored by on-going quality control techniques where applicable.

It will be necessary for laboratories to be able to demonstrate quality control procedures to ensure compliance with the EN 45001 Standard, an example of which would be compliance with the ISO/AOAC/IUPAC Guidelines on Internal Quality Control in Analytical Chemistry Laboratories (Thompson and Wood, 1995).

10.4   Internal quality control: Harmonised Guidelines For Internal Quality Control In Analytical Chemistry Laboratories

ISO, IUPAC and AOAC INTERNATIONAL have co-operated to produce agreed protocols on the 'Design, Conduct and Interpretation of Collaborative Studies' (Horwitz, 1988) and on the 'Proficiency Testing of (Chemical) Analytical Laboratories' (Thompson and Wood, 1993). The Working Group that produced these protocols has prepared a further protocol on the internal quality control of data produced in analytical laboratories. The document was finalised in 1994 and published in 1995 as the 'Harmonised Guidelines For Internal Quality Control In Analytical Chemistry Laboratories' (Thompson and Wood, 1995).

The use of the procedures outlined in the Protocol should aid compliance with the accreditation requirements specified above.

10.4.1   Basic concepts

The protocol sets out guidelines for the implementation of internal quality control (IQC) in analytical laboratories. IQC is one of a number of concerted measures that analytical chemists can take to ensure that the data produced in the laboratory are fit for their intended purpose. In practice, fitness for purpose is determined by a comparison of the accuracy achieved in a laboratory at a given time with a required level of accuracy. Internal quality control therefore comprises the routine practical procedures that enable the analytical chemist to accept a result or group of results as fit for purpose, or reject the results and repeat the analysis. As such, IQC is an important determinant of the quality of analytical data, and is recognised as such by accreditation agencies.

Internal quality control is undertaken by the inclusion of particular reference materials, called 'control materials', into the analytical sequence, and by duplicate analysis. The control materials should, wherever possible, be representative of the test materials under consideration in respect of matrix composition, the state of physical preparation and the concentration range of the analyte. As the control materials are treated in exactly the same way as the test materials, they are regarded as surrogates that can be used to characterise the performance of the analytical system, both at a specific time and over longer intervals. Internal quality control is a final check of the correct execution of all of the procedures (including calibration) that are prescribed in the analytical protocol and all of the other quality assurance measures that underlie good analytical practice. IQC is therefore necessarily retrospective. It is also required to be as far as possible independent of the analytical protocol, especially the calibration, that it is designed to test.

Ideally both the control materials and those used to create the calibration should be traceable to appropriate certified reference materials or a recognised empirical reference method. When this is not possible, control materials should be traceable at least to a material of guaranteed purity or other well characterised material. However, the two paths of traceability must not become coincident at too late a stage in the analytical process. For instance, if control materials and calibration standards were prepared from a single stock solution of analyte, IQC would not detect any inaccuracy stemming from the incorrect preparation of the stock solution.

In a typical analytical situation several, or perhaps many, similar test materials will be analysed together, and control materials will be included in the group. Often determinations will be duplicated by the analysis of separate test portions of the same material. Such a group of materials is referred to as an analytical 'run'. (The words 'set', 'series' and 'batch' have also been used as synonyms for 'run'.) Runs are regarded as being analysed under effectively constant conditions. The batches of reagents, the instrument settings, the analyst, and the laboratory environment will, under ideal conditions, remain unchanged during analysis of a run. Systematic errors should therefore remain constant during a run, as should the values of the parameters that describe random errors. As the monitoring of these errors is of concern, the run is the basic operational unit of IQC.

A run is therefore regarded as being carried out under repeatability conditions, i.e. the random measurement errors are of a magnitude that would be encountered in a 'short' period of time. In practice the analysis of a run may occupy sufficient time for small systematic changes to occur. For example, reagents may degrade, instruments may drift, minor adjustments to instrumental settings may be called for, or the laboratory temperature may rise. However, these systematic effects are, for the purposes of IQC, subsumed into the repeatability variations. Sorting the materials making up a run into a randomised order converts the effects of drift into random errors.

10.4.2 Scope of the Guidelines

The Guidelines are a harmonisation of IQC procedures that have evolved in various fields of analysis, notably clinical biochemistry, geochemistry, environmental studies, occupational hygiene and food analysis. There is much common ground in the procedures from these various fields. However, analytical chemistry comprises an even wider range of activities, and the basic principles of IQC should be able to encompass all of these. The Guidelines will be applicable in the great majority of instances, although there are a number of IQC practices that are restricted to individual sectors of the analytical community and so not included in the Guidelines.

In order to achieve harmonisation and provide basic guidance on IQC, some types of analytical activity have been excluded from the Guidelines. Issues specifically excluded are as follows:

(i) *Quality control of sampling.* While it is recognised that the quality of the analytical result can be no better than that of the sample, quality control of sampling is a separate subject and in many areas is not yet fully developed. Moreover, in many instances analytical laboratories have no control over sampling practice and quality.

(ii) *In-line analysis and continuous monitoring.* In this style of analysis there is no possibility of repeating the measurement, so the concept of IQC as used in the Guidelines is inapplicable.

(iii) *Multivariate IQC.* Multivariate methods in IQC are still the subject of research and cannot be regarded as sufficiently established for inclusion in the Guidelines. The current document regards multi-analyte data as requiring a series of univariate IQC tests. Caution is necessary in the interpretation of this type of data to avoid inappropriately frequent rejection of data.

(iv) *Statutory and contractual requirements.*

(v) *Quality assurance measures* such as pre-analytical checks on instrumental stability, wavelength calibration, balance calibration, tests on resolution of chromatography columns, and problem diagnostics are not included. For present purposes they are regarded as part of the analytical protocol, and IQC tests their effectiveness together with the other aspects of the methodology.

10.4.3 Internal quality control and uncertainty

A prerequisite of analytical chemistry is the recognition of 'fitness for purpose', the standard of accuracy that is required for an effective use of the analytical data. This standard is arrived at by consideration of the intended uses of the data, although it is seldom possible to foresee all of the potential future applications of analytical results. For this reason, in order to prevent inappropriate interpretation, it is important that a statement of the uncertainty should accompany analytical results, or be readily available to those who wish to use the data.

Strictly speaking, an analytical result cannot be interpreted unless it is accompanied by knowledge of its associated uncertainty at a stated level of confidence. A simple example demonstrates this principle. Suppose that there is a statutory requirement that a foodstuff must not contain more than 10 mg/kg of a particular constituent. A manufacturer analyses a batch and obtains a result of 9 mg/kg for that constituent. If the uncertainty on the result expressed as a half-range (assuming no sampling error) is 0.1 mg/kg (i.e. the true result falls, with a high probability, within the range 8.9–9.1) then it may be assumed that the legal limit is not exceeded. If, in contrast, the uncertainty is 2 mg/kg then there is no such assurance. The interpretation and use that may be made of the measurement thus depends on the uncertainty associated with it.

Analytical results should therefore have an associated uncertainty if any definite meaning is to be attached to them or an informed interpretation made. If this requirement cannot be fulfilled, the use to which the data can be put is limited. Moreover, the achievement of the required measurement uncertainty must be tested as a routine procedure, because the quality of data can vary, both in time within a single laboratory and between different laboratories. IQC comprises the process of checking that the required uncertainty is achieved in a run.

10.4.4 Recommendations in the Guidelines

The following recommendations represent integrated approaches to IQC that are suitable for many types of analysis and applications areas. Managers of laboratory quality systems will have to adapt the recommendations to the demands of their own particular requirements. Such adoption could be implemented, for example, by adjusting the number of duplicates and control material inserted into a run, or by the inclusion of any additional measures favoured in the particular application area. The procedure finally chosen and its accompanying decision rules must be codified in an IQC protocol that is separate from the analytical system protocol

The practical approach to quality control is determined by the frequency with which the measurement is carried out and the size and nature of each run. The following recommendations are therefore made. (The use of control charts and decision rules is covered in Appendix 1 to the Guidelines.)

In all of the following the order in the run in which the various materials are analysed should be randomised if possible. A failure to randomise may result in an underestimation of various components of error.

(i) *Short (e.g. $n < 20$) frequent runs of similar materials.* Here the concentration range of the analyte in the run is relatively small, so a common value of standard deviation can be assumed.

Insert a control material at least once per run. Plot either the individual values obtained, or the mean value, on an appropriate control chart. Analyse in duplicate at least half of the test materials, selected at random. Insert at least one blank determination.

(ii) *Longer (e.g. $n > 20$) frequent runs of similar materials.* Again a common level of standard deviation is assumed.

Insert the control material at an approximate frequency of one per ten test materials. If the run size is likely to vary from run to run it is easier to standardise on a fixed number of insertions per run and plot the mean value on a control chart of means. Otherwise plot individual values. Analyse in duplicate a minimum of five test materials selected at random. Insert one blank determination per ten test materials.

(iii) *Frequent runs containing similar materials but with a wide range of analyte concentration.* Here we cannot assume that a single value of standard deviation is applicable.

Insert control materials in total numbers approximately as recommended above. However, there should be at least two levels of analyte represented, one close to the median level of typical test materials, and the other approximately at the upper or lower decile as appropriate. Enter values for the two control materials on separate control charts. Duplicate a minimum of five test materials, and insert one procedural blank per ten test materials.

(iv) *Ad hoc analysis.* Here the concept of statistical control is not applicable. It is assumed, however, that the materials in the run are of a single type. Carry out duplicate analysis on all of the test materials. Carry out spiking or recovery tests or use a formulated control material, with an appropriate number of insertions (see above), and with different concentrations of analyte if appropriate. Carry out blank determinations. As no control limits are available, compare the bias and precision with fitness-for-purpose limits or other established criteria.

By following the above recommendations laboratories would introduce internal quality control measures which are an essential aspect of ensuring that data released from a laboratory are fit for purpose. If properly executed, quality control methods can monitor the various aspects of data quality on a run-by-run basis. In runs where performance falls outside acceptable limits, the data produced can be rejected and, after remedial action on the analytical system, the analysis can be repeated.

The Guidelines stress, however, that internal quality control is not fool-proof even when properly executed. Obviously it is subject to 'errors of both kinds', i.e. runs that are in control will occasionally be rejected and runs that are out of control occasionally accepted. Of more importance, IQC cannot usually identify sporadic gross errors or short-term disturbances in the analytical system that affect the results for individual test materials. Moreover, inferences based on IQC results are applicable only to test materials that fall within the scope of the analytical method validation. Despite these limitations, which professional experience and diligence can alleviate to a degree, internal quality control is the principal recourse available for ensuring that only data of appropriate quality are released from a laboratory. When properly executed it is very successful.

The Guidelines also stress that the perfunctory execution of any quality system will not guarantee the production of data of adequate quality. The correct procedures for feedback, remedial action and staff motivation must also be documented and acted upon. In other words, there must be a genuine commitment to quality within a laboratory for an internal quality control programme to succeed, i.e. the IQC must be part of a complete quality management system.

10.5 Proficiency testing: ISO/IUPAC/AOAC INTERNATIONAL Harmonised Protocol For Proficiency Testing Of (Chemical) Analytical Laboratories

The need for laboratories carrying out analytical determinations to demonstrate that they are doing so competently has become paramount. It may well be necessary for such laboratories not only to become accredited and to use fully validated methods, but to participate successfully in proficiency testing schemes. Thus, proficiency testing has assumed a far greater importance than previously.

10.5.1   What is proficiency testing?

A proficiency testing scheme is defined as a system for objectively checking laboratory results by an external agency. It includes comparison of a laboratory's results at intervals with those of other laboratories, the main object being the establishment of trueness.

In addition, although various protocols for proficiency testing schemes have been produced, the need now is for a harmonised protocol that will be universally accepted; the progress towards the preparation and adoption of an internationally recognised protocol is described below. Various terms have been used to describe schemes conforming to the draft protocol (e.g. external quality assessment, performance schemes, etc.), but the preferred term is 'proficiency testing'.

Proficiency testing schemes are based on the regular circulation of homogeneous samples by a co-ordinator, analysis of samples (normally by the laboratory's method of choice) and an assessment of the results. However, although many organisations carry out such schemes, there has been no international agreement on how this should be done—in contrast to the collaborative trial situation. In order to rectify this, the same international group which drew up the collaborative trial protocols was invited to prepare one for proficiency schemes (the first meeting to do so was held in April 1989). Other organisations, such as CEN, are also expected to address the problem.

10.5.2   Why proficiency testing is important

Participation in proficiency testing schemes provides laboratories with a means of objectively assessing, and demonstrating, the reliability of the data they produce. Although there are several types of schemes, they all share a common feature of comparing test results obtained by one testing laboratory with those obtained by other testing laboratories. Schemes may be 'open' to any laboratory, or participation may be invited. Schemes may set out to assess the competence of laboratories undertaking a very specific analysis (e.g. lead in blood) or more general analysis (e.g. food analysis). Although accreditation and proficiency testing are separate exercises, it is anticipated that accreditation assessments will increasingly use proficiency testing data.

10.5.3   Accreditation agencies

It is now recommended by ISO Guide 25 (ISO, 1990), the prime standard to which accreditation agencies operate, that such agencies require laboratories seeking accreditation to participate in an appropriate proficiency testing scheme before accreditation is gained.

There is now an internationally recognised protocol to which proficiency testing schemes should comply; this is the ISO/IUPAC/AOAC Harmonised Protocol described below.

10.5.4   ISO/IUPAC/AOAC International Harmonised Protocol For Proficiency Testing Of (Chemical) Analytical Laboratories

The international standardising organisations, AOAC, ISO and IUPAC have co-operated to produce an agreed 'International Harmonised Protocol For Proficiency Testing Of (Chemical) Analytical Laboratories' (Thompson and Wood, 1993). That protocol is recognised within the food sector of the European Community and also by the Codex Alimentarius Commission. The protocol makes the following recommendations about proficiency testing, all of which are important in the food sector.

10.5.5   Organisation of proficiency testing schemes

Framework.   Samples must be distributed regularly to participants who are to return results within a given time. The results will be statistically analysed by the organiser and participants will be notified of their performance. Advice will be available to poor performers, and participants will be kept fully informed of the scheme's progress. Participants will be identified by code only, to preserve confidentiality.

The scheme's structure for any one analyte or round in a series should be:

- Samples prepared;
- Samples distributed regularly;
- Participants analyse samples and report results;
- Results analysed and performance assessed;
- Participants notified of their performance;
- Advice available for poor performers, on request;
- Co-ordinator reviews performance of scheme;
- Next round commences.

Organisation.   The running of the scheme will be the responsibility of a co-ordinating laboratory/organisation. Sample preparation will either be contracted out or undertaken in-house. The co-ordinating laboratory must be of high reputation in the type of analysis being tested. Overall management of the scheme should be in the hands of a small steering committee (Advisory Panel) having representatives from the co-ordinating laboratory (who should be practising laboratory scientists), contract laboratories (if any), appropriate professional bodies and ordinary participants.

Samples.   The samples to be distributed must be generally similar in matrix to the unknown samples that are routinely analysed (in respect of matrix composition and analyte concentration range). It is essential they are of acceptable homogeneity and stability. The bulk material prepared must be effectively homogeneous so that all laboratories will receive samples that do not differ significantly in analyte concentration.

The procedure to be followed to test for sample heterogeneity is:

- Prepare whole of the bulk material in a satisfactory form.
- Divide the material into the containers.
- Select between 10 and 20 containers randomly.
- Homogenise contents of each selected container.
- From each take two test portions.
- Analyse portions in random order as one batch.
- Estimate sampling variance (one-way variance analysis, without exclusion of outliers).
- Assign σ as a precision standard at mean analyte value (i.e. the target value for the standard deviation for the proficiency test at the analyte concentration of interest).
- For effective homogeneity $\sqrt{(\text{sampling variance}/\sigma)} < 0.3$.

The co-ordinating laboratory should also show the bulk sample is sufficiently stable to ensure it will not undergo significant change throughout the duration of the proficiency test. Thus, prior to sample distribution, matrix and analyte stability must be determined by analysis after appropriate storage. Ideally the quality checks on samples referred should be performed by a different laboratory from that which prepared the sample, although it is recognised that this would probably cause considerable difficulty to the co-ordinating laboratory. The number of samples to be distributed per round for each analyte should be no more than five.

Frequency of sample distribution.   Sample distribution frequency in any one series should be not more than every 2 weeks and not less than every 4 months. A frequency greater than once every 2 weeks could lead to problems in turnaround of samples and results. If the period between distributions extends much beyond 4 months, there will be unacceptable delays in identifying analytical problems and the impact of the scheme on participants will be small. The frequency also relates to the field of application and amount of internal quality control that is required for that field. Thus, although the frequency range stated above should be adhered to, there may be circumstances where it is acceptable for a longer time-scale between sample distribution, e.g. if sample throughput per annum is very low. Advice on this respect would be a function of the Advisory Panel.

Estimating the assigned value (the 'true' result).   There are a number of possible approaches to determining the nominally 'true' result for a sample, but only three are normally considered. The result may be established from the amount of analyte added to the samples by the laboratory preparing the sample; alternatively, a 'reference' laboratory (or group of such expert laboratories) may be asked to measure the concentration of the analyte using definitive methods; or thirdly, the results obtained by the participating laboratories (or a substantial sub-group of these) may be used as the basis for the nominal 'true' result. The organisers of the scheme should provide the participants with a clear statement giving the basis for the assignment of reference values which should take into account the views of the Advisory Panel.

Choice of analytical method.   Participants can use the analytical method of their choice except when otherwise instructed to adopt a specified method. It is recommended that all methods should be properly validated before use. In situations where the analytical result is method-dependent, the true value will be assessed by using those results obtained using a defined procedure. If participants use a method which is not 'equivalent' to the defining method, then an automatic bias in result will occur when their performance is assessed.

Performance criteria.   For each analyte in a round, a criterion for the performance score may be set, against which the score obtained by a laboratory can be judged. A 'running score' could be calculated to give an assessment of performance spread over a longer period of time.

Reporting results.   Reports issued to participants should include data on the results from all laboratories together with participant's own performance score. The original results should be presented to enable participants to check correct data entry. Reports should be made available before the next sample distribution.

   Although all results should be reported, it may not be possible to do this in very extensive schemes (e.g. 700 participants determining 20 analytes in a round). Participants should, therefore, receive at least a clear report with the results of all laboratories in histogram form.

Liaison with participants.   Participants should be provided with a detailed information pack on joining the scheme. Communication with participants should be by newsletter or annual report together with a periodic open meeting; participants should be advised of changes in scheme design. Advice should be available to poor performers. Feedback from laboratories should be encouraged so that participants contribute to the scheme's development. Participants should view it as *their* scheme rather than one imposed by a distant bureaucracy.

Collusion and falsification of results. Collusion might take place between laboratories so that independent data are not submitted. Proficiency testing schemes should be designed to ensure that there is as little collusion and falsification as possible. For example, alternative samples could be distributed within a round. Also instructions should make it clear that collusion is contrary to professional scientific conduct and serves only to nullify the benefits of proficiency testing.

10.5.6 Statistical procedure for the analysis of results

The first stage in producing a score from a result x (a single measurement of analyte concentration in a test material) is to obtain an estimate of the bias, thus:

$$\text{bias} = x - X$$

where X is the true concentration or amount of analyte.

The efficacy of any proficiency test depends on using a reliable value for X. Several methods are available for establishing a working estimate of $\hat{X}$ (i.e. the assigned value).

Formation of a z-score. Most proficiency testing schemes compare bias with a standard error. An obvious approach is to form the z-score given by:

$$z = (x - \hat{X})/\sigma$$

where σ is a standard deviation; σ could be either an estimate of the actual variation encountered in a particular round ($\tilde{s}$) estimated from the laboratories' results after outlier elimination, or a target representing the maximum allowed variation consistent with valid data.

A fixed target value for σ is preferable and can be arrived at in several ways. It could be fixed arbitrarily, with a value based on a perception of how laboratories should perform. It could be an estimate of the precision required for a specific task of data interpretation; σ could be derived from a model of precision, such as the 'Horwitz Curve' (Horwitz, 1982). However, while this model provides a general picture of reproducibility, substantial deviation from it may be experienced for particular methods.

Interpretation of z-scores. If $\hat{X}$ and σ are good estimates of the population mean and standard deviation, then z will be approximately normally distributed with a mean of zero and unit standard deviation. An analytical result is described as 'well behaved' when it complies with this condition.

An absolute value of z ($|z|$) greater than three suggests poor performance in terms of accuracy. This judgement depends on the assumption of the normal distribution, which, outliers apart, seems to be justified in practice.

As z is standardised, it is comparable for all analytes and methods. Thus values of z can be combined to give a composite score for a laboratory in one round of a proficiency test.

The z-scores can therefore be interpreted as follows:

$|z| < 2$	'Satisfactory': will occur in 95% of cases produced by 'well behaved results'
$2 < |z| < 3$	'Questionable': but will occur in $\approx$5% of cases produced by 'well behaved results'
$|z| > 3$	'Unsatisfactory': will only occur in $\approx$0.1% of cases produced by 'well behaved results'

Combination of results within a round of the trial. There are several methods of combining the z-scores produced by a laboratory in one round of the proficiency test described in the Protocol. They are:

- The sum of scores, $SZ = \Sigma z$
- The sum of squared scores, $SSZ = \Sigma z^2$
- The sum of absolute values of the scores, $SAZ = \Sigma |z|$

All should be used with caution, however. It is the individual z-scores that are the critical consideration when considering the proficiency of a laboratory.

Calculation of running scores. Similar considerations apply for running scores as apply to combination scores above.

10.5.7 National proficiency schemes

A number of Member States have introduced proficiency testing schemes in the food area. Most operate according to the Harmonised Protocol and thus conform to the requirements internationally accepted for the operation of proficiency testing schemes in the food sector. The UK Ministry of Agriculture, Fisheries and Food developed a proficiency testing scheme for food analysis laboratories (The Food Analysis Performance Assessment Scheme, FAPAS). The scheme complies with the requirements of the Harmonised Protocol referred to above and thus may be regarded as a practical demonstration of the effectiveness of the Protocol.

The Food Analysis Performance Assessment Scheme. The Scheme was developed because it was appreciated that food-related legislation requires that there are data on residues and contaminants, some of which are particularly difficult to analyse. Lack of independent assessment of the data being produced in these consumer safety-related areas would hamper the work of enforcement authorities and would prejudice the recognition of results and certificates which is at the heart of the EC 'Single Market' and 'New Approach' initiatives. It would also limit the scope and reliability of food surveillance work on the UK food supply. FAPAS helps to provide the independent assessment of these data. The Scheme was also developed to assist in the effective implementation and enforcement of regulations made under UK legislation.

The Scheme was developed within the Ministry with help from representatives of all sectors (i.e. enforcement, industry/trade, and referee analysts) who are concerned with the proficiency of food analysis laboratories. In particular, representatives from the Association of Public Analysts, the British Food Manufacturing Industries Research Association, the Campden Food and Drink Research Association, the Flour Milling and Baking

Laboratory Code	MEG in mg/kg	z- Scores
1	24.50	0.16
2	24.50	0.16
3	29.20	2.13
4	26.10	0.83
5	23.70	-0.18
6	24.40	0.12
7	23.90	-0.09
8	24.60	0.20
9	26.20	0.87
10	23.50	-0.26
11	19.30	-2.02
12	23.70	-0.18
13	22.00	-0.89
14	25.50	0.58
15	17.00	-2.98
Number of Results	15	
Assigned Value	24.13	
Standard Deviation :	2.77	
Target Standard Deviation	2.39	
Min. :	17.00	
Max :	29.20	

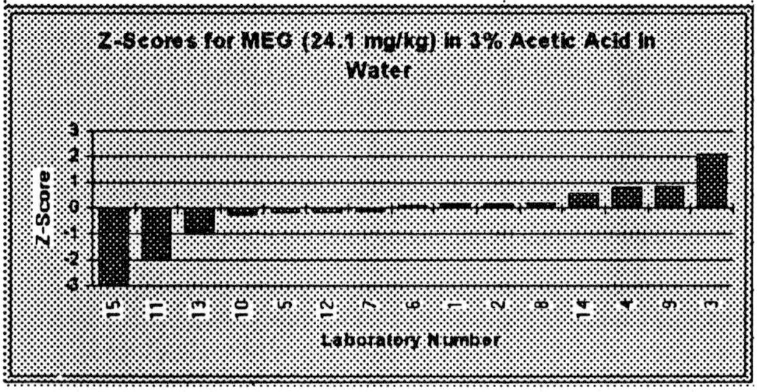

Figure 10.1 Results displayed in accordance with the Harmonised Protocol recommendations: a typical histogram of results from the specific migration series.

Research Association, the Laboratory of the Government Chemist and the Statistical Sub-Committee of the Royal Society of Chemistry Analytical Methods Committee serve on the Advisory Committee.

The Scheme has a number of series currently in operation, these being for:

- Nutritional components
- Veterinary drugs
- HPLC procedures
- Aflatoxins
- Organochlorine pesticides
- Trace elements
- Organophosphorus pesticides
- Animal feedingstuffs
- Overall migration
- Specific migration
- Alcoholic drinks by GC
- Oils and fats by GC

Participants to the scheme are from all sectors within the UK, with approximately a third of participants from non-UK laboratories.

Sectors represented include the public sector, enforcement analysts, research associations and other trade bodies, industry, education and consultants.

The results are calculated and displayed in accordance with the Harmonised Protocol recommendations.

A typical histogram of results from the specific migration series is given in Figure 10.1.

10.6 Methods of analysis

Historically, in (food) analysis laboratories, far more attention has been given to the validated method rather than the other two aspects described above. Possibly, this greater emphasis is because most organisations, be they governmental or one of the international standardising organisations working in the foodstuffs area, develop methods of analysis and incorporate them into legislation or International Standards, but do not then have any mechanism to assess how well such methods are being applied.

In addition, the development of methods of analysis for incorporation into International Standards or into foodstuff legislation was, until comparatively recently, not systematic. However, most international organisations now develop their own methods in a defined way; the most important of these is described below.

10.6.1   AOAC INTERNATIONAL (AOACI)

This organisation has required for many years that all of its methods, which are quoted worldwide, be collaboratively tested before being accepted for publication. The procedure for carrying out the necessary collaborative trials is well defined and has formed part of the internal requirements of the organisation for many years. In addition, methods are adopted by the organisation on a preliminary/tentative basis (as 'Official First Action Methods') before being adopted as 'Official Final Action Methods' by the organisation after at least two years in use by analysts on a routine basis. This procedure ensures that the methods are satisfactory on a day-to-day basis before being finally accepted by the AOACI.

The AOAC INTERNATIONAL has also developed two other procedures for 'method validation' which, though not dealing directly with the production of methods of analysis which have been validated by collaborative trial, do result in some validation. These are:

- Peer validated methods—in which a method of analysis is assessed in a very limited number of laboratories (usually three) and an assessment made of its performance characteristics, and
- Methods assessed by the AOAC Research Institute—such methods are normally proprietary products (e.g. 'test kits') for which a 'certificate' of conformance with specification is issued by the Research Institute.

10.6.2   The Codex Alimentarius Commission

This was the first international organisation working at the government level in the food sector which laid down principles for the establishment of its methods. That it was necessary for such guidelines and principles to be laid down reflects the confused and unsatisfactory situation in the development of legislative methods of analysis that existed until the early 1980s in the food sector.

The 'Principles For The Establishment Of Codex Methods Of Analysis' (FAO, 1993) are given below; other organisations which subsequently laid down procedures for the development of methods of analysis in their particular sector followed these principles to a significant degree.

Principles for the establishment of Codex methods of analysis

Purpose of Codex methods of analysis. The methods are primarily intended as international methods for the verification of provisions in Codex standards. They should be used for reference, in calibration of methods in use or introduced for routine examination and control purposes.

Methods of analysis

(A) *Definition of types of methods of analysis*

(a) Defining Methods (Type I)
Definition: A method which determines a value that can only be arrived at in terms of the method *per se* and serves by definition as the only method for establishing the accepted value of the item measured.
Examples: Howard Mould Count, Reichert-Meissl value, loss on drying, salt in brine by density.

(b) Reference Methods (Type II)
Definition: A Type II method is the one designated Reference Method where Type I methods do not apply. It should be selected from Type III methods (as defined below). It should be recommended for use in cases of dispute and for calibration purposes.
Example: Potentiometric method for halides.

(c) Alternative Approved Methods (Type III)
Definition: A Type III Method is one which meets the criteria required by the Codex Committee on Methods of Analysis and Sampling for methods that may be used for control, inspection or regulatory purposes.
Example: Volhard Method or Mohr Method for chlorides.

(d) Tentative Method (Type IV)
Definition: A Type IV Method is a method which has been used traditionally or else has been recently introduced, but for which the criteria required for acceptance by the Codex Committee on Methods of Analysis and Sampling have not yet been determined.
Examples: chlorine by X-ray fluorescence, estimation of synthetic colours in foods.

(B) *General criteria for the selection of methods of analysis*

(a) Official methods of analysis elaborated by international organisations occupying themselves with a food or group of foods should be preferred.

(b) Preference should be given to methods of analysis the reliability of which has been established in respect of the following criteria, selected as appropriate:

- specificity
- accuracy
- precision; repeatability intra-laboratory (within laboratory), reproducibility inter-laboratory (within laboratory and between laboratories)
- limit of detection
- sensitivity
- practicability and applicability under normal laboratory conditions
- other criteria which may be selected as required.

(c) The method selected should be chosen on the basis of practicability and preference should be given to methods which have applicability for routine use.

(d) All proposed methods of analysis must have direct pertinence to the Codex Standard to which they are directed.

(e) Methods of analysis which are applicable uniformly to various groups of commodities should be given preference over methods which apply only to individual commodities.

10.6.3 The European Union

The Union is attempting to harmonise sampling and analysis procedures in an attempt to meet the current demands of the national and international enforcement agencies and the likely increased problems that the open market will bring. To aid this the Union issued a Directive on Sampling and Methods of Analysis (EEC, 1985). The Directive contains a Technical Annex, in which the need to carry out a collaborative trial before it can be adopted by the Community is emphasised.

The criteria to which Community methods of analysis for foodstuffs should now conform are as stringent as those recommended by any International Organisation following adoption of the Directive. The requirements follow those described for Codex above, and are given in the Annex to the Directive. They are:

'1. Methods of analysis which are to be considered for adoption under the provisions of the Directive shall be examined with respect to the following criteria:

(i)   specificity
(ii) accuracy
(iii) precision; repeatability intra-laboratory (within laboratory), reproducibility inter-laboratory (within laboratory and between laboratories)
(iv) limit of detection
(v)   sensitivity
(vi) practicability and applicability under normal laboratory conditions
(vii) other criteria which may be selected as required.

2. The precision values referred to in 1 (iii) shall be obtained from a collaborative trial which has been conducted in accordance with an internationally recognised protocol on collaborative trials (e.g. International Organisation for Standardisation *Precision of Test Methods* (ISO 5725 (ISO, 1981)). The repeatability and reproducibility values shall be expressed in an internationally recognised form (e.g. the 95% confidence intervals as defined by ISO 5725 (ISO, 1981)). The

results from the collaborative trial shall be published or be freely available.

3. Methods of analysis which are applicable uniformly to various groups of commodities should be given preference over methods which apply to individual commodities.

4. Methods of analysis adopted under this Directive should be edited in the standard layout for methods of analysis recommended by the International Organisation for Standardisation.'

10.6.4 European Committee for Standardisation (CEN)

In Europe the International Organisation which is now developing most standardised methods of analysis in the food additive and contaminant area is the European Committee for Standardisation (CEN). Although CEN methods are not prescribed by legislation, the European Commission does place considerable importance on the work that CEN carries out in the development of specific methods in the food sector; CEN has been given direct mandates by the Commission to publish particular methods, e.g. those for the detection of food irradiation. Because of this some of the methods in the food sector being developed by CEN are described below. CEN, like the other organisations described above, has adopted a set of guidelines to which its Methods Technical Committees should conform when developing a method of analysis. The guidelines are:

'Details of the interlaboratory test on the precision of the method are to be summarised in an annex to the method. It is to be stated that the values derived from the interlaboratory test may not be applicable to analyte concentration ranges and matrices other than given in annex.

The precision clauses shall be worded as follows:

Repeatability: "The absolute difference between two single test results found on identical test materials by one operator using the same apparatus within the shortest feasible time interval will exceed the repeatability value r in not more than 5% of the cases.

The value(s) is (are): . . ."

Reproducibility: "The absolute difference between two single test results on identical test material reported by two laboratories will exceed the reproducibility value R in not more than 5% of the cases.

The value(s) is (are): . . ."

There shall be minimum requirements regarding the information to be given in an Informative Annex, this being:

- Year of interlaboratory test and reference to the test report (if available)
- Number of samples
- Number of laboratories retained after eliminating outliers
- Number of outliers (laboratories)
- Number of accepted results
- Mean value (with the respective unit)
- Repeatability standard deviation (s_r) (with the respective unit)
- Repeatability relative standard deviation (RSD$_r$) (%)
- Repeatability limit (r) (with the respective units)
- Reproducibility relative standard deviation (s_R) (with the respective unit)
- Reproducibility relative standard deviation (RSD$_R$) (%)
- Reproducibility limit (R) (with the respective unit)
- Sample types clearly described
- Notes if further information is to be given'

In addition CEN publishes its methods as either Finalised Standards or as Preliminary Standards, which are required to be reviewed after two years and then either withdrawn or converted to a Full Standard. Thus, CEN follows the same procedure as the AOACI in this regard.

10.6.5   Requirements of official bodies

Consideration of the above requirements from the four international organisations described confirms that in future all methods must be fully validated—i.e. have been subjected to a collaborative trial conforming to an international recognised protocol. In addition this, as described above, is now a legislative requirement in the food sector of the European Union.

The concept of the valid analytical method in the food sector, and its requirements, are described below:

10.6.6   Requirements for valid methods of analysis

It would be simple to say that any new method should be fully tested for the criteria given above.

However, the most 'difficult' of these is obtaining the accuracy and precision performance criteria.

Accuracy.   Accuracy is defined as the closeness of the agreement between the result of a measurement and a true value of the measureand (ISO, 1993b). It may be assessed by the use of reference materials.

However, in food analysis, there is a particular problem. In many instances, though not normally for food additives and contaminants, the numerical value of a characteristic (or criterion) in a Standard is dependent on the procedures used to ascertain its value. This illustrates the need for the (sampling and) analysis provisions in a Standard to be developed at the same time as the numerical value of the characteristics in the Standard are negotiated, to ensure that the characteristics are related to the methodological procedures prescribed.

Precision. Precision is defined as the closeness of agreement between independent test results obtained under prescribed conditions (ISO, 1992).

In a standard method the precision characteristics are obtained from a properly organised collaborative trial, i.e a trial conforming to the requirements of an International Standard (the AOAC/ISO/IUPAC Harmonised Protocol or the ISO 5725 Standard). Because of the importance of collaborative trials, and the resource that is now being devoted to the assessment of precision characteristics of analytical methods before their acceptance, they are described in detail below.

10.7 Collaborative trials

As seen above, all 'official' methods of analysis are required to include precision data. These may be obtained by subjecting the method to a collaborative trial conforming to an internationally agreed protocol.

10.7.1 What is a collaborative trial?

A collaborative trial is a procedure whereby the precision of a method of analysis may be assessed and quantified. The precision of a method is usually expressed in terms of repeatability and reproducibility values. Accuracy is not the objective.

10.7.2 IUPAC/ISO/AOAC Harmonisation Protocol

Recently there has been progress towards a universal acceptance of collaboratively tested methods and collaborative trial results and methods, no matter by whom these trials are organised. This has been aided by the publication of the IUPAC/ISO/AOAC Harmonisation Protocol on Collaborative Studies (Horwitz, 1988). That Protocol was developed under the auspices of the International Union of Pure and Applied Chemistry (IUPAC) aided by representatives from the major organisations interested in conducting collaborative studies. In particular, from the food sector, the **AOAC INTERNATIONAL**, the International Organisation for Standardisation (ISO), the International Dairy Federation (IDF), the Collaborative

International Analytical Council for Pesticides (CIPAC), the Nordic Analytical Committee (NMKL), the Codex Committee on Methods of Analysis and Sampling and the International Office of Cocoa and Chocolate were involved.

The Protocol gives a series of 11 recommendations.

10.7.3 *The components that make up a collaborative trial*

Participants. It is of paramount importance that all the participants taking part in the trial are competent, i.e. they are fully conversant and experienced in the techniques used in the particular trial, and they can be relied upon to act responsibly in following the method/protocol. It is the precision of the method that is being assessed, not the performance of the trial participants.

The number of participants must be at least eight. However, to use only laboratories who are 'experts' in the use of the method in question but will not ultimately be involved in the routine use of the method could give an exaggerated precision performance for the method.

In the case of trials organised under the UK Ministry of Agriculture, Fisheries and Food, collaborative trial programme invitations to participate are made to:

1. analytical chemists (who will be required to use the method under consideration if it reaches the legislation stage)
2. trade analysts who have a particular interest in the method under consideration
3. Research Associations, the Laboratory of the Government Chemist, etc.—i.e. those analysts who will be considered good 'general' analysts as is the case in (1) above.

It is not expected that participants will become involved in the trial unless they are likely to have to use the method on a subsequent occasion.

Sample type. The samples should, if possible, be normal materials containing the required levels of analyte. If this is not possible, then alternative laboratory-prepared samples must be used. These have the obvious disadvantage of a different analyte/sample matrix from that of the normal materials. The types of samples used should cover as wide an area as possible of sample types for which the method is to be used. Ideally for collaborative trial purposes the individual sample/sample types should be identical in appearance, so as to preserve sample anonymity.

Sample homogeneity. The bulk sample, from which all sub-samples which form the collaborative trial samples are taken, should be homogeneous, and the sub-sampling procedure such that the resulting sub-samples have an equivalent composition to that of the original bulk sample. Tests should also be carried out to determine the stability of the sample over the intended time period of the trial. The precision of the method can only be as good as the homogeneity of the samples allows it to be; there must be little or no variability due to sample heterogeneity.

Sample plan.   For the study of method performance, at least five samples covering a range of analyte concentrations representative of the commodity should be used. These samples should be duplicated, making a minimum of ten samples in all. The samples should be visually identical and given individual sample codes to preserve sample anonymity as far as possible.

However, in the case of blind duplicates it may be possible to link the pairs together. This possibility may be reduced by the use of *split level samples*, where the samples differ only slightly in concentration of analyte, and now any artificial attempt to draw these 'duplicate' samples closer together will in fact give decreased precision as it is the difference *between* the split level samples that is involved in the calculation of the precision parameter, and this will be a quantifiable amount.

Sometimes it is not feasible to have blind duplicates or split level samples and laboratories are requested to analyse samples as known duplicates. This is the least favoured option.

The method(s) to be tested.   The method(s) to be tested should have been tested for robustness (i.e. how susceptible the procedure is to small variations in method protocol/instructions) and optimised before circulation to participants. A thorough evaluation of the method at this stage can often eliminate the need for a retrial at a later stage.

Pilot study/pretrial.   These are invaluable exercises if resources permit. A pilot study involves at least three laboratories testing the method, checking the written version, highlighting any hidden problems that may be encountered during the study, and giving an initial estimation of the precision of the method. A pretrial can be used in place of or to supplement the pilot study; it is particularly useful where a new method or technique is to be used. Participants are sent one or two reference samples with which to familiarise themselves with the method before starting the trial proper. A successful pretrial is usually a prerequisite to starting the trial proper and can be used to rectify any individual problems that the laboratories are having with the method.

The trial proper.   Participants should be given clear instructions on the protocol of the trial, the time limit for return of results, the number of determinations to be carried out per sample, and how to report results: to how many decimal places, as received or on a dry matter basis, corrected or uncorrected for recovery, etc.

10.7.4 Statistical analysis

It is important to appreciate that the statistical significance of the results is wholly dependent on the quality of the data obtained from the trial. Data which contain obvious gross errors should be removed prior to statistical analysis. It is essential that participants inform the trial co-ordinator of any gross error that they know has occurred during the analysis and also if any deviation from the method as written has taken place. The statistical parameters calculated, and the outlier tests performed, are those used in the internationally agreed Protocol for the Design, Conduct and Interpretation of Collaborative Studies (Horwitz, 1988).

Analysis of data for outliers. More time has been devoted to the procedures to be used for the removal of outliers than is warranted. However, the procedure prescribed in the Harmonised Protocol is now accepted within the food sector. In it:

- Data subjected to outlier tests should be free of *known* gross errors.
- Apply Cochran's test (Horwitz, 1988) to the valid data remaining after identification of aberrant results. This measure compares the ratio of the largest variance between individual replicates with the sum of the individual variances. Because collaborative trials are only concerned with duplicates, squared differences can be used instead of variances. The calculated Cochran's value is compared with the tabulated value (one tail $p < 0.01$, but the value of $p < 0.025$ will become generally accepted in future). If the calculated value is greater than the tabulated value, then the duplicate results from that laboratory for that sample are outliers and not used in the calculation of the precision parameters.
- If the Cochran's test is positive, then further Cochran's tests are carried out on the remaining data until no outliers are detected or until at most 23.5% of the data are removed.
- After a negative Cochran's test the data are analysed for differences in the means of the duplicates from the different laboratories using Grubbs' test (Horwitz, 1988). This may be either Grubbs' Single Value Test (initially) and then the Grubbs' Paired Value Test or 'Double Grubbs''.
- In the Grubbs' Single Value Test, the means of the duplicates of each laboratory are calculated and the standard deviation of these means (s) is calculated. Standard deviations are also calculated for the means omitting the lowest mean (s_l) and the highest mean (s_h). The percentage decrease in standard deviation is calculated using s_l and s_h. The value

which gives the greatest percentage decrease is then compared with the tabulated value.

- In the Grubbs' Paired Value Test, the percentage decrease in standard deviation is calculated for the means excluding the two lowest results (s_{ll}), the two highest results (s_{hh}), and the highest and lowest results (s_{hl}). The value which gives the greatest percentage decrease is compared with the tabulated value at the 1% or 2.5% level.
- When an outlier is identified, return to the start of the scheme and retest for Cochran's and Grubbs' outliers as before.
- It is essential that no more than 23.5% of the original data should be removed as outliers—stop the outlier procedure as soon as that point is reached.

Repeatability and reproducibility parameters are then calculated from the remaining data.

Precision parameters. The most commonly quoted precision parameters are the repeatability and reproducibility.

The CEN definitions are given above. The AOACI has adopted the following definitions:

r $=$ Repeatability, the value below which the absolute difference between two single test results obtained under repeatability conditions (i.e. same sample, same operator, same apparatus, same laboratory, and short interval of time) may be expected to lie within a specific probability (typically 95%), and hence $r = 2.8 \times s_r$.

s_r $=$ Standard deviation, calculated from results generated under repeatability conditions.

RSD_r $=$ Relative standard deviation, calculated from results generated under repeatability conditions [$(s_r/\bar{x}) \times 100$], where $\bar{x}$ is the average of results over all laboratories and samples.

R $=$ Reproducibility, the value below which the absolute difference between single test results obtained under reproducibility conditions (i.e. on identical material obtained by operators in different laboratories, using the standardised test method) may be expected to lie within a certain probability (typically 95%); $R = 2.8 \times s_R$.

s_R $=$ Standard deviation, calculated from results under reproducibility conditions.

RSD_R $=$ Relative standard deviation calculated from results generated under reproducibility conditions [$(s_R/\bar{x}) \times 100$].

Assessment of the acceptability of the precision characteristics of a method of analysis. The calculated repeatability and reproducibility values can be compared with existing methods and a comparison made. If these are satisfactory then the method can be used as a validated method. If there is no method with which to compare the precision parameters, then theoretical repeatability and reproducibility values can be calculated from Horwitz's equation (Horwitz, 1982).

Horwitz trumpet and equation: $\mathrm{RSD}_R = 2^{(1 - 0.5 \log C)}$

Values are:

Concentration ratio		RSD_R
1	(100%)	2
10^{-1}		2.8
10^{-2}	(1%)	4
10^{-3}		5.6
10^{-4}		8
10^{-5}		11
10^{-6}	(mg/kg)	16
10^{-7}		23
10^{-8}		32
10^{-9}	(μg/kg)	45

Horwitz derived the equation after studying the results from many ($\sim$3000) collaborative trials. Although it represents the average RSD_R values and is an approximation of the possible precision that can be achieved, the data points from 'acceptable' collaborative trials lie within a range plus and minus two times the values derived from the equation. This idealised smoothed curve is found to be independent of the nature of the analyte or of the analytical technique that was used to make the measurement. In general the values taken from this curve are indicative of the precision that is achievable and acceptable of an analytical method by different laboratories. Its use provides a satisfactory and simple means of assessing method precision acceptability.

Summary requirements for a collaborative trial.   The critical characteristics of any collaborative trial are summarised as follows:

- the minimum number of laboratories should be eight
- the minimum number of samples should be five
- samples should not be sent to participants for analysis as known duplicates, except as a last resort
- all the original data obtained in the trial should be reproduced in the final report on the trial. A number of outlier identification procedures are given in the recommendations; although it is desirable that they should be used in the statistical analysis of the trial results it is not essential, provided that the raw collaborative trial data are available, thus enabling other organisations to re-calculate if they so desire.

10.8 Methods of analysis being developed by CEN for food additives and contaminants

CEN has established a number of Technical Committees and Working Groups in the food sector which are concerned with the development of methods of analysis; some of these are directly concerned with methods of analysis for additives and contaminants. The main activities in this area are outlined below.

The two main CEN Technical Committees which develop methods in the additives and contaminants area are CEN TC 194, methods of analysis for materials and articles, and CEN TC 275, horizontal methods. Progress in each of these Technical Committees is described below. For the methods being developed in CEN TC 194, an outline of the overall activities is given; for the methods being developed in CEN TC 275, an indication of the scope, principle and precision characteristics of some of the methods of interest being currently developed is given. The methods given are not exhaustive, but serve to illustrate the range of the work being undertaken by CEN in the areas of interest.

For the precision characteristics a repeatability value, r, and reproducibility value, R, will be given. These values are as defined by CEN, i.e.:

Repeatability, r, is the difference between two single test results found on identical test material by one operator using the same apparatus within the shortest feasible time interval, and will exceed the repeatability limit r in not more than 5% of the cases.

Reproducibility, R, is the absolute difference between two single test results on identical test material reported by two laboratories, and will exceed the reproducibility limit R on average in not more than 5% of the cases.

10.8.1 *CEN TC 194: Methods of Analysis for Materials and Articles*

Introduction. The concept of determining overall migration features in the earliest legislation for plastics, e.g. the FDA regulations of 1958 and the Italian legislation of 1963. Since then there have been constant refinements to the procedure by European national authorities and trade associations. When a European Directive appeared imminent, the need for agreed test methods was recognised in the United Kingdom and a British Standards Institution Committee was created to draft the existing procedures in standard format. Within CEN a technical committee (TC 194) was formed to prepare standards for utensils in contact with food. It was agreed by the Technical Board of CEN that Working Group 5 of TC 194 should be entitled 'Chemical Methods of Test'. Its remit is to produce all the necessary methods for determining migration, both specific and overall, to support the

European Directives. Working Group 5 of TC 194, in turn, subdivided into task groups to draft individual methods, these being:

Task Group 1 to prepare methods for overall migration.

Task Group 2 to prepare methods for all monomers subject to limitation in the EC Directive. Methods are being written for the determination of specific monomers in the food simulants and, for those monomers subject to compositional limitation, methods are being devised to determine monomers in plastics.

Task Group 3 to prepare methods for determining the release of metals from surfaces intended to come into contact with foodstuffs. Initially, methods for determining the release of lead and cadmium from silicate surfaces were being prepared. This task force will also concern itself with additives and catalyst residues in plastics where the migration of metal to foods is subject to limitation in future EC Directives.

The task groups work to a common format for their methods. Further task groups will be created to mirror the extension of the scope of the European Directives to include polymeric coatings. Thus, it is envisaged that there will be a task group for polymeric coatings upon paper, paper board and cellulose substrates, and a further task group will be established for polymeric coatings upon metal substrates.

Overall migration methods.   The procedure for determining overall migration from plastics into aqueous food simulants is to evaporate the simulants to dryness and weigh the residue. However, to obtain consistent reliable results at the low levels demanded by Directive 90/128/EEC (EEC, 1990) requires attention to detail and meticulous care by the analyst. For most plastics the overall migration into aqueous food simulants is low, that is, less than 1 milligram per square decimetre of food contact area, and well below the limit of 10 milligrams per square decimetre. Only a few plastics, such as polyamides or filled-polyolefins, give values for overall migration into aqueous food simulants in excess of the permitted limit.

The method for determining the overall migration from plastics into fat simulants is considerably more complex, as fats cannot be removed by volatilisation and an indirect procedure is required. The sample is weighed before exposure to fat. After exposure to fat, any excess fat is removed by blotting and the sample reweighed. For many plastics a weight gain will be observed as the fat penetrating the plastic exceeds the total migration from the plastic. The fat remaining in the plastic is determined by extraction with a solvent followed, usually, by gas chromatography of the methyl esters derived from the fatty triglyceride food simulant. The overall migration is then calculated by subtracting the weight of fat from the plastic after exposure to fat. Corrections may be required for some plastics for changes in moisture content.

Owing to the complexity and cost of the test the number of published results for commercial plastics in food use is not large. Nevertheless, clear conclusions can be drawn for some polymer types, for example poly(ethylene terephthalate) polymers and unplasticised poly(vinyl chloride) polymers consistently give low results, typically less than 1 mg/dm^2, whereas poly(vinyl chloride) films with monomeric plasticisers consistently give higher levels of overall migration, with virtually quantitative extraction of the monomeric plasticiser. For polyolefins the level of migration will depend upon a number of factors such as the degree of copolymerisation and the presence of certain ingredients, for example, antistatic and slip additives which are intentionally present at the surface.

The standard methods being drafted by Working Group 5 of CEN TC 194 are in many parts: the titles for the parts are given in Table 10.1.

Table 10.1

Part number	Methods which have been published or which are being developed by CEN TC 194
1	Guide to the selection of conditions and test methods for overall migration
2	Determination of overall migration into olive oil by total immersion
3	Determination of overall migration into aqueous food simulants by total immersion
4	Determination of overall migration into olive oil by cell
5	Determination of overall migration into aqueous food simulants by Tice Cell
6	Determination of overall migration into olive oil using a pouch
7	Determination of overall migration into aqueous food simulants using a pouch
8	Determination of overall migration into olive oil by article filling
9	Determination of overall migration into aqueous food simulants by article filling
10	Determination of overall migration into olive oil (modified method for use in cases where incomplete extraction of olive oil occurs)
11	Determination of overall migration at low temperature
12	Determination of overall migration from plastics at high temperature
13	Guide to the selection of conditions and determination of specific migration and residual monomer for plastics
14	Determination of acrylonitrile
15	Determination of 1,3-butadiene
16	Determination of residual isocyanate in finished product
17	Determination of monoethylene glycol
18	Determination of diethylene glycol
19	Determination of terephthalic acid
20	Determination of vinylidene chloride
21	Determination of residual vinylidene in finished product
22	Determination of the release of N-nitrosamines and N-nitrosatable substances from elastomers or rubber teats and tetines
23	Determination of the release of lead and cadmium from silicate surfaces
24	Determination of the overall migration from polymeric coatings on metal substrates
25	Determination of the overall migration from polymeric coatings on cellulosic substrates
26	Determination of the free fat on surface of food
27	Determination of the temperature at the interface between the plastic and the food
28	Determination of the molecular weight of polymer additives

Part 1 is a guide to the selection of the appropriate method and test conditions for the plastic article to be tested. It also contains general advice on how to test difficult samples and an index to the other parts of the standard. The other parts are especially constructed so that in the laboratory most samples can be tested using one part only, thereby removing the need for continual cross reference between parts. The simplest procedure, i.e. by total immersion, is divided into two parts, one for aqueous food simulants, the other for the preferred fatty food simulant, olive oil. Similarly, the procedures for testing a single surface of the plastic in a cell, or a pouch, or by filling, are split into aqueous and fat methods. As these methods have evolved they have been evaluated both by individual laboratories and by limited collaborative trials. The preliminary results emphasised the need for critical parts of the procedure to be described in great detail and for a pouch of standardised dimensions to be used in order to achieve comparable results from different laboratories.

The first ten parts of the standard have now been considered by the main CEN TC 194 committee and have been considered acceptable for conventional CEN enquiry as drafts for development (ENVs). These documents will now be edited by the CEN Editorial Committee and translated into the other official CEN languages, French and German. They will then be circulated and a period of three months allowed for comment from the national Standard Institutions. If CEN TC 194 votes in favour they will be adopted for a three-year period following which a further vote will be needed to confirm them as permanent standards.

The method for determining overall migration into the alternative fatty food simulant, a radioactively labelled synthetic mixture of fatty acid triglycerides (otherwise known as labelled HB 307), is now complete as Part 2 (Table 10.1) of the standard and will be considered by the main committee of CEN TC 194.

The CEN methods are intended as reference methods to be used in case of dispute or possible prosecution, and simplified procedures may be adopted for routine use. Also research effort in Europe is being put into finding alternative simulants for fat which will simplify the determination of overall migration.

Working Group 5 continues to draft methods for low-temperature, high-temperature and repeat-use applications and these will be published in due course. Also, to allow Directive 90/128/EEC (EEC, 1990) to be correctly applied, methods for determining the free fat on the surface of foodstuffs and for determining the temperature at the interface between plastics and foodstuffs during conditions of actual use are being developed.

CEN methods for monomers.   In addition to the drafting of methods for overall migration, many methods are being written for monomers and starting substances in Section A of Annex 11 of Directive 90/128/EEC

(EEC, 1990). Steady progress continues to be made in Task Group 2 of CEN TC 194 Working Group 5 on drafting the methods for vinylidene chloride, terephthalic acid, ethylene glycol and diethylene glycol, vinyl acetate and the isocyanates. At the request of the European Commission, a draft of a method for styrene is also being written. In order to facilitate this work, Task Group 4 has been addressing all the points of principle raised in drafting individual methods and has written the method for acrylonitrile in a format that can be followed when drafting subsequent methods. All of these methods will be circulated soon when the accompanying Part 1 has been finalised. Part 1 is a guide to the other parts of the standard and includes the procedures to be followed when exposing the plastics to the food simulants.

10.8.2   CEN TC 275 WG 1: Methods of Analysis for the Determination of Sulphite

Part 1—Optimised Monier–Williams procedure

Scope.   The Standard will specify a distillation method for quantitative determination of sulphite content in foodstuffs, in which the content of sulphite is at least 10 mg/kg. The method will be applicable in the presence of other volatile sulphur compounds. It will not be applicable to dried onions, leeks and cabbages.

Principle.   The method measures free sulphite plus reproducible portions of bound sulphites (such as carbonyl addition products) in foods. It requires heating of the test portion with a refluxing solution of hydrochloric acid to convert sulphite to sulphur dioxide with the introduction of a stream of nitrogen below the surface of the refluxing solution to sweep sulphur dioxide through a water-cooled condenser and, via a bubbler attached to the condenser, into hydrogen peroxide solution, where sulphur dioxide is oxidised to sulphuric acid. This is followed by titration of the generated sulphuric acid with standardised sodium hydroxide solution. The sulphite content is directly related to the generated sulphuric acid. Optional verification of sulphite by gravimetric determination as barium sulphate is permitted.

Precision.   Repeatability and reproducibility values obtained in recent collaborative trials are:

Sample	Concentration (mg/kg)	Repeatability, r (mg/kg)	Reproducibility, R (mg/kg)
Hominy	9.2	3.72	3.98
Fruit juice	8.1	3.81	4.54
Sea food	10.4	4.12	7.76

Part 2—Enzymatic method

Scope. The Standard will specify an enzymatic method for the quantitative determination of sulphite content in foodstuffs. The method has been proven to be applicable to wine, beer, fruit juices, dried fruits, spices, potato chips and minced meat.

Other sulphur-containing substances such as sulphate, sulphide or thiosulphate do not interfere with the determination. Carbonyl–sulphite complexes react as free sulphites. Isothiocyanates occurring (e.g. in mustard) disturb the determination of sulphite.

Principle. The method involves the oxidation of sulphite to sulphate using catalysation of the reaction by sulphite oxidase. There is formation of hydrogen peroxide at the same time.

$$SO_3^{2-} + O_2 + H_2O \xrightarrow{\text{Sulphite oxidase}} SO_4^{2-} + H_2O_2$$

There is reaction of hydrogen peroxide with NADH using catalysation of the reaction by NADH peroxidase.

$$H_2O_2 + NADH + H^+ \xrightarrow{\text{NADH peroxidase}} 2H_2O + NAD^+$$

Precision. Repeatability and reproducibility values obtained in recent collaborative trials are:

Sample	Concentration (mg/kg)	Repeatability, r (mg/kg)	Reproducibility, R (mg/kg)
Wine	75	8	16
Dried apples	800	298	311
Dried apples	960	358	374
Juice	270	37	79
Sultanas	260	45	129
Beer	4.9	0.8	1.6

10.8.3   CEN TC 275 WG 2: Methods of Analysis for the Intense Sweeteners

Determination of cyclamate and saccharin in liquid table top preparations method by HPLC

Scope. The Standard specifies a high performance liquid chromatography (HPLC) method for the determination of sodium cyclamate and saccharin in liquid table top preparations. It also allows the determination of sorbic acid in liquid table top preparations.

Principle. The method involves determination of sodium cyclamate, saccharin and sorbic acid in an appropriate dilution of a liquid table top sweetener in water by HPLC and subsequent photometric detection in the ultraviolet (UV) range. Sweeteners are identified on the basis of the retention times, and there is quantitative determination by the external standard method using peak areas or peak heights.

Precision. Repeatability and reproducibility values obtained in recent collaborative trials for a liquid table top sweetener, tested on a commercially available product, are:

Sample	Repeatability, r (g/100 ml)	Reproducibility, R (g/100 ml)
Sodium cyclamate	0.69	1.0
Saccharin	0.03	0.08
Sorbic acid	2	8

Determination of saccharin in table top preparations—spectrometric method

Scope. The Standard specifies a spectrometric method for the determination of sodium saccharin and saccharin content in solid table top preparations prepared from cyclamate–saccharin or saccharin.

Principle. The method involves preparation of the sample test solution by dissolving table top sweetener in sodium hydroxide solution and photometric determination of the sodium saccharin content at the absorption maximum of about 265 nm.

Precision. Repeatability and reproducibility values obtained in recent collaborative trials for saccharin in a table top sweetener, tested on a commercially available product, are:

Sample	Repeatability, r (g/100 g)	Reproducibility, R (g/100 g)
Table top preparation	0.42	0.85

Determination of acesulfame K in table top preparations—spectrometric method

Scope. The Standard specifies a spectrometric method for the determination of acesulfame K in solid table top preparations.

Principle. The method requires preparation of the sample test solution by dissolving table top sweetener in water and photometric determination of the acesulfame K content at the absorption maximum of about 227 nm.

Precision. Repeatability and reproducibility values obtained in recent collaborative trials for acesulfame K in a table top sweetener, tested on a commercially available product, are:

Sample	Repeatability, r (mg/100 g)	Reproducibility, R (mg/100 g)
Table top preparation	3.2	3.7

10.8.4 CEN TC 275 WG 3: Methods of Analysis for the Quantitative Determination of Pesticides and Polychlorinated Biphenyls (PCBs) in Fatty Foods

Part 1: General considerations

Scope. The Standard specifies methods of quantitative determination of residues of pesticides and PCBs in fatty food. Each method described in the standard is suitable for identifying and quantifying a definite range of those non-polar organochlorine and/or organophosphorus pesticides which occur as residues in fats and oils as well as in the fat portion of fat-containing foodstuffs, of either animal or vegetable origin. The PCB indicator congeners selected for the enforcement of maximum residue limits (MRLs) are determined along with the organochlorine pesticides.

The Standard contains the following clean-up methods that have been subjected to interlaboratory studies and/or are adopted throughout Europe:

- Method A: Liquid–liquid partitioning with acetonitrile and clean-up on a Florisil® column (AOAC).
- Method B: Liquid–liquid partitioning with dimethylformamide and clean-up on a Florisil® column (Specht).
- Method C: Column chromatography on activated Florisil® (AOAC).
- Method D: Column chromatography on partially deactivated Florisil® (Stijve).
- Method E: Column chromatography on partially deactivated aluminium oxide (Greve and Grevenstuk).
- Method F: Gel-permeation chromatography (GPC) (AOAC).
- Method G: Gel-permeation chromatography (GPC) and column chromatography on partially deactivated silica gel (Specht).
- Method H: High-performance gel-permeation chromatography (HPGPC) (MAFF).

The applicability of the eight methods A to H for residue analysis of organochlorine pesticides, PCB indicator congeners, and organophosphorus pesticides, respectively, is given in the Standard.

Principle

General. The methods described in the Standard are based on a four-stage process (in some cases two stages may be combined, in whole or in part), as follows.

Extraction. Extraction of the residues from the sample matrix by the use of appropriate solvents, so as to obtain the maximum efficiency of extraction of the residue and minimum co-extraction of any substances which can give rise to interferences in the determination. Methods for extraction of fat are recommended which are simultaneously applicable for the extraction and determination of fat and the residue analysis in the fat portion.

Clean-up. Maximum removal of interfering substances with minimal loss of analyte from the sample extract, so as to obtain a solution of the extracted residue in a solvent which is suitable for quantitative examination by the selected method of determination.

Determination. Gas chromatography (GC) with electron-capture detector (ECD), the thermionic detector (P- or N/P-mode), the flame photometric detector (FPD) or some form of mass spectrometry (MS) as appropriate.

Confirmation. Procedures to confirm the identity of observed residues, particularly in those cases where it would appear that the maximum residue limit has been exceeded.

Repeatability conditions. Each laboratory should periodically determine if its results under repeatability conditions are acceptable by analysing samples which have been spiked with appropriate standard materials at suitable concentrations, e.g. near to the maximum residue limits, or preferably by using samples with incurred residues.

Reproducibility conditions. Reproducibility conditions are defined as conditions under which test results are obtained with the same method on identical test material in different laboratories with different operators using different equipment.

The table below shows examples of acceptable differences of test results under reproducibility conditions. In this example 0.01 mg/kg is near the limit of determination. Intermediate values may be determined by interpolation from a log–log graph.

Residue level (mg/kg)	Difference ($\pm$) (mg/kg)
0.01	0.01
0.1	0.05
1.0	0.25

Part 2: Extraction of fat, pesticides and PCBs, and determination of fat content

Scope. The Standard specifies a range of analytical procedures for extracting the fat portion containing the pesticide and PCB residues from different groups of fat-containing foods.

Principle. The method involves extraction of the residues from the sample matrix by the use of appropriate solvents to obtain the maximum efficiency of extraction of the residue and minimum co-extraction of any substances that can give rise to interferences in the determination. Solvents are then removed by evaporation, and there is optional determination of the fat content by weighing the mass of the remainder.

Precision. No values are given as not appropriate.

Part 3: Clean-up methods

Introduction. The Standard comprises a range of multi-residue methods of equal status: no single method can be identified as the prime method because, in this field, methods are continuously developing. The methods selected for inclusion in this Standard have been validated and/or are widely used throughout Europe.

The residues to be analysed in the Standard are associated with the fat portion of the samples. In addition to the residues, the extracts obtained in accordance with the method given above or in accordance with the following methods contain fats and other lipids, which would interfere in the analysis. To purify the crude extracts or the fats and oils to be analysed, several methods may be used.

The Standard is based on the following clean-up methods that have been subjected to interlaboratory studies and/or are adopted throughout Europe.

- Method A: Liquid–liquid partitioning with acetonitrile and clean-up on a Florisil® column (AOAC).
- Method B: Liquid–liquid partitioning with dimethylformamide and clean-up on a Florisil® column (Specht).
- Method C: Column chromatography on activated Florisil® (AOAC).
- Method D: Column chromatography on partially deactivated Florisil® (Stijve).
- Method E: Column chromatography on partially deactivated aluminium oxide (Greve and Grevenstuk).
- Method F: Gel-permeation chromatography (GPC) (AOAC).
- Method G: Gel-permeation chromatography (GPC) and column chromatography on partially deactivated silica gel (Specht).
- Method H: High-performance gel-permeation chromatography (HPGPC) (MAFF).

Scope. The Standard specifies the details of methods A to H mentioned in the introduction for the clean-up of fats and oils or the isolated fat portion, respectively, using techniques such as liquid–liquid partition, adsorption or gel-permeation column chromatography. The scope of the methods A to H is given in detail in each method described.

Principle. The method requires removal of interfering materials from the sample extract to obtain a solution of the extracted residue in a solvent which is suitable for quantitative examination by the selected method of determination.

Part 4: Determination, confirmatory tests, miscellaneous

Scope. The Standard specifies some recommended techniques for the determination of pesticides and PCBs in fatty foodstuffs, confirmation of results and an additional clean-up procedure.

General. The methods described in this part of the Standard permit the residues present to be provisionally identified and quantified by gas chromatographic methods using selective detectors.

The procedures listed for confirmation as glass capillary GC, thin-layer chromatography (TLC), *P*-value, the GC of oxidation and other conversion products, and similar techniques, are of value. Results obtained using MS present the most definitive evidence for confirmation/identification purposes.

10.8.5 CEN TC 275 WG 5: Methods of Analysis for Mycotoxins

Methods for various mycotoxins, particularly the aflatoxins and ochratoxins, are being developed within this Working Group. Others, including patulin, will be considered in the near future. One method, for ochratoxin, is outlined below.

Quantitative determination of ochratoxin A: HPLC method for barley, maize and wheat bran

Scope and field of application. The Standard specifies a method for the determination of ochratoxin A (OTA) in barley, maize and wheat bran at concentrations above $2\,\mu g/kg$ by using high-performance liquid chromatography (HPLC).

Principle. OTA is extracted from grains with chloroform–aqueous phosphoric acid and isolated by liquid–liquid partitioning into aqueous bicarbonate solution. The solution is applied to a C_{18} cartridge, and OTA is eluted with ethyl acetate–methanol–acetic acid. The OTA is separated and identified by reversed-phase LC and quantified by fluorescence. Chromatography of OTA methyl ester derivative confirms the identification.

Precision. Repeatability and reproducibility values obtained in two recent collaborative trials are as follows:

Sample	Concentration (μg/kg)	Repeatability, r	Reproducibility, R
Barley	7.4	—	5.6
Barley	14.4	3.1	10.6
Maize	8.2	—	4.8
Maize	16.3	9.2	12.9

Sample	Concentration (μg/kg)	Repeatability, r	Reproducibility, R
Barley	3.0	1.28	1.90
Barley	2.9	1.37	1.74
Rye	4.8	2.18	3.10
Rye	2.9	1.80	2.34
Wheat bran	4.5	2.16	3.35
Wheat bran	3.8	2.23	2.58

10.8.6   CEN TC 275 WG 7: Methods of Analysis for the Determination of Nitrate and Nitrite Content

Ion-exchange chromatography (IC) method for the determination of nitrate and nitrite content of meat and meat products

Scope. The Standard specifies an IC method for the determination of nitrite and nitrate contents of meat and meat products. Inter-laboratory studies have been carried out not only with meat and meat products but also with vegetables, baby food and dairy products.

Principle. The method requires that extraction of nitrate and nitrite is carried out with hot water. Any interfering substance is removed by clarification with acetonitrile. The determination is carried out by ion-exchange chromatography (IC) and ultraviolet (UV) detection at a wavelength of 205 nm.

Precision. Repeatability and reproducibility values obtained in recent collaborative trials are as follows.

Nitrate

Sample	Concentration (mg/kg)	Repeatability, r	Reproducibility, R
Spinach	1347.1	55.6	196.9
Carrots	63.3	12.1	27.3
Corned beef	99.0	9.8	45.1
Corned beef	471.8	40.7	45.3
Baby food	83.2	7.0	16.7
Cheese	18.0	5.4	18.3

Nitrite

Sample	Concentration (μg/kg, as sodium salt)	Repeatability, r	Reproducibility, R
Corned beef	10.5	5.0	9.60
Corned beef	58.3	6.6	15.4

Flow injection method for the determination of the nitrate and/or nitrite content of vegetables and vegetable products

Scope.   The Standard specifies a continuous flow method for the quantitative determination of nitrate content of vegetables and vegetable products having a nitrate content of more than 50 mg/kg.

Principle.   Test portions are extracted with water and, after filtration, cleaned up by dialysis in the continuous flow (CF) system. Extracted nitrates are reduced on-line to nitrite by metallic cadmium, followed by reaction of nitrite with sulphanilamide and N-1-naphthylethylenediamine, resulting in the formation of a reddish purple azo dye. This dye is measured spectrometrically at a wavelength of 530 nm.

Precision.   Repeatability and reproducibility values obtained in recent collaborative trials are as follows.

Nitrate

Sample	Concentration (mg/kg)	Repeatability, r	Reproducibility, R
Beetroot	901	139	149
Beetroot	2655	246	338
Lettuce	1319	183	201
Lettuce	2738	252	252
Lettuce	4021	271	420
Endive	1981	94	209
Spinach	5197	342	573

10.8.7   CEN TC 275 WG 8: Methods of Analysis for the Determination of Food Irradiation

Various methods have been developed for establishing whether a food has been irradiated or not. Although food irradiation is a process rather than a food additive, a summary of the methods being developed by CEN TC 275 is included in this chapter because of the similarity in intent, and to demonstrate that even this most difficult estimation has been capable of being standardised. Five methods have been developed, these ensuring that detection is possible in most food likely to have been irradiated.

A: Detection of irradiated food containing bone—method by ESR spectroscopy

Scope. The Standard specifies a method for the detection of meats containing bones and fish containing bones which have been treated with ionising radiation, by analysing the electron spin resonance (ESR) spectrum or electron paramagnetic resonance (EPR) spectrum of the bones.

On the basis of results of analysis of meat containing bones from duck, frog, goose, chicken, hare, lamb, turkey, beef and pork, it is assumed that the same signals occur when other animal species are analysed.

On the basis of results obtained in the analysis of bones from eel, trout, whiting, sardine, hammerhead shark, halibut, herring, cod, salmon, mackerel and cyprinids it is assumed that the same signals occur when other fish species are analysed. The minimum detectable dose depends on the state of mineralisation of the bones, which is usually lower for small species.

Principle.   ESR spectroscopy detects paramagnetic centres (e.g. radicals). An intense external magnetic field produces a difference between the energy levels of the electron spins $m_s = +\frac{1}{2}$ and $m_s = -\frac{1}{2}$. If energy of an appropriate microwave frequency is applied (resonance frequency), the electron spins are forced to invert, and the consequent absorption of microwave energy is detected as a signal. The intensity of the signals obtained (absorption of the incident microwave) is proportional to the concentration of the paramagnetic compounds up to doses of about 6–8 kGy.

Radiation treatment produces radicals which are quite stable in solid and dry components (e.g. bones) of the food and can be detected.

Spectra are conventionally displayed as the first derivative of the absorption.

The field and frequency values depend on the experimental arrangements (sample size and sample holder), while their ratio (i.e. *g* value) is an intrinsic characteristic of the paramagnetic centre and its local co-ordination. For an identification of irradiated samples it may be helpful to measure the *g* values of the ESR signals.

Precision.   The method has been validated as a result of two Europe-wide collaborative trials with meat bones and fish bones samples: the results are summarised below.

Trial 1.   Beef bones and trout bones which were either unirradiated or irradiated to about 2 kGy, 4 kGy or 7 kGy. The following results were achieved.

Product	No. of samples	No. of false negatives	No. of false positives
Beef bone	84	0	0
Trout bone	84	5	0

Trial 2. Chicken and fish bones which were either unirradiated or irradiated to about 2 kGy, 4 kGy or 6 kGy. The following results were achieved.

Product	No. of samples	No. of false negatives	No. of false positives
Chicken bone	108	0	0
Fish bone	108	0	2

B: Detection of irradiated food containing cellulose—method by ESR spectroscopy

Scope. The Standard specifies a method for the detection of foods containing cellulose which have been treated with ionising radiation, by analysing the electron spin resonance (ESR) spectrum or electron paramagnetic resonance (EPR) spectrum of the food.

On the basis of results of analysis of stones, shells or dry parts of food containing cellulose from berries (chilled or frozen), pistachio nuts, French prunes, peanuts, hazelnuts, coconuts, macadamia nuts, almonds, brazil nuts, pecan nuts and walnuts, it may be assumed that the same signals may also be encountered in similar foods.

Principle. ESR spectroscopy detects paramagnetic substances (e.g. radicals). An intense external magnetic field produces a difference between the energy levels of the electron spins $m_s = +\frac{1}{2}$ and $m_s = -\frac{1}{2}$. If energy of an appropriate microwave frequency is applied (resonance frequency), the electron spins are forced to invert and the consequent absorption of microwave energy is detected as a signal. The intensity of the signals obtained (absorption of the incident microwave) is proportional to the concentration of the paramagnetic compounds up to doses of about 6–8 kGy. Radiation treatment produces radicals, which can be detected in solid and dry parts of the food.

Spectra are conventionally displayed as the first derivative of the absorption.

The field and frequency values depend on the experimental arrangements (sample dimension and sample holder), while their ratio (i.e. *g* value) is an intrinsic characteristic of the sample.

Precision. The method has been validated as a result of two Europe-wide collaborative trials with pistachio shells samples; the results are summarised below.

Trial 1. Pistachio shells which were either unirradiated or irradiated to about 2 kGy, 4 kGy or 7 kGy. The following results were achieved.

Product	No. of samples	No. of false negatives	No. of false positives
Pistachio shells	84	15	2

Trial 2. Pistachio shells which were either unirradiated or irradiated to about 4 kGy or 6 kGy. The following results were achieved.

Product	No. of samples	No. of false negatives	No. of false positives
Pistachio shells	68	0	1

C: Detection of irradiated food containing silicate minerals—method by thermoluminescence

Scope. The Standard specifies a method of detecting whether food has been treated by ionising radiation by thermoluminescence analysis of contaminating silicate minerals. The method is applicable to those foodstuffs from which a sufficient amount of silicate minerals can be isolated. The method has been successfully tested in inter-laboratory tests with herbs and spices and their mixtures. From other studies, including an inter-laboratory test on fresh fruits and vegetables, it may be concluded that the method, after suitable modification, is applicable to a large variety of foods including seafood. At present, this standard pertains only to the detection of the irradiation treatment of herbs and spices and their mixtures.

Principle. Silicate minerals contaminating food products store energy by charge-trapping processes as a result of exposure to ionising radiation. Releasing such energy, by controlled heating of isolated silicate minerals, gives rise to measurable thermoluminescence (TL) glow curves.

Silicate minerals are therefore isolated from the food products, mostly by a density separation step. In order not to obscure the TL, the isolated silicate minerals should be as free as possible of organic constituents of the food. A first glow of the separated mineral extracts is recorded (glow 1). Since various amounts or types of minerals (quartz, feldspar, etc.) exhibit very variable integrated TL intensities after irradiation, a second TL glow (glow 2) of the same sample after exposure to a fixed dose of radiation is necessary to normalise the TL response.

The TL glow ratio thus obtained is used to indicate the irradiation treatment of the food since the population of irradiated samples on principle yields higher TL glow ratios than that of unirradiated samples. Glow shape parameters offer additional evidence for identifying irradiated foods. This method of TL analysis relies only on the silicate minerals which can be separated from various foods and is in principle not influenced by the kind of food product.

Precision. The procedure as described in the European Standard is based on inter-laboratory studies with herbs and spices and their mixtures. Fourteen participants tested 18 different herbs and spices or mixtures 3 or 9 months after irradiation with a dose of about 6 kGy or 10 kGy, respectively. Of a total of 317 samples examined, 99.1% were correctly identified. Only three irradiated samples were classified as unirradiated. None of the unirradiated samples were classified as irradiated.

D: Detection of irradiated food containing fat by GC analysis of hydrocarbons

Scope. The Standard specifies a method for the identification of irradiation treatment of food which contains fat. It is based on the gas-chromatographic detection of radiation-induced hydrocarbons (HC). The method has been tested in inter-laboratory tests on raw chicken, pork and beef. Some slight modifications might be necessary if it is applied to other fatty foodstuffs.

Principle. During irradiation, chemical bonds are broken in primary and secondary reactions. In triglycerides breaks occur mainly in the α- and β-positions with respect to the carbonyl groups resulting in the respective C_{n-1} and the $C_{n-2:1}$ HC. To predict these chief radiolytic products, the fatty acid composition of the samples must be known.

For detection of HCs the fat is isolated from the sample by melting it out or by solvent extraction. The HC fraction is obtained by adsorption chromatography prior to separation using gas chromatography and detection with a flame ionisation detector or a mass spectrometer.

Precision. The method has been validated as a result of three Europe-wide collaborative trials with meat bones and fish bones samples; the results are summarised below.

Trial 1. Chicken samples irradiated with about 5 kGy and three unirradiated ones were tested about 2 weeks and 6–8 weeks after irradiation. Radiation-induced HCs were detected in all the irradiated samples.

Trial 2. Chicken meat samples which were either unirradiated or given doses of approximately 0.5, 3.0 and 5.0 kGy were tested one and six months after irradiation. The following results were achieved.

Time after irradiation	No. of samples	No. of false negatives	No. of false positives
1 month	119	7	2
6 months	120	8	0

Trial 3. Chicken, pork and beef samples which were unirradiated or irradiated with 0.8 kGy, 2.8 kGy or 7 kGy (mean doses) and tested three and six months after irradiation. The following results were achieved.

Species	Time after irradiation	No. of samples	No. of false negatives	No. of false positives
Chicken	3 months	160	0	0
Chicken	6 months	126	1	0
Pork	3 months	153	1	3
Pork	6 months	140	1	4
Beef	3 months	149	2	2
Beef	6 months	136	1	0

E: Detection of irradiated food containing fat by GC-analysis of 2-alkylcyclobutanones

Scope.   The Standard specifies a method of identifying food containing fat which has been treated with ionising radiation by mass spectrometry of radiation-induced 2-alkylcyclobutanones after gas-chromatographic separation.

The method has been tested in inter-laboratory tests on raw chicken, pork, and liquid whole egg. Some slight modifications may be necessary if it is applied to other foodstuffs.

Principle.   During irradiation, the acyl–oxygen bond in triglycerides is cleaved and this reaction results in the formation of 2-alkylcyclobutanones containing the same number of carbon atoms as the parent fatty acid; the alkyl group is located in ring position 2. Thus, if the fatty acid composition is known, the 2-alkylcyclobutanones formed can be predicted. The two cyclobutanones which are presently available are 2-dodecylcyclobutanone (DCB) and 2-tetradecylcyclobutanone (TCB) which are formed from palmitic and stearic acid respectively, during irradiation. To date, there is no evidence that the cyclobutanones can be detected in unirradiated foods. The 2-alkylcyclobutanones are extracted using *n*-hexane or *n*-pentane along with the fat. The extract is then fractionated by using adsorption chromatography prior to separation using gas chromatography and detection with a mass spectrometer.

Precision.   The method has been validated as a result of three Europe-wide collaborative trials with meat bones and fish bones samples; the results are summarised below.

Trial 1.   Laboratories quantified 2-dodecylcyclobutanone in 15 coded samples of chicken which were either not irradiated or irradiated at doses of approximately 0.5, 3.0 and 5.0 kGy, one and six months after irradiation. The results are summarised below.

Time after irradiation	No. of samples	No. of false negatives	No. of false positives
1 month	74	0	0
6 months	60	2	0

Trial 2. Laboratories used 2-dodecylcyclobutanone and 2-tetradecyl-cyclobutanone to detect nine coded samples of chicken and liquid whole egg, while eight laboratories analysed pork. The samples were either unirradiated or given doses of 1.0 kGy and 3.0 kGy. The results are summarised below.

Sample	No. of samples	No. of false negatives	No. of false positives
Chicken	99	1	0
Liquid whole egg	99	0	0
Pork	72	0	0

10.9 Conclusions

The need for a laboratory to demonstrate that it is producing reliable and acceptable results has become of major importance. This is aided by the use of proficiency testing schemes, validated methods and internal quality control; these are essential aspects of ensuring that data released from a laboratory are fit for purpose. All three aspects have now been described in appropriate International Protocols and Guidelines. In particular:

- If properly executed, quality control methods can monitor the various aspects of data quality on a run-by-run basis. In runs where performance falls outside acceptable limits, the data produced can be rejected and, after remedial action on the analytical system, the analysis can be repeated.
- The International Harmonised Protocol For Proficiency Testing Of (Chemical) Analytical Laboratories has now been adopted by a number of international organisations. It has been shown to be a simple, transparent procedure which participants readily support.
- Methods of analysis may now be fully validated by procedures which have been adopted by a number of international organisations. A number of particular methods have been developed by CEN and these will become of increasing importance.

References

EEC (1985) Council Directive 85/591/EEC Concerning the Introduction of Community Methods of Sampling and Analysis for the Monitoring of Foodstuffs Intended for Human Consumption. *Official Journal*, **L372**, 31/12/85.

EEC (1989) Council Directive 89/397/EEC on the Official Control of Foodstuffs. *Official Journal*, **L186**, 30/6/89.

EEC (1990) Commission Directive 90/128/EEC Relating to Plastics Materials and Articles Intended to Come into Contact with Foodstuffs. *Official Journal*, **L349**, 13/12/90.

EEC (1993) Council Directive 93/99/EEC on the Subject of Additional Measures Concerning the Official Control of Foodstuffs. *Official Journal*, **L290**, 24/11/93.

FAO (1993) *Procedural Manual of the Codex Alimentarius Commission*, 8th edn. FAO/WHO Food Standards Programme, Food and Agriculture Organisation, Rome.

Horwitz, W. (1982) Evaluation of analytical methods for regulation of foods and drugs. *Anal. Chem.*, **54**, 67A–76A.

Horwitz, W. (1988) Protocol for the design, conduct and interpretation of method performance studies. *Pure Appl. Chem.*, **60**, 855–864. (Revision in press.)

ISO (1981) *Precision of Test Methods*. International Organisation for Standardisation, Geneva (revised 1986 with further revision in preparation).

ISO (1990) *Accreditation of Laboratories*, ISO Guide 25, 3rd edn. International Organisation for Standardisation, Geneva.

ISO (1992) *Terms and Definitions used in Connection with Reference Materials*, ISO Guide 30. International Organisation for Standardisation, Geneva.

ISO (1993) *International Vocabulary for Basic and General Terms in Metrology*, 2nd edn. International Organisation for Standardisation, Geneva..

Thompson, M. and Wood, R. (1993) The International Harmonised Protocol for the Proficiency Testing of (Chemical) Analytical Laboratories. *Pure Appl. Chem.*, **65**, 2123–2144 (also published in *J. AOAC International* (1993, **76**, 926–940).

Thompson, M. and Wood, R. (1995) Harmonised guidelines for internal quality control in analytical chemistry laboratories. *Pure Appl. Chem.* **67**, 649–666.

Index